*Nanomaterials for the
Life Sciences
Volume 9*
Carbon Nanomaterials

*Edited by
Challa S. S. R. Kumar*

Related Titles

Kumar, C. S. S. R. (ed.)
Nanotechnologies for the Life Sciences
10 Volume Set
2011
ISBN: 978-3-527-31301-3

Kumar, C. S. S. R. (Ed.)
Nanomaterials for the Life Sciences (NmLS)
Book Series, 10 Volumes

Vol. 6
Semiconductor Nanomaterials
2010
ISBN: 978-3-527-32166-7

Vol. 7
Biomimetic and Bioinspired Nanomaterials
2010
ISBN: 978-3-527-32167-4

Vol. 8
Nanocomposites
2010
ISBN: 978-3-527-32168-1

Vol. 9
Carbon Nanomaterials
2011
ISBN: 978-3-527-32169-8

Vol. 10
Polymeric Nanomaterials
2011
ISBN: 978-3-527-32170-4

*Nanomaterials for the
Life Sciences
Volume 9*

Carbon Nanomaterials

*Edited by
Challa S. S. R. Kumar*

WILEY-VCH Verlag GmbH & Co. KGaA

The Editor

Dr. Challa S. S. R. Kumar
CAMD
Louisiana State University
6980 Jefferson Highway
Baton Rouge, LA 70806
USA

■ All books published by **Wiley-VCH** are carefully produced. Nevertheless, authors, editors, and publisher do not warrant the information contained in these books, including this book, to be free of errors. Readers are advised to keep in mind that statements, data, illustrations, procedural details or other items may inadvertently be inaccurate.

Library of Congress Card No.: applied for

British Library Cataloguing-in-Publication Data
A catalogue record for this book is available from the British Library.

Bibliographic information published by the Deutsche Nationalbibliothek
The Deutsche Nationalbibliothek lists this publication in the Deutsche Nationalbibliografie; detailed bibliographic data are available on the Internet at <http://dnb.d-nb.de>.

© 2011 WILEY-VCH Verlag GmbH & Co. KGaA, Weinheim

All rights reserved (including those of translation into other languages). No part of this book may be reproduced in any form – by photoprinting, microfilm, or any other means – nor transmitted or translated into a machine language without written permission from the publishers. Registered names, trademarks, etc. used in this book, even when not specifically marked as such, are not to be considered unprotected by law.

Composition Toppan Best-set Premedia Ltd., Hong Kong
Printing and Binding Strauss GmbH, Mörlenbach
Cover Design Schulz Grafik-Design, Fußgönheim

Printed in the Federal Republic of Germany
Printed on acid-free paper

ISBN: 978-3-527-32169-8

Contents

Preface XV
List of Contributors XIX

Part One Overview of Synthesis, Characterization, and Applications in Biomedicine 1

1   Carbon Nanomaterials: Synthetic Approaches 3
        Jean-Philippe Tessonnier
1.1   Introduction 3
1.2   General Concepts on the Synthesis of Carbon (Nano-)Materials 4
1.2.1   Uncatalyzed Synthesis of Carbon (Nano-)Materials 4
1.2.2   Catalyzed Synthesis of Carbon (Nano-)Materials 5
1.3   Synthesis from Solid Precursors 6
1.3.1   Nanodiamonds 6
1.3.1.1 Turning Graphite into Diamond 7
1.3.1.2 Explosive Detonation Synthesis 7
1.3.2   Fullerenes, Nanohorns, Single- and Multi-Wall Carbon Nanotubes 9
1.4   Catalytic Chemical Vapor Deposition 10
1.4.1   Definitions 10
1.4.2   Mechanistic Aspects 12
1.4.3   Single- and Multi-Wall Carbon Nanotubes 14
1.4.3.1 Floating Catalyst CCVD 15
1.4.3.2 Immobilized Catalyst CCVD 16
1.4.4   Aligned Carbon Nanotubes 17
1.4.5   Carbon Nanotubes Synthesized from Biocompatible Catalysts 18
1.4.6   Metal- and PAH-Induced Toxicity of Carbon Nanotubes 19
1.5   Purification Techniques 20
1.6   Importance of Defects and Curvature for Further Functionalization 22
1.7   Functionalization: Creating Anchoring Points for Bioactive Molecules 23
1.7.1   Functionalization by Oxidation 24
1.7.2   Functionalization by Coupling Reactions 25

1.7.3	Noncovalent Functionalization 26
1.8	Conclusion and Outlook 26
	References 26

2 Nanocarbons: Characterization Tools 35
Dang Sheng Su
2.1 Introduction 35
2.2 Diffraction Techniques 36
2.3 Imaging 37
2.3.1 Electron Microscopy 37
2.3.1.1 Electron–Specimen Interactions 38
2.3.1.2 Scanning Electron Microscopy 39
2.3.1.3 Transmission Electron Microscopy 41
2.3.1.4 Scanning Transmission Electron Microscopy 49
2.3.2 Scanning Probe Microscopy 49
2.3.2.1 Scanning Tunneling Microscopy 49
2.3.2.2 Atomic Force Microscopy 52
2.4 Spectroscopy 53
2.4.1 Energy-Dispersive X-Ray Spectroscopy 53
2.4.2 Electron Energy-Loss Spectroscopy 55
2.4.3 X-Ray Absorption Spectroscopy 57
2.4.4 X-Ray Photoelectron Spectroscopy 58
2.4.5 Raman Spectroscopy 62
2.4.6 Infrared Spectroscopy 64
2.5 Summary 66
References 66

3 Synthesis, Characterization, and Biomedical Applications of Graphene 69
Albert Dato, Velimir Radmilovic und Michael Frenklach
3.1 Introduction 69
3.2 Synthesis of Graphene 70
3.2.1 Chemical Exfoliation 71
3.2.2 Epitaxial Growth 71
3.2.3 Substrate-Free Gas-Phase Synthesis 72
3.2.4 Chemical Vapor Deposition 72
3.2.5 Arc Discharge of Graphite Electrodes 72
3.2.6 Liquid-Phase Production 72
3.3 Characterization of Graphene 73
3.3.1 Raman Spectroscopy 73
3.3.2 Transmission Electron Microscopy 75
3.3.3 Electron Diffraction 75
3.3.4 Electron Energy Loss Spectroscopy 76
3.3.5 Elemental Analysis 77
3.4 Biomedical Applications of Graphene 78

3.4.1	Biocompatible Graphene Paper	79
3.4.2	Drug Delivery	79
3.4.3	Biodevices	81
3.4.4	Imaging of Soft Materials	82
3.5	Conclusions	83
	References	83
4	**Carbon Nanohorns and Their Biomedical Applications**	**87**
	Shuyun Zhu and Guobao Xu	
4.1	Introduction	87
4.2	Structure and Properties	88
4.3	Functionalization	90
4.3.1	Covalent Functionalization	91
4.3.2	Noncovalent Functionalization	92
4.4	Biomedical Applications	93
4.4.1	Toxicity Assessment of SWCNHs	95
4.4.2	SWCNHs Used in Drug-Delivery Systems	95
4.4.3	SWCNHs Used in Magnetic Resonance Analysis	100
4.4.4	Biosensing Applications of SWCNHs	101
4.5	Conclusions	103
	Acknowledgments	104
	References	104
5	**Bio-Inspired Magnetic Carbon Materials**	**111**
	Elby Titus, José Gracio, Duncan P. Fagg, Manjo K. Singh and Antonio C. M. Sousa	
5.1	Introduction	111
5.2	Allotropic Forms of Carbon	112
5.3	Magnetism in Diamond	113
5.3.1	Biomedical Applications of Magnetic Diamond	113
5.4	Magnetism in Graphite	115
5.4.1	Biomedical Applications of Magnetic Graphite	116
5.5	Magnetism in Carbon Nanotubes/Fullerenes	117
5.5.1	Biomedical Applications of Magnetic Carbon Nanotubes/Fullerenes	120
5.6	Magnetism in Graphene	124
5.6.1	Biomedical Applications of Magnetic Graphene	125
5.7	Conclusion	126
	References	126
6	**Multi-Walled Carbon Nanotubes for Drug Delivery**	**133**
	Nicole Levi-Polyachenko	
6.1	Introduction	133
6.2	Gene Therapy	138
6.3	Antibacterial Therapy	140

6.4	Wound Healing *142*
6.5	Chemotherapy *145*
6.5.1	Hyperthermic Drug Delivery Using CNTs *146*
6.5.2	Drug Transport Using CNTs *150*
6.6	Summary and Future Perspectives *154*
	References *155*

7	**Carbon Nanotube-Based Three-Dimensional Matrices for Tissue Engineering** *161*
	Izabela Firkowska and Michael Giersig
7.1	Introduction *161*
7.2	Carbon Nanotubes *162*
7.3	Carbon Nanotubes for Matrix Enhancement *164*
7.4	Cellular Responses to CNT-Based Matrices *166*
7.5	CNT Engineering into Three-Dimensional Matrices *166*
7.5.1	Vertically Aligned CNT-Based Matrices *166*
7.5.2	Three-Dimensional Cavity Network of Interconnected Nanotubes *170*
7.5.3	Freestanding MWNT-Based Matrix *175*
7.5.3.1	Modification of the MWNT-Based Matrix Surface with Bioactive Calcium Phosphate Nanoparticles *178*
7.6	Summary *180*
	References *182*

8	**Electrochemical Biosensors Based on Carbon Nanotubes** *187*
	Jonathan C. Claussen, Jin Shi, Alfred R. Diggs, D. Marshall Porterfield and Timothy S. Fisher
8.1	Introduction *187*
8.2	CNT Properties *188*
8.2.1	Mechanical *188*
8.2.2	Electrical *188*
8.2.3	Chemical/Electrochemical *189*
8.3	Electrochemical Biosensing *190*
8.4	CNT-Based Electrode Fabrication *190*
8.4.1	Adsorption *190*
8.4.2	Covalent Bonding *192*
8.4.3	Polymer Entrapment *192*
8.4.4	Aligned Arrays *195*
8.4.4.1	Nanoelectrodes *197*
8.4.5	Hybrid (CNT/Metal Nanoparticle) Electrodes *197*
8.5	Applications *199*
8.5.1	Nonenzymatic Biosensing *199*
8.5.1.1	Nicotinamide Adenine Dinucleotide (NADH) *199*
8.5.1.2	Homocysteine *201*
8.5.1.3	Dopamine *202*

8.5.1.4	Indole Acetic Acid (IAA)	202
8.5.2	Enzymatic Biosensing	203
8.5.2.1	Glucose	203
8.5.2.2	Glutamate	205
8.5.2.3	Ethanol	207
8.6	Conclusions	209
	References	210

9 Single-Walled Carbon Nanotube Biosensors 217
Jeong-O Lee and Hye-Mi So
- 9.1 Introduction 217
- 9.2 The Sensing Mechanisms of Nanotube Biosensors 218
- 9.3 The Immobilization of Biomolecules on SWNTs 221
- 9.3.1 Covalent Binding 221
- 9.3.2 Noncovalent Binding 222
- 9.3.3 Other Immobilization Methods (Metal Particles, etc.) 223
- 9.4 Various Receptors for Nanotube Biosensors 224
- 9.4.1 Aptamers 224
- 9.4.2 Fragment Antibodies 227
- 9.4.3 Enzymes and Proteins 229
- 9.4.4 Other Receptor Types 230
- 9.5 The Application of Nanotube Biosensors to Pathogen Detection 231
- 9.6 The Future of Nanotube Biosensors 234
- References 235

10 Environmental Impact of Fullerenes 239
Naohide Shinohara
- 10.1 Introduction 239
- 10.2 Methods Used to Prepare Fullerene Suspensions 239
- 10.2.1 Solubility of Fullerene 239
- 10.2.2 Aqueous Suspensions of Fullerenes 246
- 10.2.3 Toxicity of Aqueous Fullerene Suspensions as a Factor of the Dispersion Method 246
- 10.3 Toxicological Data Relating to Fullerenes 247
- 10.3.1 Toxicological Effects of C_{60} on Fish 247
- 10.3.2 Toxicological Effects of C_{60} on Invertebrates 250
- 10.3.3 Toxicological Effects of C_{60} on Algae 251
- 10.3.4 Toxicological Effects of C_{60} on Bacteria and Soil Microbes 251
- 10.3.5 Toxicological Effects of C_{60} on Other Organisms 254
- 10.4 Possible Emission Sources of C_{60} 254
- 10.5 The Environmental Fate of C_{60} 262
- 10.6 Fullerenes in the Environment 265
- 10.7 Conclusion 265
- References 266

11 Computational Tools for the Biomedical Application of Carbon Nanomaterials 271
Leela Rakesh

- 11.1 Introduction 271
- 11.2 Simulation Methods 278
- 11.2.1 Background 278
- 11.2.2 Molecular Modeling 280
- 11.3 Results and Discussions 281
- 11.3.1 Branched PEGylated DPCC Functionalized PTX Physical Loading on SWNTs 287
- 11.3.2 Interaction of Irinotecan-co-br-PEG-DPCC-unco-SWNT and with ssDNA (Adenine-Thymine (AT)) 292
- 11.3.3 Interaction of Nanotube with 20-Base Pair Guanine–Thymine -ssDNA in the Presence of Calcitriol 296
- 11.4 Future Perspectives 298
- 11.5 Executive Summary 300

Acknowledgments 301
References 301

Part Two Overview of Applications in Cancer 307

12 Carbon Nanotubes for Cancer Therapy 309
William H. Gmeiner

- 12.1 Introduction 309
- 12.1.1 Limitations of Current Therapy Options 309
- 12.1.2 Developing Nanomaterials for Cancer Treatment 310
- 12.1.3 CNTs: Physical Properties, Manufacture, and Chemical Modifications 311
- 12.2 Hyperthermia for Cancer Treatment 312
- 12.2.1 Current Ablative Technologies 315
- 12.2.2 Use of CNTs for Hyperthermia Treatment 316
- 12.3 CNTs for Drug Delivery 320
- 12.3.1 Localization of CNTs to Malignant Tissues 321
- 12.3.2 Drug Delivery Using CNTs 322
- 12.4 Imaging Using CNTs 323
- 12.5 CNT-Related Toxicity 324
- 12.6 Summary and Future Perspective 325

Acknowledgments 326
Abbreviations 326
References 327

13 Cancer Treatment with Carbon Nanotubes, Using Thermal Ablation or Association with Anticancer Agents 333
Roger G. Harrison, Luís F. F. Neves, Whitney M. Prickett and David Luu

- 13.1 Introduction 333

13.2	Use of Nanotubes as Heated Particles *334*
13.3	Use of Anticancer Agents Associated with Nanotubes *338*
13.4	Summary *343*
13.5	Future Perspective *344*
	Acknowledgments *345*
	References *345*

14 Carbon Nanotubes for Targeted Cancer Therapy *349*
Reema Zeineldin

14.1	Introduction *349*
14.2	Cancer *350*
14.3	Conventional Cancer Chemotherapy versus Nanocarrier-Mediated Drug Delivery *352*
14.3.1	Challenges with Chemical Compounds as Therapeutic Agents *352*
14.3.2	Advantages of Nanocarriers as Drug-Delivery Vehicles *352*
14.4	Carbon Nanotubes as Drug-Delivery Vehicles *353*
14.5	Cellular Uptake of CNTs *354*
14.6	Functionalization of CNTs with Polyethylene Glycol *355*
14.7	Targeting of Cancers *357*
14.7.1	Passive Targeting *357*
14.7.2	Active Targeting *358*
14.7.3	Trafficking of Targeted Drug-Delivery Vehicles *358*
14.8	Targeted Cancer Therapy Employing CNTs and a Critique of Current Studies *359*
14.8.1	erbB Family Members *360*
14.8.2	Folate Receptor α *363*
14.8.3	Biotin Receptor *365*
14.8.4	Integrins *366*
14.8.5	Markers for Lymphomas or Leukemias *367*
14.8.6	Disialoganglioside (GD2) *368*
14.9	Summary and Future Perspective *368*
	Acknowledgments *371*
	References *371*

15 Application of Carbon Nanotubes to Brain Tumor Therapy *381*
Dongchang Zhao and Behnam Badie

15.1	Introduction *381*
15.2	The Current Challenge of Brain Tumor Therapy *382*
15.2.1	Current Status of Clinical Practice in Brain Tumor Therapy *382*
15.2.2	The Progress of Investigational Therapies for Brain Tumors *382*
15.2.2.1	Targeted Molecular Therapy *382*
15.2.2.2	Anti-Angiogenic Therapy *383*
15.2.2.3	Immunotherapy *383*
15.2.2.4	Gene Therapy *384*
15.3	The Characteristics of CNTs for Biological Applications *385*
15.3.1	Single-Walled and Multi-Walled Carbon Nanotubes *385*

15.3.2	Functionalization of CNTs	385
15.3.2.1	Covalent Surface Modification	385
15.3.2.2	Noncovalent Surface Modification	386
15.3.3	CNT Delivery System	386
15.3.3.1	Delivery of Antibodies and Peptides	387
15.3.3.2	Delivery of siRNA	387
15.3.3.3	Delivery of DNA Molecules	387
15.3.3.4	Delivery of CpG	388
15.3.3.5	Delivery of Vaccines	388
15.3.3.6	Delivery of Chemical Drugs	388
15.4	Strategies of Application of CNTs to Brain Tumor Therapy	389
15.4.1	CNTs Targeting Brain Tumor-Macrophages	389
15.4.1.1	Internalization of CNTs by BV2 Microglia Cells *in vitro*	389
15.4.1.2	Preferential Uptake of CNTs by Macrophages in a Glioma Model	390
15.4.1.3	Phosphatidylserine-Coated CNTs Targeting Microglia/Macrophages	391
15.4.2	CNTs Targeting Tumor Cells and Preliminary Efforts Towards *In Vivo* Cancer Therapy	392
15.4.2.1	CNTs Actively Targeting Tumor Cells	392
15.4.2.2	CNTs Passively Targeting Tumor Cells	392
15.4.2.3	CNTs Thermal Effects on Tumor Cells	393
15.5	Toxicity Issues of CNTs in Brain Tumor Therapy	394
15.6	Conclusions and Future Directions	395
	Acknowledgments	395
	References	395

16 Carbon Nanotubes in Cancer Therapy, including Boron Neutron Capture Therapy (BNCT) *403*
Amartya Chakrabarti, Hiren Patel, John Price, John A. Maguire and Narayan S. Hosmane

16.1	Introduction	403
16.2	Carbon Nanotubes in the Treatment of Cancer	403
16.2.1	Drug Delivery	404
16.2.2	Imaging and Probing	406
16.2.3	Photothermal and Photoacoustic Therapy	407
16.3	BNCT and Its Development through Nanotechnology	409
16.3.1	BNCT: A Brief Overview	409
16.3.2	Liposomes	411
16.3.3	Dendritic Macromolecules	411
16.3.4	Magnetic Nanoparticles	413
16.4	The Role of Carbon Nanotubes in BNCT	413
16.5	Summary and Future Outlook	415
	References	415

17	**Fullerenes in Photodynamic Therapy** *419*
	Sulbha K. Sharma, Ying-Ying Huang, Pawel Mroz, Tim Wharton,
	Long Y. Chiang and Michael R. Hamblin
17.1	Introduction *419*
17.2	Photodynamic Therapy *420*
17.2.1	Traditional Photosensitizers *421*
17.2.2	Photophysics and Photochemistry in PDT *422*
17.2.3	Anticancer Mechanism of PDT *424*
17.2.3.1	Cellular Effects *424*
17.2.3.2	*In Vivo* Effects *425*
17.2.4	Antimicrobial Mechanism of PDT *425*
17.3	Fullerenes as Photosensitizers *425*
17.3.1	Photophysics of Fullerenes *426*
17.3.2	Photochemistry of Fullerenes *427*
17.3.3	Interactions of Fullerenes with DNA *429*
17.3.4	Drug-Delivery Strategies for Fullerenes *430*
17.3.5	Strategies to Overcome the Unfavorable Spectral Absorption of Fullerenes *432*
17.3.5.1	Covalent Attachment of Light-Harvesting Antennae to Fullerenes *433*
17.3.5.2	Two-Photon PDT *434*
17.4	Anticancer Effects of Fullerenes *436*
17.4.1	*In Vitro* PDT with Fullerenes *436*
17.4.2	*In Vivo* PDT with Fullerenes *438*
17.5	Fullerenes for Antimicrobial Photoinactivation *439*
17.5.1	Photoinactivation of Viruses *439*
17.5.2	Photoinactivation of Bacteria and Other Pathogens *440*
17.6	Summary and Future Perspectives *441*
	Acknowledgments *442*
	References *442*

Index *449*

Preface
Vol 9 – Carbon Nanomaterials for the Life Sciences

Since the discovery of the C_{60} molecule, the family of carbon nanomaterials has been steadily growing. Carbon nanotubes, the most famous, were the first to be explored for their potential in biological and biomedical applications. Since then, a number of other types of carbon nanomaterials ranging from graphene, fullerenes, carbon nanoparticles, and carbon nanohorns have also been gaining importance thanks to development of methods for their chemical modification and bio-functionalization in order to conjugate them with proteins, carbohydrates, nucleic acids and any other biomolecule. This opened up an entirely new and exciting research direction involving Carbon Nanomaterials and biology especially in targeting the cell's behavior at the sub cellular or molecular level. This book, Carbon Nanomaterials for the Life Sciences, is the 9th volume in the series on Nanomaterials for the Life Sciences and covers the latest advances in synthesis and bio-functionalization of carbon nanomaterials along with their applications in drug delivery, biosensing, tissue engineering, and cancer therapy. The book is broadly divided into two parts. In the first part, there are eleven chapters covering various aspects of synthesis, characterization and different biomedical applications. In the second part there are six chapters with their combined focus on applications in cancer treatment and diagnosis.

The book begins with a chapter on synthesis of carbon nanomaterials particularly in the context of their biofunctionalization and applications in both biochemistry and biology along with a perspective on reducing their toxicity. In the second chapter, some of the most important techniques applicable for the characterization of nanocarbons starting with classic diffraction techniques, followed by imaging techniques using electron microscopy and scanning probe microscopy are described. In addition, some important spectroscopic methods which are essential for determination of electronic structure and functionalities are also included. The third chapter is completely dedicated graphene exploring its synthesis, characterization, and biomedical applications. In this chapter evidence is provided to show its promising application in drug delivery, bio devices, and molecular imaging. The fourth chapter brings out the importance of carbon nanohorns (CNHs), a more recent addition to the list of carbon allotropes. Even though the field of carbon nanohorns is still in its infancy, a number of biomedical applications have been explored due to their unique structures and properties and the chapter four captures the most recent ones. The theme of the fifth chapter is exploring magnetic properties of carbon nanomaterials (diamond, graphite and carbon nanotubes) and the chapter provides both fundamental explanations for magnetism as well as the latest developments related to their synthesis, characterization and applications in biomedicine.

The focus of the sixth chapter is multi-walled nanotubes (MWNT). MWNTs by definition are many nested tubes of multiple concentric sidewalls with diameters of a few nanometers up to hundreds of nanometers and lengths in the micron range. In this chapter application of MWNT for a broad array of drug delivery applications including delivery of agents to modify genetic function, transdermal delivery, antibiotic agents and delivery of chemotherapeutic agents directly via intracellular transport or indirectly via hyperthermia to induce intracellular transport are presented. Moving from drug delivery applications, the seventh chapter provides reader with up to date information on investigations into suitability of carbon nanotubes as biomaterial for tissue engineering highlighting a broad range of issues starting from fabrication, properties and performance of three dimensional carbon nanotube-based matrices. Highlight of this chapter is the attention on the biocompatibility studies proving the capability of carbon nanotubes to support long-term survival of osteoblast and fibroblast cells.

Yet another fascinating application of carbon nanomaterials in general and carbon nanotubes in particular is biosensing taking advantage of their electrochemistry. In the eight chapter, therefore, various properties of CNTs including mechanical, electrical, and chemical/electrochemical properties that make them enticing for biosensing were discussed. This was followed description of latest results regarding various types of electrochemical sensors were highlighted including those based on amperometric, enzymatic systems. Continuing to elaborate on biosensors, the ninth chapter describes the use of various types of single-walled carbon nanotube field effect transistors (SWNT-FETs) in biomedical applications. It will be obvious from reading the chapter that it is possible to develop highly sensitive biosensors for diverse purposes by using SWNT-FETs and suitable receptors.

The final two chapters in the first section deal with two conceptually different topics. In the first chapter, tenth in the book, readers are provided an opportunity to understand fullerenes from the point of view of environmental hazard assessment. In the last chapter utilization of computational simulations tools for the development of PEGylated-lipophilic polymers with or without functionalized SWNT as drug delivery carriers for anticancer therapeutics is presented. This chapter is extremely important as it demonstrates the necessity of computer modeling to identify biocompatible and safe nanomaterials and nanotools for early detection of diseases.

With around 1.5 million people diagnosed and 1/3 rd of them dead in year 2009, Cancer continues to be one of the top deadliest diseases of humankind. Adding to the problem is several types of cancers such as pancreatic, glioblastoma multiforme remain essentially incurable. Successful nanotechnology-based approaches, for both sensitive and timely diagnosis as well as effective therapy of a number of primary and metastatic tumors, are providing a glimmer of hope for cancer patients. The second section in this book is completely dedicated to similar efforts albeit focusing only on carbon nanomaterials. In this section, the first chapter explains several important advances for the use of carbon nanotubes, especially in the photothermal treatment, drug delivery and tumor imaging for cancer treatment in pre-clinical models. Continuing to stress on one of the most important properties of carbon nanotubes, their strong absorbance of light in the near-infrared range (700-1400 nm), leading to their heating, the next chapter (Chapter 13) summarizes progress on the use of carbon nanotubes to treat cancer through heating of the nanotubes by near-infrared or radiofrequency field radiation. The chapter also covers therapeutic approaches taking advantage of complexation of the nanotubes with anticancer agents. The next chapter in this section is devoted to targeted cancer therapy using carbon nanotubes. CNTs for drug delivery have the potential to practically realize the concept of "magic bullet." The chapter provides latest review on the topic. This is the first review of its kind focusing on advantages of CNT based nanocarriers in targeted cancer drug delivery including details about CNT functionalization and cellular uptake.

In the 15th chapter, readers will find most recent review on application of CNTs to brain tumor therapy. It highlights critical characteristics of CNTs for biological applications. It also brings out current challenges in brain tumor therapy with special emphasis on the use of CNTs to target tumor macrophages with a perspective on toxicity issues. The penultimate chapter summarizes the role of carbon nanotubes in several forms of cancer therapy. It also brings out the essential role of CNTs in boron neutron capture therapy (BNCT), a technology that appears promising. Form BNCT, the last chapter takes the readers to photo dynamic therapy; a non-surgical, minimally invasive approach for the treatment of solid tumors and many non malignant diseases. In this chapter, the ability of fullerenes to absorb visible light leading to generation of reactive single oxygen (ROS) that can efficiently inactivate viruses and kill cancer cells and pathogenic microbial cells is highlighted. Thanks to fullerenes the field of PDT got a new lease of life even though fullerenes do have some disadvantages in their optical absorption spectrum compared to currently well established materials. However, according to the authors of the chapter, rational strategies to overcome these deficiencies do exist and more will likely be developed.

Overall, this book is the first of its kind summarizing the literature to date on application of Carbon Nanomaterials in life sciences. It is invaluable for the reader as contributors for each of the chapters are renowned experts in their particular fields. I am grateful for their contributions. Thank you for reading!

Challa S. S. R. Kumar
Baton Rouge, USA

List of Contributors

Behnam Badie
City of Hope National Medical Center
Division of Neurosurgery
1500 East Duarte Road
Duarte, CA 91010
USA

Amartya Chakrabarti
Northern Illinois University
Department of Chemistry and Biochemistry
DeKalb, IL 60115
USA
Southern Methodist University
Department of Chemistry
Dallas, TX 75275
USA

Long Y. Chiang
University of Massachusetts
Department of Chemistry
Lowell, MA 01854
USA

Jonathan C. Claussen
Purdue University
Birck Nanotechnology Center
Bindley Biosciences Center
Department of Agricultural and Biological Engineering
West Lafayette, IN 47907
USA

Albert Dato
University of California
at Berkeley
Applied Science and Technology Graduate Group
230 Bechtel Engineering Center
Berkeley, CA 94720-1708
USA

Alfred R. Diggs
Purdue University
Birck Nanotechnology Center
Bindley Biosciences Center
Department of Agricultural and
Biological Engineering
West Lafayette, IN 47907
USA

Duncan P. Fagg
University of Aveiro
Center for Mechanical
Technology & Automation
Nanotechnology Research
Division
3810-193 Aveiro
Portugal

Izabela Firkowska
Freie Universität Berlin
Department of Physics
Arnimalle 14
14195 Berlin
Germany

Timothy S. Fisher
Purdue University
Birck Nanotechnology Center
School of Mechanical Engineering
West Lafayette, IN 47907
USA

Michael Frenklach
University of California
at Berkeley
Applied Science and Technology
Graduate Group
230 Bechtel Engineering Center
Berkeley, CA 94720-1708
USA

Michael Giersig
Freie Universität Berlin
Department of Physics
Arnimalle 14
14195 Berlin
Germany

William H. Gmeiner
Wake Forest University School
of Medicine
Medical Center Boulevard
Department of Cancer Biology
Winston-Salem, NC 27157
USA

José Gracio
University of Aveiro
Center for Mechanical
Technology & Automation
Nanotechnology Research
Division
3810-193 Aveiro
Portugal

Michael R. Hamblin
Massachusetts General Hospital
Wellman Center for
Photomedicine
Boston, MA 02114
USA
Harvard Medical School
Department of Dermatology
Boston, MA 02115
USA
Harvard-MIT
Division of Health Sciences and
Technology
Cambridge, MA 02139
USA

Roger G. Harrison
University of Oklahoma
School of Chemical, Biological
and Materials Engineering
100 E. Boyd
Norman, OK 73019
USA

Narayan S. Hosmane
Northern Illinois University
Department of Chemistry and
Biochemistry
DeKalb, IL 60115
USA

Ying-Ying Huang
Massachusetts General Hospital
Wellman Center for
Photomedicine
Boston, MA 02114
USA
Harvard Medical School
Department of Dermatology
Boston, MA 02115
USA
Aesthetic, Esthetic and Plastic
Center of Guangxi Medical
University
Nanning 530022
China

Jeong-O Lee
Korea Research Institute of
Chemical Technology
NanoBio Fusion Research Center
Shinseongno 19, Eusung-Gu
Daejeon 305-343
Korea

Nicole Levi-Polyachenko
Wake Forest University Health
Sciences
Department of Plastic and
Reconstructive Surgery
Winston-Salem, NC 27157
USA

David Luu
University of Oklahoma
School of Chemical, Biological
and Materials Engineering
100 E. Boyd
Norman, OK 73019
USA

John A. Maguire
Southern Methodist University
Department of Chemistry
Dallas, TX 75275
USA

Pawel Mroz
Massachusetts General Hospital
Wellman Center for
Photomedicine
Boston, MA 02114
USA
Harvard Medical School
Department of Dermatology
Boston, MA 02115
USA

List of Contributors

Luís F. F. Neves
University of Oklahoma
School of Chemical, Biological
and Materials Engineering
100 E. Boyd
Norman, OK 73019
USA

Hiren Patel
Northern Illinois University
Department of Chemistry and
Biochemistry
DeKalb, IL 60115
USA

D. Marshall Porterfield
Purdue University
Birck Nanotechnology Center
Bindley Biosciences Center
Department of Agricultural and
Biological Engineering
Weldon School of Biomedical
Engineering
West Lafayette, IN 47907
USA

John Price
Northern Illinois University
Department of Chemistry and
Biochemistry
DeKalb, IL 60115
USA

Whitney M. Prickett
University of Oklahoma
School of Chemical, Biological
and Materials Engineering
100 E. Boyd
Norman, OK 73019
USA

Velimir Radmilovic
University of California
at Berkeley
Applied Science and Technology
Graduate Group
230 Bechtel Engineering Center
Berkeley, CA 94720-1708
USA

Leela Rakesh
Central Michigan University
Department of Mathematics
Center for Polymer Fluid
Dynamics & Applied
Mathematics, Science of
Advanced Materials
Mount Pleasant
Michigan, MI 48859
USA

Sulbha K. Sharma
Massachusetts General Hospital
Wellman Center for
Photomedicine
Boston, MA 02114
USA

Jin Shi
Purdue University
Birck Nanotechnology Center
Bindley Biosciences Center
Weldon School of Biomedical
Engineering
West Lafayette, IN 47907
USA

Naohide Shinohara
National Institute of Advanced
Industrial Science and
Technology (AIST)
Research Institute of Science for
Safety and Sustainability (RISS)
16-1 Onogawa, Tsukuba
Ibaraki 305-8569
Japan

Manjo K. Singh
University of Aveiro
Center for Mechanical
Technology & Automation
Nanotechnology Research
Division
3810-193 Aveiro
Portugal

Hye-Mi So
National NanoFab Center
Daejeon 305-806
Korea

Antonio C. M. Sousa
University of Aveiro
Center for Mechanical
Technology & Automation
Nanotechnology Research
Division
3810-193 Aveiro
Portugal

Dang Sheng Su
Fritz-Haber Institute of the
Max-Planck Society
Department of Inorganic
Chemistry
Faradayweg 4-6
14195 Berlin
Germany
Institute of Metal Research,
Chinese Academy of Science
Catalysis and Materials Division
Wenhua Road
110016 Shenyang
China

Jean-Philippe Tessonnier
Fritz Haber Institute of the
Max-Planck Society
Department of Inorganic
Chemistry
Faradayweg 4-6
14195 Berlin
Germany

Elby Titus
University of Aveiro
Center for Mechanical
Technology & Automation
Nanotechnology Research
Division
3810-193 Aveiro
Portugal

Tim Wharton
Lynntech Inc.
College Station, TX 77840
USA

Guobao Xu
Chinese Academy of Sciences
Changchun Institute of Applied
Chemistry
State Key Laboratory of
Electroanalytical Chemistry
5625 Renmin Street
Changchun, Jilin 130022
China

Reema Zeineldin
Massachusetts College of
Pharmacy and Health Sciences
Department of Pharmaceutical
Sciences
19 Foster Street
Worcester, MA 01608
USA

Dongchang Zhao
City of Hope National Medical
Center
Division of Neurosurgery
1500 East Duarte Road
Duarte, CA 91010
USA

Shuyun Zhu
Chinese Academy of Sciences
Changchun Institute of Applied
Chemistry
State Key Laboratory of
Electroanalytical Chemistry
5625 Renmin Street
Changchun, Jilin 130022
China
Graduate University of the
Chinese Academy of Sciences
Beijing 100864
China

Part One
Overview of Synthesis, Characterization, and Applications in Biomedicine

1
Carbon Nanomaterials: Synthetic Approaches
Jean-Philippe Tessonnier

1.1
Introduction

The strength of carbon nanomaterials lies in the fact that most of their properties can be modified and adjusted, depending on the target application(s). The morphology, surface chemistry, as well as the physical properties of carbon nanomaterials are all susceptible to change by tuning the synthesis procedures, or by employing post-treatments. As the possibilities are almost unlimited, not only the materials scientists but also (bio-)chemists and biologists have at their disposal an immense playground to invent new materials for electronics, transportation, energy production and storage, or medicine. In the latter case, carbon nanomaterials – such as fullerenes, nanohorns, and nanotubes – have already contributed to the development of new biosensors, diagnostic tools, drugs, and therapies [1–4].

Although nanomaterials such as fullerenes or carbon nanotubes (CNTs) are remarkably popular and already available from several suppliers – as in the case of any other chemical compound – their synthesis remains a dynamic field of research. Today, improvements are continuously being made to increase the selectivity of desired nanomaterialsas, as well as their purity and quality. In the case of CNTs, much progress has been made in controlling their diameter, length, and chirality – all of which may be critically important, depending on the desired application(s). In contrast to organic molecules, carbon nanomaterials do not exhibit a textbook structure; rather, they contain impurities, structural defects and heteroatoms, all of which may have possible implications for the target application. The aim of this chapter is to bridge the gap between materials scientists, biologists and/or biochemists, by showing not only how carbon nanomaterials are produced, but also highlighting the consequences that specific synthesis parameters have on the functionalization and application(s) of these materials in both biochemistry and biology.

1.2
General Concepts on the Synthesis of Carbon (Nano-)Materials

1.2.1
Uncatalyzed Synthesis of Carbon (Nano-)Materials

The uncatalyzed synthesis of carbon materials or nanomaterials always proceeds through the temperature-driven rearrangement of carbon atoms provided by a precursor. Precursors in any physical state – whether gaseous (carbon monoxide, hydrocarbons), liquid (thermoplastic polymers, hydrocarbons), or solid (wood, sugars, polymers) – can be employed [5]. Common examples are activated carbon produced from the pyrolysis of wood, or candle soot resulting from the incomplete combustion of wax. The thermal decomposition of the precursor proceeds through pyrolysis for gases, carbonization for liquids, and decomposition in the case of solids.

The structure of the resulting material depends on both the nature of the precursor and the temperature of treatment [5–8]. The variety of carbon materials which exists will not be discussed in more detail at this point, as most of these materials are not employed in the life sciences. However, a basic knowledge of the mechanisms involved during graphitization is important to better understand those parameters that are crucial during functionalization (see Sections 1.6 and 1.7). Carbon materials are composed of small crystallites – also termed basic structural units (BSUs) – that are crosslinked in either a three-dimensional (3-D) or a two-dimensional (2-D) network [5, 8]. In the latter case, the material can be further graphitized. After synthesis, the crystallites (sp^2-hybridized carbon) are interconnected with disorganized carbon, at which stage a large proportion of heteroelements is still present in the material (Figure 1.1).

Thermal treatment, performed in a vacuum or in an inert gas (annealing) at 1500 K, removes the heteroelements, while simultaneously any five-membered rings are converted into six-membered rings and the BSUs grow to form larger crystallites (Figure 1.1). At approximately 1900 K, the carbon atoms linking the BSUs become organized into large polycyclic molecules. However, the graphene sheets are not fully parallel (as in graphite), due to the existence of rows of sp^3 carbon (defects). Finally, at 2500 K all of the carbon atoms become hybridized sp^2,

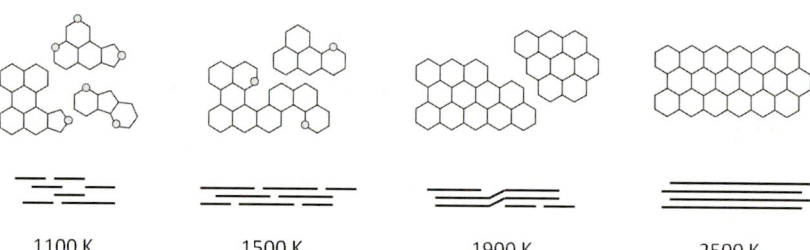

Figure 1.1 Elementary steps occurring during the thermal treatment of graphitizable materials. The circles mark the presence of heteroatoms.

and are integrated into graphene sheets. The stacking is then considered to be perfect, such that synthetic graphite is obtained.

1.2.2
Catalyzed Synthesis of Carbon (Nano-)Materials

The decomposition of gaseous hydrocarbons on transition metals to form solid carbon deposits is a well-known phenomenon in heterogeneous catalysis. In petrochemistry, these reactions lead to coking of the catalyst and trigger its deactivation by encapsulating the active sites [9, 10]. Indeed, during the 1950s, several groups utilized such knowledge of catalyst deactivation to synthesize carbon filaments [11, 12]. Subsequently, during in the 1970s and 1980s, small vapor-grown carbon fibers (VGCFs) with diameters below 100 nm were synthesized via the decomposition of hydrocarbons on transition metal nanoparticles [13, 14]. As a consequence, the VGCFs were introduced into polymers for purposes of their reinforcement, or to render them electrically conductive (e.g., to dissipate electrostatic electricity in fuel pipes) [15]. Very soon, carbon fibers of different sizes were obtained by varying the catalyst or the reaction conditions. In 1991, Iijima described in his groundbreaking publication the structure of carbon nanotubes [16]. The subsequent enthusiasm for CNTs led many research groups to attribute their discovery to Iijima, although they were actually already observed in the 1950s [11, 12].

The mechanisms involved in the catalyzed synthesis of carbon materials differ radically from those associated with the pyrolysis of hydrocarbons. Typically, the hydrocarbons are decomposed to the elementary carbon atoms on the metallic surface of the catalyst, and then recombined to form solid carbon. Most carbon is directly incorporated into crystallites (sp^2 carbon) and the amount of sp^3 carbon is significantly lower than for materials prepared by pyrolysis (Figure 1.2) [17, 18].

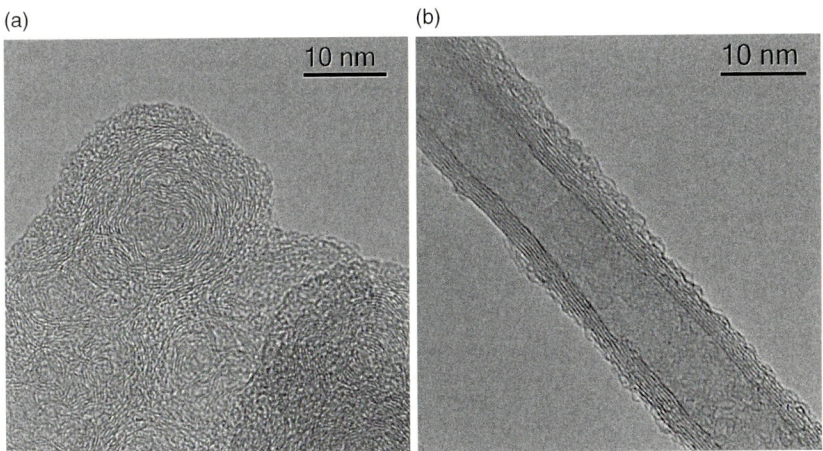

Figure 1.2 HRTEM images. (a) Soot particles; (b) Multi-wall carbon nanotube. Note that the crystallites are curved and shorter in the case of soot.

In addition, the morphology of the recombined carbon depends on the shape of the catalyst. For example, a planar surface will lead to graphene sheets, while a particle would produce carbon (nano-)fibers or single- and multi-wall CNTs (see Section 1.4). The catalyst, by definition, also acts on the kinetics of the solid carbon formation, by lowering the activation energy of the reaction. Consequently, the synthesis can also be conducted at a much lower temperature than the uncatalyzed pyrolysis, below 1200 K.

1.3
Synthesis from Solid Precursors

Carbon exhibits three natural allotropes for which all the atoms are in the same hybridization state: diamond (sp^3 carbon); graphite (sp^2); and carbyne (sp) [8, 19]. Starting from these allotropes, several polytypes (or families of carbon materials) can be distinguished which have carbon atoms of various hybridization states coexisting in the same material. For example, activated carbon and carbon black are made of crystallites (sp^2 carbon) interconnected with sp^3 carbon atoms (see Section 1.2.1). At a very early stage, attempts were made to identify paths which linked the different polytypes and allotropes, there being a twofold aim to: (i) understand carbon materials formation in spatial dust; and (ii) be able to transform carbon materials on demand. In many of these fundamental experiments, solid carbon (usually graphite) was exposed to extreme conditions so as to dissociate and recombine carbon atoms and, indeed, several carbon nanomaterials were created, notably nanodiamonds, fullerenes, and CNTs.

1.3.1
Nanodiamonds

Nanodiamonds (NDs) constitute an emerging class of materials for biological, and in particular for medical, applications [20–24]. Although NDs have been known since the late 1950s [25], very few reports exist relating to their potential applications in the life sciences [21, 22]. Importantly, NDs have one major advantage over other carbon nanomaterials such as fullerenes and nanotubes, namely that their biocompatibility has already been demonstrated [22]. *Diamondoids*, which are subnanometer clusters of similar structure, are used in drugs [26]. For example, adamantane – the smallest of these clusters – is an important building block for drugs used to treat Parkinson's disease and Alzheimer's disease. Recently, a new synthetic technique has allowed the creation of ultrananocrystalline diamond (UNCD) films [27], which have been used to produce biosensors or to coat biomedical devices and implants. Unfortunately, this technique is incapable of producing NDs in sufficiently large amounts, and has therefore been restricted to niche applications [23].

Typically, ND nanoparticles exhibit sizes between 5 and 25 nm, and are therefore more suited than UNCD films to a broad range of applications in the life sciences, for example as drug and gene carriers [21, 24]. Two different ND synthetic techniques which were developed during the 1960s are currently used on an industrial scale, namely shock-wave synthesis and explosive detonation. In both cases, high pressures and temperatures are generated in order to convert a solid carbon precursor into diamond.

1.3.1.1 Turning Graphite into Diamond

The high-pressure high-temperature (HPHT) synthesis, which was developed during the late 1950s [25], is based on the initial proposal of Rossini and Jessup (in 1938), and later by Bridgman, that there is a thermodynamic pressure and temperature equilibrium line between graphite and diamond in the phase diagram of carbon [28]. Therefore, it should be possible to directly transform graphite into diamond if the pressure and temperature are sufficiently high, and if the formed diamond can be quenched in order to avoid its retransformation into graphite. The research group at General Electric managed to circumvent the technical difficulties encountered, and developed an instrument that was capable of pressing graphite at up to 20 GPa and heating it at temperatures of up to 4000 K [29, 30]. In the early experiments, small diamonds were produced from a graphite bar at 13 GPa and 3300 K within a few milliseconds [29]. However, it was shown later that diamonds could also be produced under milder conditions if Group VIII transition metals were mixed with the graphite powder. The best results were obtained when using metal alloys, for example Ni–Fe–Mn, with diamonds being produced at 5 GPa and 1750 K [28]. As large amounts of metal or metal alloys were typically employed, interest was centered on the so-called "catalyst–solvent" effect, although because of the harsh synthesis conditions involved, studies of the growth mechanism with *in situ* techniques remained extremely challenging. To date, the role of neither the metal nor metal alloy has been fully elucidated, although several mechanisms have been proposed (see Refs [28, 31]). The HPHT technique is better suited to the synthesis of large diamonds (micron to millimeter scale), with diamonds of up to 2 carats (0.4 g) currently being produced using this method [32].

During the 1960s, DuPont developed, in parallel, a procedure to recreate the HPHT conditions by using shock waves. For this, a steel plate is covered with explosive (e.g., trinitrotoluene; TNT) and placed in front of a physical mixture of iron and graphite. After detonation, the steel plate collides with the iron–graphite mixture at an approximate velocity of 5 km s^{-1}; this produces a peak pressure of 30 GPa and causes the temperature to increase to about 1000 K within a few milliseconds [32]. Shock-wave synthesis typically leads to high-quality, monocrystalline and polycrystalline particles (up to 60 μm) that are then milled to obtain NDs of approximately 25 nm diameter [22].

1.3.1.2 Explosive Detonation Synthesis

Explosive detonation synthesis was developed during the 1960s in the former USSR [22, 23]. The process involves the creation of an explosion in a hermetic tank, with the explosive producing the HPHT conditions required to grow NDs, while simultaneously providing the necessary carbon atoms. An additional carbon precursor is not needed. In this process, TNT or any other explosive is typically mixed with cyclotrimethylenetrinitramine (hexogen) in an oxygen-deficient ratio, after which the explosive is ignited in the closed tank which has previously been filled with an inert gas. Within a fraction of a second, the pressure in the tank can increase to 20–30 GPa, while the temperature may reach (locally) 3000 K. The excess carbon atoms in the gas phase are condensed into ND particles, typically of 5 nm diameter. Because of their extremely small size, NDs produced by detonation synthesis are often referred to as ultra-dispersed diamonds (UDDs) or ultra-nanocrystalline diamonds (UNCDs). In order to prevent the transformation of UNCDs into graphite, or their oxidation with CO, CO_2 and H_2O that has been generated by the explosion, the temperature must be rapidly lowered, at a rate of at least 3000 K min^{-1}. The result of the detonation is a mixture called "diamond soot"; this contains 40–80% diamonds, 35–45% graphite-like particles, and 1–5% metal and metal oxide impurities. The diamonds are then separated and purified by chemical treatment, with strong oxidants such as chromic acid [23] or gas-phase HNO_3 [33] being used to oxidize and dissolve the impurities. In general, depending on the purification procedure used, commercial samples contain 90–99% diamond and 1–10% graphite impurities.

Because of their size, UNCDs are perhaps more suited to biological applications than are NDs produced by shock-wave synthesis. However, it must be emphasized that UNCDs synthesized by detonation are composed of a diamond core covered by few shells of onion-like graphite, as well as a layer of disorganized carbon (Figure 1.3) [33, 34]. Aleksenskii *et al.* demonstrated that oxidizing treatments

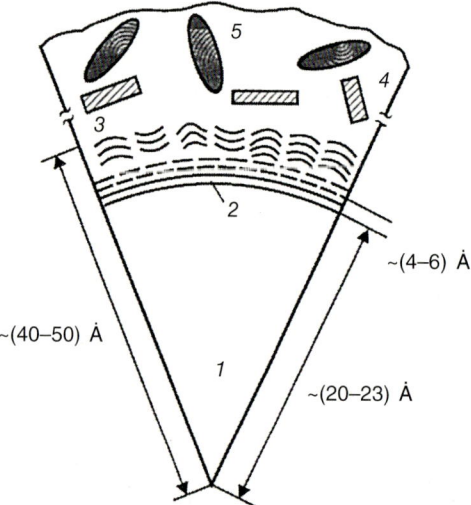

Figure 1.3 Schematic representation of the UNCD structure. The diamond core (1) is covered with few onion-like graphite shells (2), as well as a thick disorganized layer containing curled graphene sheets (3), graphite-like nanoparticles (4), and metal oxide inclusions (5). Reproduced from Ref. [33].

under harsh conditions (e.g., with 70% HNO_3 in an autoclave at 523 K) is sufficient to remove the disorganized carbon layer, as well as most of the onion-like shells [33]. These oxidizing treatments also lead to the creation of many oxygen-containing functional groups on the UNCD surface. Both, the presence of onion-like graphite and/or of functional groups interfere during the synthesis of UNCD-based composites, as well as during biological applications [22] (see also Section 1.7).

1.3.2
Fullerenes, Nanohorns, Single- and Multi-Wall Carbon Nanotubes

Fullerenes were discovered in 1985, as a result of Kroto's experiments to understand the formation of carbon clusters [35]. For this, carbon atoms were evaporated from a graphite disk by laser ablation, and then allowed to condense and cluster in the cold part of the set-up. Kroto optimized the reaction conditions to obtain C_{60} as a major product. It is interesting to note that theoreticians predicted the existence of fullerene (and even some of its properties) long before its discovery [19, 36]. Unfortunately, laser ablation produced only minute amounts of C_{60}. A second breakthrough occurred in 1990, when Krätschmer and Huffman developed a set-up to produce carbon clusters and soot, with the aim of identifying bands in the absorption spectra of interstellar dust [37]. The carbon atoms were evaporated by creating an electric arc (arc discharge synthesis) between two graphite electrodes in a chamber filled with 100–200 Torr helium, such that carbon deposits containing soot and fullerenes were condensed on the cold parts of the chamber. This set-up allowed the gram-scale production of fullerenes, and permitted further study of their physical and chemical properties. Shortly thereafter it was found that, by introducing slight variations in the reaction conditions (e.g., using a DC current instead of AC, or different He pressures), both single- and multi-wall CNTs could be grown, using the same procedure [16, 38]. At a later date, single-wall carbon nanohorns were also obtained [39, 40].

Although fullerenes synthesized by arc discharge are typically mixed with other carbon materials, their purification is relatively easy. Fullerenes are soluble in various organic solvents [41], and so can be easily separated from other carbon materials such as soot by Soxhlet extraction (e.g., using toluene). Following evaporation of the solvent, a mixture of C_{60} with fullerenes of higher mass is obtained, and this can be further purified using liquid chromatography [19]. Alternatively, the fullerene/soot mixture can simply be filtered over a charcoal/silica gel plug, with a relatively high yield. However, in contrast to liquid-phase chromatography, this method might not be capable of separating C_{60} from small organic molecules, such as polycyclic aromatic hydrocarbons (PAHs) [19]. As PAHs are known to be both carcinogenic and mutagenic, it is important to take great care in the purification of C_{60} before their further use in biological applications. After purification, it is also important to consider possible reactions with oxygen when storing C_{60} in air. It has been shown that oxygen is adsorbed strongly onto the C_{60} surface and,

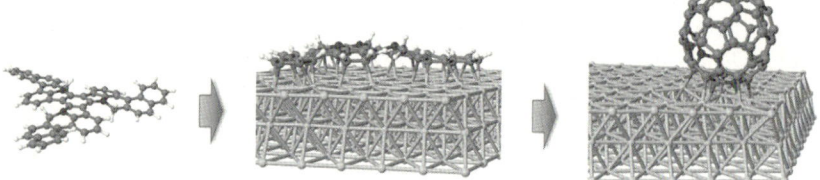

Figure 1.4 Molecular structure of the $C_{57}H_{33}N_3$ molecule during its cyclodehydrogenation on a platinum(111) surface to produce C_{60}. Adapted from Ref. [45].

depending on temperature, can form oxygen-containing functional groups such as peroxo species, which are also known to be cytotoxic [42, 43].

Although arc-discharge synthesis typically produces high-quality carbon nanomaterials, with very few structural defects, the method is difficult to scale up for mass production. Consequently, alternative synthesis techniques have been investigated and developed. In the case of fullerenes, wet chemistry [44], as well as the platinum-catalyzed cyclodehydrogenation of organic polyaromatic precursors offer interesting and scalable solutions (Figure 1.4) [45]. For CNTs, catalytic chemical vapor deposition (CCVD) offers the best results.

1.4
Catalytic Chemical Vapor Deposition

1.4.1
Definitions

Catalytic chemical vapor deposition (CCVD) is, as its name clearly indicates, a variation of the chemical vapor deposition (CVD) technique. Although the link seems obvious, many research groups have investigated CCVD via a purely catalytic approach, forgoing the knowledge acquired over more than 60 years with CVD. It is, however, essential to first grasp the fundamental physical and chemical phenomena that govern CVD in order to obtain a complete picture of CCVD. The rediscovery of CVD over the past few years has created new opportunities to synthesize single-wall carbon nanotubes (SWNTs) and multi-wall carbon nanotubes (MWNTs), without using a metal catalyst, and thus facilitating applications in both biology and medicine (see Section 1.4.5) [46–52].

The CVD of carbon was first employed during the 1880s, and has been widely developed since the 1950s [32, 53]. The process consists of the thermal decomposition (pyrolysis) of a gaseous carbon precursor (usually a hydrocarbon, such as methane, ethylene or acetylene) on the surface of a substrate (1.1):

$$C_xH_y \rightleftarrows xC + \frac{y}{2}H_2 \qquad (1.1)$$

1.4 Catalytic Chemical Vapor Deposition

This reaction obeys thermodynamic and kinetic rules, which must be taken into account in order to obtain homogeneous carbon coatings. For hydrocarbon precursors, carbon deposition is thermodynamically favored at high temperature, above 1000 K. For example, CVD with methane as a precursor is performed at a temperature between 1373 and 1673 K [32]. Then, by adjusting the reaction parameters such as temperature, total pressure, gas velocity and nature, and the partial pressure of the carbon precursor, it is possible to control the deposition rate and quality of the carbon coating [54]. As the gas-phase pyrolysis of hydrocarbons leads to the formation of carbon blacks (soot-like particles), it is crucial that the reaction occurs on the surface of the substrate and not in the gas phase, in order to obtain homogeneous coatings.

For CCVD, a transition metal – generally Fe, Ni, Co or one of their alloys – is used to catalyze the decomposition of the carbon precursor. Consequently, the temperature of the reaction can be significantly decreased; typically, CCVD is carried out between 700 and 1273 K. Because the reaction is catalyzed, other carbon-containing precursors may also be employed; for example, carbon can be deposited by CO disproportionation [e.g., the Boudouard reaction; Eq. (1.2)]:

$$2CO \rightleftarrows C + CO_2 \tag{1.2}$$

In contrast to hydrocarbons, the rate of the CO disproportionation reaction reaches a maximum at 900 K, after which CO_2 formation is thermodynamically favored and the reaction rate decreases [55, 56]. It must be emphasized that, as for CVD, the deposited carbon reproduces the shape of the metal substrate; that is, flat particles lead to flat graphene sheets, while round nanoparticles produce nanofilaments (nanotubes and nanofibers). The elementary steps involved in the formation of the nanofilaments are discussed in Section 1.4.2.

As no consensus has yet been reached, the scientific literature proposes confusing definitions for CNTs. The most common definitions derive from research groups who classify as a nanotube any carbon filament with an external diameter less than 100 nm and having a tubular shape – that is, with a cylindrical pore at its center. Unfortunately, however, a purely morphological definition does not take the structure of the nanofilament into account. Indeed, as can be seen in Figure 1.5, it implies that materials with a completely different structure, and thus with different physical and chemical properties, will share the same name [18]. In contrast, Dresselhaus and Endo defined a CNT as a nested coaxial array of SWNTs, with each nanotube being formed by a graphene sheet rolled into a cylinder of nanometer-size diameter [15]. In order to avoid any ambiguity in this chapter, and in the scientific literature in general, it is suggested that reference is made to Dresselhaus and Endo's definition, as this also meets the recent recommendations of the International Organization for Standardization (ISO) [57].

Some examples of carbon nanomaterials produced using CCVD are shown in Figure 1.5. Vapor-grown carbon (nano)fibers (VGCFs; also referred to as VGCNFs) are obtained by feeding methane and iron carbonyl (as catalyst precursor) into a horizontal reactor maintained at 1373 K (Figure 1.5a). Typically, the VGCFs

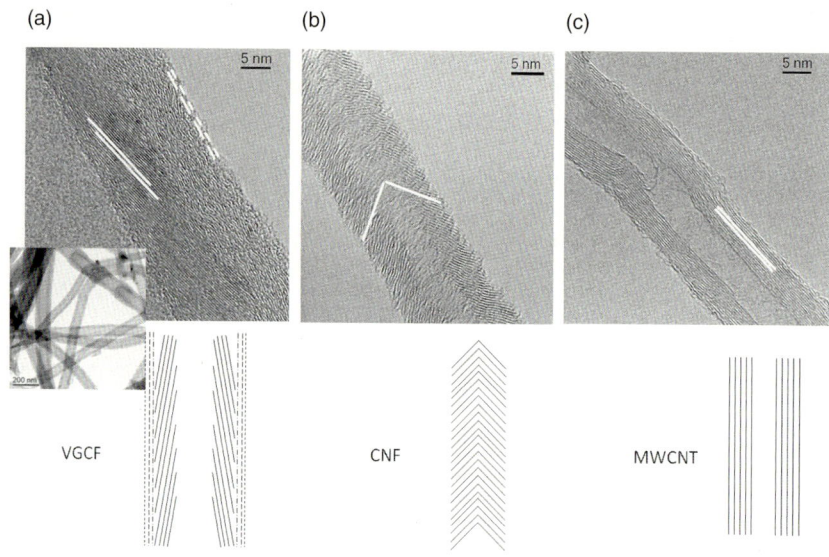

Figure 1.5 HRTEM images highlighting the structural differences of (a) a VGCF, (b) a CNF, and (c) a MWNT. The inset in panel (a) shows a low-magnification TEM image of the same material (VGCF).

produced are significantly larger than the MWNTs, with 80–90 nm instead of 10–15 nm external diameter, and they also show a dual structure [18]. The core, which is grown catalytically, exhibits a fishbone-like stacking of the graphene sheets. However, because of the high temperature required for the reaction, a thick layer of pyrolytic carbon is also formed by CVD on the external walls of the VGCFs. The structures of commercial fishbone carbon nanofibers (CNFs) and MWNTs, obtained by CCVD of ethylene on Ni and FeCo nanoparticles, are shown in Figure 1.5b and c, respectively.

1.4.2
Mechanistic Aspects

The reaction mechanism involved in CCVD has been investigated since the 1970s, mainly as the result of research into heterogeneous catalyst deactivation. Indeed, the formation of carbon deposits represents a major issue for many important petrochemical processes [10]. In the case of the steam-reforming reaction, the hydrocarbons are converted to CO and H_2. In this case, the reaction takes place under conditions where it competes thermodynamically with the Boudouard reaction (Eq. 1.2), or with the pyrolysis of the hydrocarbon [9]. Solid carbon, as a byproduct, can poison the catalyst by encapsulating the metal particles, leading to a complete deactivation of the catalyst material. In addition, carbon nanofilaments can grow inside the pores of the catalyst support and lead to its fragmentation.

1.4 Catalytic Chemical Vapor Deposition

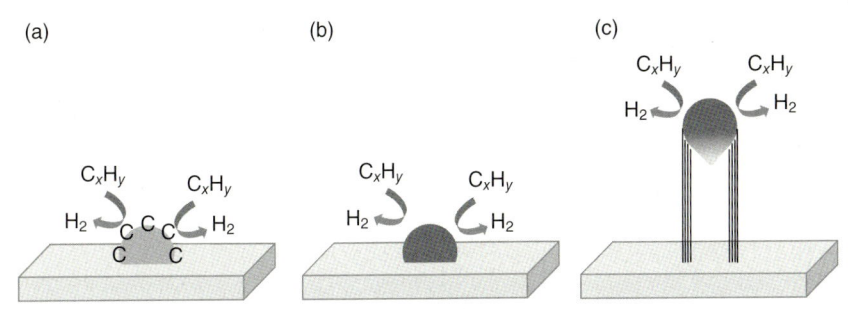

Figure 1.6 The three steps involved in the VLS mechanism. (a) Decomposition of the carbon-containing precursor on the surface of the catalyst particle; (b) Diffusion of carbon atoms through the particle as a solid solution; (c) Precipitation of carbon at the metal–support interface and formation of a nanofiber or a nanotube.

The first carbon filaments were observed, using electron microscopy, during the 1950s [11, 12, 19]. Some time later the pioneering studies of Baker on the formation of carbon filaments on nickel (a steam-reforming catalyst), as well as on Group VIII metals and metal alloys, led to the establishment of a growth mechanism which is still widely accepted today [58, 59]. In fact, Baker suggested that the vapor–liquid–solid (VLS) theory developed by Wagner and Ellis to explain the growth of Si whiskers [60] might be transposed to also explain the growth of carbon filaments [58]. The VLS mechanism consists of three successive steps (Figure 1.6) [58]. First, a carbon precursor present in the gas phase is adsorbed onto the surface of the catalyst particle, and then dissociates to form elementary carbon atoms. The carbon atoms then dissolve in the metal and diffuse though the particle. Finally, the carbon atoms precipitate at the metal–support interface to form a carbon nanofilament.

Whilst the VLS mechanism is still widely employed to explain the growth of CNFs, as well as of SWNTs and MWNTs [61–63], its validity is often discussed, with the main problem relating to the second step. Carbon diffusion on/through the catalyst particle would be expected to be the rate-limiting step of the reaction. Baker showed that the activation energy calculated from experimental data was in good agreement with values obtained for carbon dissolution in metal single crystals [58, 59, 64]. The results of recent kinetic experiments also support this mechanism in the case of CNFs and MWNTs grown on iron [63, 65, 66]. However the driving force which would push carbon atoms to diffuse through the whole particle remains unclear [67–72]. In addition, the VLS mechanism implies the formation of a metal carbide. Iron carbide (Fe_3C) is often observed using transmission electron microscopy (TEM) and X-ray diffraction (XRD) after growth. However fresh Fe_3C does not catalyze the CNT growth [73]. It appears, therefore, that iron carbide is either the deactivated catalyst, or is formed during cooling to room temperature.

Consequently, it was proposed by several groups during the late 1970s and 1980s that carbon diffusion should rather take place on the surface of the catalyst particle than through the particle [74, 75] and, indeed, many studies now support this hypothesis. More recent *ab initio* molecular dynamics simulations have shown that the surface diffusion of carbon atoms on a 1 nm Fe particle, and their incorporation into the forming graphene sheet, occurs significantly faster than does carbon diffusion in the bulk of the particle [76]. Raty *et al.* failed to observe any carbon atoms plunging into the surface of the metallic particle in their calculations, while Hofmann *et al.* demonstrated, both experimentally and theoretically, that the activation energy for surface diffusion is significantly lower than for bulk diffusion in the case of Ni, Fe, and Co [77]. Helveg *et al.* provided the first time-resolved *in-situ* TEM images of a MWNT growing from a Ni particle [78]. In this case, the lattice fringes of the Ni were observed throughout the experiment, thus proving that CNTs can grow from solid Ni particles. Hofmann *et al.* studied the chemical and physical state of nickel and iron particles extensively, using both *in-situ* TEM and *in-situ* X-ray photoelectron spectroscopy (XPS) [79–81]. It was concluded by these authors that CNTs grow from solid particles, and that a carbide formation could be excluded. Recently, Wirth *et al.* conducted an extensive reinvestigation of the reaction kinetics over a large range of temperatures and pressures [82], and showed the growth rate to depend both on dissociation of the hydrocarbon at the surface of the catalyst, and on surface diffusion of the atomic carbon. Thus, it appears that the diffusion of carbon through the particle, though plausible at high temperature (especially for iron), is in fact an exception [83–85]. Typically, CNTs probably grow following a three-step CCVD mechanism that involves dissociation of the carbon precursor, surface diffusion of the carbon atoms, and their precipitation in the form of a nanotube.

1.4.3
Single- and Multi-Wall Carbon Nanotubes

Both, SWNTs and MWNTs are typically synthesized with similar catalysts, with the diameter of the metal catalyst particle being the critical factor defining whether SWNTs or MWNTs are produced [86–92]. Large particles of 10 nm or more lead to MWNTs, whereas SWNTs are synthesized from particles smaller than 2 nm. In addition, the diameter of the CNT is most often directly proportional to the parent catalyst particle. Nasibulin *et al.* demonstrated that, for SWNTs, the particle is 1.6-fold larger than the corresponding nanotube, independent of the synthesis conditions [93]. In the case of SWNTs, it must also be highlighted that their electronic properties depend on their diameter, and on their chirality. A SWNT is formed by a graphene sheet rolled along an axis termed chiral vector C_h [13], which is defined by $C_h = n\vec{a}_1 + m\vec{a}_2$ (Figure 1.7), where a_1 and a_2 are the two vectors. In the (n,m) notation $(n,0)$ and $(0,m)$ denote zigzag nanotubes, whereas (n,n) are armchair SWNTs. All other (n,m) correspond to chiral nanotubes. A SWNT exhibits metallic properties when its $|n - m|$ is a multiple of 3. In all other cases, the SWNT is semiconducting, and therefore, an armchair SWNT is always

1.4 Catalytic Chemical Vapor Deposition | 15

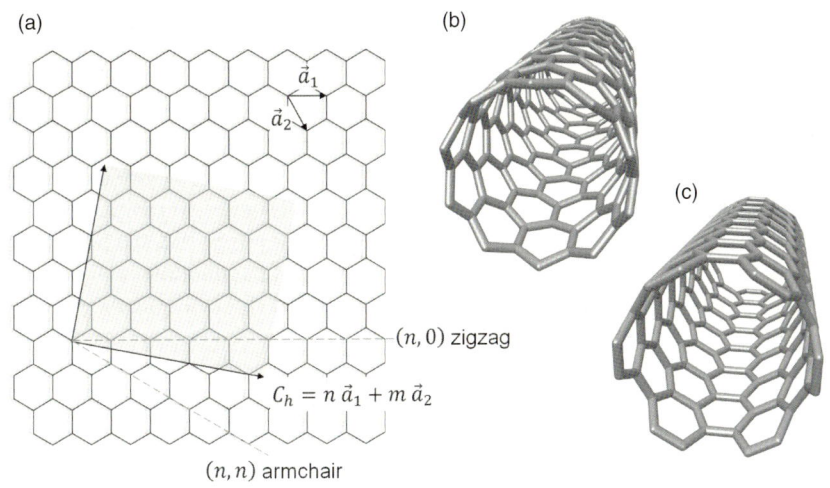

Figure 1.7 Construction of a SWNT by rolling a graphene sheet along the chiral vector C_h. (a) Depending on (n,m), the SWNT can be zigzag, armchair or chiral; (b) Structure of a (5,0) zigzag SWNT; (c) Structure of a (6,0) armchair SWNT.

metallic. The number of (n,m) possibilities depends on the SWNT diameter (Figure 1.7). Control of the catalyst particle size is, therefore, of crucial importance.

1.4.3.1 Floating Catalyst CCVD

In the case of the floating catalyst CCVD technique, the catalyst precursor – usually a metallocene [94] or a metal carbonyl [56] – is directly vaporized in the reaction chamber at the same time as the carbon-containing precursors. The organometallic compound is immediately decomposed to form catalyst nanoparticles which remain suspended in the gas stream. In fact, two different reactions take place simultaneously: (i) the catalyst particle grows by CVD; and (ii) the SWNTs grow on the catalyst by CCVD, with the reaction conditions being very carefully controlled.

In the HiPCO process (developed at Rice University and based on this concept), iron carbonyl $Fe(CO)_5$ is injected along with carbon monoxide in a reactor maintained at 1300 K [56]. The iron particles are maintained at a diameter of 1–2 nm by utilizing a reverse CVD reaction (i.e., the reaction of Fe with CO to reform the volatile $Fe(CO)_5$), and in the meantime the SWNTs grow from the suspended Fe particles by CO disproportionation. In order to increase the CCVD reaction rate, the reaction is performed at 20 bar pressure. Compared to other processes, the HiPCO system produces high-quality SWNTs, with minimal carbon impurities (the main impurity is iron, but this can be removed with mild acid treatment). It has been shown recently that the floating catalyst CCVD also offers a degree of chirality control when tuning the composition of bimetallic catalyst nanoparticles, thus opening a new route to the selective synthesis of semi-conducting or metallic SWNTs [94].

1.4.3.2 Immobilized Catalyst CCVD

Although floating catalyst CCVD offers excellent results, the complexity of the reaction and the relatively low yields typically obtained favor alternative processes. Of these, heterogeneous catalysts, where the active phase is immobilized on a substrate or in a matrix, represent the best option. Thus, by tuning the properties of the catalyst support as well as the catalyst synthesis parameters, it is possible to control the particle size and enhance the catalytic activity of the active phase.

During catalyst preparation, two steps have a major influence on the average particle size and its statistical distribution – namely, the catalyst synthesis, and its thermal pretreatment (e.g., calcination, reduction). A catalyst is usually composed of an active phase (the transition metal) as well as a substrate on which the active phase is deposited (the catalyst support), or a matrix in which the active phase is dispersed. A strong affinity between the metal and the support or matrix is important at each step of the catalyst preparation. In the case that the metal–substrate interaction is low, the aggregation of metal atoms becomes easy, and large metal particles will be formed. Such aggregation can take place during impregnation of the support with the solution containing the metal salt precursor, during drying, during the calcination step (when the metal salt is decomposed into a metal oxide precursor), or during reduction (when the metal oxide is converted into the final metallic active phase). At the relatively high temperatures needed for calcination and/or reduction, the metal atoms and nanoparticles may become mobile and sinter to form larger particles. The latter are usually unselective in CNT synthesis, and typically become encapsulated with one or several layers of carbon. In addition, sintering may lead to a broader distribution of particle diameters. Subsequently, the final product may contain CNTs of various diameters, ranging from a few nanometers to more than 30 nm, and consequently aggregation and sintering must be avoided. In general, high-surface-area oxides, such as silica, alumina or magnesia, are employed as supports, as their strong interaction with the metal particles hinder their diffusion and sintering [95]. Several groups have shown that alumina provides the best results with most metals and metal alloys [96–98]. For example, Fe/Al_2O_3 and $FeCo/Al_2O_3$ are employed as catalysts in the industrial production of MWNTs [66, 99]. Mattevi et al. showed that, for Fe/Al_2O_3, Fe exists in 2+ and 3+ states at the interface with the support [98], and concluded that the Fe particles were chemically bonded to the alumina. This strong interaction led to smaller particles, with a narrower size distribution, than when Fe was supported on silica. In addition, the strain at the interface may well induce a reconstruction of the Fe particle, and this might account for the higher catalytic activity of Fe/Al_2O_3 compared to Fe/SiO_2. Another clear difference observed when replacing low-interaction supports (e.g., silica) with a high-interaction counterpart (e.g., alumina) lies in the change from a tip-growth to a root-growth mechanism. The CNT growth mechanism (see Section 1.3.2) implies carbon precipitation at the rear of the particle, at the interface with the support; as a result, the particle will be lifted and the growth catalyzed by the particle located at the tip of the nanotube (tip growth). However, if the metal–support interaction is strong, the catalyst particle will remain anchored to the support surface, and growth will proceed with the particle located at the root of the CNT (root growth). Gohier *et al.* demonstrated that the mechanism could also switch from tip growth to root growth simply by decreasing the metal particle size [100], which explains why SWNTs usually follow the root growth mechanism. It should also be noted that, after reduction, the catalyst can be further modified by exposing it to defined gas compositions and temperatures. For instance, Zhou *et al.* showed that, under defined conditions, the metal particles would undergo a degree of structural reconstruction [101]; this allowed the growth of either platelet CNFs, fishbone CNFs or MWNTs, using the same catalyst, by simply varying the pretreatment conditions. The morphology (diameter, length), structure (alignment of the walls), existence of structural defects, homogeneity of the CNTs (same size), yield and formation of carbon-containing impurities are all parameters that depend on the growth conditions.

Alternatively, the metal can also be dispersed in mixed-oxide crystal structures, for example in perovskite type ABO_3 and spinel-type AB_2O_4 oxides [102]. The A or B crystallographic sites are then occupied by Fe, Ni, or Co cations. In this case, the metal remains typically dispersed in the oxide matrix during calcinations; however, during the reduction step the metal cations are reduced and metal nanoparticles begin to nucleate. Zhao *et al.* showed that this technique would lead to the creation of homogeneous MWNTs [103].

1.4.4
Aligned Carbon Nanotubes

In general, CNT arrays are synthesized for applications in electronics (e.g., for field emission [4]), and therefore most studies have in the past focused on the growth of SWNT arrays on coated silicon wafers, although the techniques developed can also be used to grow MWNTs and/or CNFs. In addition, other supports can be employed, including either monolithic (plates) or in a powder form (alumina grains) [104]. Notably, the catalysts developed to synthesize randomly oriented CNTs can also be generally used to grow CNT arrays. For this, the Si wafers are first coated with a layer of silica or alumina, after which the metal is either deposited as a film by evaporation or by the CVD of ferrocene. A final thermal treatment with hydrogen or ammonia leads to dewetting of the support and the formation of metal nanoparticles [98]. The diameter of the nanoparticles—and consequently of the CNTs—is directly proportional to the thickness of the metal film. The critical parameter governing the growth of aligned CNTs is the density of the catalyst particles on the support. For example, if the density of the metal nanoparticles is low, then the SWNTs will grow in random directions [98]. The density of the SWNTs must reach approximately 5×10^{11} per cm^2 in order to become vertically aligned [105]. In this situation, the neighboring SWNTs will then support each other in a concerted step, and growth will proceed as a "forest," orthogonal to the support. Parameters such as the temperature, pressure, and nature of the precursor each have a major influence on the quality of the mat, and on the length of the nanotubes [82]. In general, plasma-enhanced CVD (PECVD) provides better results because the plasma dissociates the hydrocarbon precursor to more reactive species. By using PECVD, Hofmann *et al.* were able to grow aligned nanofilaments at 393 K, which made it possible to synthesize forests on thermally unstable films such as plastics. However, it was noted that below 543 K, the structure of the nanofilament was essentially amorphous because insufficient energy was given to the system to anneal any defects [106]. Reasonably organized CNFs were obtained at temperatures in excess of 663 K. In order to increase the life time of the catalyst, oxidizing compounds such as water were added to the feed, in order to prevent any coke formation that could poison the catalyst. This "super-growth" technique led to very long SWNTs [107], with arrays of more than 0.5 mm being obtained within a few minutes [108]. Because of the high density of SWNTs, the growth rate can decrease due to diffusion limitations of the reactant gas molecules to the metal nanoparticles (root growth). However, this problem can be circumvented by patterning the catalyst so that CNT-free channels exist where the gas molecules can diffuse freely [109]. It was also shown that the geometry of the reactor would play a significant role. Typically, a cold wall shower head reactor offers the best results, because the reactants are fed directly orthogonal to the catalyst [110].

1.4.5
Carbon Nanotubes Synthesized from Biocompatible Catalysts

Although Fe, Ni and Co are the best catalysts to grow CNTs, it has been shown that many other elements can be used to grow CNTs [111]. Recently, several groups have demonstrated the possibility of growing SWNTs on oxides, such as silica [49, 50], alumina [49], or zirconia [52]. The nucleation and growth of the SWNTs was either attributed to CVD on solid silica nanoparticles [50] or at a molten silica particle of diameter <2 nm [49]. It was also suggested that the SiO_2 particles were first transformed into SiC by a carbothermal reaction, and that the SiC particles then catalyze the CVD [48]. In any case, this observation constitutes a breakthrough in CNT synthesis. Both, SWNTs and SWNT arrays can be grown on virtually any support, the only limiting factor being the formation of nanoparticles. In these preliminary studies, silica nanoparticles were formed either by the sputtering of SiO_2 (instead of thermal growth under oxidative conditions) [50] or by scratching the surface with a Si wafer [50] or a diamond blade [49].

Liu *et al.* showed that it is possible to grow short SWNTs on these transition metal-free supports [51], because the growth velocity was significantly lower ($8\,\text{nm}\,\text{s}^{-1}$ instead of $4\text{--}20\,\mu\text{m}\,\text{s}^{-1}$ for metal catalysts). Consequently, Liu *et al.* were able to control accurately the length of the SWNTs – a technique which in future might be of great interest for biological applications. The fact that CNTs are shorter and do not contain any metal impurities would favor their phagocytosis and reduce their pathogenicity [112].

1.4.6
Metal- and PAH-Induced Toxicity of Carbon Nanotubes

Since the evaluation of CNT-related toxicity is a time-demanding process, most investigations are ongoing. In the past, preliminary results have often been contradictory depending on the CNT products, their purification, and their functionalization. However, there appears to be some level of agreement with regards to the cytotoxicity of the remaining catalyst particles. In 2007, Pisanic *et al.* showed that iron oxide nanoparticles hindered the growth of neurons [113]. Both, SWNT and MWNT products contain various amounts of transition metals as well as oxide impurities from the catalyst supports. Depending on the synthesis process used and their possible purification, CNTs contain between 1% and 30% of inorganic impurities [114, 115], most of which can be easily removed by washing with acids and bases (see Section 1.5). The presence of carbon impurities is usually neglected, except in the case of SWNTs prepared by arc discharge, which can contain up to 30–50% carbon soots in the raw material [19]. Commercially available CCVD-prepared SWNTs and MWNTs are expected to be free of carbon impurities – an assumption which is generally based on images recorded using scanning electron microscopy (SEM) and transmission electron microscopy (TEM). At very high magnification, however, the CNT's surface is rarely clean (see Figure 1.8), with large amounts of carbonaceous debris being observed. Such debris includes large polycyclic molecules (most likely graphene sheets fragments) formed during the CNT synthesis. Although PAHs exhibit a similar structure, they are significantly

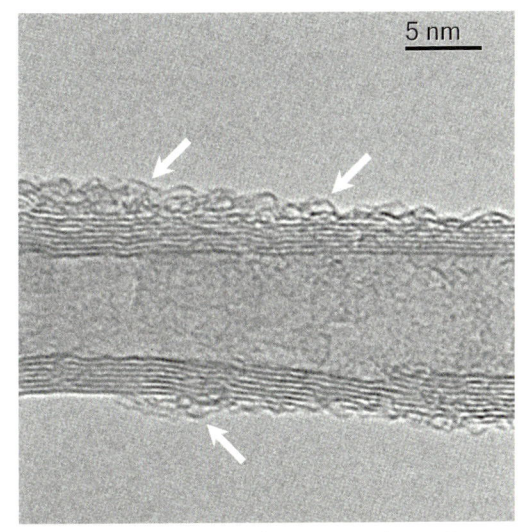

Figure 1.8 HRTEM image of a commercial MWNT. Most MWNTs exhibit various amounts of carbonaceous debris dispersed on their surface. In some cases, the debris can almost completely cover the nanotube's surface (arrows in the image).

smaller, and therefore it is very difficult to observe them even when using high-resolution TEM (HRTEM). The PAHs are known to be carcinogenic and mutagenic and, if present on the CNT surface, they may have a significant influence on the results of toxicity studies. Plata *et al.* investigated the PAHs present in different CNT products and also in the gas stream during CNT synthesis [116, 117], and in both cases concluded that whilst some PAHs were observed, their concentrations were lower than were found in the air. Nonetheless, these results must be regarded with great caution. In the case of PAHs present in the gas phase, only the CCVD reaction performed with ethylene as carbon precursor was investigated. Although the reaction conditions are relevant to industrial CNT production, higher PAH concentrations may be identified for bench-scale CNT synthesis when other precursors are employed. When PAHs were present in commercial CNT products, Plata *et al.* used organic solvents to extract them. Although extraction with toluene is an effective and well-known purification technique for fullerenes, it has not been shown as successful for extracting PAHs from CNTs, as the PAHs interact strongly with the CNT walls via van der Waals interactions. Consequently, noncovalent functionalization techniques have been developed, in which desired functions are grafted onto pyrenes (see Section 1.7). Subsequently, the interaction between the arene and the CNT was shown to be strong, with the functional groups anchored irreversibly onto the CNTs walls, such that a total extraction of the PAHs would not be possible. Finally, the high affinity of PAHs for CNTs was confirmed by Yang *et al.*, who suggested that, with time, PAHs present in the air might be adsorbed onto CNTs, thus increasing their potential toxicity [118].

1.5
Purification Techniques

The purification of CNTs involves the removal of any inorganic impurities as well as disordered carbon present in the raw product. Inorganic impurities mainly consist of the remaining catalyst, although additional impurities derived either from the catalyst preparation or from the CNT synthesis and handling may be identified. Although, in general, Fe, Ni or Co, as well as alumina, silica and magnesia, are found (Figure 1.9) [114, 115], Cr, Cu, Mo or Mn have also been found occasionally in the raw material. Washing with mineral acids such as HCl, HNO_3 and H_2SO_4 represents a cheap and efficient method to remove most of the inorganic impurities. On some occasions, the acid treatment is followed by a base treatment, which is more efficient for dissolving the catalyst support. To date, the reaction conditions reported have varied widely with regards to the concentration of acid, and the reaction temperature and reaction time. Typically, the acid washing is conducted with concentrated HNO_3, at between 373 and 403 K, and for 2 to 20 h. Nitric acid is used most often because it is a strong acid and also a good oxidant; this allows any metal impurities to be dissolved and most of the disordered carbon to be oxidized in a single step. Unfortunately, this method has one major disadvantage, in that its efficiency varies for different CNT products, and so the condi-

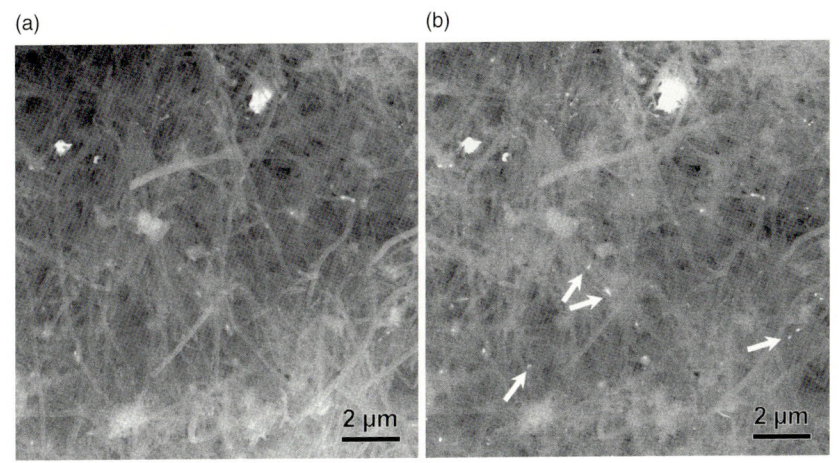

Figure 1.9 SEM images of commercial CNTs obtained from (a) secondary electrons and (b) back-scattered electrons. The back-scattered electron signal is a function of the atomic number of the elements in the sample. Remaining catalyst particles appear as bright spots in the image, making their detection easier (arrows).

tions must be optimized. Whilst weak conditions might not be effective for removing all of the impurities, a strong treatment would functionalize the CNT surface and create a wide variety of oxygen-containing functional groups [119]. In the worst case, the CNT might even be destroyed. Notably, SWNTs are more reactive than MWNTs because of their curvature, and are easily damaged under strong oxidizing conditions. Salzmann et al. have developed a simple technique to remove most of the carbonaceous debris covering SWNTs, without destroying them [120, 121]. Initially, the debris and the SWNT surface are oxidized with HNO_3, after which the oxidized SWNTs are washed with sodium hydroxide solution. Because of the high pH, the surface acidic groups will be deprotonated, while both the debris and the CNTs will become negatively charged. The consecutive electrostatic repulsion is sufficiently strong to detach the debris from the CNT surface, and it can be removed by washing with distilled water, followed by simple filtration and rinsing. After purification, the CNTs can be annealed at high temperature in vacuum or in inert gas, in order to remove the functional groups and heal any defects.

Recently, much effort has been expended in developing new catalysts capable of growing SWNTs with desired chiralities. This task is extremely challenging, and, instead several research groups have developed parallel techniques to separate metallic from semiconducting SWNTs. Details of the various procedures employed, as well as their advantages and drawbacks, have been reported [105]. Arnold et al. have shown that it is possible to separate not only metallic from semiconducting SWNTs, but also SWNTs of different sizes, by using density-gradient ultracentrifugation. This technique is nondestructive and, to date, has provided the best results [122].

1.6
Importance of Defects and Curvature for Further Functionalization

The reactivity of CNTs upon further functionalization with heteroatoms as well as with organic and organometallic molecules is not clear. Most of the exposed CNT surface area is constituted by the basal plane of graphite, which is known to be poorly reactive. All of the carbon atoms are hybridized sp^2, and the π electrons are delocalized over the entire nanotubes; this not only increases the stability of the structure but also provides the reason why CNTs are chemically stable in most aggressive media. In contrast, nanodiamonds are composed only of sp^3 carbon atoms, with their surface being terminated by C–H bonds; consequently, hydrogen can be replaced relatively easily with desired functional groups or molecules [22]. Fortunately, however, the CNT surface is not perfect and flat (as in graphite), and a variety of topological defects exist as well as vacancies (Figure 1.10a) [123]. During their synthesis, heteroatoms such as nitrogen or boron can also be inserted into the backbone of the CNT, and it is possible to take advantage of these structural defects to anchor the desired molecules. Very often, CNTs are first oxidized with HNO_3 in order to create a variety of oxygen-containing groups on their surface; these groups may then be employed as anchoring points to graft the desired molecules [115, 124–128]. Recently, Tessonnier et al. developed a concept based on the defect-mediated functionalization of MWNTs [129]. In this case, carbon atoms located close to vacancies are terminated with hydrogen, in order to replace the missing carbon atom. These potentially reactive C–H bonds may be regarded as possible anchoring points for the direct grafting of desired groups or molecules by C–C coupling reactions.

In the case of SWNTs, the small diameter induces a strong curvature (as occurs with fullerenes). Subsequently, the strain caused by the curvature leads to a pyramidalization of carbon atoms [126], such that the sp^2 hybridized orbitals are no longer in the same plane, but rather form an angle of 11.6°, which is very close to the angle expected for sp^3 carbon atoms. It is, therefore, not surprising that C_{60} is very reactive towards addition reactions, where sp^2 carbon atoms become sp^3. Nyogi et al. calculated the pyramidalization angle as well as the π orbital misalignment angle as a function of the SWNT diameter and chirality (Figure 1.10b) [126]. These authors showed that, for small SWNTs, the pyramidalization angle is close enough to the value obtained for C_{60} to expect a similar chemistry. Whilst SWNTs are certainly less reactive than C_{60}, cycloaddition reactions would be expected to take place. In contrast, the curvature of the MWNT basal plane and the corresponding strain are too low to expect a similar chemistry to occur. Cycloaddition reactions might take place on a Stone–Wales defect, as their local curvature is similar to that of C_{60}. However, grafting directly onto a defect-free sidewall remains a major challenge, and defects would be required in this case.

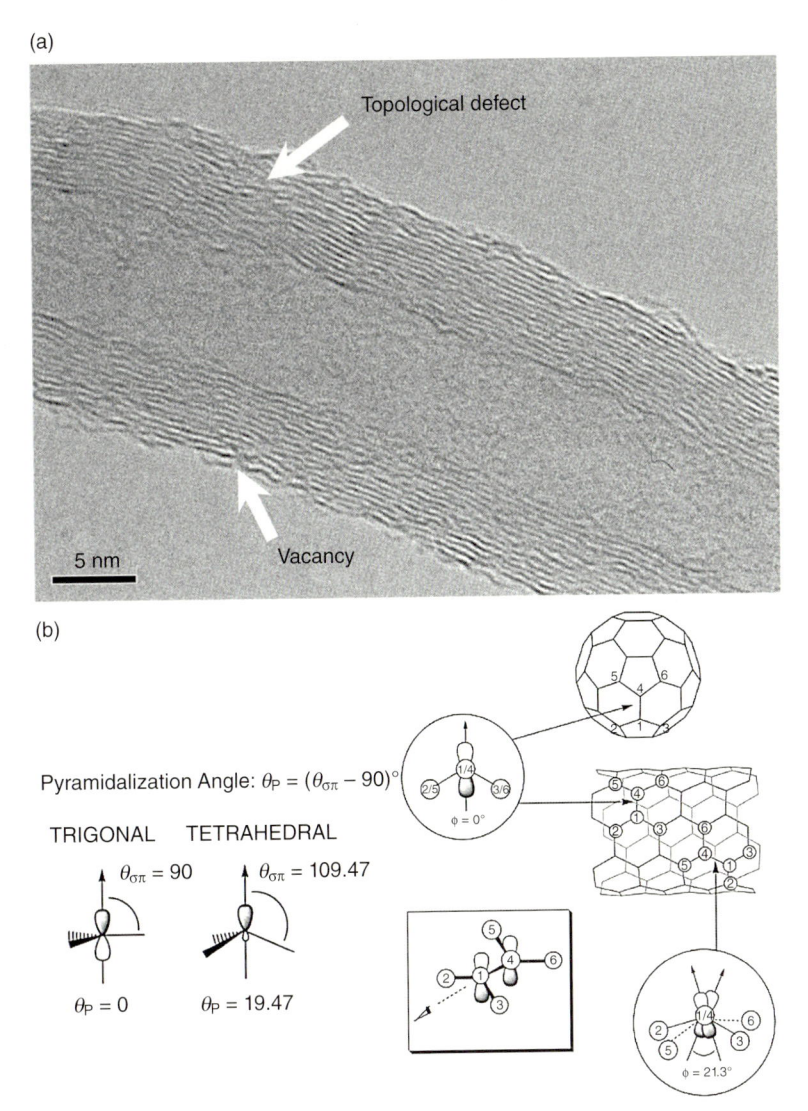

Figure 1.10 (a) HRTEM image of a MWNT showing the presence of topological defects as well as vacancies; (b) Pyramidalization angle and misalignment of π orbitals in a (5,5) SWNT. Reproduced from Ref. [124].

1.7
Functionalization: Creating Anchoring Points for Bioactive Molecules

A broad variety of functionalization techniques has been developed for the further application of carbon nanomaterials in materials science as well as in the life sciences [115, 124–128]. Attention will be focused here on procedures developed to functionalize CNTs, bearing in mind that similar techniques can be employed for nanodiamonds and fullerenes. Functionalization is an important step, as it allows CNTs to be opened, to be dispersed in solutions, to graft desired functions onto their surface, or for them to be coupled with another material. It is possible to distinguish three different families of functionalization techniques, namely oxidations, organic chemistry coupling reactions, and noncovalent functionalizations.

1.7.1
Functionalization by Oxidation

The oxidation reactions are the most popular because they permit the simultaneous purification of CNTs, the opening of their caps, and the functionalization of their surfaces. Various oxidants are employed, including HNO_3, H_2SO_4, H_2O_2, aqua regia (HCl:HNO_3, 3:1), $KMnO_4$, O_2, O_3, and H_2O. Of these oxidants, HNO_3 is certainly the most widely employed, because the treatment is relatively simple, efficient, and less dangerous than when using other oxidants. In addition, HNO_3 is expected to mainly create carboxylic acid groups, on which desired molecules can later be grafted by amidation or esterification. Although this belief is still widespread, especially in the life sciences, it has been shown that oxidizing treatments—and especially HNO_3 treatments—are unselective. Typically, a variety of oxygen-containing groups is created simultaneously (Figure 1.11a), all of which show different acidic and even basic properties. All of these groups must be taken into account when predicting the dispersion of CNTs in solutions, and the properties of the synthesized material. For example, it has been shown that NR_2-MWNTs synthesized with oxidized MWNTs are significantly less basic than when untreated MWNTs are employed, because the acidic oxygen-containing groups interact with basic amines [129, 130].

The effect of the oxidizing treatment varies significantly, depending on the CNT material used (SWNTs or MWNTs), the density of any structural defects, and the reaction conditions. The oxidation process starts on defects and at the tips of the CNTs, where the curvature—and thus reactivity—is high. Under strongly oxidizing conditions (high temperature and concentration), the reaction may also proceed at the basal planes for SWNTs, such that new defects will be created and at some point the SWNTs might be cut or even fully destroyed. In contrast, MWNTs are far less reactive, with minimal changes generally observed using Raman spectroscopy; this means that the oxidation takes place only on existing defects. Tessonnier *et al.* have recently shown that both the number and the nature of the oxygen-containing groups created with HNO_3 change, depending on the graphitic character of the nanomaterial, and thus its density of defects [131]. This observation might provide new opportunities for the selective functionalization of CNTs via oxidation. Alternatively, any undesired groups could be removed by thermal treatment in a vacuum or an inert gas (see Section 1.1). The oxygen-containing groups each exhibit different thermal stabilities; typically, carboxylic acids will be lost between 400 and 650 K, while phenols are stable up to 800 K (Figure 1.11b). However, some of the groups might be reformed upon exposure to the air [132].

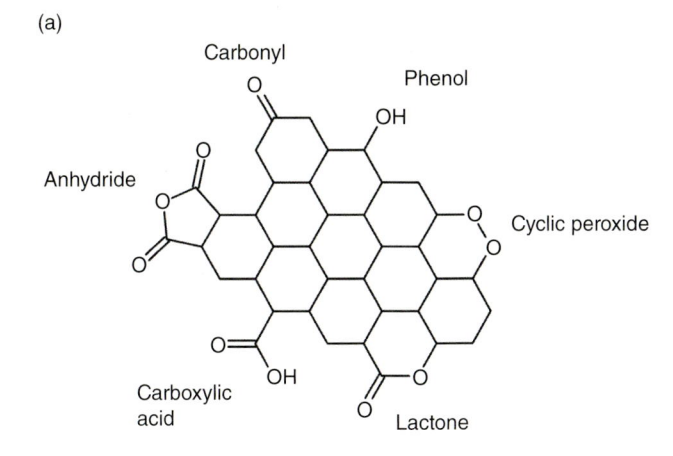

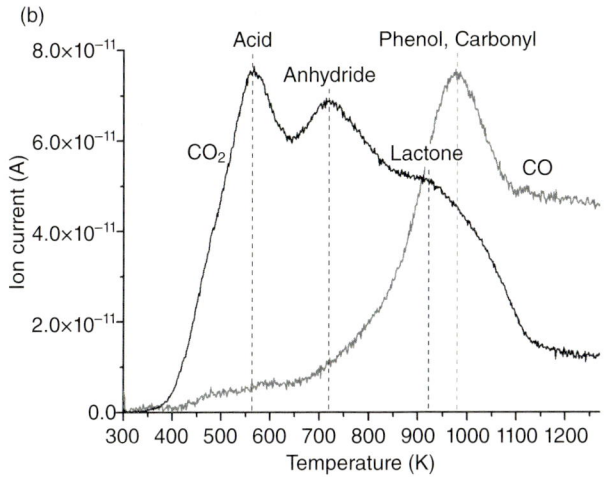

Figure 1.11 (a) The most common oxygen-containing functional groups which can coexist on the surface of oxidized CNTs; (b) The groups are lost as CO or as CO_2 during heating in inert gas. Thus, it is possible to quantify them by using temperature-programmed desorption (TPD).

1.7.2
Functionalization by Coupling Reactions

Grafting using organic chemistry coupling reactions represents the most exciting and promising functionalization technique. These procedures do not cause damage to the CNTs; rather, they are selective and the desired groups are bonded covalently to the CNT backbone by C–C bonds, thus reducing the risk of leaching. Many different reactions have been identified, the most interesting being cycloaddition, for example Diels–Alder reactions. As with oxidation, the CNT carbon atoms must be prone to react, either because of strain in SWNTs or because they are located next to defects (localized C–C double bond).

1.7.3
Noncovalent Functionalization

Noncovalent functionalization relies on strong van der Waals interactions between the CNT surface and small PAHs, such as pyrene. The attachment takes place by π–π stacking, and the anchoring has been shown to be irreversible. As there are no covalent bonds between the addend and the CNT, its electronic structure will not be modified.

1.8
Conclusion and Outlook

The synthesis of carbon nanomaterials is an ever-developing field of research, with many materials already having exhibited encouraging results for major applications in the life sciences, such as cancer diagnosis and therapy. Although most studies have focused on CNTs (SWNTs, MWNTs, arrays), very promising results have also been obtained with nanodiamonds, while fullerenes might receive additional attention if their large-scale production issues were to be resolved. A strong collaboration between materials scientists and biologists is crucial to further optimize carbon nanomaterials, depending on their target applications. Knowledge regarding the real state of a surface and possible impurities in the raw material must be shared. The presence of metal impurities and carbonaceous debris (possibly PAHs), along with the limited industrial production of high-quality SWNTs and SWNTs arrays, clearly reduces the real potential of these materials. The possible CNT toxicity must first be investigated and its causes understood in order to modify the material accordingly, by acting directly on its synthesis. Until recently, transition-metal nanoparticles were believed to be a prerequisite for growing CNTs, with the presence of metal impurities in the product being unavoidable. More recently, however, SWNTs have been grown directly on silica and other oxides without the need for any metal, thus eliminating one potential source of toxicity. Similar progresses are foreseeable, not only to reduce the surface carbonaceous debris, but also to design CNTs with controlled length and chirality, depending on the target application.

References

1 Ke, P.C. and Larcom, L.L. (2007) Carbon nanotubes in cancer therapy and diagnosis, in *Nanomaterials for Cancer Diagnosis* (ed. C. Kumar), Wiley-VCH Verlag GmbH, Weinheim, pp. 232–84.

2 Ye, J.-S. and Sheu, F.-S. (2007) Carbon nanotube-based sensor, in *Nanomaterials for Biosensors* (ed. C. Kumar), Wiley-VCH Verlag GmbH, Weinheim, pp. 27–55.
3 Bianco, A., Wu, W., Pastorin, G., Klumpp, C., Lacerda, L., Partidos, C.D., Kostarelos, K. and Prato, M. (2007) Carbon nanotube-based vectors for delivering immunotherapeutics and drugs, in *Nanomaterials for Medical Diagnosis and Therapy* (ed. C. Kumar), Wiley-VCH Verlag GmbH, Weinheim, pp. 85–142.
4 Endo, M., Strano, M. and Ajayan, P. (2008) Potential applications of carbon nanotubes, in *Carbon Nanotubes: Advanced Topics in the Synthesis, Structure, Properties and Applications* (eds A. Jorio, G. Dresselhaus and M.S. Dresselhaus), Springer, Berlin, pp. 13–61.
5 Radovic, L.R. (2009) Physicochemical properties of carbon materials: a brief overview, in *Carbon Materials for Catalysis* (eds P. Serp and J.L. Figueiredo), John Wiley & Sons, Inc., Hoboken, pp. 1–44.
6 Radovic, L.R. and Rodriguez-Reinoso, F. (1997) Carbon materials in catalysis, in *Chemistry and Physics of Carbon*, vol. 25 (ed. P.A. Thrower), Marcel-Dekker, New York, pp. 243–358.
7 Schlögl, R. (1999) Carbons, in *Preparation of Solid Catalysts* (eds G. Ertl, H. Knözinger and J. Weitkamp), Wiley-VCH Verlag GmbH, Weinheim, pp. 150–240.
8 Schlögl, R. (2007) Carbons, in *Handbook of Heterogeneous Catalysis* (eds G. Ertl, H. Knözinger, F. Schüth and J. Weitkamp), Wiley-VCH Verlag GmbH, Weinheim, pp. 357–427.
9 Bartholomew, C.H. (1982) Carbon deposition in steam reforming and methanation. *Catalysis Reviews – Science and Engineering*, **24**, 67–112.
10 Moulijn, J.A., van Diepen, A.E. and Kapteijn, F. (2007) Deactivation and regeneration, *Handbook of Heterogeneous Catalysis* (eds G. Ertl, H. Knözinger, F. Schüth and J. Weitkamp), Wiley-VCH Verlag GmbH, Weinheim, pp. 1829–46.
11 Boehm, H.P. (1997) The first observation of carbon nanotubes. *Carbon*, **35**, 581–4.
12 Monthioux, M. and Kuznetsov, V.L. (2006) Who should be given the credit for the discovery of carbon nanotubes? *Carbon*, **44**, 1621–3.
13 Dresselhaus, M. and Avouris, P. (2001) Introduction to carbon materials research, in *Carbon Nanotubes: Synthesis, Structure, Properties and Applications* (eds M.S. Dresselhaus, G. Dresselhaus and P. Avouris), Springer, Berlin, pp. 1–9.
14 Tibbetts, G.G., Lake, M.L., Strong, K.L. and Rice, B.P. (2007) A review of the fabrication and properties of vapor-grown carbon nanofiber/polymer composites. *Composites Science and Technology*, **67**, 1709–18.
15 Dresselhaus, M. and Endo, M. (2001) Relation of carbon nanotubes to other carbon materials, in *Carbon Nanotubes: Synthesis, Structure, Properties and Applications* (eds M.S. Dresselhaus, G. Dresselhaus and P. Avouris), Springer, Berlin, pp. 11–28.
16 Iijima, S. (1991) Helical microtubules of graphitic carbon. *Nature*, **354**, 56–8.
17 Muller, J.-O., Su, D.S., Wild, U. and Schlögl, R. (2007) Bulk and surface structural investigations of diesel engine soot and carbon black. *Physical Chemistry Chemical Physics*, **9**, 4018–25.
18 Tessonnier, J.-P., Rosenthal, D., Hansen, T.W., Hess, C., Schuster, M.E., Blume, R., Girgsdies, F., Pfänder, N., Timpe, O., Su, D.S. and Schlögl, R. (2009) Analysis of the structure and chemical properties of some commercial carbon nanostructures. *Carbon*, **47**, 1779–98.
19 Dresselhaus, M.S., Dresselhaus, G. and Eklund, P.C. (1996) *Science of Fullerenes and Carbon Nanotubes*, Academic Press, San Diego.
20 Berger, M. (2009) *Nano-Society: Pushing the Boundaries of Technology*, The Royal Society of Chemistry, Cambridge.
21 Chen, M., Pierstorff, E.D., Lam, R., Li, S.-Y., Huang, H., Osawa, E. and Ho, D. (2009) Nanodiamond-mediated delivery of water-insoluble therapeutics. *ACS Nano*, **3**, 2016–22.
22 Schrand, A.M., Hens, S.A.C. and Shenderova, O.A. (2009) Nanodiamond particles: properties and perspectives for bioapplications. *Critical Reviews in Solid State and Materials Sciences*, **34**, 18–74.

23 Shenderova, O.A. and McGuire, G. (2006) Nanocrystalline diamond, in *Nanomaterials Handbook* (ed. Y. Gogotsi), CRC Press, Boca Raton, pp. 203–37.
24 Zhang, X.-Q., Chen, M., Lam, R., Xu, X., Osawa, E. and Ho, D. (2009) Polymer-functionalized nanodiamond platforms as vehicles for gene delivery. *ACS Nano*, 3, 2609–16.
25 Bundy, F.P., Hall, H.T., Strong, H.M. and Wentorf, R.H. (1955) Man-made diamonds. *Nature*, 176, 51–5.
26 Schwertfeger, H., Fokin, A.A. and Schreiner, P.R. (2008) Diamonds are a chemist's best friend: diamondoid chemistry beyond adamantane. *Angewandte Chemie International Edition*, 47, 1022–36.
27 Gruen, D.M. (1999) Nanocrystalline diamond films. *Annual Review of Materials Science*, 29, 211–59.
28 Mallika, K., DeVries, R.C. and Komanduri, R. (1999) On the low pressure transformation of graphite to diamond in the presence of a "catalyst-solvent". *Thin Solid Films*, 339, 19–33.
29 Bundy, F.P. (1962) Direct conversion of graphite to diamond in static pressure apparatus. *Science*, 137, 1057–8.
30 Hall, H.T. (1960) Ultra-high-pressure, high-temperature apparatus: the "belt". *The Review of Scientific Instruments*, 31, 125–31.
31 Pavel, E. (1998) Combinative mechanism of HP-HT catalytic synthesis of diamond. *Physica B*, 245, 288–92.
32 Pierson, H.O. (1993) *Handbook of Carbon, Graphite, Diamond and Fullerenes: Properties, Processing and Applications*, Noyes Publications, Park Ridge.
33 Aleksenskii, A.E., Baidakova, M.V., Vul', A.Y. and Siklitskii, V.I. (1999) The structure of diamond nanoclusters. *Physics of the Solid State*, 41, 668–71.
34 Aleksenskii, A.E., Baidakova, M.V., Vul', A.Y., Davydov, V.Y. and Pevtsova, Y.A. (1997) Diamond-graphite phase transition in ultradisperse-diamond clusters. *Physics of the Solid State*, 39, 1007–15.
35 Kroto, H.W., Heath, J.R., O'Brien, S.C., Curl, R.F. and Smalley, R.E. (1985) C_{60}: buckminsterfullerene. *Nature*, 318, 162–3.
36 Boyd, D.B. and Slanina, Z. (2001) Introduction and foreword to the special issue commemorating the thirtieth anniversary of Eiji Osawa's C60 paper. *Journal of Molecular Graphics and Modelling*, 19, 181–4.
37 Kratschmer, W., Lamb, L.D., Fostiropoulos, K. and Huffman, D.R. (1990) Solid C_{60}: a new form of carbon. *Nature*, 347, 354–8.
38 Ajayan, P.M. and Iijima, S. (1992) Smallest carbon nanotube. *Nature*, 358, 23–23.
39 Yudasaka, M., Iijima, S. and Crespi, V. (2008) Single-wall carbon nanohorns and nanocones, in *Carbon Nanotubes: Advanced Topics in the Synthesis, Structure, Properties and Applications* (eds A. Jorio, G. Dresselhaus and M.S. Dresselhaus), Springer, Berlin, pp. 605–29.
40 Yamaguchi, T., Bandow, S. and Iijima, S. (2004) Synthesis of carbon nanohorn particles by simple pulsed arc discharge ignited between pre-heated carbon rods. *Chemical Physics Letters*, 389, 181–5.
41 Ruoff, R.S., Tse, D.S., Malhotra, R. and Lorents, D.C. (1993) Solubility of fullerene (C_{60}) in a variety of solvents. *Journal of Physical Chemistry*, 97, 3379–83.
42 Bensch, W., Werner, H., Bartl, H. and Schlögl, R. (1994) Single-crystal structure of C_{60} at 300 K. Evidence for the presence of oxygen in a statically disordered model. *Journal of the Chemical Society, Faraday Transactions*, 90, 2791–7.
43 Werner, H., Schedel-Niedrig, T., Wohlers, M., Herein, D., Herzog, B., Schlögl, R., Keil, M., Bradshaw, A.M. and Kirschner, J. (1994) Reaction of molecular oxygen with C_{60}: spectroscopic studies. *Journal of the Chemical Society, Faraday Transactions*, 90, 403–9.
44 Scott, L.T., Boorum, M.M., McMahon, B.J., Hagen, S., Mack, J., Blank, J., Wegner, H. and de Meijere, A. (2002) Rational chemical synthesis of C_{60}. *Science*, 295, 1500–3.
45 Otero, G., Biddau, G., Sanchez-Sanchez, C., Caillard, R., Lopez, M.F., Rogero, C.,

Palomares, F.J., Cabello, N., Basanta, M.A., Ortega, J., Mendez, J., Echavarren, A.M., Perez, R., Gomez-Lor, B. and Martin-Gago, J.A. (2008) Fullerenes from aromatic precursors by surface-catalysed cyclodehydrogenation. *Nature*, **454**, 865–8.

46 Rummeli, M.H., Kramberger, C., Gruneis, A., Ayala, P., Gemming, T., Buchner, B. and Pichler, T. (2007) On the graphitization nature of oxides for the formation of carbon nanostructures. *Chemistry of Materials*, **19**, 4105–7.

47 Rummeli, M.H., Schaffel, F., Kramberger, C., Gemming, T., Bachmatiuk, A., Kalenczuk, R.J., Rellinghaus, B., Buchner, B. and Pichler, T. (2007) Oxide-driven carbon nanotube growth in supported catalyst CVD. *Journal of the American Chemical Society*, **129**, 15772–3.

48 Bachmatiuk, A., Börrnert, F., Grobosch, M., Schäffel, F., Wolff, U., Scott, A., Zaka, M., Warner, J.H., Klingeler, R., Knupfer, M., Büchner, B. and Rümmeli, M.H. (2009) Investigating the graphitization mechanism of SiO_2 nanoparticles in chemical vapor deposition. *ACS Nano*, **3**, 4098–104.

49 Huang, S., Cai, Q., Chen, J., Qian, Y. and Zhang, L. (2009) Metal-catalyst-free growth of single-walled carbon nanotubes on substrates. *Journal of the American Chemical Society*, **131**, 2094–5.

50 Liu, B., Ren, W., Gao, L., Li, S., Pei, S., Liu, C., Jiang, C. and Cheng, H.-M. (2009) Metal-catalyst-free growth of single-walled carbon nanotubes. *Journal of the American Chemical Society*, **131**, 2082–3.

51 Liu, B., Ren, W., Liu, C., Sun, C.-H., Gao, L., Li, S., Jiang, C. and Cheng, H.-M. (2009) Growth velocity and direct length-sorted growth of short single-walled carbon nanotubes by a metal-catalyst-free chemical vapor deposition process. *ACS Nano*, **3**, 3421–30.

52 Steiner, S.A., Baumann, T.F., Bayer, B.C., Blume, R., Worsley, M.A., MoberlyChan, W.J., Shaw, E.L., Schlögl, R., Hart, A.J., Hofmann, S. and Wardle, B.L. (2009) Nanoscale zirconia as a nonmetallic catalyst for graphitization of carbon and growth of single- and multiwall carbon nanotubes. *Journal of the American Chemical Society*, **131**, 12144–54.

53 Tada, M. and Iwasawa, Y. (2007) Supported catalysts from chemical vapor deposition and related techniques, in *Handbook of Heterogeneous Catalysis* (eds G. Ertl, H. Knözinger, F. Schüth and J. Weitkamp), Wiley-VCH Verlag GmbH, pp. 539–55.

54 Hitchman, M.L. and Jensen, K.F. (1993) *Chemical Vapor Deposition: Principles and Applications*, Academic Press, London, p. 677.

55 Anisimov, A.S., Nasibulin, A.G., Jiang, H., Launois, P., Cambedouzou, J., Shandakov, S.D. and Kauppinen, E.I. (2010) Mechanistic investigations of single-walled carbon nanotube synthesis by ferrocene vapor decomposition in carbon monoxide. *Carbon*, **48**, 380–8.

56 Nikolaev, P. (2004) Gas-phase production of single-walled carbon nanotubes from carbon monoxide: a review of the HiPCO process. *Journal of Nanoscience and Nanotechnology*, **4**, 307–16.

57 ISO (2008) ISO/TS 27687:2008; Nanotechnologies–Terminology and Definitions for Nano-Objects: Nanoparticle, Nanofibre and Nanoplate, International Organization for Standardization, Geneva.

58 Baker, R.T.K., Barber, M.A., Harris, P.S., Feates, F.S. and Waite, R.J. (1972) Nucleation and growth of carbon deposits from the nickel catalyzed decomposition of acetylene. *Journal of Catalysis*, **26**, 51–62.

59 Baker, R.T.K., Harris, P.S., Thomas, R.B. and Waite, R.J. (1973) Formation of filamentous carbon from iron, cobalt and chromium catalyzed decomposition of acetylene. *Journal of Catalysis*, **30**, 86–95.

60 Wagner, R.S. and Ellis, W.C. (1964) Vapor-liquid-solid mechanism of single crystal growth. *Applied Physics Letters*, **4**, 89–90.

61 Bajwa, N., Li, X., Ajayan, P.M. and Vajtai, R. (2008) Mechanisms for catalytic CVD growth of multiwalled carbon nanotubes. *Journal of Nanoscience and Nanotechnology*, **8**, 6054–64.

62 Dupuis, A.-C. (2005) The catalyst in the CCVD of carbon nanotubes – a review. *Progress in Materials Science*, **50**, 929–61.

63 Serp, P. (2009) Carbon nanotubes and nanofibers in catalysis, in *Carbon Materials for Catalysis* (eds P. Serp and J.L. Figueiredo), John Wiley & Sons, Inc., Hoboken, pp. 309–72.

64 Baker, R.T.K. and Harris, P.S. (1978) The Formation of Filamentous Carbon, in *Chemistry and Physics of Carbon*, vol. 14 (ed. P.A. Thrower), Marcel Dekker, New-York, pp. 83–165.

65 Kim, K.-E., Kim, K.-J., Jung, W.S., Bae, S.Y., Park, J., Choi, J. and Choo, J. (2005) Investigation on the temperature-dependent growth rate of carbon nanotubes using chemical vapor deposition of ferrocene and acetylene. *Chemical Physics Letters*, **401**, 459–64.

66 Pirard, S.L., Douven, S., Bossuot, C., Heyen, G. and Pirard, J.-P. (2007) A kinetic study of multi-walled carbon nanotube synthesis by catalytic chemical vapor deposition using a Fe-Co/Al_2O_3 catalyst. *Carbon*, **45**, 1167–75.

67 Rostrup-Nielsen, J. and Trimm, D.L. (1977) Mechanisms of carbon formation on nickel-containing catalysts. *Journal of Catalysis*, **48**, 155–65.

68 Tibbetts, G.G., Devour, M.G. and Rodda, E.J. (1987) An adsorption-diffusion isotherm and its application to the growth of carbon filaments on iron catalyst particles. *Carbon*, **25**, 367–75.

69 Louchev, O.A., Laude, T., Sato, Y. and Kanda, H. (2003) Diffusion-controlled kinetics of carbon nanotube forest growth by chemical vapor deposition. *Journal of Chemical Physics*, **118**, 7622–34.

70 Sacco, A., Thacker, P., Chang, T.N. and Chiang, A.T.S. (1984) The initiation and growth of filamentous carbon from [alpha]-iron in H_2, CH_4, H_2O, CO_2, and CO gas mixtures. *Journal of Catalysis*, **85**, 224–36.

71 Kock, A.J.H.M., de Bokx, P.K., Boellaard, E., Klop, W. and Geus, J.W. (1985) The formation of filamentous carbon on iron and nickel catalysts: II. Mechanism. *Journal of Catalysis*, **96**, 468–80.

72 Alstrup, I. (1988) A new model explaining carbon filament growth on nickel, iron, and Ni–Cu alloy catalysts. *Journal of Catalysis*, **109**, 241–51.

73 Baker, R.T.K., Alonzo, J.R., Dumesic, J.A. and Yates, D.J.C. (1982) Effect of the surface state of iron on filamentous carbon formation. *Journal of Catalysis*, **77**, 74–84.

74 Oberlin, A., Endo, M. and Koyama, T. (1976) Filamentous growth of carbon through benzene decomposition. *Journal of Crystal Growth*, **32**, 335–49.

75 Yang, R.T. and Chen, J.P. (1989) Mechanism of carbon filament growth on metal catalysts. *Journal of Catalysis*, **115**, 52–64.

76 Raty, J.-Y., Gygi, F. and Galli, G. (2005) Growth of carbon nanotubes on metal nanoparticles: a microscopic mechanism from ab initio molecular dynamics simulations. *Physical Review Letters*, **95**, 096103.

77 Hofmann, S., Csányi, G., Ferrari, A.C., Payne, M.C. and Robertson, J. (2005) Surface diffusion: the low activation energy path for nanotube growth. *Physical Review Letters*, **95**, 036101.

78 Helveg, S., Lopez-Cartes, C., Sehested, J., Hansen, P.L., Clausen, B.S., Rostrup-Nielsen, J.R., Abild-Pedersen, F. and Norskov, J.K. (2004) Atomic-scale imaging of carbon nanofibre growth. *Nature*, **427**, 426–9.

79 Hofmann, S., Sharma, R., Ducati, C., Du, G., Mattevi, C., Cepek, C., Cantoro, M., Pisana, S., Parvez, A., Cervantes-Sodi, F., Ferrari, A.C., Dunin-Borkowski, R., Lizzit, S., Petaccia, L., Goldoni, A. and Robertson, J. (2007) In situ observations of catalyst dynamics during surface-bound carbon nanotube nucleation. *Nano Letters*, **7**, 602–8.

80 Wirth, C.T., Hofmann, S. and Robertson, J. (2009) State of the catalyst during carbon nanotube growth. *Diamond and Related Materials*, **18**, 940–5.

81 Hofmann, S., Blume, R., Wirth, C.T., Cantoro, M., Sharma, R., Ducati, C., Hävecker, M., Zafeiratos, S., Schnoerch, P., Oestereich, A., Teschner, D., Albrecht, M., Knop-Gericke, A., Schlögl, R. and Robertson, J. (2009) State of transition metal catalysts during carbon nanotube growth. *Journal of Physical Chemistry C*, **113**, 1648–56.

82 Wirth, C.T., Zhang, C., Zhong, G., Hofmann, S. and Robertson, J. (2009) Diffusion- and reaction-limited growth of carbon nanotube forests. *ACS Nano*, **3**, 3560–6.

83 Jiang, A., Awasthi, N., Kolmogorov, A.N., Setyawan, W., Börjesson, A., Bolton, K., Harutyunyan, A.R. and Curtarolo, S. (2007) Theoretical study of the thermal behavior of free and alumina-supported Fe-C nanoparticles. *Physical Review B*, **75**, 205426.

84 Harutyunyan, A.R., Mora, E., Tokune, T., Bolton, K., Rosen, A., Jiang, A., Awasthi, N. and Curtarolo, S. (2007) Hidden features of the catalyst nanoparticles favorable for single-walled carbon nanotube growth. *Applied Physics Letters*, **90**, 163120–3.

85 Harutyunyan, A.R., Awasthi, N., Jiang, A., Setyawan, W., Mora, E., Tokune, T., Bolton, K. and Curtarolo, S. (2008) Reduced carbon solubility in Fe nanoclusters and implications for the growth of single-walled carbon nanotubes. *Physical Review Letters*, **100**, 195502.

86 Flahaut, E., Bacsa, R., Peigney, A. and Laurent, C. (2003) Gram-scale CCVD synthesis of double-walled carbon nanotubes. *Chemical Communications*, 1442–3.

87 Murakami, Y., Chiashi, S., Miyauchi, Y., Hu, M., Ogura, M., Okubo, T. and Maruyama, S. (2004) Growth of vertically aligned single-walled carbon nanotube films on quartz substrates and their optical anisotropy. *Chemical Physics Letters*, **385**, 298–303.

88 Ago, H., Nakamura, K., Imamura, S. and Tsuji, M. (2004) Growth of double-wall carbon nanotubes with diameter-controlled iron oxide nanoparticles supported on MgO. *Chemical Physics Letters*, **391**, 308–13.

89 Saito, T., Ohshima, S., Xu, W.-C., Ago, H., Yumura, M. and Iijima, S. (2005) Size control of metal nanoparticle catalysts for the gas-phase synthesis of single-walled carbon nanotubes. *Journal of Physical Chemistry B*, **109**, 10647–52.

90 Yamada, T., Namai, T., Hata, K., Futaba, D.N., Mizuno, K., Fan, J., Yudasaka, M., Yumura, M. and Iijima, S. (2006) Size-selective growth of double-walled carbon nanotube forests from engineered iron catalysts. *Nature Nanotechnology*, **1**, 131–6.

91 Yu, H., Zhang, Q., Luo, G. and Wei, F. (2006) Rings of triple-walled carbon nanotube bundles. *Applied Physics Letters*, **89**, 223106–3.

92 Dervishi, E., Li, Z., Watanabe, F., Xu, Y., Saini, V., Biris, A.R. and Biris, A.S. (2009) Thermally controlled synthesis of single-wall carbon nanotubes with selective diameters. *Journal of Materials Chemistry*, **19**, 3004–12.

93 Nasibulin, A.G., Pikhitsa, P.V., Jiang, H. and Kauppinen, E.I. (2005) Correlation between catalyst particle and single-walled carbon nanotube diameters. *Carbon*, **43**, 2251–7.

94 Chiang, W.-H., Sakr, M., Gao, X.P.A. and Sankaran, R.M. (2009) Nanoengineering $NixFe_{1-x}$ catalysts for gas-phase, selective synthesis of semiconducting single-walled carbon nanotubes. *ACS Nano*, **3**, 4023–32.

95 Moisala, A., Nasibulin, A.G. and Kauppinen, E.I. (2003) The role of metal nanoparticles in the catalytic production of single-walled carbon nanotubes – a review. *Journal of Physics: Condensed Matter*, **15**, S3011–35.

96 Colomer, J.-F., Bister, G., Willems, I., Konya, Z., Fonseca, A., Van Tendeloo, G. and Nagy, J.B. (1999) Synthesis of single-wall carbon nanotubes by catalytic decomposition of hydrocarbons. *Chemical Communications*, 1343–4.

97 Su, M., Zheng, B. and Liu, J. (2000) A scalable CVD method for the synthesis of single-walled carbon nanotubes with high catalyst productivity. *Chemical Physics Letters*, **322**, 321–6.

98 Mattevi, C., Wirth, C.T., Hofmann, S., Blume, R., Cantoro, M., Ducati, C., Cepek, C., Knop-Gericke, A., Milne, S., Castellarin-Cudia, C., Dolafi, S., Goldoni, A., Schloegl, R. and Robertson, J. (2008) In-situ X-ray photoelectron spectroscopy study of catalyst–support interactions and growth of carbon nanotube forests. *Journal of Physical Chemistry C*, **112**, 12207–13.

99 Philippe, R., Serp, P., Kalck, P., Kihn, Y., Bordère, S., Plee, D., Gaillard, P.,

Bernard, D. and Caussat, B. (2009) Kinetic study of carbon nanotubes synthesis by fluidized bed chemical vapor deposition. *AIChE Journal*, **55**, 450–64.

100 Gohier, A., Ewels, C.P., Minea, T.M. and Djouadi, M.A. (2008) Carbon nanotube growth mechanism switches from tip- to base-growth with decreasing catalyst particle size. *Carbon*, **46**, 1331–8.

101 Zhou, J.-H., Sui, Z.-J., Li, P., Chen, D., Dai, Y.-C. and Yuan, W.-K. (2006) Structural characterization of carbon nanofibers formed from different carbon-containing gases. *Carbon*, **44**, 3255–62.

102 Melezhik, A.V., Sementsov, Y.I. and Yanchenko, V.V. (2005) Synthesis of fine carbon nanotubes on coprecipitated metal oxide catalysts. *Russian Journal of Applied Chemistry*, **78**, 938–44.

103 Zhao, M.-Q., Zhang, Q., Jia, X.-L., Huang, J.-Q., Zhang, Y.-H. and Wei, F. (2010) Hierarchical composites of single/double-walled carbon nanotubes interlinked flakes from direct carbon deposition on layered double hydroxides. *Advanced Functional Materials*, **20**, 677–85.

104 Philippe, R., Caussat, B., Falqui, A., Kihn, Y., Kalck, P., Bordère, S., Plee, D., Gaillard, P., Bernard, D. and Serp, P. (2009) An original growth mode of MWNTs on alumina supported iron catalysts. *Journal of Catalysis*, **263**, 345–58.

105 Joselevich, E., Dai, H., Liu, J., Hata, K. and Windle, H.A. (2008) Carbon nanotube synthesis and organization, in *Carbon Nanotubes: Advanced Topics in the Synthesis, Structure, Properties and Applications* (eds A. Jorio, G. Dresselhaus and M.S. Dresselhaus), Springer, Berlin, pp. 101–64.

106 Hofmann, S., Ducati, C., Robertson, J. and Kleinsorge, B. (2003) Low-temperature growth of carbon nanotubes by plasma-enhanced chemical vapor deposition. *Applied Physics Letters*, **83**, 135–7.

107 Hata, K., Futaba, D.N., Mizuno, K., Namai, T., Yumura, M. and Iijima, S. (2004) Water-assisted highly efficient synthesis of impurity-free single-walled carbon nanotubes. *Science*, **306**, 1362–4.

108 Yun, Y., Shanov, V., Tu, Y., Subramaniam, S. and Schulz, M.J. (2006) Growth mechanism of long aligned multiwall carbon nanotube arrays by water-assisted chemical vapor deposition. *Journal of Physical Chemistry B*, **110**, 23920–5.

109 Zhong, G., Iwasaki, T., Robertson, J. and Kawarada, H. (2007) Growth kinetics of 0.5 cm vertically aligned single-walled carbon nanotubes. *Journal of Physical Chemistry B*, **111**, 1907–10.

110 Yasuda, S., Futaba, D.N., Yamada, T., Satou, J., Shibuya, A., Takai, H., Arakawa, K., Yumura, M. and Hata, K. (2009) Improved and large area single-walled carbon nanotube forest growth by controlling the gas flow direction. *ACS Nano*, **3**, 4164–70.

111 Yuan, D., Ding, L., Chu, H., Feng, Y., McNicholas, T.P. and Liu, J. (2008) Horizontally aligned single-walled carbon nanotube on quartz from a large variety of metal catalysts. *Nano Letters*, **8**, 2576–9.

112 Liu, Z., Tabakman, S., Welsher, K. and Dai, H. (2009) Carbon nanotubes in biology and medicine: *in vitro* and *in vivo* detection, imaging and drug delivery. *Nano Research*, **2**, 85–120.

113 Pisanic, T.R., II, Blackwell, J.D., Shubayev, V.I., Finones, R.R. and Jin, S. (2007) Nanotoxicity of iron oxide nanoparticle internalization in growing neurons. *Biomaterials*, **28**, 2572–81.

114 Hou, P.-X., Liu, C. and Cheng, H.-M. (2008) Purification of carbon nanotubes. *Carbon*, **46**, 2003–25.

115 Pillai, S.K., Ray, S.S. and Moodley, M. (2008) Purification of multi-walled carbon nanotubes. *Journal of Nanoscience and Nanotechnology*, **8**, 6187–207.

116 Plata, D.E.L., Hart, A.J., Reddy, C.M. and Gschwend, P.M. (2009) Early evaluation of potential environmental impacts of carbon nanotube synthesis by chemical vapor deposition. *Environmental Science and Technology*, **43**, 8367–73.

117 Plata, D.L., Gschwend, P.M. and Reddy, C.M. (2008) Industrially synthesized single-walled carbon nanotubes: compositional data for users,

environmental risk assessments, and source apportionment. *Nanotechnology*, **19**, 185706.

118 Yang, K., Zhu, L. and Xing, B. (2006) Adsorption of polycyclic aromatic hydrocarbons by carbon nanomaterials. *Environmental Science and Technology*, **40**, 1855–61.

119 Datsyuk, V., Kalyva, M., Papagelis, K., Parthenios, J., Tasis, D., Siokou, A., Kallitsis, I. and Galiotis, C. (2008) Chemical oxidation of multiwalled carbon nanotubes. *Carbon*, **46**, 833–40.

120 Salzmann, C.G., Llewellyn, S.A., Tobias, G., Ward, M.A.H., Huh, Y. and Green, M.L.H. (2007) The role of carboxylated carbonaceous fragments in the functionalization and spectroscopy of a single-walled carbon-nanotube material. *Advanced Materials*, **19**, 883–7.

121 Shao, L., Tobias, G., Salzmann, C.G., Ballesteros, B., Hong, S.Y., Crossley, A., Davis, B.G. and Green, M.L.H. (2007) Removal of amorphous carbon for the efficient sidewall functionalisation of single-walled carbon nanotubes. *Chemical Communications*, 5090–2.

122 Arnold, M.S., Green, A.A., Hulvat, J.F., Stupp, S.I. and Hersam, M.C. (2006) Sorting carbon nanotubes by electronic structure using density differentiation. *Nature Nanotechnology*, **1**, 60–5.

123 Charlier, J.C. (2002) Defects in carbon nanotubes. *Accounts of Chemical Research*, **35**, 1063–9.

124 Hirsch, A. and Vostrowsky, O. (2007) Functionalization of carbon nanotubes, in *Functional Organic Materials*, vol. 1 (eds T.J.J. Müller and U.H.F. Bunz), Wiley-VCH Verlag GmbH, Weinheim, pp. 1–57.

125 Balasubramanian, K. and Burghard, M. (2005) Chemically functionalized carbon nanotubes. *Small*, **1**, 180–92.

126 Niyogi, S., Hamon, M.A., Hu, H., Zhao, B., Bhowmik, P., Sen, R., Itkis, M.E. and Haddon, R.C. (2002) Chemistry of single-walled carbon nanotubes. *Accounts of Chemical Research*, **35**, 1105–13.

127 Peng, X. and Wong, S.S. (2009) Functional covalent chemistry of carbon nanotube surfaces. *Advanced Materials*, **21**, 625–42.

128 Tasis, D., Tagmatarchis, N., Bianco, A. and Prato, M. (2006) Chemistry of carbon nanotubes. *Chemical Reviews*, **106**, 1105–36.

129 Tessonnier, J.-P., Villa, A., Majoulet, O., Su, D.S. and Schlögl, R. (2009) Defect-mediated functionalization of carbon nanotubes as a route to design single-site basic heterogeneous catalysts for biomass conversion. *Angewandte Chemie International Edition*, **48**, 6543–6.

130 Villa, A., Tessonnier, J.-P., Majoulet, O., Su, D.S. and Schlögl, R. (2009) Amino-functionalized carbon nanotubes as solid basic catalysts for the transesterification of triglycerides. *Chemical Communications*, 4405–7.

131 Tessonnier, J.-P., Rosenthal, D., Girgsdies, F., Amadou, J., Begin, D., Pham-Huu, C., Su, D.S. and Schlogl, R. (2009) Influence of the graphitisation of hollow carbon nanofibers on their functionalisation and subsequent filling with metal nanoparticles. *Chemical Communications*, 7158–60.

132 Burg, P. and Cagniant, D. (2008) Characterization of carbon surface chemistry, in *Chemistry and Physics of Carbon*, vol. 30 (ed. L.R. Radovic), Taylor & Francis (CRC Press), Boca Raton, pp. 129–75.

2
Nanocarbons: Characterization Tools
Dang Sheng Su

2.1
Introduction

The use of nanocarbons in the life sciences and biology strongly depends on their physico-chemical characteristics. These properties can be classified into: (i) morphology (shape, size, structural architecture); (ii) nanostructure (detailed atomic arrangement, defects); (iii) chemistry (elements, composition); (iv) electronic structure (nature of the bonding between atoms); and (v) functionalization (acid–base properties, polarity). It is of essence to establish a structure–property correlation for nanocarbons to determine how the synthetic process might be optimized, in order to obtain nanocarbons with the desired texture and functionality for applications.

Many techniques have been established for the characterization of nanomaterials, and also for nanocarbons. For example, atoms, ions, electrons, neutrons, and photons can each be used as primary probes to excite secondary effects such as electrons, X-rays, ions and light from the irradiated or illuminated regions. The recording, monitoring and analysis of the primary probe after interaction with the specimen, and of the secondary effects as a function of the different variables (energy, intensity, time, angle, etc.), represent the underlying principles of the characterization techniques. In general, techniques for characterization can be categorized as diffraction, imaging, and spectroscopy.

The development of characterization tools has been rapid, such that today certain dedicated technologies such as aberration–correction for transmission electron microscopy (TEM) have been commercialized. In this chapter, some of the most important characterization tools for nanocarbons are introduced, together with a description of the results obtained with these new technical development during the past few years. The initial discussion is centered on classic diffraction techniques, and this is followed by imaging techniques using electron microscopy and scanning probe microscopy. Some important spectroscopic methods that are essential when determining the electronic structure and functionalities of nanocarbons are also described.

Nanomaterials for the Life Sciences Vol.9: Carbon Nanomaterials. Edited by Challa S. S. R. Kumar
Copyright © 2011 WILEY-VCH Verlag GmbH & Co. KGaA, Weinheim
ISBN: 978-3-527-32169-8

2.2
Diffraction Techniques

Diffraction techniques are used routinely in chemistry, biology, and materials science to determine the shape and symmetry of a molecule, and the unit cell structure of crystalline samples. For nanomaterials and nanocarbons with dimensions of only a few nanometers, the translation symmetry of the corresponding bulk materials is limited, or even lost, with the spectrum peak or diffraction spot becoming broad or diffuse due to such size effects. Nonetheless, diffraction techniques are still widely used for the characterization of nanomaterials, as they are comparably simple in operation, and provide rapid results. Diffraction techniques provide the "fingerprints" of the phase composition of nanomaterials, as compared to the structural data of bulk materials that are available for almost all types of material, as for example, listed in the Powder Diffraction File (PDF). Electrons, neutrons, and X-rays can be used for diffraction at a sample [referred to here as electron diffraction (ED), neutron diffraction (ND), and X-ray diffraction (XRD), respectively]. X-rays and neutrons can be used to probe large volumes of materials, providing an averaged information of nanomaterials, whereas ED can be used to probe a volume of nanometer scale (nanodiffraction), and is the only technique capable of providing structural information relating to individual nanoparticles.

Today, XRD or ED in a transmission electron microscope represent the most frequently used diffraction techniques. Each is based on the elastic scattering of X-rays or electrons, respectively, from structures that have a certain short- or long-range order. If an X-ray or electron beam interacts with a crystalline sample, then the elastically scattered X-rays or electrons will be coherent. The diffracted beams arise from the strong constructive interference from all atoms in the materials. In crystalline solids, the atoms are arranged periodically in three-dimensional (3-D) repeated unit cells, and this leads to a corresponding periodic distribution of the scattering centers. The structure factor, $F(\theta)$, is defined in terms of the sum of the atomic scattering factor from all of the atoms in the unit cell, multiplied by the phase factor that takes account of the difference in phase between waves scattered from atoms on different planes with the Miller indices (h, k, l). The recorded or observed diffraction intensity is then proportional to the module of the structure form factor $I(\theta) \sim |F(\theta)|^2$.

The analysis of the angular distribution and intensities of these diffracted waves therefore provide information on the atomic identities and arrangements in the materials—that is, the phase and structure of the studied sample. Diffracted beams can only be observed at an angle given by the Bragg equation:

$$m\lambda = 2d_{hkl} \sin\theta$$

where λ is the wavelength of the incident X-ray or electron beam, d_{hkl} is the lattice distance of the (h, k, l) plane, and m is an integer.

For graphite and diamond with long-range periodic order, the diffraction pattern is sharply defined. However, for amorphous materials such as glassy and amorphous carbon, the diffraction pattern shows diffuse maxima which are related to the average interatomic spacing between the atoms. A radial profile through the diffraction pattern essentially gives a radial distribution function of the atoms surrounding a particular scattering center in the amorphous sample. The effect of the finite sizes of sample is seen as the broadening of the peaks in XRD, as explained by the Scherrer equation.

The most-often used X-ray technique is that of *powder diffractometry*. For this, a monochromatic beam is incident at an angle θ on a specimen with a minimum thickness of about 20 μm. The detector is then set to receive reflections at an angle θ that varies over the angular range of interest, either by keeping the incident beam direction fixed and rotating the specimen and detector, or by keeping the specimen fixed and rotating the incident beam and detector in opposite scenes. Neutrons emerge from a high-flux nuclear reactor. Single-wavelength beams are achieved through the use of crystal monochromators. Neutron diffraction is geometrically similar to XRD, is very much a bulk technique, and is not widely used for the characterization of nanocarbons due to its limited benefits for the analysis of nanostructures. Electron diffraction is mostly obtained in a transmission electron microscope (as discussed later in the chapter).

2.3
Imaging

2.3.1
Electron Microscopy

Electron microscopy is one of the most important techniques used to conduct morphological and structural investigations of nanocarbons. Depending on the electrons that are used or collected to form an image, the technique is divisible into either TEM, where the transmitted electron beam is used to study the bulk structure of a thin specimen, and scanning electron microscopy (SEM), where the secondary or backscattered electron is used to analyze both the surface and sub-surface of nanocarbons.

In both techniques, an electron beam must first be formed for the illumination. The beam can be generated in either of two ways: (i) by thermionic emission, where the thermal energy is used to overcome the surface potential barrier of a solid surface, allowing the extraction of electrons from the conduction band of an emitter (tungsten or lanthanum hexaboride, LaB_6, filament); or (ii) by applying an extremely high electric field to reduce the surface potential barrier of an emitter (i.e., ZnO tip), which is referred to as a field emission gun (FEG). A "cold emission" is obtained when electrons are extracted only through an applied high field at room temperature, but this requires a very high vacuum as the surface of the tip must be very clean. Alternatively, the tip can be heated to a moderate temperature (Schottky or thermally assisted field emission), which requires a lower field strength and a lower vacuum compared to the cold emitter. The emitted electrons

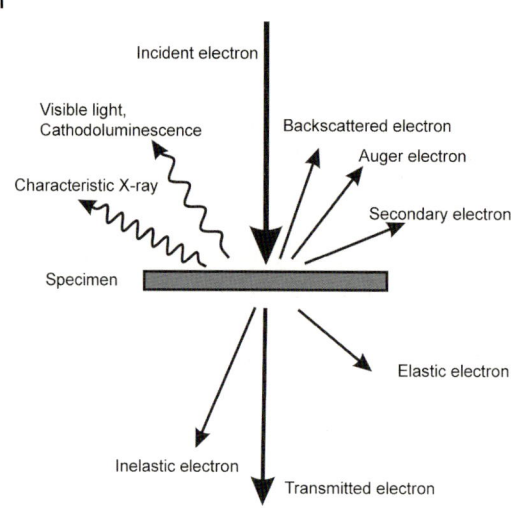

Figure 2.1 Schematic diagram of the electron-beam interaction in a thin specimen.

are collimated and focused using a Wehnelt cylinder to form a beam. The accelerating voltage is typically 1–30 kV for SEM, and 100–300 kV for TEM.

2.3.1.1 Electron–Specimen Interactions

When an electron passes through an atom, it will be scattered through either a coulombic or electrostatic interaction. Elastic scattering occurs when the incident electrons interact with the potential field formed by the nuclei of the atoms (electron–nuclei interaction). While this process involves essentially no energy loss, the electron can be declined by momentum transfer. Inelastic scattering occurs when interactions between the incident electrons and the atomic electrons occur, such that the scattered electrons lose energy (electron–electron interaction). The probability of electrons undergoing any type of interaction with atoms is determined by the interaction cross-section.

The result of the interaction of an incident electron beam with a thin specimen is shown in Figure 2.1. Whilst a variety of electrons, photons, and other signals can be generated, there are three types of transmitted electron: (i) non-scattered electrons; (ii) elastically scattered electrons; and (iii) inelastically scattered electrons. The transmitted and elastically scattered electrons are used to form an image or diffraction pattern, while the inelastically scattered electrons can be analyzed to provide spectroscopic information, for example, by using electron energy-loss spectroscopy. There are also three types of electron that can be emitted from the electron-entrance surface of the specimen:

- *Secondary electrons* with energies less than 50 eV that escape from the specimen arising from ionized electrons associated with atoms close to the surface of the solid. Secondary electrons are abundant and are used extensively for imaging in SEM.

- *Auger electrons*, produced by the decay of the excited atoms.
- *Backscattered electrons*, that have energies close to those of the incident electrons. These are beam electrons that are reflected from the sample by Rutherford backscattering from the nucleus, and are also used for imaging in SEM.

The de-excitation of atoms that are excited by the primary electrons also produces X-rays as well as light (*cathodoluminescence*). The energies and wavelengths of the emitted X-rays are characteristic of the involved atoms, and can be used to provide both qualitative and quantitative information of the elements present in the regions of interest. Depending on which probe or signals are to be used, a variety of microscopes (e.g., scanning or transmission electron microscope, Auger electron microscope) or techniques (energy-dispersive X-ray spectroscopy, electron energy-loss spectroscopy) have been developed. Scanning and transmission electron microscopy will be discussed in the following section, while spectroscopic methods will be introduced later.

2.3.1.2 Scanning Electron Microscopy

Scanning electron microscopy is regarded as the first choice for obtaining the overall information relating to nanocarbons, before other techniques are used. Indeed, SEM is a high-routine characterization tool in nanomaterials research, as it is relatively easily operated and rapidly provides information regarding morphological and compositional information when an analysis of energy or wavelength of the characteristic X-ray is performed. A SEM image is obtained by scanning the sample surface with a beam of electrons in a raster scan pattern. The electrons interact with the atoms that make up the sample, producing signals (as described above) that contain information relating to the sample's surface topography, composition, and other properties.

The basic layout of the SEM instrument is shown in Figure 2.2. The source of electrons is an electron gun, that can be of either the FEG or thermionic type. Two or more condenser lenses are used to demagnify the crossover of the electron beam produced by the gun. The objective lens then focuses the electrons to form a very narrow probe with diameters between 1 and 10 nm on the specimen, with the angular spread of the beam being limited by the objective aperture. The focused beam, on passing through the optical axis, scans the specimen surface in a two-dimensional (2-D) raster, while an appropriate detector monitors the signals that are emitted from each point on the surface. Simultaneously, by using the same scan generator, a beam scans across the recording monitor. The intensity of each pixel on the monitor is directly related to the emission intensity of the selected interaction at the corresponding point on the specimen surface. The magnification is the ratio of the dimensions of the raster on the display device and the raster on the specimen. Higher magnification micrographs can be obtained by reducing the size of the raster on the specimen.

The operational mode of the SEM system depends on the types of signal used for the imaging or for analysis. Secondary electron imaging (SEI) is the most common or standard detection mode. Due to the very narrow electron beam, SEM

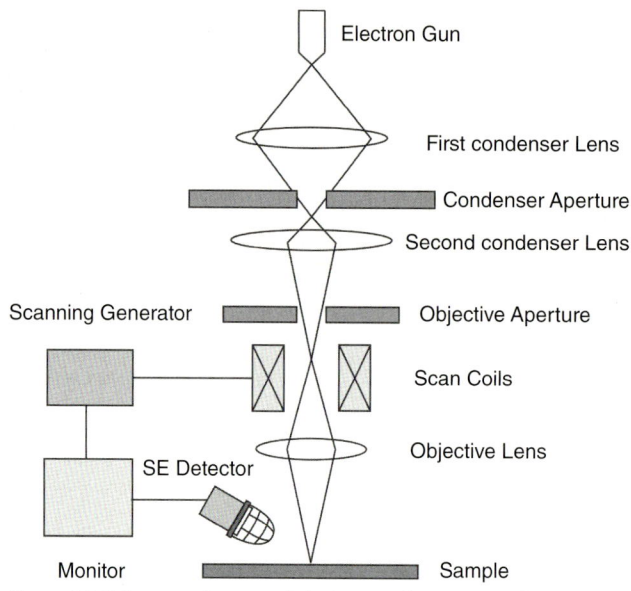

Figure 2.2 Schematic diagram of the layout of a scanning electron microscope.

micrographs have a large depth of field and yield a characteristic 3-D appearance that is useful for understanding the surface structure of a sample. As the detected secondary electrons mainly originate from the top few nanometers of the specimen surface, the emission diameter will be only slightly larger than the probe diameter. The lateral resolution for SEI is of the order of 1–5 nm.

Backscattered electrons (BSEs) have very high energies and low yields compared to secondary electrons. The intensity of the BSE signal is related to the atomic number (Z) of the specimen, and BSE images can therefore provide information regarding the distribution of different elements in the sample. The resolution in BSE images is typically worse than 10 nm, due to the larger penetration depth. A lower acceleration voltage (<1 kV) with high-brightness LaB_6 and FEG sources can be used to increase the details of surface and obviate sample charging in nonconducting or poorly conducting samples.

Recently, techniques have been developed to detect the transmitted electrons (TE) from a thin specimen in a SEM system, so as to form TE images [2]. This provides the possibility of obtaining a "TEM" image in a SEM system. The SEI or BSE images, together with TE image, can be obtained from the same area of one sample. A BSE image of Ni nanoparticles on a carbon support is shown in Figure 2.3, where the Ni nanoparticles exhibit a bright contrast due to the atomic number of Ni being larger than that of carbon. The TE image reveals that the supports have a tubular structure.

Environmental scanning electron microscopy (ESEM) allows the imaging of uncoated nonconducting samples, or of specimens which are in the hydrated state in a much degraded vacuum in the SEM chamber. ESEM allows also the study of

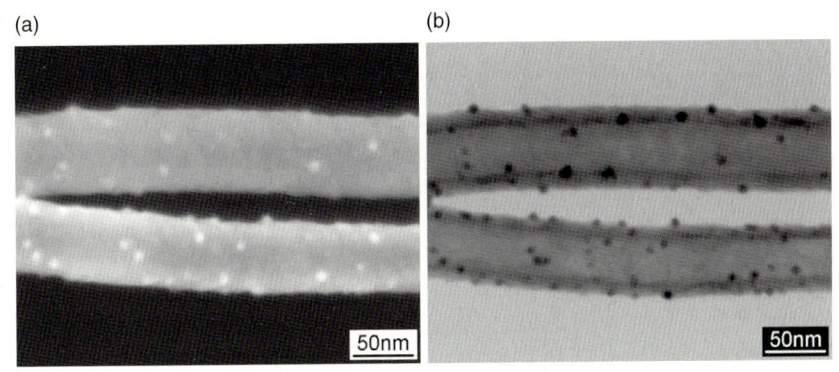

Figure 2.3 SEM images of a 1 wt% Ni/CNT sample prepared by incipient wetness impregnation. The same area was acquired in the BSE (a) and TE (b) modes [1].

in situ reactions at higher temperatures under a variable pressure, although the image resolution is degraded due to interaction between the incident and scattered electrons with the gaseous atmosphere [2].

2.3.1.3 Transmission Electron Microscopy

The transmission electron microscope is, without doubt, the "eyes" of nanoscience and nanotechnology. All of the earlier reports of the discovery of carbon nanotubes (CNTs), carbon nanofibers (CNFs), and any other nanocarbons with extraordinary morphologies, have been evidenced via TEM images. In the transmission electron microscope, an image is formed due to the electrons being transmitted through an ultra-thin specimen, magnified, and recorded using a charge-coupled device (CCD) camera, as frequently used in a modern TEM system.

The basic construction of a TEM system is shown in Figure 2.4. The electrons emitted from a filament are accelerated by creating a potential between the filament and the anode. Condenser lenses and apertures are then used to focus the electron beam to a certain size and to illuminate the specimen. After interacting with the sample, the transmitted electrons are focused onto the back focal plane of the objective lens, from where they are propagated and enter the first and second intermediate lenses. The projector lens forms the final enlarged image or diffraction pattern on the fluorescent screen, or on the CCD camera chip. Modern TEM systems are normally equipped with energy-dispersive X-ray (EDX) detectors and/or an electron energy loss spectroscopy (EELS) system located on that side of the microscope column near the specimen stage, or below the view screen, respectively (Figure 2.4).

The transmitted electrons can also be used to form a diffraction pattern containing similar information as XRD, but from a very small volume of the chosen area of a specimen [this is referred to as selected area electron diffraction (SAED)]. This is achieved by projecting the back focus-plane of the objective lens to the

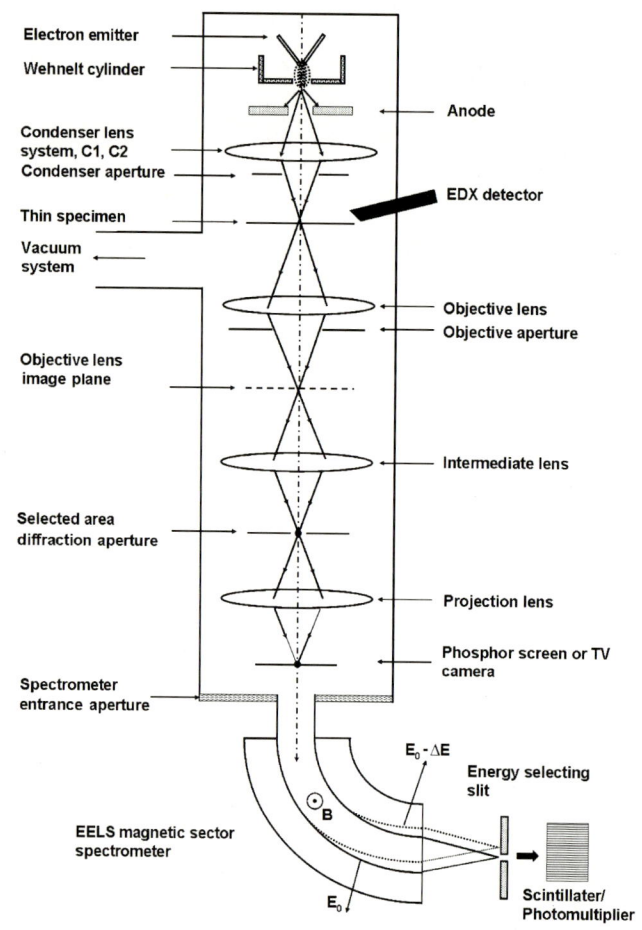

Figure 2.4 Schematic diagram of the layout of an analytical transmission electron microscope.

viewing screen CCD chips. The most important feature of a modern electron microscope is its imaging versatility: typically, images at lower, medium and high magnification with various contrast mechanics can be obtained, providing comprehensive information regarding a sample, from its general morphology to details at atomic resolution. The TEM system is capable of imaging at a significantly high resolution that enables the examination of fine details as small as columns of atoms. High-resolution imaging is the most frequently used method for examining the very fine details of nanocarbons. A high-resolution TEM image of five-wall CNTs, with the fullerene and tungsten atoms clearly resolved, is shown in Figure 2.5.

2.3.1.3.1 **TEM Imaging** Currently, there are three basic contrast mechanisms which contribute to TEM images.

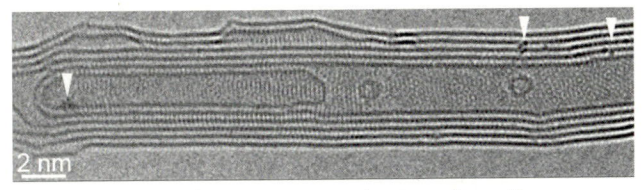

Figure 2.5 High-resolution TEM image showing a five-wall MWNT containing two fullerenes in the hollow core. A few W clusters were observed (indicated with white arrowheads) in the MWNT [3].

Mass–Thickness Contrast The physics of mass–thickness contrast is the incoherent Rutherford scattering of electrons. The cross-section of Rutherford scattering is proportional to the atomic number Z – that is, the mass or the density ρ – as well as thickness t. For a simple and qualitative interpretation, regions of a sample with high-Z (thus high mass) will scatter more electrons than regions with low-Z of the same thickness. Similarly, thicker regions will scatter more electrons into a large angle than the thinner regions of the same average Z, due to multiple scattering. Electrons scattered into large angles in a TEM system cannot be used to form an image. Specimen regions, which are thicker or of higher density, appear then dark in the image; this difference in the intensity of the image is termed the "mass-thickness contrast."

Diffraction Contrast Diffraction is the major contrast mechanism in crystalline specimens, especially at medium magnifications. It arises from the various amplitudes of the undiffracted beam and diffracted beams, resulting in intensity variations in the image formed by the different beams. If the objective aperture is centered on the undiffracted (transmitted) beam, then there will be a *bright-field* image, in which region of the specimen which is in the Bragg position for strong diffraction to occur will appear dark. Correspondingly, if the objective aperture is centered on any one of diffracted beams that used to form the image, this is referred to as a *dark-field* image, in which only the region exciting the Bragg beam appears bright, while the rest appears dark. Diffraction contrast is very useful when studying the defect structure of graphitic carbon with long-range ordering.

Images obtained using TEM at medium magnification contain, in general, both contrast mechanisms. Such images first give the impression of how the prepared nanocarbon appears. A quantitative analysis of such images will then provide valuable information about the specimen. A TEM image of multi-walled CNTs (MWNTs), with the inner and outer diameter distributions estimated from many images, is shown in Figure 2.6.

Phase Contrast High-resolution TEM represents one of the most powerful methods for the direct atomic structural analysis of nanocarbons. In discussions

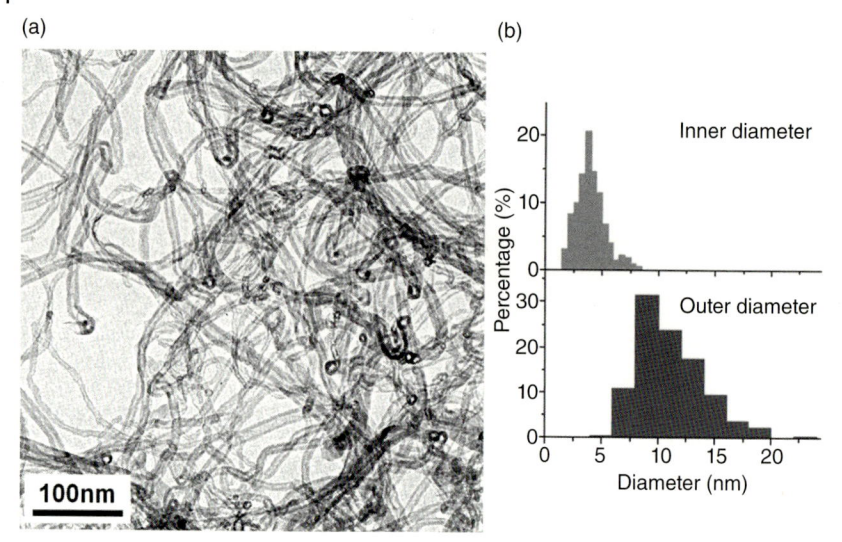

Figure 2.6 (a) TEM image of MWNTs; (b) Statistic distribution of the inner and outer diameters of CNTs, measured on TEM images.

of high-resolution images in a TEM system, it is more convenient to treat electrons as a wave (the dualistic property of particles). An electron wave is scattered by interactions with the inner potential of a specimen, and these interactions result in both phase and amplitude changes in the electron wave. The contrast of images on the atomic level is due to phase contrast, caused by a small phase-shift in the diffracted beams by the scattering, and by objective lens aberrations. For a sufficiently thin specimen, the "weak phase object" (WPO) approximation, which assumes that the electron wave is only phase-modulated (phase contrast) and not amplitude-modulated, can be applied. The image intensity is then correlated to the projected electrostatic potential of the sample, leading to atomic structural information. However, this can be modified by the aberration and defocus effects of the objective lens.

In the WPO approximation, the image intensity $I(x,y)$ at the image plane of the objective lens results from a 2-D Fourier synthesis of the diffraction beams, modified by a phase-contrast transfer function (CTF) factor, $\sin \chi$, given by Scherzer, as:

$$I(x,y) = 1 - i\sigma V(x,y) * FT \sin \chi$$

where $*$ represents a convolution integral and FT the Fourier transform. $\sigma = \pi / \lambda E$. Here, $V(x, y)$ is the interaction constant which is a function of the electron wavelength, and energy is the specimen potential. The phase-contrast imaging of a high-resolution image is controlled by $\sin\chi$, which contains the basic phase-contrast sinusoidal terms modified by an attenuation envelope function, $P(\theta)$, which is essentially due to the partial coherence of the electron beam:

$$\sin\chi = P(\theta)\sin(2\pi/\lambda)\left(\Delta f\theta^2/2 - C_s\theta^4/4\right)$$

where θ is the radian scattering angle, C_s is the spherical aberration coefficient of the objective lens, Δf is the objective lens defocus value and $P(\theta)$ depends on the coherence condition of the incident beam. The CTF is a quantitative measure of the trustworthiness of the lens in recording a reliable image. Directly interpretable structure images are recorded near the Scherzer defocus, defined as:

$$\Delta f(S) = -1.2 C_s^{1/2} \lambda^{1/2}$$

The point resolution corresponds to the first zero in the CFT. Thus, under the WPO approximation, near $\Delta f(S)$, the image can be directly related to the 2-D projected potential of the specimen with dark regions corresponding to columns of heavier atoms.

2.3.1.3.2 Aberration-Correction As discussed above, image quality and the resolution is strongly governed by the lens aberration. Spherical aberration occurs when the power of a lens varies with radial distance from the optical axis (the spherical aberration coefficient C_s is larger than zero). As a consequence, a point in the object becomes a disc in the image plane (Figure 2.7). Spherical aberration degrades the TEM imaging performance in several ways. Notably, it causes a general blurring of the image as the discs from adjacent object points overlap in the image. It also causes a phenomenon known as *delocalization*, whereby electron waves interfere with one another over an extended area in the image (Fresnel fringes), appearing in images as an extension of the perimeter of a sample beyond

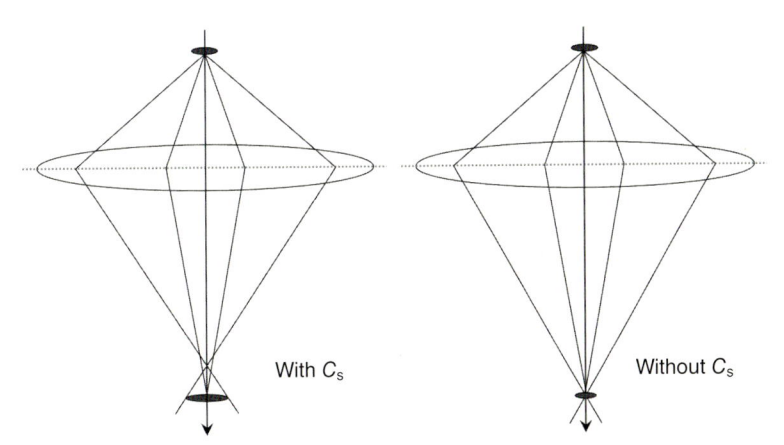

Figure 2.7 Schematic illustration of optical ray diagram showing the effect of spherical aberration of an objective lens. Due to the spherical aberration, a point in the object becomes a disc in the image plane. With aberration correction (C_s), a point in the object is imaged as a point in the image plane.

the actual surface or interface [4]. This delocalization effect leads to the difficulty for a straightforward interpretation of the structure. Time-consuming image reconstruction process could be used to retrieval the "real" structure of the studied samples if a set of images at varying defocus are recorded (the defocus series). Recently, spherical aberration correction techniques applied in TEM have been well developed and commercialized. In such so-called C_s-corrected TEM, a combination of multipole correctors produce a negative spherical aberration to compensate the lens aberration. As the C_s can be reduced to a minimum, the C_s-corrected TEM exhibits a higher resolution than the uncorrected TEM. Most importantly, delocalization effect is corrected so that an artifact-free imaging of a sample is possible.

The power of C_s-corrected high-resolution TEM (HRTEM) is demonstrated in Figure 2.8, which shows a visualization of the carbon network of individual molecules. A fullerene molecule functionalized with pyrrolidine, attached to a single-walled CNT (SWNT) is clearly visible. The new technical improvements of C_s-corrected TEM, combined with image simulation, highlight the newfound ability to provide an atomic resolution analysis of noncrystalline materials, particularly those made from light elements such as organic or biological molecules which, previously, were believed difficult to visualize with HRTEM. Keeping the accelerating voltage of TEM as low as possible is beneficial for minimizing any knock-on damage of soft matter, provided that the resolution can be compensated by the C_s correction [5].

2.3.1.3.3 **Tomography** In the TEM system, 2-D information is generated by projecting a 3-D structure. A 2-D projection of many materials with nanoarchitecture can lead to the overlap of many features, so that it may be difficult to draw conclusions relating to their 3-D structure. For example, a projected image of CNTs with supported nanoparticles cannot provide information on the location of the particles – are they *outside* or *inside* the CNTs? Recently, electron tomography in association with TEM has been developed to obtain 3-D information of the studied specimen. Although the methodology used originated during the late 1960s, principally for the analysis of biological molecules, it is now a standard technique for analyzing macromolecules, with nanometer resolution. Electron tomography involves the recording of multiple images of a single sample, tilted through a series of angles, such that a set of images (a "tilt series") can be collected. In the same way as in established X-ray tomography, the information from the tilt series of 2-D projects is then analyzed to yield a detailed 3-D construction of the structure. The intensity in the micrograph should be a monotonic function of the amount of material projected parallel to the electron beam; in other words, the diffraction contrast must be excluded.

Figure 2.9a shows a typical 2-D image of a CNT with supported Ni nanoparticles. From such a bright-field image, it is not possible to determine the exact location of the Ni nanoparticles and, in addition – due to the very small size of the Ni nanoparticles – the difference in mass–thickness contrast between Ni and carbon is very weak. In the sections through the reconstructed volume (a longitudinal section is

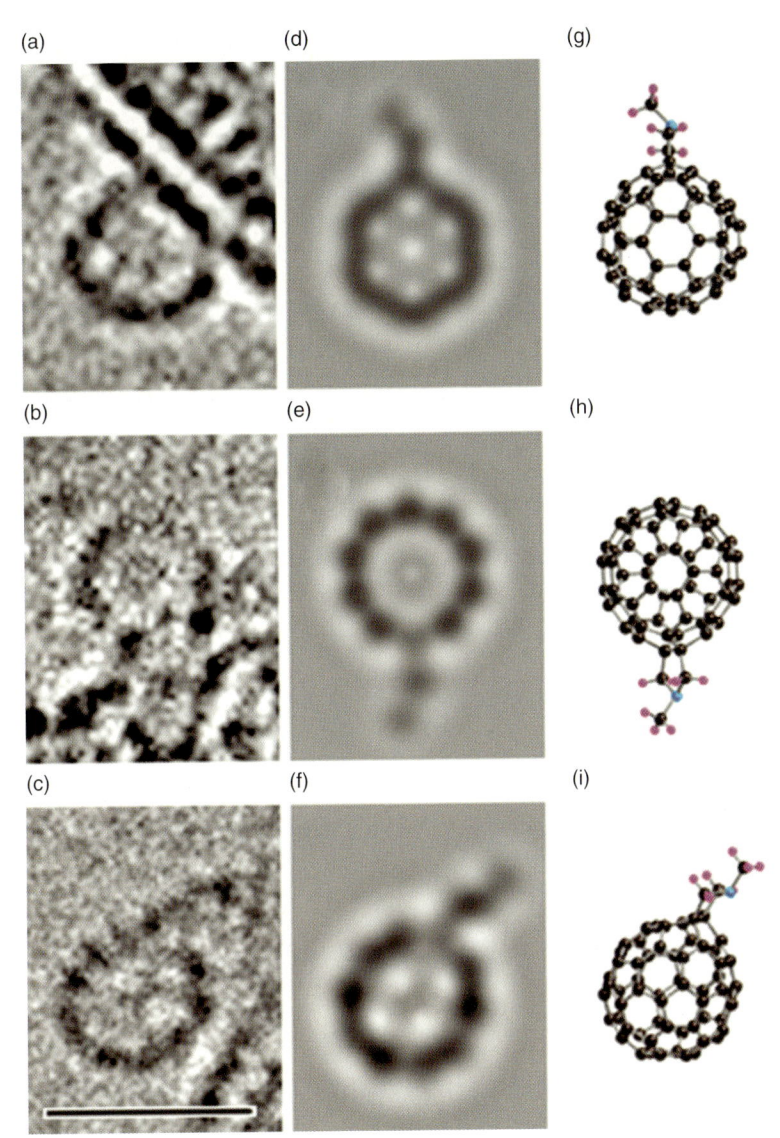

Figure 2.8 (a–c) HRTEM images of functionalized fullerene molecules with pyrrolidine (C60-C$_3$NH$_7$) attached to the surface of a SWNT. The intramolecular structures are clearly visible for each fullerene; (d–f) Image simulations of C60 fullerene derivates for various orientations to be compared. The corresponding atomic models are also shown (g–i) [5].

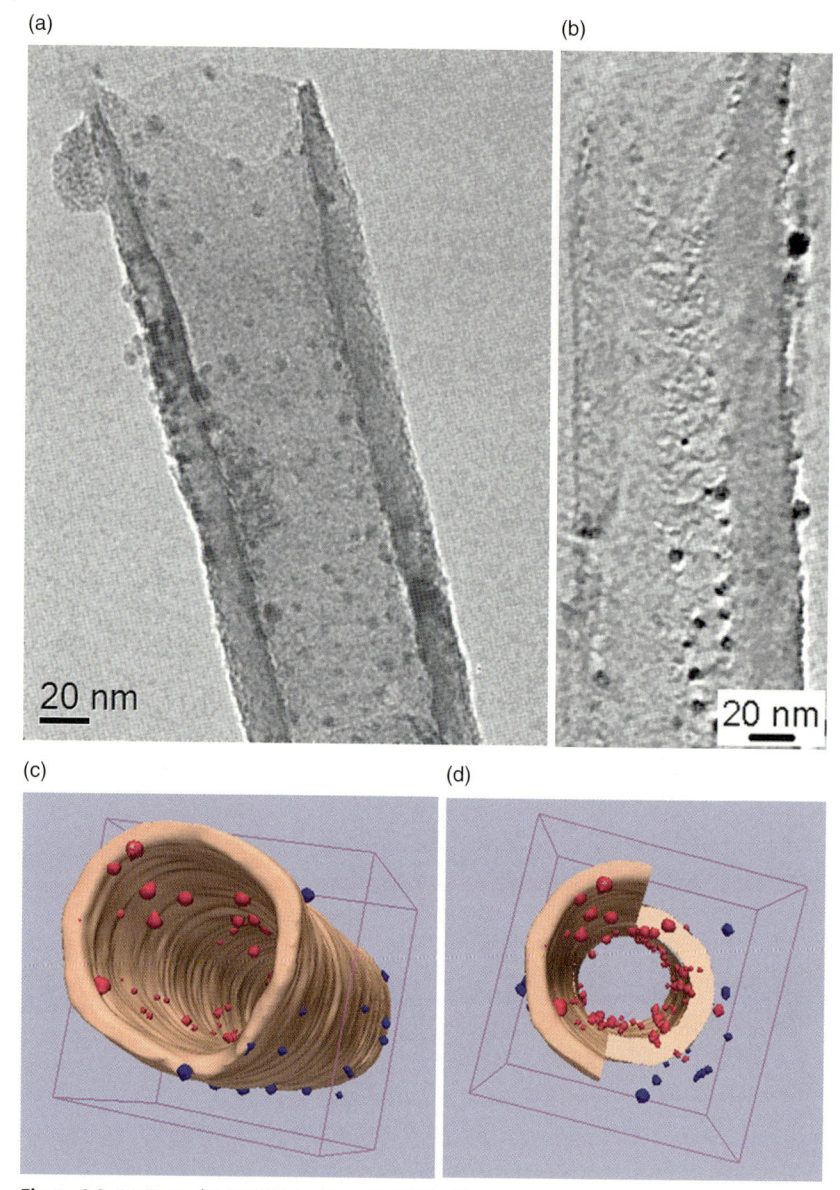

Figure 2.9 (a) Typical 2-D TEM image of a CNT with Ni nanoparticles, used to reconstruct the 3-D image; (b) Longitudinal section through the reconstructed volume; (c and d) Modeling of the reconstructed volume (CNT shown in pink; Ni particles inside the tube shown in red; Ni particles on the external surface shown in blue) [1].

shown in Figure 2.9b), the individual analysis of their size is facilitated by an increase in the signal-to-noise ratio due to the redundancy of information coming from several images. By using the 3-D positions of these particles with respect to the inner and outer surfaces of the tube (as obtained by modeling), the Ni nanoparticles inside or outside the CNTs can be clearly distinguished (Figure 2.9c and d). Thus, some 75% of the Ni particles are seen to be located inside the tube.

2.3.1.4 Scanning Transmission Electron Microscopy

In modern TEM systems, an electron detector can be located below a thin TEM specimen. When a small probe, produced by a FEG and a condenser-objective lens system, is scanned across the specimen and the signal is detected and imaged on a monitor or frame store, this construction and operation mode is termed scanning transmission electron microscopy (STEM). In this case, the resolution is determined mainly by the probe diameter, which is usually of the order of 1 nm. By using a corrector of spherical aberration in the probe-forming lens, sub-angstrom resolution can be demonstrated. STEM offers high-resolution nanoanalytical capabilities for both EDX and EELS, together with bright-field, dark-field, and high-angle annular dark field (HAADF) imaging. Bright-field STEM imaging uses an axial detector to detect electrons scattered through a relatively small angle, and the images contain diffraction contrast. In contrast, dark-field STEM imaging is essentially incoherent and employs an annular detector to detect electrons scattered through higher angles.

The most important method of high-resolution STEM is Z-contrast imaging. This utilizes the fact that high-angle scattering (with scattering angles >30 mrad) follows Rutherford's law – that is, the scattering cross-section is proportional to Z^2. The use of a HAADF-STEM signal does not have the complexity of conventional bright-field scattering in HRTEM associated with diffraction complications. The incoherent HAADF-STEM images are directly interpretable. When using HAADF techniques, single heavy-metal atoms on nanocarbon support can be imaged; hence, the method is very useful for detecting metallic clusters on carbon supports.

2.3.2
Scanning Probe Microscopy

In scanning probe microscopy, the sample surface is probed by monitoring the interaction between a localized probe and a sample surface. Conceptually, the process is fundamentally different from conventional microscopy, since neither light nor electron beams are used. However, these techniques are excellent for achieving lateral resolution and in manipulating the samples at an atomic resolution.

2.3.2.1 Scanning Tunneling Microscopy

The scanning tunneling microscope is the respective instrument for imaging surfaces at the atomic level, without illumination and without lenses. The

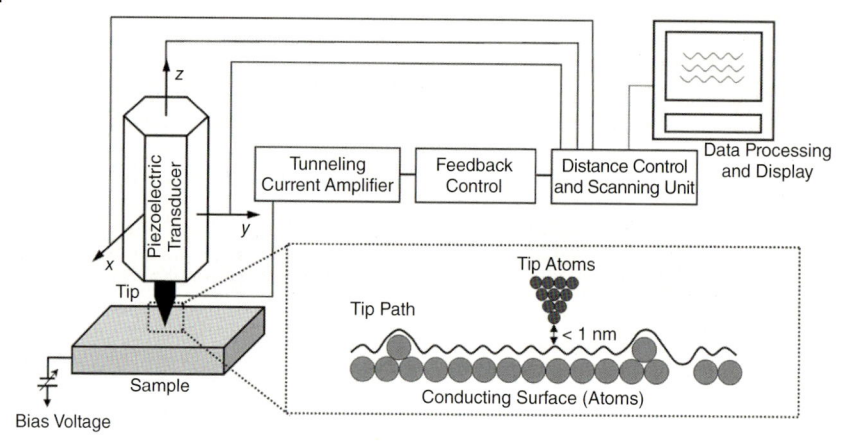

Figure 2.10 Schematic diagram of a STM set-up with the principle of tip and sample interaction.

fundamental principle of scanning tunneling microscopy (STM) is rather simple. An atomically sharp metal tip (anode) is brought into such close proximity (<1 nm) to a sample surface (cathode) that an overlap occurs between the tip and the surface electron wave function. If a small bias voltage (V_t) is applied to the sample, the electrons can tunnel elastically from filled tip states into sample tip states, or *vice versa*, depending on the polarity of V_t. This "vacuum" tunneling establishes a small tunneling current (I_t) within the nano-ampere range, which is a function of tip position, the applied voltage, and the atomic species present on the surface. It is this current that is used to generate an STM image. Notably, STM cannot be used to image insulating materials and, in principle, no vacuum is required except when studying the adsorption of species on the surface.

A schematic illustration of the STM set-up is shown in Figure 2.10. The system consists of a sharp conducting tip, a piezoelectric-controlled height, a (x, y) scanner, a vibration isolation system, and a computer. The tip is often made from tungsten or platinum–iridium. The image is acquired by monitoring the current as the tip scans across the surface laterally, under the control of a piezoelectric driver. The resolution of an image is limited by the radius of curvature of the scanning tip, with a lateral resolution of 0.1 nm and a depth resolution of 0.01 nm being common. With this resolution, individual atoms within materials are routinely imaged and manipulated. When using STM, a surface can be scanned in two different modes: (i) constant height; or (ii) constant current. In constant-height mode, the tip moves in a horizontal plane above the sample, and the tunneling current varies as a function of the surface topography and the local surface electronic states of the sample. The tunneling current measured at each location on the sample surface constitutes the data set, and thus the topographic image. In constant-current mode, a feedback system is used to keep the tunneling current constant by adjusting the height of the scanner at each measurement point. In this mode, the motion of the scanner therefore constitutes the data set.

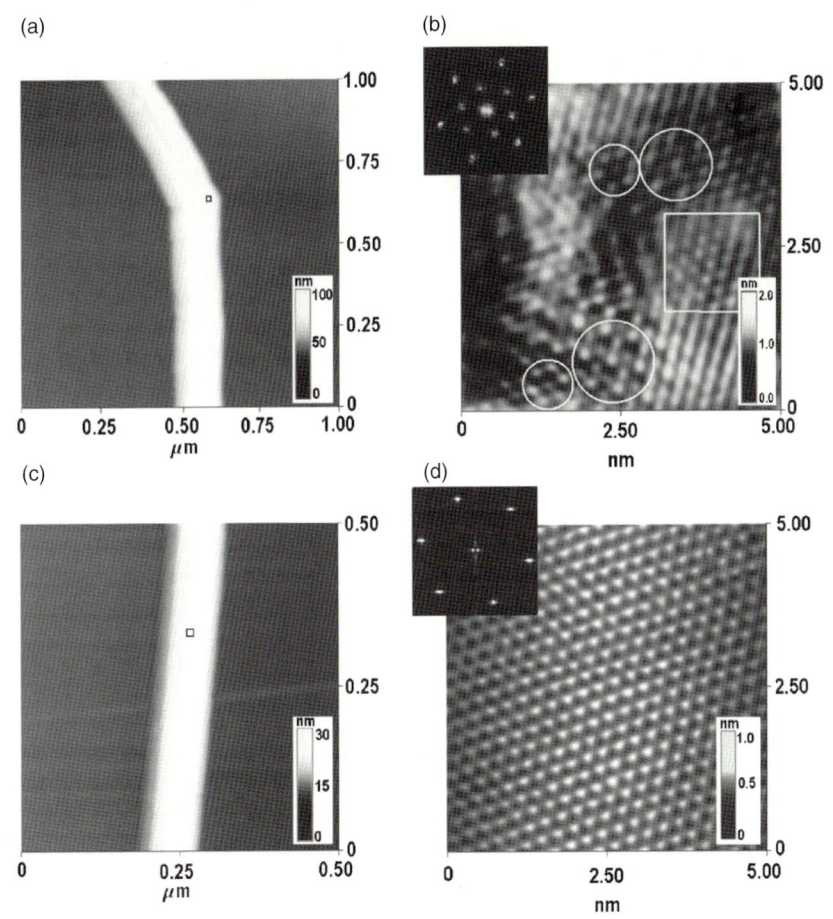

Figure 2.11 (a) STM image of a MWNT bend junction with a bend angle of 30°; (b) Atomic resolution achieved in the junction region, showing complex interference patterns of coexisting superstructures. For the sake of comparison the atomic resolution image (d) of a defect-free MWNT (c) is also presented. The 2-D Fourier transforms of current images are shown as insets of the atomic resolution images recorded at 0.1 V bias [6].

The STM system with atomic resolution is ideally suited for characterization of the surface structure, such as individual adsorbate atoms, defect sites, the defect structure, or just morphology in general. The STM image of an as-grown MWNT bend junction is shown in Figure 2.11a. An atomic-resolution image (Figure 2.11b), achieved at the junction region (indicated by the small rectangle in Figure 2.11a), revealed that specific configurations of the defects lead to nanotube bend junctions. For comparison, an atomic-resolution image obtained on the defect-free region of a straight MWNT is shown in Figure 2.11d; here, the triangular lattice of graphite is revealed, which indicates that the two outermost graphite layers exhibit *ABAB* stacking [6].

2.3.2.2 Atomic Force Microscopy

The atomic force microscope represents another important form of scanning probe microscopy for imaging, measuring, and manipulating matter at the nanoscale. While STM is limited to the surfaces of conducting specimens, atomic force microscopy (AFM) can be used to study nonconducting materials, such as insulators and semiconductors, as well as electrical conductors. In AFM, the magnitudes of atomic forces rather than the tunneling currents are monitored as a function of the probe position on the sample surface. In simple terms, information in AFM is gathered by "feeling" the surface with a mechanical probe.

The conceptual illustration and principle of AFM is shown schematically in Figure 2.12. The system consists of a cantilever with a sharp tip (probe), at the end of which is a piezoelectric scanner that is used to scan the specimen surface, a laser diode, a control system, and other units. When the tip is brought into the proximity of a sample surface, the forces between the tip and the sample lead to the cantilever being deflected according to Hooke's law. Depending on the situation, the forces that are measured in AFM include mechanical contact force, van der Waals forces, electrostatic repulsive forces, and magnetic forces. Additional quantities can be measured simultaneously by using specialized types of probe. Typically, the deflection is measured using a laser spot that is reflected from the top surface of the cantilever into an array of photodiodes. Other detection methods used include optical interferometry, capacitive sensing, and/or piezoresistive AFM cantilevers.

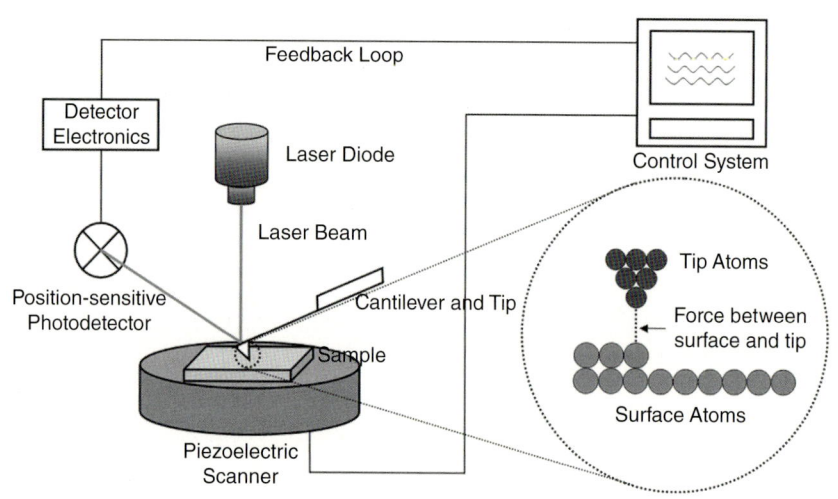

Figure 2.12 Schematic illustration of an AFM set-up with the principle of tip and sample interaction.

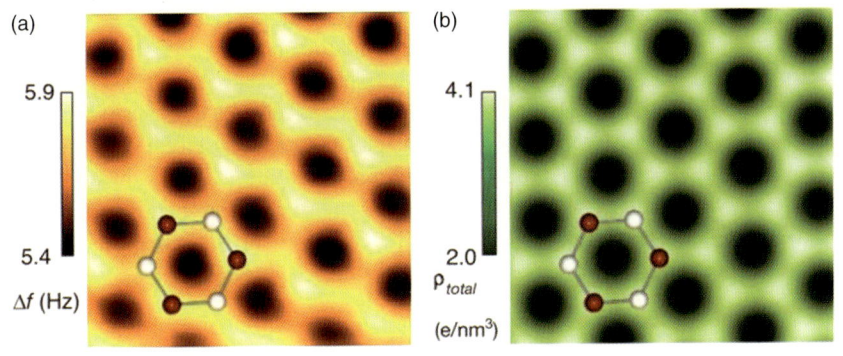

Figure 2.13 (a) Experimental image of graphite in constant-height dynamic AFM modes; (b) Calculated total charge density as a good approximation for a repulsive AFM image [7].

The atomic force microscope can be operated in several modes, depending on the application. Topographic AFM data sets are generated by operating in either constant-height or constant-force mode. In constant-height mode, the scanner height is fixed during the scan, and the spatial variation of the cantilever deflection is recorded. This mode is used to record real-time images of change surfaces, and also often for recording atomic-scale images of atomically flat surfaces, where the cantilever deflections – and thus the variations in applied force – are small. In constant-force mode, the deflection of the cantilever can be used as the input to a feedback circuit that moves the scanner up and down in the z direction, responding to the topography by keeping the cantilever deflection constant. In this mode, data sets can be collected which are generated from the scanner motion in the z direction; this mode is generally preferred for most applications.

The experimental AFM and theoretical images of a graphite surface are shown in Figure 2.13. Here, the hexagonal unit cells are clearly visible in the AFM image [7]. In general, AFM can be used to study 3-D nanotopography, morphology, homogeneity, dispersability, and the purity of nanocarbons. In addition, with the cantilever tip of this instrument, the mechanical (Young's modulus) and electrical (V–I characteristic) properties of nanocarbons can be determined.

2.4 Spectroscopy

2.4.1 Energy-Dispersive X-Ray Spectroscopy

Energy-dispersive X-ray spectroscopy (EDS, also EDX) is an analytical technique used for the element analysis and chemical characterization of a sample. The underlying principle is the analysis of X-ray signals as a function of X-ray energy.

As discussed above, the interaction between a high-energy electron and an inner shell electron from an atom results in the ejection of a bound inner-shell electron; this leaves the atom in an excited state with an electron-shell vacancy. De-excitation by transition from the outer shell produces either an Auger electron or X-rays that are characteristic of the elements in the sample, and can be used for composition analysis. The characteristic X-ray peaks sit on a continuous X-ray background emitted by the deceleration of fast electrons by the nuclei of atoms (Bremsstrahlung). The relation between the energy of the characteristic X-ray and the investigated element is given by Moseley's law:

$$E = A(Z-1)^2$$

where A is constant and Z is the atomic number. Currently, EDS is the most convenient and commonly used method for chemical composition analysis. If the wavelength of X-ray is analyzed, the technique is termed wavelength-dispersive X-ray analysis (WDX). A typical EDS spectrum is displayed in Figure 2.14.

The EDS set-up consists of a beam source, an X-ray detector, a plus processer, and an analyzer. Although free-standing EDS systems exist, most such systems are found associated with SEM and TEM set-ups. In these cases, the beam source serves as the electron gun of the microscope. As the windows in front of the X-ray detector can absorb low-energy X-rays, the conventional EDS system cannot detect precisely the presence of elements with an atomic number less than 5. Recently, a new type of EDS detector – the silicon drift detector (SDD) – has been developed and commercialized. The SDD detector consists of a high-resistivity silicon chip, where electrons are driven to a small collecting anode. The fitting of EDS with a SDD detector allows rapid X-ray recording and analysis, with a

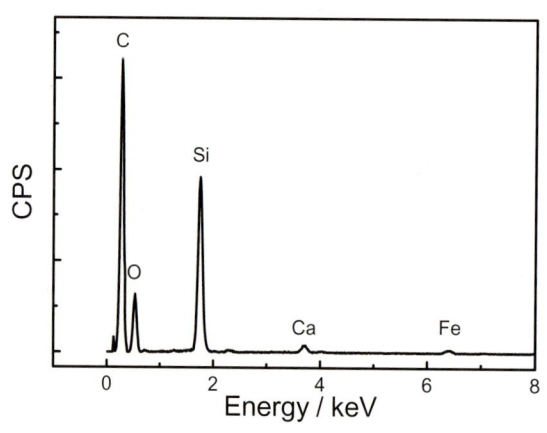

Figure 2.14 EDS spectrum of a SiO_2/CNT hybrid, showing the element-characteristic signals of carbon from CNT, oxygen and silica from the support, and of calcium and iron from the catalyst for CNT growth.

2.4.2
Electron Energy-Loss Spectroscopy

Electron energy loss spectroscopy (EELS), in association with TEM or STEM, is based on an analysis of the inelastic scattering suffered by the electron beam, by measuring the energy distribution of the transmitted electrons. The various energy losses observed in a typical EELS spectrum are shown in Figure 2.15a, in which gives scattered electron intensity is plotted as a function of the decrease in kinetic energy (the energy loss, E) of the transmitted electrons, and essentially represents

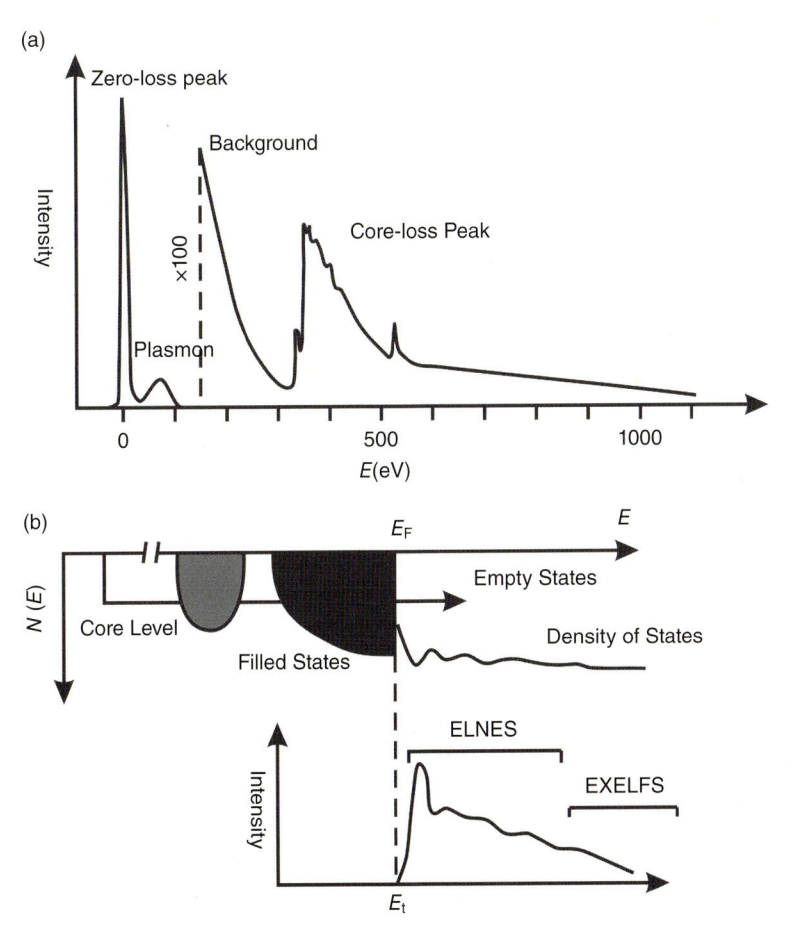

Figure 2.15 (a) Schematic diagrams of an EELS spectrum; (b) Relationship between the empty DOS and ELNES intensity in the ionization edge fine structure [8].

the response of the electrons in the solid to the disturbances introduced by the incident electrons. In a specimen of thickness less than the mean free path for inelastic scattering (ca. 100 nm at 100 keV), by far the most intense feature is the zero-loss peak at 0 eV energy loss, which contains all the elastically and quasi-elastically (e.g., phonon) scattered electron components. The energy resolution of EELS is defined as the full width at half-maximum (FWHM) of the zero-loss peak, which is usually limited by the energy spread in the electron source. For an FEG source, the energy resolution is typically about 0.8–1 eV. A monochromator attached to a FEG can increase the energy resolution down to 0.1 eV or even smaller, but at a huge expense in terms of a loss in beam intensity. The EELS set-up is shown schematically in Figure 2.4.

The low-loss region of the EELS spectrum, which extends from 0 to about 50 eV, corresponds to the excitation of electrons in the outermost atomic orbitals; these are often delocalized due to interatomic bonding, and extend over several atomic sites. The dominant feature in the low-loss spectrum arises from the collective oscillations of the valence electrons, known as *plasmons*. The energy of the plasmon peak is governed by the density of the valence electrons. In a thicker specimen, there will be additional peaks at multiples of the plasmon energy, corresponding to the excitation of more than one plasmon. A further feature in the low-loss spectra of insulators are peaks, known as *interband transitions*, which correspond to the excitation of valence electrons to low-energy unoccupied electronic states above the Fermi level. In a more detailed analysis, the low-loss spectrum may be related to the dielectric response function of the materials, which allows a correlation with optical measurements, including reflectivity and band gap determination in insulators and semiconductors.

The high-loss region of the EELS spectrum extends from about 50 eV to several thousand eV, and corresponds to the excitation of electrons from localized inner orbitals on a single atomic site to extended, unoccupied electron energy levels above the Fermi level. In solids, the unoccupied electronic state near the Fermi level may be appreciably modified by chemical bonding, leading to a complex density of states (DOS), as shown in Figure 2.15b. This is reflected in the electron energy loss near-edge structure (ELNES) within the first 30–40 eV above the edge threshold. The ELNES provides information on the local structure and bonding. Beyond the near-edge region, superimposed on the gradually decreasing tail of the core-loss edge, a region of weaker, extended oscillations is observed; this is known as the extended energy loss fine structure (EXELFS). The period of the oscillations can be used to determine bond distance, while the amplitude reflects the coordination number of the particular atom. The intensity or area under the ionization edge is proportional to the number of atoms, thus allowing a quantitative analysis of element. EELS is particularly sensitive to the detection and quantification of light elements ($Z < 11$), as well as transition metals and rare earth elements.

The major advantage of EELS is in identifying and quantifying the chemical composition and structural details of light elements, such as carbon and nitrogen. In the case of carbon, ELNES of carbon K-edges can reflect the differences in structure and bonding of various carbon materials by the intensities and shapes of the 1s-π^* and -σ^* peaks. For example, Figure 2.16 displays a series of carbon

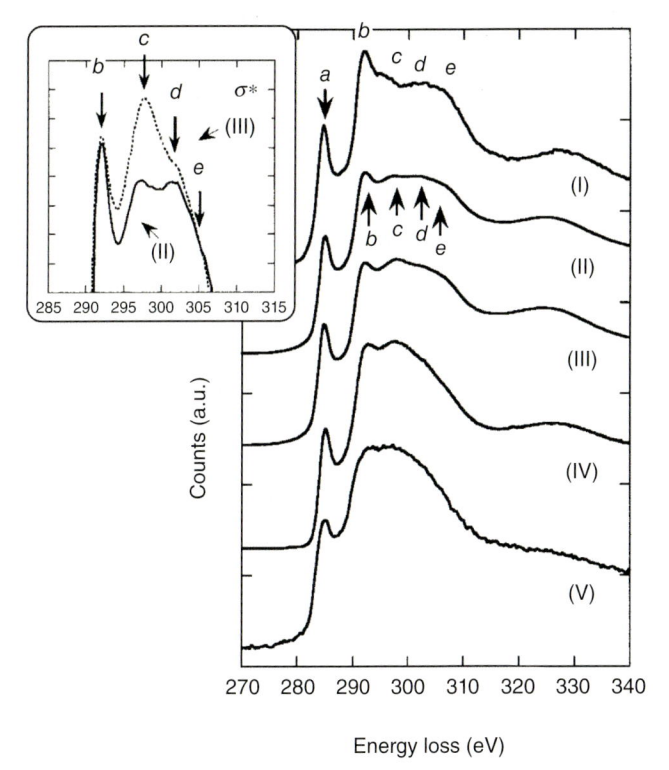

Figure 2.16 Spectra I–V: A series of C K-edge ELNES spectra corresponding to the various carbon nanostructures. I, a MWNT consisting of well-stacked graphitized layers of 10–16 nm diameter; II, coupled double-layers in an aggregate of nanohorns of about 3 nm diameter; III, an aggregate of various-sized fullerenes (1–3 nm); IV, double-walled CNTs with well-controlled diameters ~0.7 nm; V, an isolated nanohorn consisting of a curved, free-standing single graphene layer. For further details, see the text [9].

K-edge ELNES spectra corresponding to the various carbon nanostructures [9]. Each spectrum clearly shows two energy loss features. The first sharp peak (marked *a*) at around 285 eV corresponds to an electronic transition from carbon 1s to π^*, characteristic of the sp^2-bonded carbon, while four peaks (*b–e*) at 290–310 eV indicate transitions to σ^* states. As the curvature radius of the examined carbon nanostructures from I to V decreases, the peaks *c*, *d*, and *e* merge together and grow wider, indicating the electronic states of carbon nanostructures governed by the curvature and/or the interlayer coupling between adjacent graphene layers.

2.4.3
X-Ray Absorption Spectroscopy

X-ray absorption spectroscopy (XAS) is the counterpart of EELS for determining the local geometric and/or electronic structure of matter, but using X-ray photons instead of electrons as the probe. The experiment is usually performed at synchrotron radiation sources, which provide intense and tunable X-ray beams. The XAS data are obtained by tuning the photon energy using a crystalline monochromator to a range where core electrons can be excited (0.1–100 keV photon energy). Three main regions are found on a spectrum generated by XAS data. The dominant feature is called the "rising edge," and is sometimes referred to as XANES (X-ray absorption near-edge structure) or NEXAFS (near-edge X-ray absorption fine structure), which involves strong intensity oscillations within about 50 eV of the absorption edge onset. The extended X-ray absorption fine structure (EXAFS) region is at higher energies involving weaker-intensity oscillations. Both, XANES and/or NEXAFS are directly related to the unoccupied density of state, have characteristic features and energy positions, and can be used to determine the bonding environments of atoms in a solid. EXAFS oscillation corresponds to the scattering of the ejected photoelectrons of neighboring atoms, and can be used to determine bond lengths and coordination numbers in terms of a radial distribution function around the ionized atoms. The electron equivalents of the both techniques in EELS are shown in Figure 2.15b.

2.4.4
X-Ray Photoelectron Spectroscopy

X-ray photoelectron spectroscopy (XPS) is a surface-sensitive analysis technique that is widely used to determine the surface composition, and the chemical and electronic states of the surface atoms of the material under investigation. This method, which is also known as electron spectroscopy for chemical analysis (ESCA), is widely used as a fingerprinting tool by using empirically derived tables for the chemical shift to analyze the data [10]. When a sample is irradiated by a beam of soft, usually monochromatic X-rays (200–2000 eV), photoelectrons are emitted from the atoms in the specimen surface as a consequence of the photoelectric effect. The XPS spectra are obtained by measuring the kinetic energy distribution and number of the ejected photoelectrons that escape from the top 1–10 nm of the material being analyzed. From the measured kinetic energies of the emitted photoelectrons, the binding energy (BE) can be determined (the BE of each core-level electron is characteristic of the atom). Core-level electrons are not directly involved in chemical bonding, but they are influenced by the surrounding chemical environment of the photoemitting atoms. Changes in the chemical environment of the photoemitting atoms lead to a shift of the peak position in the spectrum. Thus, the observed BE chemical shift provides information about the electronic configuration of the atoms under investigation.

Normally, XPS measurements are performed under ultrahigh vacuum (UHV) conditions. The basic XPS instrument consists of a two-chamber apparatus. The first (preparation) chamber is used for sample introduction; notably, any contamination of samples will generate additional XPS signals that make a straightforward analysis difficult. The second chamber is a high-vacuum chamber used for XPS measurements, and contains the X-ray source, the detection system, and the

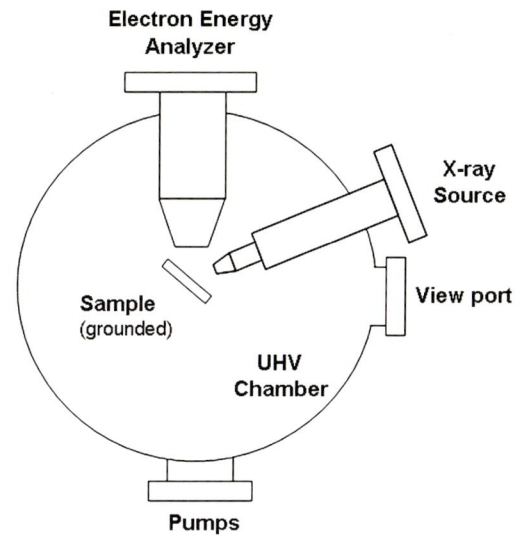

Figure 2.17 Schematic diagram of an XPS set-up.

sample holder with a view-window for monitoring (Figure 2.17). The amount of available information rapidly decays with the degrading vacuum. The BE of various photoelectron peaks (1s, 2s, 2p, etc.) are well tabulated [10], and therefore XPS is a suitable tool for elemental identification. The elements can also be quantified by measuring the integrated photoelectron peak intensities and using a standard set of sensitivity factors so as to obtain a surface atomic composition.

Currently, XPS is the most widely used analytical technique for characterizing the chemical configuration of the carbon–heteroatom bonds on the surface of carbon materials. The C1s XPS spectrum for graphitic samples has an asymmetric peak line-shape [12], with tailing on the high-energy side of the spectrum. Any disruptions of the perfect graphite structure, whether by defects, heteroatoms, or functional groups, will introduce components into the high binding energy side of the C1s spectrum. Figure 2.18 reports the C1s spectra of a CNF sample and the C1s of highly oriented pyrolytic graphite (HOPG). The line shape of the C1s spectrum for HOPG is considered to be that of almost pure, defect-free graphite, for which the line width is mainly determined by the spectral resolution. The BE of the main "graphitic" peak may vary slightly (the value reported for HOPG is about 284.6 eV). Any features above 285 eV are assigned to defective sp^2 graphitic structure and the sp^3 carbon species. At a BE higher than that of sp^3 carbon, features due to carbon–heteroatom bonds can be found. Compared to the HOPG sample, CNFs are characterized by a less intense and more asymmetric C1s peak shifted to a high BE value, with a larger FWHM. The difference between the two spectra at higher BE is significant, as clearly illustrated in Figure 2.18b. The distinct feature at around 288.8 eV is attributed to C–O heterobonds [13, 14]. The chemical configuration of the carbon–heteroatom bonds could not be obtained in

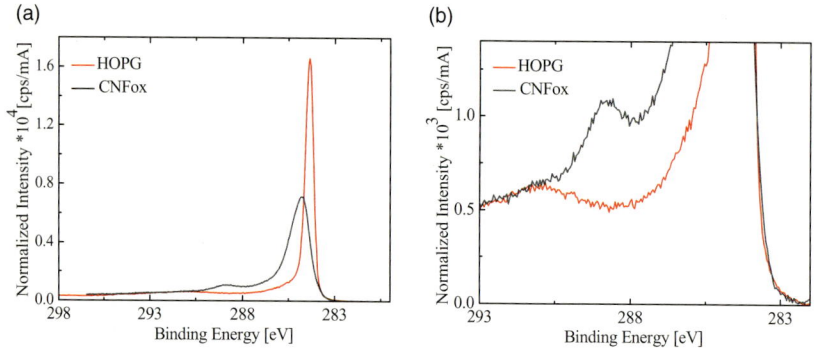

Figure 2.18 (a) C1s XPS spectra for HOPG (red line) and oxidized CNFs (black line); (b) Details of the C1s spectrum in the high BE region.

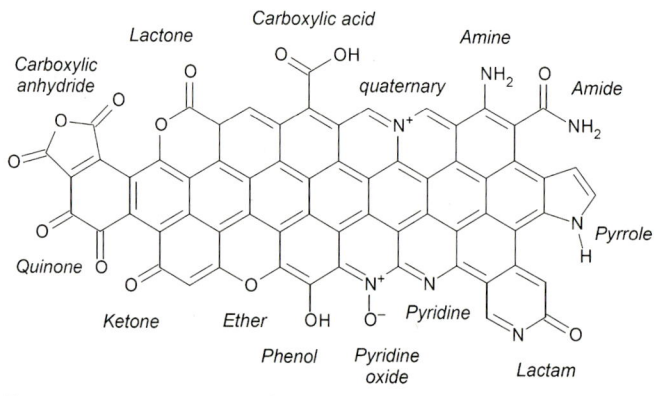

Figure 2.19 Various oxygen and nitrogen species and functional groups on the prismatic sites of a sp^2 carbon basal plane.

straightforward fashion through an analysis of the C1s spectrum. Rather, this is assessed by measuring and analyzing the corresponding core-level spectrum, which contains more detailed information.

Oxygen and nitrogen are two major elements used for the functionalization of nanocarbons (Figure 2.19). Due to the high variety of chemical configurations of the carbon–oxygen or carbon–nitrogen heterobonds, the N1s and O1s XPS peaks are usually very broad. Deconvolution of the N1s and O1s spectra into several components represents the most common approach to analyzing the O and N species on carbon. Four different BE regions can be identified in the experimental N1s core level spectra of carbon materials [11], as shown in Figure 2.20a:

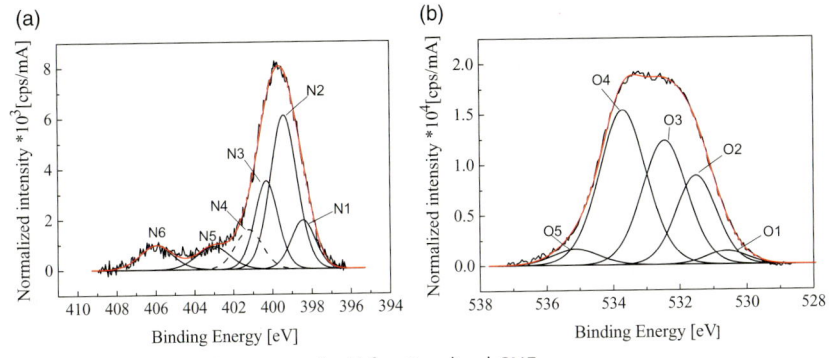

Figure 2.20 (a) N1s XPS spectrum for N-functionalized CNFs; (b) O1s XPS spectrum for O-functionalized CNFs. The red line indicates the smoothed spectrum of the experimental data.

- The first region, ranging from 398–399 eV; this refers to the pyridine-like N atoms (N1), and considers all nitrogen electronic configurations contributing to the π system with one p-electron.

- The region between 399–400 eV; this is referred to as the region of the N–H bond in amide/amine groups (N2).

- The third region between 400–400.5 eV; this refers to N atoms in pyrrole-like configuration and donating two p-electrons to the π system. Species such as pyridone, lactam and pyrrole functional groups are present in this region (N3).

- A fourth BE region at around 2.5 eV; this is higher than pyridine-like N atom configuration, and is usually referred to quaternary nitrogen (N4) due to intra- or inter-molecular hydrogen bonding. At higher BEs (402–406 eV), the N–O bonds are found (N5 and N6).

As shown in the O1s XPS data in Figure 2.20b, the component at the lowest BE side, 530.7 eV (O1), is usually attributed to a highly conjugated form of carbonyl oxygen such as quinine or pyridone groups; the components in the region between 531.1–531.8 eV are assigned to a carbon–oxygen double bonds (O2); and the components in the region between 532.4–533.6 eV are assigned to carbon–oxygen single bonds in ether-like (O3) and in hydroxyl group configuration (O4). However, there is still no well-accepted assignment of the O1s spectral feature to carbon–oxygen bonds, due to the fact that the BE chemical shift for a certain functionality is influenced by the neighboring functional groups. In addition, the adsorbed water molecules on nanocarbon interacting with the O species gives rise to a chemical shift, or to additional contributions to O1s signal, rendering their assessments complicated. For instance, detailed studies of the adsorption of water carried on polycrystalline graphite have led to the identification of a very broad peak centered at 533 eV [15], although in other studies the signal O5 above 535 eV was assigned to adsorbed water and/or oxygen [16, 17]. Other complementary techniques, such as XPS coupled with temperature-programmed desorption (TPD) mass spectroscopy, must be used for the analysis of oxygen species in carbon materials.

2.4.5
Raman Spectroscopy

Raman spectroscopy is used to study vibrational, rotational, and other low-frequency modes in a system. The basis of Raman spectroscopy is the inelastic scattering of monochromatic light, which interacts with phonons or other excitations in the system. For example, in contrast to elastic Rayleigh scattering, a small fraction of the scattered light (ca. 1 in 10 million photons) can be inelastically scattered by an excitation, which gives a frequency different from the incident photons (this is termed Stokes or anti-Stokes scattering). The difference in energy (frequency) then provides information about the phonon mode.

Experimentally, Raman spectroscopy is quite simple, especially when applied to bulk samples, and the instrumentation is widely available. A schematic set-up of a Raman spectrometer is shown in Figure 2.21. In operation, a laser is focused through a lens on the sample surface; the scattered light is then passed into a spectrometer and dispersed on a charge-coupled device (CCD) detector. Although the choice of laser wavelength depends on the required application, the most common choice is visible light (632.8 nm red from a HeNe laser, or 514.5 nm green from an Ar ion laser), ultraviolet (e.g., 325 nm from a He:Cd laser), or near infrared (e.g., 785 nm from a diode laser).

Raman spectroscopy provides a fast, nondestructive, and preparation-free diagnosis of carbon materials in terms of their electronic structure (sp^2, sp^3 hybridization, or a mixture of them), the ordering degree of sp^2 bound carbon atoms, and the nature of nanocarbons (SWNTs instead of MWNTs, or single graphene instead of graphite). Practically all carbon allotropes are Raman active; notably, fullerenes, CNTs and crystalline graphite feature sharp Raman lines that make them clearly

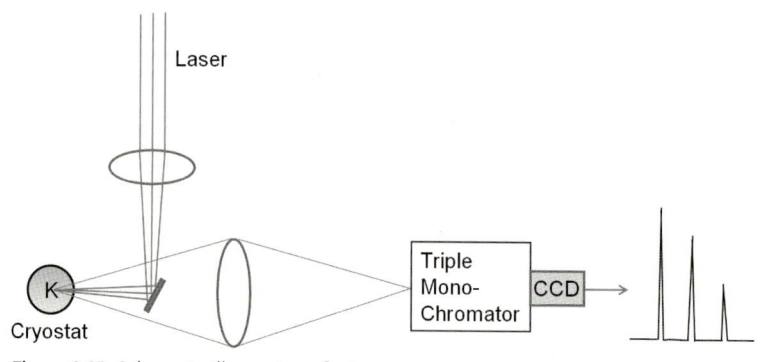

Figure 2.21 Schematic illustration of a Raman spectroscopy set-up.

distinguishable from the other carbon allotropes. For sp^3-bonded carbon (e.g., diamond), the Raman spectrum has a single shape line at 1335 cm^{-1}, whereas for sp^2-bonded carbon (e.g., graphite) the Raman spectrum has a single peak at 1575 cm^{-1}. This is the only active Raman mode of the infinite graphite lattice (called the G-band, after crystalline graphite, corresponding to the in-plane vibration of sp^2 carbon atoms and being a doubly degenerated phonon mode of the E$_{2g}$ symmetry mode of graphite). However, the peak at 1355 cm^{-1} is present in nearly all sp^2-bonded nanocarbons. This peak is called the D-band, after *d*isordered graphite, caused by the breakdown of solid-state Raman select rules which prevent its appearance in the spectrum of graphite with perfect structure. The intensity of the D-band in a Raman spectrum increases when the amount of disordered sp^2-bonded carbon increases, and when the crystalline size of graphite decreases. It should be noted that all carbon forms contribute to the Raman spectra, typically giving rise to a two-band feature with D-band and G-band. The ratio of D-band to G-band (the I_D/I_G ratio) has been widely used to quantitatively characterize the graphitization of the sample.

The Raman spectra of MWNTs and SWNTs are shown in Figure 2.22. Here, the major features are the sharp G-band peaks in the frequency bunch (between 1500 and 1600 cm^{-1}) and the D-band features at around 1350 cm^{-1}. A second-order observed mode between 2450 and 2650 cm^{-1} is assigned to the first overtone of the D-mode, and often called G'-mode. For SWNTs, there are some second-order modes between 1700 and 1800 cm^{-1}.

Raman scattering can probe the elastic vibrations of an entire nanoparticle, which most often shows a distinct size-dependence. This effect is used widely to identify SWNTs, which have characteristic features in the low-frequency peak <300 cm^{-1} (or bunches of peaks for polydispersed samples when the resonating conditions are met) (Figure 2.22). These features are assigned to the A$_{1g}$ radial

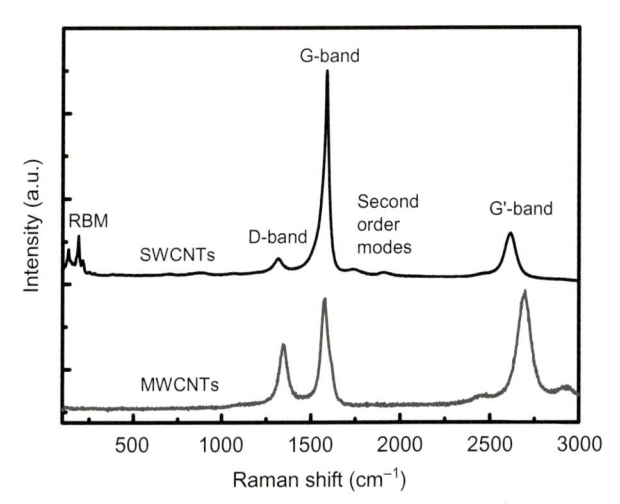

Figure 2.22 Raman spectra of MWNT and SWNT samples.

breathing mode (RBM) of the tubes [18]. The RBM feature is a unique phonon mode, which is very useful when confirming that bulk samples contain either SWNTs or double-walled CNTs. The frequency of the RBM mode depends essentially on the diameter of the tube through the relation:

$$\omega_{RBM} = \frac{A}{d} + B$$

where the parameters A and B are determined experimentally. For an isolated SWNT on a Si/SiO_2 substrate, the experimental value of A is found to be $248\,nm\,cm^{-1}$ and $B = 0\,cm^{-1}$, while for SWNT bundles, the CNT/CNT interactions lead to values of $A = 234\,nm\,cm^{-1}$ and $B = 10\,cm^{-1}$. By using a crystal of identical and infinite nanotubes, and taking into account the van der Waals interactions between tubes with a Lennard–Jones potential, Alvarez et al. [19] found the calculated value of RBM parameters as $A = 232\,nm\,cm^{-1}$ and $B = 6.5\,cm^{-1}$. Moreover, the resonant Raman effect was used to study SWNTs, while the Kataura plot could be used to determine the metallic and semi-conducting nanotubes. Graphene comprises one monolayer of carbon atoms packed into a 2-D honeycomb lattice. Typically, a Raman spectrum of graphene has a G-band at $1580\,cm^{-1}$ and a 2-D band (G′-band) at $2670\,cm^{-1}$, originating from a two-phonon double-resonance Raman process. The obvious difference between the Raman features of single-layered graphene and graphite is the 2-D band. Indeed, the presence of a sharp and symmetric 2-D band is widely used to identify single-layer graphene.

Typical Raman frequencies and all symmetries of carbon are listed in Table 2.1. Raman spectroscopy can also provide information concerning the strains, alignments and purities of CNTs or nanocarbons; these data are available in reviews of the Raman spectroscopy of new carbon materials [20, 21].

2.4.6
Infrared Spectroscopy

Infrared (IR) spectroscopy is another technique used to analyze the energy of molecular vibrations in the IR region of the electromagnetic spectrum. An IR

Table 2.1 Important symmetry modes and Raman frequencies from CNTs.

Notation		Position (cm^{-1})
A_{1g}	RBM	$A/d + B$
D	D-band	~1340
A_{1g}	G-band	1500–1605
E_{1g}	G-band	1585
E_{2g}	G-band	1591
E'_{2g}		1620
$2 \times D$	G′-band	2500–2700

RBM = radial breathing mode.

spectrometer consists of an IR source that emits light throughout the whole frequency range; the beam is split, with one beam passing either through the sample (transmission) or alternatively reflected, while the other beam is used as a reference beam. The IR spectrum covers a wavenumber range of 400–5000 cm^{-1}. Analysis of the transmitted light reveals how much energy was absorbed at each wavelength; this is achieved using a monochromatic beam which changes in wavelength over time, or by using a Fourier transform instrument to measure all of the wavelengths at once. From this, a transmittance or absorbance spectrum can be produced, showing at which IR wavelengths the sample absorbs. In Fourier transform infrared (FTIR) spectroscopy, the transmitted beam recombines with the reference beam after a path difference has been introduced. Performing a Fourier transform on this signal data results in a spectrum which is identical to that from conventional (dispersive) IR spectroscopy.

For carbon and carbonaceous materials, different functional groups have characteristic vibration frequencies that arise due to the stretching, bending, rocking, and twisting of bonds, and this allows the particular functional group to be identified. However, these may change slightly when the functional group is incorporated into carbon. The IR spectral regions of various chemical bonds that can be found in organic materials are shown in Figure 2.23. Not all possible vibrations are exhibited in IR spectroscopy, as a dipole moment must be created during the vibration. In molecules possessing a center of symmetry, however, the vibrations symmetrical about the center of symmetry are IR-inactive. Generally, IR-inactive vibrations are Raman active, which makes these techniques complementary for the analysis of functional groups or bonds.

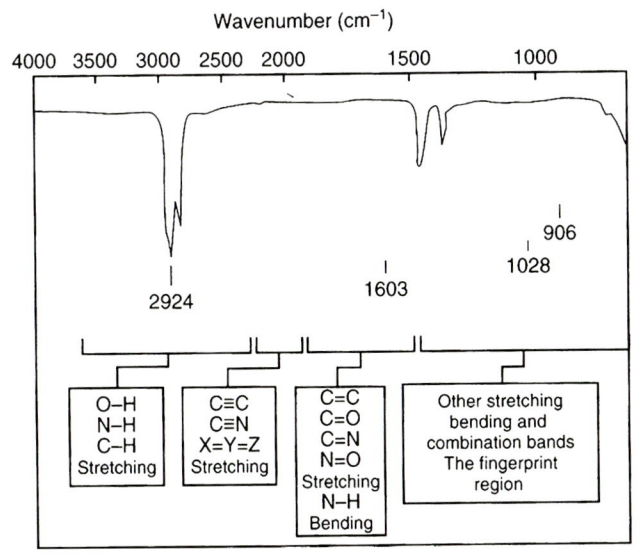

Figure 2.23 Diagram showing the IR spectral regions of different chemical bonds in nanocarbons [22].

2.5 Summary

In this chapter, an attempt has been made to cover some of the most important techniques applicable to the characterization of nanocarbons. Depending on the information required, various characterization tools – both bulk- and surface-sensitive – are available to provide comprehensive information. Among these, imaging techniques such as TEM and STEM have, on the basis of aberration correction and high-brightness FEG techniques, undergone a remarkable development during recent years, to a point where information at the atomic level concerning defects or heteroatoms can be investigated with a precision previously unavailable. The improvement of energy resolution down to 0.1 eV, by using a monochromator in a TEM system, has allowed EELS to provide the near-edge structures in the high-energy loss region comparable to those obtained with XAS, where the one is bulk-sensitive and the other surface-sensitive. When combined with other spectroscopic methods, such as Raman or IR spectroscopy, these methods have opened new horizons for the characterization of nanocarbons, by not only guiding the synthetic strategy but also establishing a correction of structure–property performances. It should be mentioned that many other characterization tools are available, including secondary ion mass spectrometry, Rutherford backscattering spectrometry, nuclear magnetic resonance, Auger electron spectroscopy, and some ultrahigh vacuum surface techniques, all of which may provide useful information on nanocarbon materials. It should be also pointed out that additional analytical chemistry methods are required to characterize some of the specialized properties of nanocarbons that cannot be assessed when using the tools discussed in this chapter. For instance, a Boehm titration may be needed to determine the acidity of nanocarbons, or a zeta-potential measurement to determine the surface charge. The hydrophilic or hydrophobic properties of carbon materials can be determined by contact-angle measurements.

References

1 Tessonnier, J.P., Ersen, O., Weinberg, G., Pham-Huu, C., Su, D.S. and Schlögl, R. (2009) Selective deposition of metal nanoparticles inside or outside multiwalled carbon nanotubes. *ACS Nano*, **3**, 2081.

2 Bogner, A., Jouneau, P.-H., Thollet, G., Basset, D. and Gauthier, C. (2007) A history of scanning electron microscopy developments: towards "wet-STEM" imaging. *Micron*, **38**, 390.

3 Jin, C.H., Lan, H.P., Suenaga, K., Peng, L.M. and Iijima, S. (2008) Metal atom catalyzed enlargement of fullerenes. *Physical Review Letters*, **101**, 176102.

4 Su, D.S., Jacob, T., Hansen, T.W., Wang, D., Schlögl, R., Freitag, B. and Kujawa, S. (2008) Surface chemistry of Ag particles: identification of oxide species by aberration-corrected TEM and by DFT calculations. *Angewandte Chemie International Edition*, **47**, 5005.

5 Liu, Z., Suenaga, K. and Iijima, S. (2007) Imaging the structure of an individual C60 fullerene molecule and its deformation process using HRTEM with

atomic sensitivity. *Journal of the American Chemical Society*, **129**, 6666.
6. Tapasztó, L., Nemes-Incze, P., Osváth, Z., Darabont, A., Lambin, P. and Biró, L.P. (2006) Electron scattering in a multiwall carbon nanotube bend junction studied by scanning tunneling microscopy. *Physical Review B*, **74**, 235422.
7. Hembacher, S., Giessibl, F.J., Mannhart, J. and Quate, C.F. (2003) Revealing the hidden atom in graphite by low-temperature atomic force microscopy. *Proceedings of the National Academy of Sciences of the United States of America*, **100**, 12539.
8. Williams, D.B. and Carter, C.B. (1996) *Transmission Electron Microscopy: A Textbook for Materials Science*, Plenum Press, New York, pp. 743.
9. Suenaga, K., Sandré, E., Colliex, C., Pickard, C.J., Kataura, H. and Iijima, S. (2001) Electron energy-loss spectroscopy of electron states in isolated carbon nanostructures. *Physical Review B*, **63**, 165408.
10. Vincent Crist, B. (2007) A review of XPS data-banks. *XPS Reports*, **1**, 1.
11. Arrigo, R., Hävecker, Schlögl, R. and Su, D.S. (2008) Dynamic surface rearrangement and thermal stability of nitrogen functional groups on carbon nanotubes. *Chemical Communications*, **44**, 4891.
12. Castle, J.E., Chapman-Kpodo, H., Proctor, A. and Salvi, A.M. (2000) Curve-fitting in XPS using extrinsic and intrinsic background structure. *Journal of Electron Spectroscopy and Related Phenomena*, **106**, 65.
13. Sherwood, P.M.A. (1996) Surface analysis of carbon and carbon fibres for composites. *Journal of Electron Spectroscopy and Related Phenomena*, **81**, 319.
14. Wild, U., Pfänder, N. and Schloegl, R. (1997) The application of XPS for integral analysis of filter samples. *Fresenius' Journal of Analytical Chemistry*, **357**, 420.
15. Marchon, B., Carrazza, J., Heinemann, H. and Somorjai, G.A. (1988) TPD and XPS studies of O_2, CO_2, and H_2O adsorption on clean polycrystalline graphite. *Carbon*, **26**, 507.
16. Desimoni, E., Casella, G.I., Cataldi, T.R.I., Salvi, A.M., Rotunno, T. and Croce, E.D. (1992) Remarks on the surface characterization of carbon fibers. *Surface and Interface Analysis*, **18**, 623.
17. Schlögl, R., Loose, G. and Wesemann, M. (1990) On the mechanism of the oxidation of graphite by molecular oxygen. *Solid State Ionics*, **43**, 183.
18. Rao, A.M., Richter, E., Bandow, S., Dresselhaus, G. and Dresselhaus, M.S. (1997) Diameter-selective Raman scattering from vibrational modes in carbon nanotubes. *Science*, **275**, 187.
19. Alvarez, L., Righi, A. and Guillard, T. (2000) Resonant Raman study of the structure and electronic properties of single-wall carbon nanotubes. *Chemical Physics Letters*, **316**, 186.
20. Saito, R. and Fantini, J.J. (2008) Excitonic states and resonance Raman Spectroscopy of single-wall carbon nanotubes. *Topics in Applied Physics*, **111**, 251–86.
21. Dresselhaus, M.S., Jorio, A., Hofmann, M., Dresselhaus, G. and Saito, R. (2010) Perspectives on carbon nanotubes and graphene Raman spectroscopy. *Nano Letters*, **10** (3), 751–8.
22. Brydson, R.M. and Hammond, C. (2005) Generic methodologies for nanotechnology: classification and fabrication, in *Nanoscale Science and Technology* (eds R.W. Kelsall, I.W. Hamley and M. Geoghegan), John Wiley & Sons Ltd, p. 103.

3
Synthesis, Characterization, and Biomedical Applications of Graphene

Albert Dato, Velimir Radmilovic and Michael Frenklach

3.1
Introduction

Graphene, a single atomic layer of sp^2-bonded carbon atoms tightly packed in a two-dimensional (2-D) honeycomb lattice (Figure 3.1), is a novel material that possesses unique properties [1–9]. For instance, electrons traveling through graphene can traverse submicron distances without being scattered, even at room temperature [1–3]. Electrons propagating through graphene also lose their effective rest mass, and thus mimic relativistic particles moving through space at close to the speed of light [1–3]. Defect-free graphene is the strongest material ever measured [5], is highly pliable but fractures like glass at high strains [6], and has a large theoretical surface area [7]. Furthermore, graphene exhibits a room-temperature thermal conductivity that is superior to the conductivities of single-walled carbon nanotubes (SWNTs) and diamond, which is the best bulk crystalline thermal conductor [8].

The remarkable properties of graphene have generated great interest throughout the scientific community. Graphene research has intensified since the material was first isolated in 2004 [9], and this has resulted in the development of numerous graphene-synthesis techniques and characterization methods. The first reports on the biomedical applications of graphene emerged in 2008; subsequently, graphene and graphene-based materials have been shown to be biocompatible and potentially used in a myriad of biomedical applications, including drug delivery, biodevices, and molecular imaging.

The aim of this chapter is three-fold: first, the current avenues to graphene synthesis will be discussed (Section 3.2), and this will be followed by an overview of the most widely utilized graphene characterization methods (Section 3.3). Finally, a review will be provided of the biomedical applications of graphene developed to date (Section 3.4).

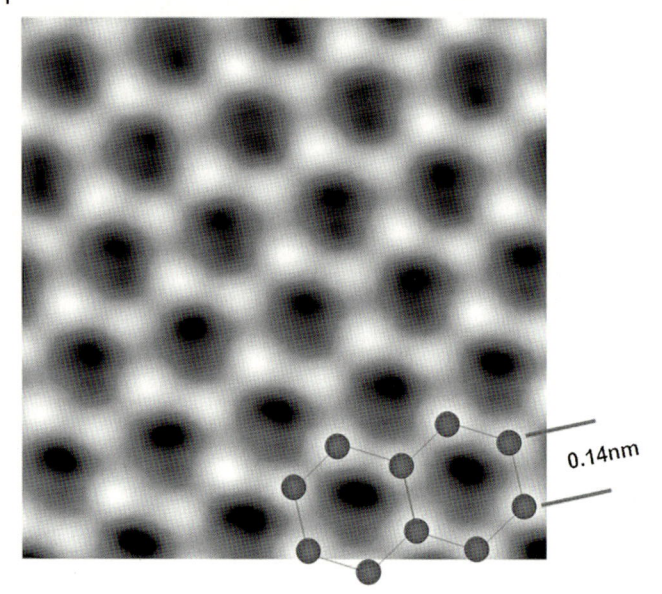

Figure 3.1 An atomic-resolution transmission electron microscopy image of graphene. The individual carbon atoms appear white in the image. The graphene sheet was synthesized using the substrate-free gas-phase method. Reproduced with permission from Ref. [4]; © 2009, Royal Society of Chemistry Publishers.

3.2
Synthesis of Graphene

Stable three-dimensional (3-D) graphite crystals are composed of stacked graphene layers that are weakly coupled by van der Waals forces. The existence of a single layer of graphite was originally thought to be impossible, as both theoretical and experimental studies had shown that free-standing 2-D materials were thermodynamically unstable [1, 10–12]. For example, the melting temperature of thin films rapidly decreases with their thickness, while the decomposition or segregation of films into islands was observed at thicknesses of dozens of atomic layers [1, 12]. Thus, a free-standing 2-D graphene sheet was theorized to be unstable with respect to the formation of curved carbon structures such as soot, nanotubes, and fullerenes [1, 9]. However, this theory was disproved in 2004, when free-standing graphene was created [9].

Graphene was first isolated through the simple mechanical exfoliation of highly oriented pyrolytic graphite (HOPG) [9]. The technique involves the repeated peeling of HOPG with Scotch tape until atomically thin sheets are obtained. To create individual graphene sheets through the mechanical exfoliation of HOPG is an unwieldy and time-intensive process, however, and although sheets that were hundreds of microns thick were prepared in this way, only small quantities of graphene that were suitable for research purposes could be obtained. That the method could not be scaled-up for commercial applications has clearly provided the driving power behind the quest for alternative methods of graphene synthesis.

3.2.1
Chemical Exfoliation

Graphene has been synthesized through the chemical exfoliation of bulk graphite. The exfoliation process begins with the oxidative treatment of graphite to transform it into graphite oxide [13–16]. The latter material consists of oxidized graphene layers which have epoxide and hydroxyl groups attached to their basal planes, and carbonyl and carboxyl groups attached at the edges [13]. These oxygen functionalities render the individual sheets of graphite oxide hydrophilic, such that water molecules can readily intercalate into the interlayer galleries of graphite oxide [14]. The sonication of graphite oxide results in an exfoliation of the bulk material into individual layers of graphene oxide (GO) which, once obtained, are subjected to a chemical reduction so as to obtain individual graphene sheets [13–16]. For example, graphene has been produced through the direct dispersion of GO in hydrazine [16]. The high throughput and low cost of the oxidation–exfoliation–reduction of graphite has led to it becoming one of the most widely used methods for graphene synthesis.

Graphene flakes have also been fabricated via the thermal and chemical exfoliation of graphite [17]. In this case, graphite was thermally exfoliated by brief heating to 1273 K in forming gas, and the resultant graphite flakes were then ground and intercalated with oleum. Tetrabutylammonium hydroxide (TBA, 40% solution in water) was then inserted into the oleum-intercalated graphite in an *N,N*-dimethylformamide solution. Sonication of the TBA-inserted oleum-intercalated graphite was then carried out in solution to form a homogeneous suspension. The graphene was obtained by centrifugation of the suspension (24 000g for 3 min), such that predominantly single-layer (graphene) was retained in the supernatant. Large pieces of material were then removed from the supernatant and thermally annealed in H_2 at 1073 K.

3.2.2
Epitaxial Growth

Multilayered, rotationally disordered graphene sheets have been grown through the thermal decomposition of single-crystal 6H- and 4H-SiC(0001) substrates at 1573 K and pressures below 1.3 kPa [18, 19]. Growth of these sheets was found to depend on the surface of the SiC crystal [18]; typically, growth on the Si-terminated face was slow and terminated at relatively short times at high temperatures. In contrast, growth on the C-terminated face was not self-limiting and resulted in relatively thick sheets of between five and 100 layers.

3.2.3
Substrate-Free Gas-Phase Synthesis

Graphene was first synthesized without the use of substrates or graphite through the substrate-free gas-phase method [20]. In this case, free-standing graphene was produced by sending liquid ethanol droplets directly into atmospheric-pressure microwave-generated argon plasmas. Over a time scale on the order of 10^{-1} s, the ethanol droplets evaporated and dissociated in the plasma, forming graphene. An extensive characterization has revealed that the synthesized graphene exhibited a highly ordered structure that was free of any detrimental oxygen functionalities [4] (Figure 3.1). In contrast to alternative graphene synthesis methods, this single-step approach is capable of a rapid and continuous production of high-quality graphene, under ambient conditions.

3.2.4
Chemical Vapor Deposition

Large-area graphene layers were created using chemical vapor deposition (CVD) on Ni films [21, 22]. The first part of the CVD process involved the electron-beam evaporation of Ni onto SiO_2/Si substrates, followed by thermal annealing to generate Ni films with microstructures that consisted of single-crystalline grains with sizes ranging from 1 to 20 μm. This polycrystalline film was then exposed to a highly diluted hydrocarbon flow under ambient pressure at 1273 K, which resulted in the formation of graphene films with thicknesses ranging from one to ten layers. Recently, CVD has also been employed to grow graphene on Cu foils at temperatures of up to 1273 K, using a mixture of methane and hydrogen [23]. Graphene films were grown directly on Cu surfaces via a surface-catalyzed process.

3.2.5
Arc Discharge of Graphite Electrodes

Graphene was prepared through the direct current arc discharge between graphite electrodes in a water-cooled stainless steel chamber filled with a mixture of hydrogen and helium [24]. This process resulted in the deposition of sheets containing two to four graphene layers on the inner walls of the chamber. The presence of H_2 during the arc-discharge process was speculated to terminate the dangling carbon bonds with hydrogen, thus preventing the formation of closed structures, such as polyhedral particles and nanotubes. In addition to synthesizing graphene, both p- and n-doped sheets were also obtained by injecting boron and nitrogen into the arc-discharge chamber, respectively [24].

3.2.6
Liquid-Phase Production

Graphene was synthesized via the sonication of graphite powder in a water–surfactant solution [25]. Sonication was found to break apart the bulk graphite into few-layer graphene flakes, which were then coated by the surfactant sodium dodecylbenzene sulfonate. The coulombic repulsion between the surfactant-coated sheets was found to prevent their re-aggregation into 3-D graphite structures. Subsequently, graphene sheets were obtained after centrifugation of the sonicated suspension and decanting of the supernatant.

Graphene flakes with controlled thicknesses have also been produced through a similar procedure that was followed by density gradient ultracentrifugation [26]. The use of sodium cholate as surfactant was found to promote the exfoliation of graphite during sonication. The resulting graphene–surfactant complexes were found to have buoyant densities that depended on the thicknesses of the graphene flakes. Consequently, graphene sheets with varying thicknesses could easily be isolated using density gradient ultracentrifugation.

3.3
Characterization of Graphene

Before graphene was first isolated, several research groups had attempted to produce 2-D sheets through the exfoliation of HOPG [27, 28]. There was, however, a low rate of success in locating atomically thin graphene sheets, because the latter formed a minority among a multitude of thicker flakes, which usually consisted of between tens and hundreds of graphene layers [27, 28]. In 2004, atomic force microscopy (AFM) was first recognized as one of the few characterization techniques capable of detecting, imaging, and measuring graphene layers. However, the low throughput of AFM systems meant that the search for graphene on a substrate using this technique was essentially impossible. Nonetheless, a breakthrough was made when graphene was shown to become visible under an optical microscope when deposited on a SiO_2 substrate with a specific oxide thickness of 300 nm [9]. As a consequence, graphene could be detected when AFM was used to scan regions on substrates where this optical effect occurred. Although today, AFM is still used to characterize graphene, further investigations into the material have resulted in the development of numerous specific graphene-characterization techniques.

3.3.1
Raman Spectroscopy

Raman spectroscopy provides "fingerprint spectra" that allow distinctions to be made between a broad range of carbon materials. Indeed, Raman spectroscopy is one of the most widely used graphene-characterization methods, because it allows for the unambiguous and nondestructive identification of the graphene layers [29–31]. The electronic structure of graphene is uniquely captured in its Raman spectrum, and single-layer, bi-layer (two layers) and few-layer (less than ten layers) graphene sheets have been shown to have unique Raman "fingerprints" that reflect changes in the electronic structure and electron–phonon interactions in the materials [29–31].

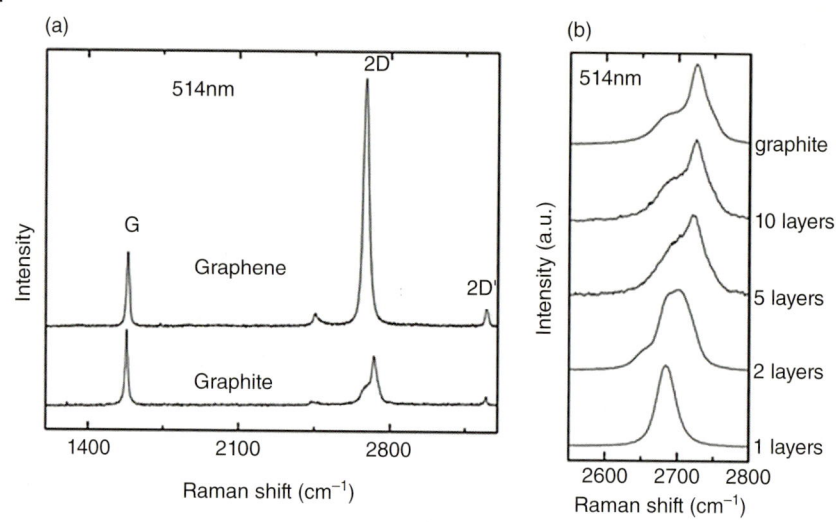

Figure 3.2 (a) Raman spectra of graphene and graphite; (b) Comparison of the 2D peaks of single-layer, bi-layer, and few-layer graphenes. Reproduced with permission from Ref. [31]; © 2007, Elsevier.

The two most intense features in the Raman spectra of graphene and graphite are the G and 2D peaks located at 1580 and ~2700 cm^{-1}, respectively [29–31] (Figure 3.2a). The number of graphene layers in a sheet can be determined from the position, shape, and intensity of the 2D peak. For example, the graphite 2D peak is positioned at a Raman shift greater than 2700 cm^{-1} (Figure 3.2a); the peak is also asymmetric because it consists of two components that are approximately one-fourth and one-half the intensity of the G peak, respectively (Figure 3.2a). Single-layer graphene exhibits a single, sharp 2D peak that is symmetric and down-shifted to 2680 cm^{-1}. Furthermore, the 2D peak of a single-layer graphene sheet is about fourfold more intense than its G peak (Figure 3.2a). Bi-layer graphene has a much broader, asymmetric 2D peak than single-layer graphene, because it consists of four components; furthermore, the bi-layer 2D peak is up-shifted [29, 30] to 2700 cm^{-1} (Figure 3.2b). Few-layer sheets consisting of five or more graphene layers exhibit similar 2D peaks to that of graphite [29, 30] (Figure 3.2b). The ability to distinguish between single-, bi-, and few-layer graphene sheets from the shape, position, and intensity of the 2D peak has led to Raman spectroscopy becoming an invaluable graphene-characterization technique.

Raman spectroscopy can also be used to determine the presence of defects and disorder in graphene sheets. Carbon structures such as graphite, graphene, graphene oxide, and soot exhibit common Raman D and G peaks, positioned near 1350 and 1580 cm^{-1}, respectively [29–34]. The shapes and intensities of these peaks yield information about the degree of disorder in the structures. Away from its edges, a defect-free graphene sheet exhibits a single, sharp G peak [29–31] at ~1580 cm^{-1}. The D peak appears when defects become present in a graphitic sample (in fact, it was assigned the letter "D" because it represents "defects" or "disorder") [32, 33]. As the degree of defects and disorder in a carbonaceous material increases, so too do the intensity of its D peak and the ratio of the intensities of the D and G peaks (I_D/I_G) [29–34]. The D peak also broadens and remains near $\sim$1350 cm^{-1}, while the G peak broadens and displaces to higher values in excess of 1600 cm^{-1} as the sample becomes more disordered [33, 34]. Although pristine graphene does not exhibit a D peak, the edges of the sheets are always seen as defects. A D peak will appear in the spectra of a defect-free graphene sheet if the laser spot includes its edges [31]. Disordered graphene created by alternative synthetic approaches have shown strong and broadened D peaks, wide and up-shifted G peaks, and I_D/I_G values that approached or exceeded unity [14, 24]. Thus, Raman spectroscopy can be used not only to determine the number of layers in a graphene sheet, but also to probe the sheet's quality.

3.3.2
Transmission Electron Microscopy

In addition to using TEM to image graphene, AFM [9, 16, 17] and scanning tunneling microscopy [19, 35] (STM) have each been employed to image and characterize this material. Unfortunately, however, both techniques suffer from slow data acquisition rates [36, 37], and any regions containing monolayer sheets must be detected optically by using special substrates prior to characterization. STM is also especially affected by sample cleanliness and conductivity (for details, see: http://www.siliconfareast.com/stm.htm) [37]. Although, in the past, graphene has also been imaged using scanning electron microscopy (SEM) [14, 16], the number of layers in a sheet cannot be determined unambiguously when using this method. To date, no other technique is available for imaging graphene which is better and faster than TEM.

Although, because it is atomically thin, graphene appears transparent and featureless in TEM images [38], the number of layers in a sheet can be determined by analyzing its folded regions [29, 38]. The folded regions occur locally parallel to the electron beam, and single-layer graphene has been found to exhibit one dark line, similar to TEM images of SWNTs [38, 39] (Figure 3.3a). Both, bi-layer and few-layer graphene sheets exhibit multiple dark lines, similar to the TEM images of double- and multi-walled CNTs, respectively [38–40] (Figure 3.3a). More recent technological advancements in TEM have resulted in the development of aberration-corrected electron microscopes capable of attaining sub-angstrom resolutions at low accelerating voltages. Notably, these advanced microscopes have been used to image the C atoms on a graphene sheet [4, 36, 41] (see Figure 3.1).

3.3.3
Electron Diffraction

Electron diffraction patterns can be used to distinguish between single- and bi-layer graphene [29]. The diffraction pattern for graphite exhibits an inner hexagon

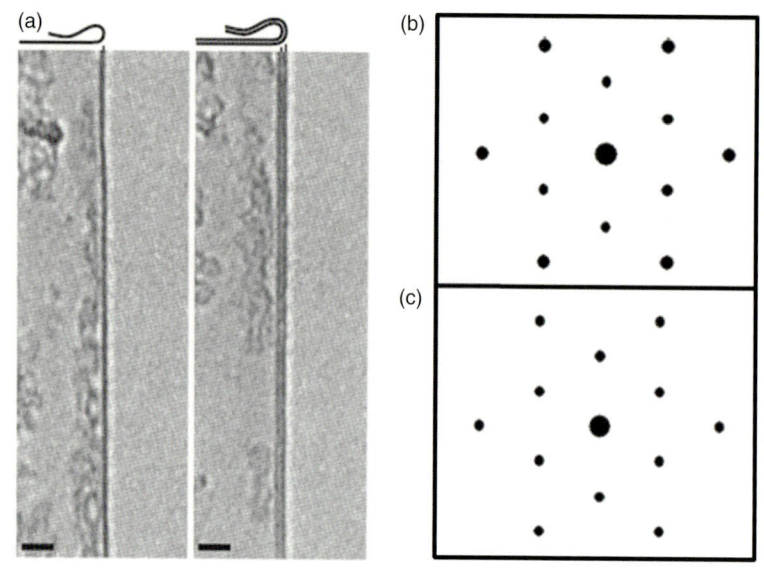

Figure 3.3 (a) TEM images of the folded regions of single-layer and bi-layer graphene; (b, c) Simulated electron diffraction patterns of bi-layer and single-layer graphene, respectively. (a) Reproduced with permission from Ref. [38]; © 2007, Macmillan Publishers Ltd.; (b, c) Reproduced with permission from Ref. [20]; © 2008, American Chemical Society.

corresponding to indices (0–110) (2.13 Å spacing) and an outer hexagon corresponding to indices (1–210) (1.23 Å spacing) [29]. The intensities of the (1–210) diffraction spots of an A-B stacked bi-layer graphene sheet were demonstrated to have twice the intensities of the (0–110) spots [29] (Figure 3.3b). The intensities of the spots in the inner and outer hexagons of single-layer graphene have been shown to be equivalent [29, 38] (Figure 3.3c); this phenomenon has been demonstrated in both experimental [20, 29, 38] and computational simulations [20].

3.3.4
Electron Energy Loss Spectroscopy

The technique of electron energy loss spectroscopy (EELS) provides information regarding the electronic structure of specimen atoms, revealing details about the nature of these atoms and their bonding. The technique has been used to unambiguously distinguish between different carbon materials, such as diamond [42, 43], graphite [42–44], and amorphous carbon [42–44]. Carbon atoms have hybridized s and p orbitals which are termed σ and π, respectively. Graphite contains sp^2 bonds in the basal planes and van der Waals bonding between the planes. Thus, the main features in the EELS spectra of graphite and graphene in the

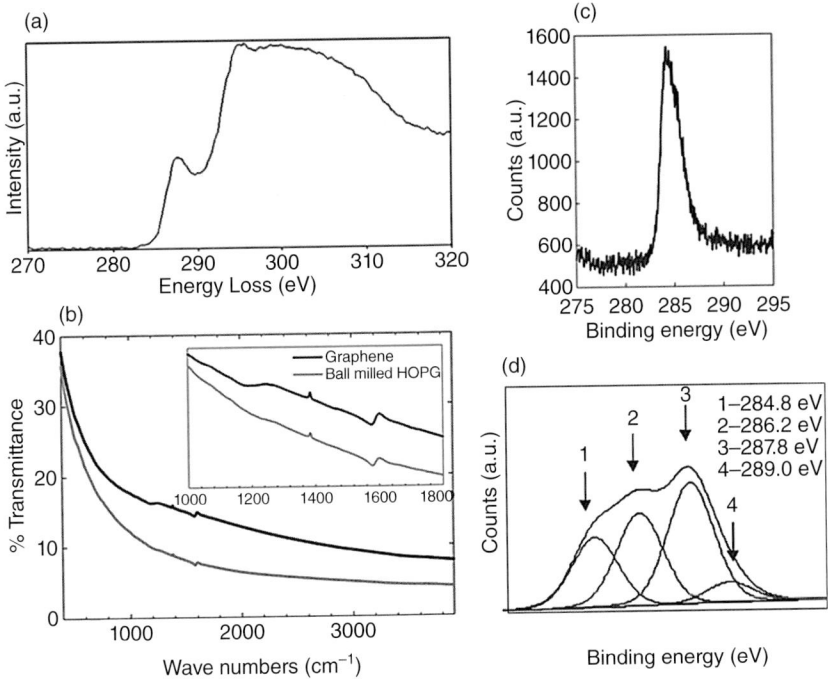

Figure 3.4 (a) EELS spectrum of a single-layer graphene sheet; (b) FT-IR spectra of HOPG and oxygen-free graphene; (c) XPS spectrum of clean and highly ordered graphene; (d) XPS spectrum of GO, exhibiting peaks corresponding to C atoms bonded to oxygen. (a) Reproduced with permission from Ref. [20]; © 2008, American Chemical Society; (b, c) Reproduced with permission from Ref. [4]; © 2009, Royal Society of Chemistry Publishers; (d) Reproduced with permission from Ref. [13]; © 2006, Royal Society of Chemistry Publishers.

carbon K-edge region are a peak at 285 eV that corresponds to electron transitions from the K-shell to empty π^* states, and a peak at 291 eV that corresponds to transitions from the K-shell to σ^* states [20, 42–44]. Amorphous carbon and graphite have similarities in their bonding [43], and thus the EELS spectrum of amorphous carbon is similar to that of graphite, except that it exhibits rounded edges instead of peaks at 285 eV and 291 eV [42–44]. Thus, EELS can be used to determine whether an atomically thin carbon sheet is amorphous or graphitic. An EELS spectrum obtained from single-layer graphene is shown in Figure 3.4a.

3.3.5
Elemental Analysis

The presence of oxygen functionalities on a graphene sheet can detrimentally affect its unique properties [16, 17, 45]. TEM imaging, EELS, Raman spectroscopy, and electron diffraction can each provide definitive information regarding the structure, number of layers, and degree of disorder in a graphene sheet. However, conventional TEM images and diffraction patterns cannot be used to assess the presence of functional groups, and EELS and Raman spectroscopy only provide an ambiguous identification of any detrimental elements. Therefore, a detailed elemental analysis is required to completely determine the purity of a graphene sheet.

The presence of functional groups on graphene has been determined using Fourier transform infrared spectroscopy (FT-IR). HOPG is a pure carbon material, and its FT-IR spectrum is shown in Figure 3.4b. As discussed in Section 3.2.1, HOPG can be oxidized and exfoliated into single-layer graphene oxide sheets. The prominent features in the FT-IR spectrum of electrically-insulating graphene oxide [13] include absorption bands corresponding to C–O stretching at 1053 cm^{-1}, C–OH stretching at 1226 cm^{-1}, phenolic O–H deformation vibration at 1412 cm^{-1}, C=C ring stretching at 1621 cm^{-1}, C=O carbonyl stretching at 1733 cm^{-1}, and O–H stretching vibrations at 3428 cm^{-1}. Additionally, one CH_3– and two CH_2– peaks occur at 2960, 2922, and 2860 cm^{-1}, respectively [46]. Therefore, the purity of a graphene sheet can be determined by the presence or absence of these features in its FT-IR spectrum. The FT-IR spectra of pure, oxygen-free graphene have exhibited similar features to the spectrum of HOPG [4] (Figure 3.4b).

X-ray photoelectron spectroscopy (XPS) has also been used to determine the presence of functional groups on graphene [4, 13, 16, 17]. The C 1s XPS spectra of HOPG and graphene have exhibited a single peak at a binding energy of ~284.8 eV corresponding to C–C bonds [4, 13, 16, 17] (Figure 3.4c). In contrast, XPS spectra for GO have exhibited additional peaks corresponding to C–O, C=O, and O–C=O bonding at ~286.2, ~287.8, and ~289.0 eV, respectively [13, 16, 17] (Figure 3.4d).

Elemental analysis by combustion has been utilized to determine the purity of graphene [4, 16], with the method accurately determining the mass composition of C, O, H, and other elements. Graphene created via the mechanical exfoliation of high-purity HOPG is composed of 99.99% carbon by mass, while sheets synthesized from alternative synthesis methods have exhibited residual oxygen functionalities [16]. Graphene obtained via the substrate-free gas-phase method has been shown to exhibit a purity approaching that of HOPG, with 99% C, 0.9% H, and 0.1% O by mass [4].

3.4
Biomedical Applications of Graphene

Graphene has been envisioned for use in numerous applications. For example, it has been suggested as a candidate to replace Si in integrated circuits [1], and could potentially be used in individual ultrahigh-frequency transistors, as well as flexible and transparent electronic devices [22]. Other proposed electronic applications for graphene include supercapacitors, batteries, interconnects, and field emitters [2]. Graphene has also been envisioned for use as fillers in composite materials to create electrically conducting plastics [47], and as a sensor capable of detecting individual gas molecules [48]. The first novel biomedical applications of graphene began to appear in 2008; a review of biomedical graphene applications developed to date is provided in the following subsections.

3.4.1
Biocompatible Graphene Paper

Thin papers made from both graphene [15, 49] and GO [50] have been fabricated via the vacuum filtration of graphene sheets through membrane filters. These graphene-based papers have exhibited a combination of superior thermal, mechanical, and electrical properties [15, 49, 50]. The biocompatibility of graphene paper was investigated by culture of the mouse fibroblast cell line (L-929) [49], which has been used to assess the cytotoxicity of surfaces used for cell growth [49] and also to study the biocompatibility of CNTs [51]. Subsequently, L-929 cells were found to adhere and proliferate on graphene paper (Figure 3.5), indicating that the material was biocompatible.

3.4.2
Drug Delivery

Graphene has been proposed as an ideal material for the attachment and delivery of drugs, such as anti-cancer medications. The reasons for this is that graphene is atomically thin, possesses a large theoretical surface area, and both sides of a single sheet are accessible for drug binding. Unfortunately, however, graphene is

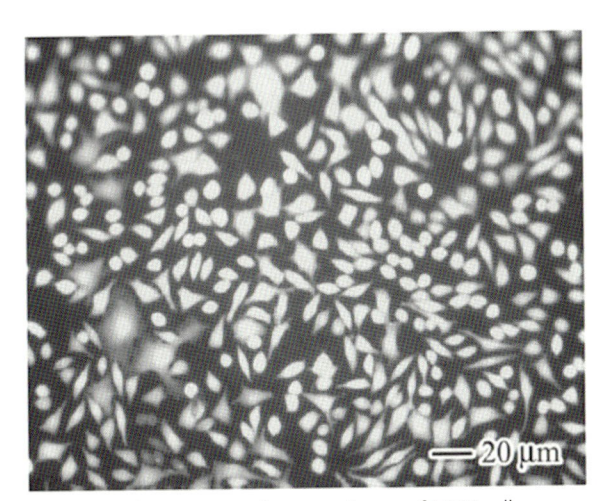

Figure 3.5 Fluorescence microscopy image of L-929 cells growing on graphene paper. Reproduced with permission from Ref. [49]; © 2008, Wiley-VCH Verlag GmbH & Co. KGaA.

insoluble in water. Although graphene can be oxidized to form water-soluble GO, the material has been found to aggregate in biological solutions rich in salts and proteins, such as cell media and serum [52, 53], which would make it unsuitable for drug-delivery purposes. The functionalization of nanoscale GO (NGO) with branched, biocompatible poly(ethylene glycol) (PEG) has been shown to result in NGO possessing high aqueous solubility and stability in biological solutions [52, 53]. Such oxidized and functionalized NGO sheets were created by first sonicating GO into nanoscale pieces with lateral dimensions of between 5 and 50 nm; the PEG was then attached to the carboxylic acid groups on the NGO sheets (Figure 3.6). No obvious toxic effects were identified when the biocompatibility of the NGO–PEG sheets was studied [52, 53].

The drug-attachment and -delivery capabilities of NGO–PEG were subsequently investigated [52, 53]. The NGO–PEG sheets were found to readily complex with

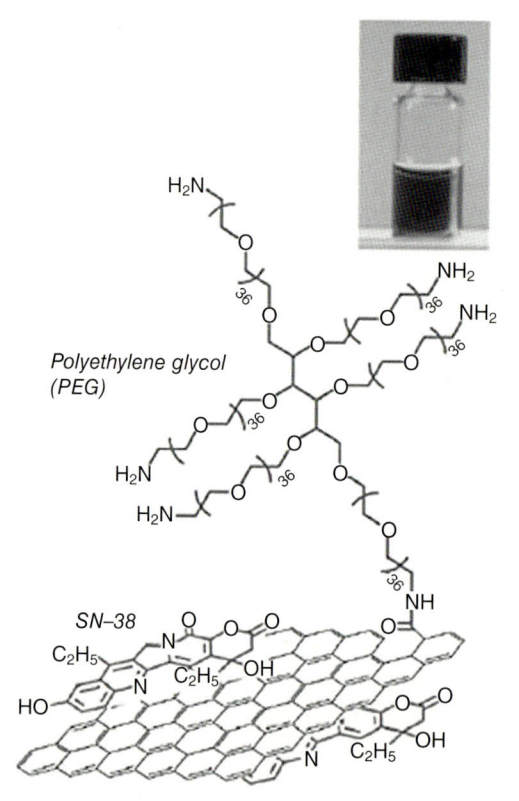

Figure 3.6 A schematic of a NGO–PEG sheet loaded with SN-38. The inset shows a photograph of an aqueous solution of NGO–PEG–SN-38. Reproduced with permission from Ref. [52]; © 2008, American Chemical Society.

the aromatic molecule SN-38 [52], a highly potent cancer-killing drug, the direct clinical use of which has been hindered by its water-insoluble nature [52, 54]. The prepared NGO–PEG–SN-38 sheets (Figure 3.6) were shown to be water-soluble, and also highly effective in killing human colon cancer cells. Other types of insoluble aromatic drug molecules have also been successfully attached to the NGO–PEG sheets, including camptothecin analogues and Iressa, a potent epidermal growth factor receptor inhibitor [52].

Further research into the attachment and delivery of anti-cancer drugs on NGO sheets has yielded promising results. When the loading and release of the anti-cancer drug doxorubicin hydrochloride on NGO was investigated, it was found that the weight ratio of the loaded drug to a single NGO sheet could reach 200% [55]. NGOs were thus shown to have much higher loading ratios than alternative nanocarriers, such as nanoparticles, which have typical ratios of less than 100% [55]. NGO has also been functionalized with sulfonic acid groups and subsequently covalently bound with folic acid [56], and this enabled the sheets to specifically target human breast cancer cells with folic acid receptors. Moreover, when the functionalized NGO sheets were loaded with two potent anti-cancer drugs, doxorubicin and camptothecin [56], they specifically targeted breast cancer cells and demonstrated a higher cytotoxicity than did NGO sheets loaded with a single drug.

3.4.3
Biodevices

The fabrication of novel hybrid biodevices through the attachment of microscale and nanoscale biological systems to graphene has been achieved [57]. Graphene–DNA hybrids were fabricated by tethering single-stranded DNA to oxidized graphene sheets [57]. The DNA was found to tether preferentially onto the thicker layers, folds, and wrinkles of graphene, such that the attachment of DNA resulted in a 128% increase in the conductivity of the sheets. The attachment of fluorescent cDNA to the DNA–graphene hybrids resulted in a 71% increase in conductivity, while the de-hybridization of cDNA from the sheets restored their original conductivity (Figure 3.7a). These results demonstrated the high sensitivity of the graphene–DNA hybrid nanostructures, and suggested that they might function as label-free DNA detectors or molecular transistors.

In the same study, chemically modified graphene sheets were shown to be capable of detecting bacteria [57]. For this, positively charged graphene-amine (GA) sheets were prepared by immersing oxidized graphene sheets in a solution containing ethylenediamine. When a single negatively charged *Bacillus cereus* cell was attached to the GA sheet (Figure 3.7b), it exhibited a 42% increase in conductivity, which was attributed to the p-type characteristic of GA. Thus, the attachment of negatively charged bacteria was proposed as being equivalent to a negative potential gating, which increased the hole density – and hence the conductivity – of the GA sheets.

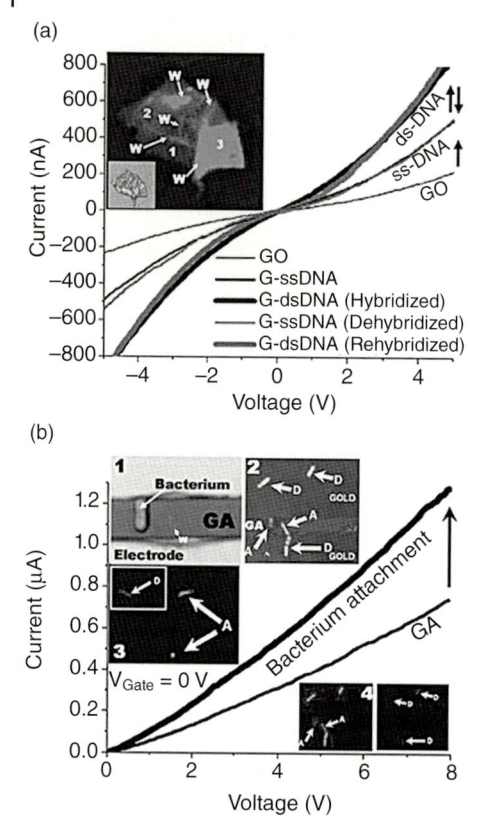

Figure 3.7 (a) Current–voltage characteristics of a graphene–DNA hybrid biodevice; (b) Current–voltage characteristics of GA sheets with and without bacterium attachment. Reproduced with permission from Ref. [57]; © 2008, American Chemical Society.

3.4.4
Imaging of Soft Materials

For TEM, graphene represents the ideal support film, as it is atomically thin, chemically inert, consists of light atoms, and possesses a highly ordered structure. Additionally, the material is electrically and thermally conductive, and structurally stable [58]. Taken together, these properties have enabled the direct atomic-resolution TEM imaging of the organic molecular coatings of citrate-capped gold nanoparticles [58]. As shown in Figure 3.8, the molecular coating of a gold nanoparticle was imaged by masking the known crystalline reflections of the graphene support and the gold nanoparticle. These results suggested that atomically thin graphene support films could also be used to directly image a diverse range of molecular coatings on nanoparticles, such as DNA, proteins, and antibody–antigen pairs.

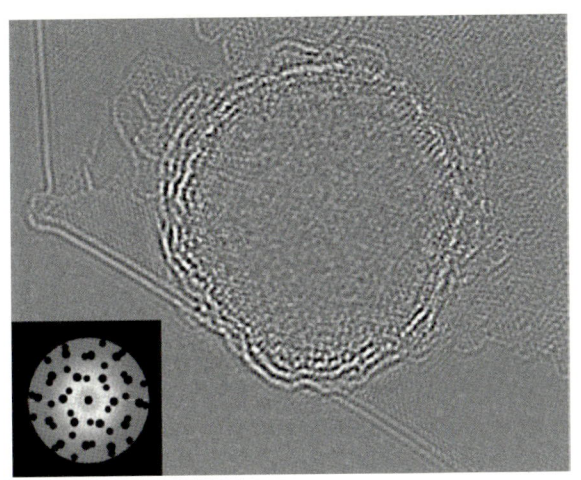

Figure 3.8 Atomic-resolution direct TEM image of the citrate molecules coating a gold nanoparticle supported by a graphene sheet. The image was obtained by masking the reflections of the nanoparticle and underlying graphene sheet, as shown in the inset. Reproduced with permission from Ref. [58]; © 2009, American Chemical Society.

3.5
Conclusions

Graphene research continues to grow at a rapid pace, with new methods for obtaining graphene continuing to be discovered, and other current synthetic approaches being improved. Simpler and better graphene synthesis and characterization methods may greatly accelerate the development of novel graphene applications. The successful utilization of biocompatible graphene sheets in drug delivery, biodevices, and molecular imaging suggests that graphene is a promising material for biomedical applications.

References

1 Geim, A.K. and Novoselov, K.S. (2007) The rise of graphene. *Nature Materials*, **6**, 183–91.

2 Geim, A.K. (2009) Graphene: status and prospects. *Science*, **324**, 1530–4.

3 Service, R.F. (2009) Carbon sheets an atom thick give rise to graphene dreams. *Science*, **324**, 875–7.

4 Dato, A., Lee, Z., Jeon, K.-J., Erni, R., Radmilovic, V., Richardson, T.J. and Frenklach, M. (2009) Clean and highly ordered graphene synthesized in the gas phase. *Chemical Communications*, 6095–7.

5 Lee, C., Wei, X.D., Kysar, J.W. and Hone, J. (2008) Measurement of the elastic properties and intrinsic strength of monolayer graphene. *Science*, **321**, 385–8.

6 Booth, T.J., Blake, P., Nair, R.R., Jiang, D., Hill, E.W., Bangert, U., Bleloch, A., Gass, M., Novoselov, K.S., Katsnelson, M.I. and Geim, A.K. (2008) Macroscopic graphene membranes and their

extraordinary stiffness. *Nano Letters*, **8**, 2442–6.

7 Rao, C.N.R., Sood, A.K., Subrahmanyam, K.S. and Govindaraj, A. (2009) Graphene: the new two-dimensional nanomaterial. *Angewandte Chemie International Edition*, **48**, 7752–77.

8 Balandin, A.A., Ghosh, S., Bao, W.Z., Calizo, I., Teweldebrhan, D., Miao, F. and Lau, C.N. (2008) Superior thermal conductivity of single-layer graphene. *Nano Letters*, **8**, 902–7.

9 Novoselov, K.S., Geim, A.K., Morozov, S.V., Jiang, D., Zhang, Y., Dubonos, S.V., Grigorieva, I.V. and Firsov, A.A. (2004) Electric field effect in atomically thin carbon films. *Science*, **306**, 666–9.

10 Mermin, N.D. (1968) Crystalline order in 2 dimensions. *Physical Review*, **176**, 250–4.

11 Venables, J.A., Spiller, G.D.T. and Hanbucken, M. (1984) Nucleation and growth of thin films. *Reports on Progress in Physics*, **47**, 399–459.

12 Evans, J.W., Thiel, P.A. and Bartelt, M.C. (2006) Morphological evolution during epitaxial thin film growth: formation of 2D islands and 3D mounds. *Surface Science Reports*, **61**, 1–128.

13 Stankovich, S., Piner, R.D., Chen, X.Q., Wu, N.Q., Nguyen, S.T. and Ruoff, R.S. (2006) Stable aqueous dispersions of graphitic nanoplatelets via the reduction of exfoliated graphite oxide in the presence of poly(sodium 4-styrenesulfonate). *Journal of Materials Chemistry*, **16**, 155–8.

14 Stankovich, S., Dikin, D.A., Piner, R.D., Kohlhaas, K.A., Kleinhammes, A., Jia, Y., Wu, Y., Nguyen, S.T. and Ruoff, R.S. (2007) Synthesis of graphene-based nanosheets via chemical reduction of exfoliated graphite oxide. *Carbon*, **45**, 1558–65.

15 Li, D., Müller, M.B., Gilje, S., Kaner, R.B. and Wallace, G.G. (2008) Processable aqueous dispersions of graphene nanosheets. *Nature Nanotechnology*, **3**, 101–5.

16 Tung, V.C., Allen, M.J., Yang, Y. and Kaner, R.B. (2009) High-throughput solution processing of large-scale graphene. *Nature Nanotechnology*, **4**, 25–9.

17 Li, X.L., Zhang, G.Y., Bai, X.D., Sun, X.M., Wang, X.R., Wang, E. and Dai, H.J. (2008) Highly conducting graphene sheets and Langmuir-Blodgett films. *Nature Nanotechnology*, **3**, 538–42.

18 Hass, J., Feng, R., Li, T., Li, X., Zong, Z., de Heer, W.A., First, P.N., Conrad, E.H., Jeffrey, C.A. and Berger, C. (2006) Highly ordered graphene for two dimensional electronics. *Applied Physics Letters*, **89**, 143106.

19 Berger, C., Song, Z.M., Li, X.B., Wu, X.S., Brown, N., Naud, C., Mayou, D., Li, T.B., Hass, J., Marchenkov, A.N., Conrad, E.H., First, P.N. and de Heer, W.A. (2006) Electronic confinement and coherence in patterned epitaxial graphene. *Science*, **312**, 1191–6.

20 Dato, A., Radmilovic, V., Lee, Z., Phillips, J. and Frenklach, M. (2008) Substrate-free gas-phase synthesis of graphene sheets. *Nano Letters*, **8**, 2012–16.

21 Reina, A., Jia, X.T., Ho, J., Nezich, D., Son, H.B., Bulovic, V., Dresselhaus, M.S. and Kong, J. (2009) Large area, few-layer graphene films on arbitrary substrates by chemical vapor deposition. *Nano Letters*, **9**, 30–5.

22 Kim, K.S., Zhao, Y., Jang, H., Lee, S.Y., Kim, J.M., Ahn, J.H., Kim, P., Choi, J.Y. and Hong, B.H. (2009) Large-scale pattern growth of graphene films for stretchable transparent electrodes. *Nature*, **457**, 706–10.

23 Li, X., Cai, W., An, J., Kim, S., Nah, J., Yang, D., Piner, R., Velamakanni, A., Jung, I., Tutuc, E., Banerjee, S.K., Colombo, L. and Ruoff, R.S. (2009) Large-area synthesis of high-quality and uniform graphene films on copper foils. *Nature*, **324**, 1312–14.

24 Subrahmanyam, K.S., Panchakarla, L.S., Govindaraj, A. and Rao, C.N.R. (2009) Simple method of preparing graphene flakes by an arc-discharge method. *Journal of Physical Chemistry C*, **113**, 4257–9.

25 Lotya, M., Hernandez, Y., King, P.J., Smith, R.J., Nicolosi, V., Karlsson, L.S., Blighe, F.M., De, S., Wang, Z., McGovern, I.T., Duesberg, G.S. and Coleman, J.N. (2009) Liquid phase production of graphene by exfoliation of graphite in surfactant/water solutions.

Journal of the American Chemical Society, **131**, 3611–20.

26 Green, A.A. and Hersam, M.C. (2009) Solution phase production of graphene with controlled thickness via density differentiation. *Nano Letters*, **9**, 4031–6.

27 Zhang, Y., Small, J.P., Amori, M.E.S. and Kim, P. (2005) Electric field modulation of galvanomagnetic properties of mesoscopic graphite. *Physical Review Letters*, **94**, 176803.

28 Shioyama, H. (2001) Cleavage of graphite to graphene. *Journal of Material Science Letters*, **20**, 499–500.

29 Ferrari, A.C., Meyer, J.C., Scardaci, V., Casiraghi, C., Lazzeri, M., Mauri, F., Piscanec, S., Jiang, D., Novoselov, K.S., Roth, S. and Geim, A.K. (2006) Raman spectrum of graphene and graphene layers. *Physical Review Letters*, **97**, 187401.

30 Casiraghi, C., Pisana, S., Novoselov, K.S., Geim, A.K. and Ferrari, A.C. (2007) Raman fingerprint of charged impurities in graphene. *Applied Physics Letters*, **91**, 233108.

31 Ferrari, A.C. (2007) Raman spectroscopy of graphene and graphite: disorder, electron-phonon coupling, doping and nonadiabatic effects. *Solid State Communications*, **143**, 47–57.

32 Ferrari, A.C. and Robertson, J. (2000) Interpretation of Raman spectra of disordered and amorphous carbon. *Physical Review B*, **61**, 14095–107.

33 Cuesta, A., Dhamelincourt, P., Laureyns, J., Martinezalonso, A. and Tascon, J.M.D. (1994) Raman microprobe studies on carbon materials. *Carbon*, **32**, 1523–32.

34 Gruber, T., Zerda, T.W. and Gerspacher, M. (1994) Raman studies of heat-treated carbon blacks. *Carbon*, **32**, 1377–82.

35 Marchini, S., Gunther, S. and Wintterlin, J. (2007) Scanning tunneling microscopy of graphene on Ru(0001). *Physical Review B*, **76**, 075429.

36 Girit, Ç.Ö., Meyer, J.C., Erni, R., Rossell, M.D., Kisielowski, C., Yang, L., Park, C.-H., Crommie, M.F., Cohen, M.L., Louie, S.G. and Zettl, A. (2009) Graphene at the edge: stability and dynamics. *Science*, **323**, 1705–8.

37 Meyer, J.C., Girit, C.O., Crommie, M.F. and Zettl, A. (2008) Imaging and dynamics of light atoms and molecules on graphene. *Nature*, **454**, 319–22.

38 Meyer, J.C., Geim, A.K., Katsnelson, M.I., Novoselov, K.S., Booth, T.J. and Roth, S. (2007) The structure of suspended graphene sheets. *Nature*, **446**, 60–3.

39 Gass, M.H., Bangert, U., Bleloch, A.L., Wang, P., Nair, R.R. and Geim, A.K. (2008) Free-standing graphene at atomic resolution. *Nature Nanotechnology*, **3**, 676–81.

40 Iijima, S. (1991) Helical microtubules of graphitic carbon. *Nature*, **354**, 56–8.

41 Meyer, J.C., Kisielowski, C., Erni, R., Rossell, M.D., Crommie, M.F. and Zettl, A. (2008) Direct imaging of lattice atoms and topological defects in graphene membranes. *Nano Letters*, **8**, 3582–6.

42 Berger, S.D., McKenzie, D.R. and Martin, P.J. (1988) EELS analysis of vacuum arc-deposited diamond-like films. *Philosophical Magazine Letters*, **57**, 285–90.

43 Duarte-Moller, A., Espinosa-Magana, F., Martinez-Sanchez, R., Avalos-Borja, M., Hirata, G.A. and Cota-Araiza, L. (1999) Study of different forms of carbon by analytical electron microscopy. *Journal of Electron Spectroscopy and Related Phenomena*, **104**, 61–6.

44 Chu, P.K. and Li, L. (2006) Characterization of amorphous and nanocrystalline carbon films. *Materials Chemistry and Physics*, **96**, 253–77.

45 Mkhoyan, K.A., Contryman, A.W., Silcox, J., Stewart, D.A., Eda, G., Mattevi, C., Miller, S. and Chhowalla, M. (2009) Atomic and electronic structure of graphene-oxide. *Nano Letters*, **9**, 1058–63.

46 Jeong, H.K., Noh, H.J., Kim, J.Y., Jin, M.H., Park, C.Y. and Lee, Y.H. (2008) X-ray absorption spectroscopy of graphite oxide. *Europhysics Letters*, **82**, 67004.

47 Stankovich, S., Dikin, D.A., Dommett, G.H., Kohlhaas, K.M., Zimney, E.J., Stach, E.A., Piner, R.D., Nguyen, S.T. and Ruoff, R.S. (2006) Graphene-based composite materials. *Nature*, **442**, 282–6.

48 Schedin, F., Geim, A.K., Morozov, S.V., Hill, E.W., Blake, P., Katsnelson, M.I. and Novoselov, K.S. (2007) Detection of individual gas molecules adsorbed on graphene. *Nature Materials*, **6**, 652–5.

49 Chen, H., Muller, M.B., Gilmore, K.J., Wallace, G.G. and Li, D. (2008)

Mechanically strong, electrically conductive, and biocompatible graphene paper. *Advanced Materials*, **20**, 3557–61.

50 Dikin, D.A., Stankovich, S., Zimney, E.J., Piner, R.D., Dommett, G.H.B., Evmenenko, G., Nguyen, S.T. and Ruoff, R.S. (2007) Preparation and characterization of graphene oxide paper. *Nature*, **448**, 457–60.

51 Correa-Duarte, M.A., Wagner, N., Rojas-Chapana, J., Morsczeck, C., Thie, M. and Giersig, M. (2004) Fabrication and biocompatibility of carbon nanotube-based 3D networks as scaffolds for cell seeding and growth. *Nano Letters*, **4**, 2233–6.

52 Liu, Z., Robinson, J.T., Sun, X. and Dai, H. (2008) PEGylated nanographene oxide for delivery of water-insoluble cancer drugs. *Journal of the American Chemical Society*, **130**, 10876–7.

53 Sun, X., Liu, Z., Welsher, K., Robinson, J.T., Goodwin, A., Zaric, S. and Dai, H. (2008) Nano-graphene oxide for cellular imaging and drug delivery. *Nano Research*, **1**, 203–12.

54 Tanizawa, A., Fujimori, A., Fujimori, Y. and Pommier, Y. (1994) Comparison of topoisomerase I inhibition, DNA damage, and cytotoxicity of camptothecin derivatives presently in clinical trials. *Journal of the National Cancer Institute*, **86**, 836–42.

55 Yang, X., Zhang, X., Liu, Z., Ma, Y., Huang, Y. and Chen, Y. (2008) High-efficiency loading and controlled release of doxorubicin hydrochloride on graphene oxide. *Journal of Physical Chemistry C*, **112**, 17554–8.

56 Zhang, L., Xia, J., Zhao, Q., Liu, L. and Zhang, Z. (2010) Functional graphene oxide as a nanocarrier for controlled loading and targeted delivery of mixed anticancer drugs. *Small*, **6**, 537–44.

57 Mohanty, N. and Berry, V. (2008) Graphene-based single-bacterium resolution biodevice and DNA transistor: interfacing graphene derivatives with nanoscale and microscale biocomponents. *Nano Letters*, **8**, 4469–76.

58 Lee, Z., Jeon, K.-J., Dato, A., Erni, R., Richardson, T.J., Frenklach, M. and Radmilovic, V. (2009) Direct imaging of soft-hard interfaces enabled by graphene. *Nano Letters*, **9**, 3365–9.

4
Carbon Nanohorns and Their Biomedical Applications

Shuyun Zhu and Guobao Xu

4.1
Introduction

During recent years, nanostructured materials have attracted much attention in the materials sciences, and a considerable number of applications, including electronic devices and energy-related applications, have been identified [1–4]. Since the discovery of the C_{60} molecule [5], the family of carbon nanostructures has been steadily growing. Following the development of multi-walled carbon nanotubes (MWNTs) [6] and single-walled carbon nanotubes (SWNTs) [7], carbon onions [8] and carbon cones [9], an additional – and highly attractive – addition to the family has been that of carbon nanohorns (CNHs).

As a new carbon allotrope within the family of elongated CNTs, CNHs were first observed by Iijima *et al.* in 1996, in the carbon soot resulting from the CO_2 laser ablation of graphite [10]. Subsequently, this new structure was fully characterized by the same research group in 1999 [11]. Initially, CNHs were prepared in considerable amounts via the CO_2 laser ablation of a graphite target, at room temperature under an argon atmosphere, with both high yield (75%) and high purity (95%) [11]. To date, two different types of CNHs – termed "dahlia-like" and "bud-like" – have been observed. In the "dahlia-like" form, the single-walled carbon nanohorns (SWCNHs) protrude from the particle surface, whereas in the "bud-like" form the horn-shaped structure appears to develop inside the particle itself. On examination, the diameters of the "dahlia-like" SWCNHs were found to range from 2 to 5 nm, and the lengths from 40 to 50 nm. Typically, about 2000 SWCNHs were assembled to form a spherical aggregate of 80–100 nm diameter, while the internal cone angle of the "dahlia-like" SWCNH was found to be about 20°. Previously, the nature and pressure of the buffer gas used in the above method were shown to play key roles in the purity and morphology of the product. For example, "dahlia-like" SWCNH aggregates were produced with a yield of 95% when Ar was used, whereas the yield of "bud-like" SWCNH aggregates was only 70–80% when He or N_2 were used [12]. It was also shown that the CO_2 laser ablation of a graphite rod in a Ne atmosphere produced "Ne-dahlia" SWCNHs, with aggregate diameter

---

Nanomaterials for the Life Sciences Vol.9: Carbon Nanomaterials. Edited by Challa S. S. R. Kumar
Copyright © 2011 WILEY-VCH Verlag GmbH & Co. KGaA, Weinheim
ISBN: 978-3-527-32169-8

of 50 nm [13]. At this point, the practical daily production capacity of 1 kg that could be achieved was about 100-fold greater than that for previous methods, simply by changing some of the experimental conditions [14]. Since the main disadvantage of the laser ablation route is its high cost, an arc discharge method has also been employed for the bulk preparation of CNHs, though this has led to variations not only in the product purity and morphology but also of the aggregate diameters. For example, CNH aggregates with diameter of 30–150 nm were synthesized by using a new cavity arc jet method under open-air conditions [15]. A convenient and inexpensive method for producing CNHs in open air, using a welding arc torch, was also presented [16]. In this case, SWCNH particles with mean size of 50 nm and a higher purity of 90% were synthesized by using a simple pulsed arc discharge [17]. An arc-in-liquid method was also used to prepare SWCNHs [18–20]. Whilst CNH production has included SWCNHs, double-wall carbon nanohorns (DWCNHs), and multiwall carbon nanohorns (MWCNHs), the majority of interest has until now been focused on SWCNHs, as detailed in the following subsections.

4.2
Structure and Properties

From a structural aspect, pristine CNHs aggregate in a dahlia-like shape, with a layer of conical single-layered tips sticking out in all directions from the spherical superstructure (Figure 4.1), and with the surrounding tips capped by highly stretched five-membered rings. Notably, the graphene sheets of SWCNHs are less perfect than those of the SWNTs, due to the presence of many defects, such as

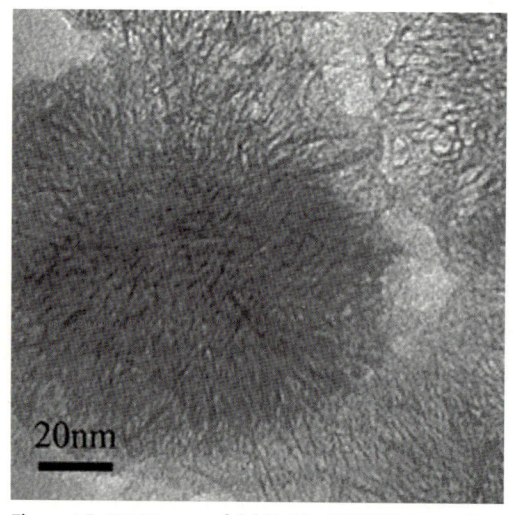

Figure 4.1 TEM image of dahlia-like SWCNHs. Reproduced from Ref. [21].

pentagons and heptagons, within the hexagonal network. Data acquired using thermogravimetry-differential thermal analysis (TG-DTA) have indicated that the as-grown CNHs consist of 70% tubular, 15% defective (on tips), 12% graphitic, and 2.5% amorphous carbon [22]. The Raman spectra of pristine SWCNHs exhibit different characteristics compared to those of graphite, glassy carbon, nanosoot, and diamond-like amorphous carbon [23]. Typically, the Raman spectrum for a SWCNH has a unique character, namely that two bands with almost equal scattering strengths are detected at 1593 and 1341 cm^{-1}. The Raman peak observed at 1593 cm^{-1} (the "G" band) has been assigned to E_{2g}-like vibrations in the sp^2-bonded carbon atoms, whereas the Raman peak at 1341 cm^{-1} (the "D" band) may be related to A_{1g}-symmetry modes associated with the M point in a graphitic Brillouin zone. The D band is attributed to a loss of the basal plane lattice due to the fullerene-terminated tips of the nanohorns, rather than to random lattice defects.

The structural features of pristine SWCNHs have also been studied using X-ray diffraction (XRD) [24]. The results of XRD analyses have indicated a wider spacing (~0.4 nm) for the van der Waals distance of the aggregated SWCNHs, this being greater than the interlayer spacing of graphite (0.335 nm). Thus, SWCNH aggregates should have both microporosity and mesoporosity originating from the above specific structure, and consequently the porosity of CNHs has been investigated in a series of studies. One such study of CNHs through N_2 adsorption revealed the existence of micropores with a pore volume of 0.11 ml g^{-1} [25]. A nanohorn aggregate presents two groups of pores: (i) the inter-nanohorn pores – the interstitial pores between individual nanohorns in the spherical aggregate; and (ii) the intra-nanohorn pores – the hollow interior spaces of the individual nanohorns (see Figure 4.1). Access to this latter set of pores requires the nanohorns to be subjected to an opening treatment. To date, numerous attempts have been made to obtain intra-nanohorn pores. For instance, the partial oxidation of CNHs during a high-temperature treatment in the presence of O_2 led to the formation of windows on the walls of the CNHs, such that the internal pores became accessible. Heat treatment at 693 K then resulted in an almost perfect opening of the intraparticle pores, leading to a micropore volume of 0.47 ml g^{-1} and a specific surface area of 1010 $m^2 g^{-1}$. Both, the number and size of the windows was also found to be increased as the heating temperature was raised [26]. The volume of the interstitial pores was much smaller than that of the intratube space, which was made available for adsorption after the nanohorns had been opened by an oxygen treatment. Consequently, the micropore structure of pristine SWCNHs consisted of interstitial channels between the nanohorns, while both interstitial and internal nanotube spaces formed the microporosity in oxidized SWCNHs (oxSWCNHs). The interstitial pore volume of the SWCNHs was 0.11 ml g^{-1}, and this was preserved after treatment in oxygen; the volume of the open intratube space in oxSWCNHs was 0.36 ml g^{-1} [27]. Furthermore, SWCNH particles were oxidized at different temperatures in O_2 to control the size and number of the nanowindows [28]. Later, the use of a slow rate of temperature increase (1 °C min^{-1} or less) and a low oxygen concentration (21% in air) was shown to result in hole opening, thus avoiding the formation of C-dust [29]. An investigation of water adsorption in oxSWCNHs by heat treatment in oxygen suggested that oxygen-containing functional groups, such as carboxyl (COOH), hydroxyl (OH), and carbonyl (CO), existed at the defect sites of the holes [30]. It had been confirmed previously that an oxidation treatment of SWCNHs increased the quantity of oxygen-containing functional (OCF) groups. The heat treatment of oxSWCNHs under a H_2 gas flow (300 ml min^{-1}) at 1200 °C for 3 h resulted in opened SWCNHs with reduced oxygen functionalities (termed hydrogen gas-treated oxSWCNHs; hSWCNHs) [31]; the subsequent compression of pristine SWCNHs led to a remarkable increase in microporosity [32].

Acids also have been used for pore opening; for example, the treatment of dahlia-like SWCNHs with H_2SO_4, an H_2O_2/H_2SO_4 mixture [33], or HNO_3 [34, 35], followed by heat treatment, resulted in an enhanced microporosity. Notably, treatment with H_2O_2/H_2SO_4 led to remarkable increases in both the total surface area and the micropore volume. In particular, the latter was increased almost 5.5-fold (from 0.11 to 0.60 ml g^{-1}) after successive H_2O_2/H_2SO_4 and heat treatments, whereas the mesopore volume showed a dramatic decrease (from 0.28 to 0.06 ml g^{-1}). Hence, the micropore volume was increased remarkably, from 28% to 91%. Alternatively, treatment with HNO_3 led to a considerable effect on both the microporosity and mesoporosity of the SWCNH assemblies. Typically, the HNO_3 treatment of SWCNHs caused a development in microporosity due to increases in both the internal and interstitial pore volumes. Recently, the thermal hole-closing processes of SWCNHs before and after heat treatment were investigated [36]. The results obtained indicated that all of the holes formed at ≤400 °C were closed by heat treatment at 1200 °C, whereas thermal hole closing was ineffective for holes formed at >500 °C. Later on, phenomena related to the behavior of these holes were reported in detail [37]. When the initial hole size was greater than 0.7–0.9 nm, temporary oxygen bridges associated with quick closing were not observed; however, when the hole was less than 0.7–0.9 nm the holes were closed quickly, and remained closed.

4.3
Functionalization

Chemical modification of the external surfaces of SWCNHs is a rapidly growing field, as such modification would be expected not only to advance the manipulation of SWCNHs by introducing a desired solubility, but also to make a fundamental contribution to the study of their solution properties. The dissolution of SWCNHs would enable full spectroscopic investigations to be conducted, as well as comprehensive studies of their properties with regards to nanotechnological and biomedical applications. Soluble SWCNHs have two major advantages, in that they are easy to handle, and are compatible with other materials. The dissolution of SWCNHs in organic or aqueous media depends heavily on the functionalization methodology and procedure used. Notably, different functionalization strategies have been described that have included the covalent attachment of organic fragments, as well as noncovalent interactions based on π–π stacking

4.3.1
Covalent Functionalization

The covalent functionalization of SWCNHs was first reported via a chemical methodology applied extensively to other carbon-rich nanostructured materials (i.e., fullerenes and CNTs), namely the 1,3-dipolar cycloaddition of azomethine yields generated *in situ* upon the thermal condensation of aldehydes and α-amino acids (Figure 4.2) [38]. In this case, the functionalized SWCNHs were soluble in both organic solvents and in water, while the nature of the α-amino acid used to generate the azomethine ylides governed the solubility of the modified nanohorns.

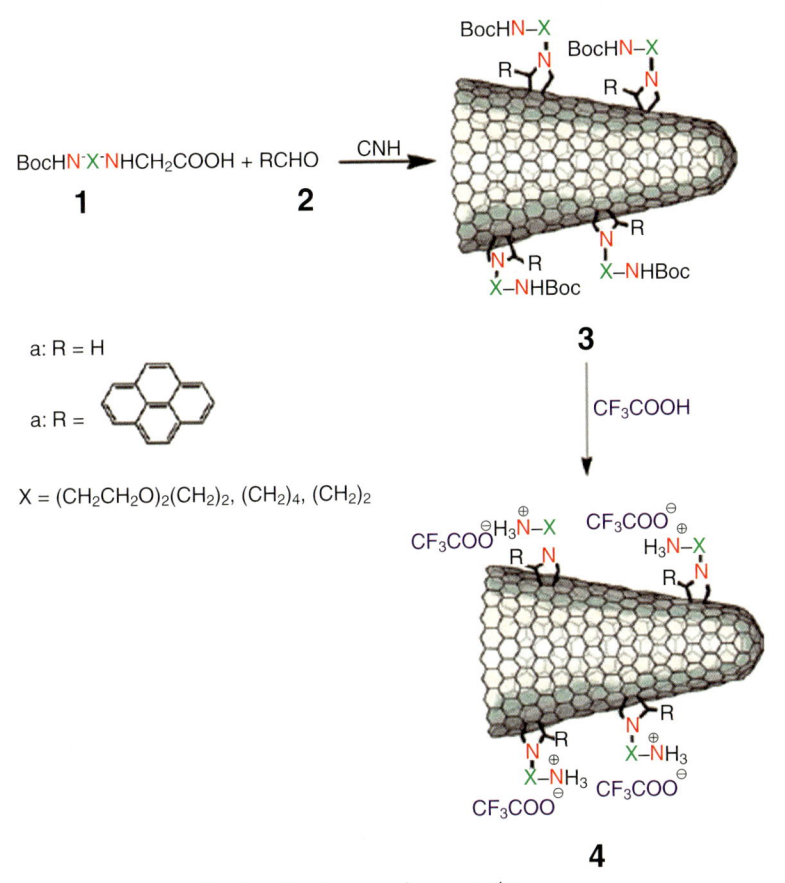

Figure 4.2 Functionalization of carbon nanohorns via the 1,3-dipolar cycloaddition of azomethine ylides. Reproduced from Ref. [38].

With this type of chemistry, a large number of pyrrolidine moieties could be introduced at locations near the conical tip of CNHs [39]. The functionalized CNH hybrids, where the pyrrolidines carried different groups, were shown to be stable in solution, as no precipitation was observed during long-term (months) storage in the dark. Later, Cioffi et al. also described the functionalization of SWCNHs via a 1,3-dipolar cycloaddition [40], where the SWCNH derivatives demonstrated good solubility in common organic solvents. Previously, the 1,3-dipolar cycloaddition reaction of azomethine ylides onto SWCNHs had been investigated via theoretical calculations based on Austin Model 1 (AM1), density functional theory (DFT), and own N-layered integrated molecular orbital and molecular mechanics (ONIOM) methods [41]. An alternative approach was reported which described the covalent functionalization of pristine SWCNHs and oxSWCNHs by direct nucleophilic addition, or via the amidation of carboxyl groups, respectively, with a monoprotected diamine derivative [42]. The functionalized SWCNH derivatives exhibited reasonable solubility in organic solvents. Subsequently, a new method for the covalent functionalization of CNHs by means of anionic polymerization, using a "grafting-to" approach, was described [43]. When, recently, the chemical functionalization of oxSWCNHs with a polymer (i.e., polyethylene oxide; PEO) was reported [44], the synthesized CNHs–PEO material was shown to be soluble in a variety of solvents, including water, tetrahydrofuran (THF), $CHCl_3$ and dimethylformamide (DMF). Another strategy for the covalent functionalization of CNHs, utilizing in situ-generated aryl diazonium compounds, was also reported [45]. In this case, the covalent grafting of aryl moieties onto the skeletons of the CNHs enhanced the solubility. Recently, an efficient strategy to produce multifunctionalized CNHs via a simple and rapid microwave-induced method, in combination with two different addition reactions, was reported [46]. The novel CNH derivatives created proved to be soluble in both water and in many common solvents.

Functionalization procedures of CNHs, based on the opening of their conical and highly strained ends, have also been reported. The first report involved the attachment of various organic amines, alcohols and thiols [47], such that soluble CNH-based amide, ester, and thioester materials were easily produced. Alternatively, oxidized CNHs were covalently functionalized with porphyrin (H_2P), and this resulted in the formation of CNH–H_2P nanohybrids [48]. Such nanohybrids were found to remain dissolved in DMF, THF and toluene for several weeks; moreover, a lack of any significant precipitation of the solubilized material provided strong evidence that the hybrid material was very stable. The CNH–H_2P nanohybrid synthesized via this method was also used to construct photoelectrochemical solar cells [49], while others functionalized CNHs with pyrene chromophores to form dispersible nanohybrids that were soluble in dichloromethane [50].

4.3.2
Noncovalent Functionalization

Although the covalent attachment of various addends either onto the graphite-like sidewalls of CNHs (e.g., via pyrrolidine moieties) or at the conical-shaped tip (e.g., via the formation of amides, esters, etc.) leads to significant solubilization and dispersion of the functionalized material, the resultant perturbation of the continuous π-electronic network of the CNHs represents a significant drawback, especially with regards to applications based on nanoelectronics. Thus, in order to overcome such drawbacks, a variety of supramolecular approaches utilizing noncovalent π–π stacking interactions between the sidewalls of CNHs with aromatic organic materials and/or synergistic electrostatic interactions has been developed. Based on this concept, a bifunctional molecule bearing a pyrene moiety (π–π stacking interactions with CNHs) on one end and silane (covalent bonding with hydroxyl groups of an oxide surface) on the other end, was used to form a pattered assembly of a CNH monolayer on an oxide surface [51]. The same methodology was employed for the synthesis of a CNH/nanoparticle binary nanomaterial, using a pyrene-succinimidyl ester bifunctional molecule [52]. Notably, the noncovalent method used here to construct SWCNH-based binary nanomaterials preserved the original structural and functional features of SWCNHs. This, in turn, provided an opportunity to investigate the various properties of SWCNHs in solution, namely their molecular recognition and catalytic capabilities. Subsequently, tetracationic water-soluble porphyrin (H_2P^{4+}) was immobilized onto the skeleton of CNHs by π–π stacking interactions, but without disrupting their π-electronic networks [53]. In another study, a combination of π–π stacking and electrostatic interaction was used to integrate an anionic porphyrin (H_2P^{2-}) with CNHs, mediated by positive pyrene (pyr^+) units [54]. The resultant water-soluble CNHs–pyr^+–H_2P^{2-} nanoensemble was characterized using optical spectroscopy and electron microscopy. The strong fluorescence emission of H_2P^{2-} was significantly quenched by the CNHs, suggesting an effective energy and electron transfer between the photoexcited porphyrin and extended π-electronic network of the nanohorns. Similarly, water-soluble CNH-tetrathiafulvalene (TTF) nanoensembles were prepared by utilizing positively charged pyrene as an assembly medium [55]. Recently, a simple method for the noncovalent polymer functionalization of CNHs using a block polyelectrolyte was reported [56]. The protocol for the noncovalent functionalization utilized the amphiphilic poly[sodium (2-sulfamate-3-carboxylate) isoprene-b-styrene] block polyelectrolyte. The hydrophobic polystyrene block was anchored onto the nanohorn surface through hydrophobic interactions, while the polyelectrolyte block stabilized the formed hybrid nanoassembly through electrosteric interactions. The nanohybrid colloid was shown to be water-soluble and stable for several months.

Such functionalization of SWCNHs may not only open new doors for the use of nanohorns in the materials sciences, but also increase their solubility in water with regards to biomedical applications, such as drug- and gene-delivery systems.

4.4
Biomedical Applications

On the basis of their unique structures and properties, SWCNHs have been employed for a wide variety of applications, including adsorption [57], methane storage [58], nanoparticles and catalyst support [59], fuel cells [60], supercapacitors [61], gas sensors [62], and electrochemical detection [63]. Details of such applications in the field of biomedicine are introduced in the following subsections.

Structurally, SWCNHs are composed of a graphite carbon atom structure similar to that of SWNTs. However, the main advantage of the CNHs is that, upon aggregation, they form secondary particles (80–100 nm in size) that in turn provide an extremely large surface area and facilitate gas/liquid penetration into the structure. Moreover, the SWCNH is prepared at room temperature by CO_2 laser ablation without catalyst. Hence, pure samples of SWCNH are easily available, which favors the study of their properties and applications. As SWCNH is made of graphene sheet, it is stable enough for any use in the physiological environment. Furthermore, the hollow space inside SWCNHs and the interstitial spaces among SWCNHs provide suitable sites to incorporate and protect drugs. Therefore, SWCNH aggregates can be regarded as potentially good drug carriers possessing certain advantages over other drug carriers. Even compared to CNTs, SWCNHs still have some advantages in application for the field of biomedicine:

- Although CNTs appear to show minimal toxicity, this aspect has not yet been clarified because the toxicity tests are influenced by the contaminations with metal catalysts used in the production processes. The toxicological action of defects on the CNTs, introduced by the purification processes to remove the metal catalysts, is also unclear. In contrast, SWCNHs are highly pure, and essentially do not contain any metals. Consequently, investigations into the toxicity of SWCNHs are difficult to misinterpret.

- The CNTs assemble to form bundles of micrometer length, whereas thousands of SWCNHs assemble to form a spherical aggregate with diameters of up to 100 nm. Materials with such large diameters demonstrate enhanced permeability and retention under passive tumor-targeting conditions, and tend to accumulate at the tumor tissues. The SWCNHs may also remain in the tumors for longer periods, due to their appropriate size when injected intratumorally, and inhibit tumor growth.

- The holes on the SWCNH walls are easily opened by heat treatment in O_2 gas, while the number and size of the holes can be changed by controlling the heat-treatment conditions. The inside spaces are also sufficiently wide for the large molecules to be incorporated.

- The SWCNHs have a large diameter and a short length. By comparison, CNTs have narrow diameters (1–2 nm) and long lengths (micrometers), so that both the incorporation and release of drugs are difficult; consequently, the drugs are often attached to the outside walls. In contrast, oxSWCNHs have large diameters (2–5 nm) and shorter length (40–50 nm), and the drugs can be easily incorporated into the inside walls of oxSWCNHs and released slowly from within. Hence, SWCNHs with opened holes are well suited to drug incorporation and controlled release, and will have potential applications in the field of biomedicine.

4.4.1
Toxicity Assessment of SWCNHs

In order to avoid potential health hazards caused by occupational exposure to SWCNHs, and to promote their industrial and biomedical application, the toxicity of the SWCNHs should be investigated proactively from various aspects. For example, to investigate the pulmonary toxicity of SWCNHs [64], mice were exposed to 30 μg of surfactant-suspended SWCNHs or an equal volume of vehicle control by pharyngeal aspiration. The SWCNH-exposed mice did not exhibit any overt clinical symptoms of distress from the time of anesthesia recovery until the time of sacrifice at 24 h or seven days post-exposure. A whole-lung microarray analysis indicated that SWCNH-exposed mice did not have any robust changes in gene expression, while histology showed no evidence of granuloma formation or fibrosis after SWCNH aspiration. Together, these results suggested that SWCNHs were a relatively innocuous agent when delivered to mice *in vivo*, using aspiration as a delivery mechanism. Elsewhere, the toxicities of as-grown SWCNHs were investigated comprehensively *in vivo* and *in vitro* and compared to those of CNTs and fullerenes [65]. Two types of bacterial genotoxicity test were first conducted to assess the carcinogenic potential of SWCNHs. Subsequently, dermal and ocular reactions were investigated on the skin and eyes, respectively, as these superficial organs had the greatest risk of exposure to SWCNHs. A peroral administration test for SWCNHs was also conducted, since oral ingestion was a likely uptake pathway for SWCNH exposure. As there is a risk of inhaling fluffy SWCNHs floating in the air, the effects of intratracheal instillation of SWCNHs on the lungs as a surrogate for inhalational exposure was investigated to evaluate pulmonary toxicity. The results of all of these tests confirmed that as-grown SWCNHs had low toxicities, and implied that SWCNHs should have a negligible impact on the living body.

4.4.2
SWCNHs Used in Drug-Delivery Systems

The significant characteristics of the carbon nanomaterials include their stability, inertness, and large surface area, all of which suggest the potential utility of these materials as carriers in drug-delivery systems (DDSs). Previously, CNTs have been used as the shells in DDSs because they are stable both chemically and mechanically, material incorporation into their inner hollow spaces is easy, and their outside walls can be chemically modified to achieve the desired targeting effect. To date, many reports have been made of the application of SWCNHs for DDSs.

To investigate the potential use of CNHs as drug-delivery carriers, the *in vitro* binding and release of the anti-inflammatory glucocorticoid dexamethasone (DEX) by as-grown SWCNHs and oxSWCNHs were investigated [66]. Adsorption analyses using [^{3}H] -DEX indicated that 200 mg DEX could be loaded onto oxSWCNHs; this was about sixfold larger for as-grown SWCNHs. The adsorption profiles showed the oxSWCNHs to have a higher affinity for DEX than as-grown SWCNHs. The controlled release of drugs from a drug–carrier complex is a cardinal property of a DDS. Hence, when the cumulative release of DEX from a DEX–oxSWCNHs complex in phosphate-buffered saline (PBS) at pH 7.4 and at 37 °C was investigated, during the first few days the amount of DEX released was almost proportional to the incubation time. However, the release rate then gradually declined, such that about 50% of the total bound DEX was released from the complex within a period of two weeks. In contrast, in the cell culture medium about 50% of the DEX was released during the first 8 h, and the remainder into the RPMI1640 after 24 h. The biological integrity of the DEX released from the DEX–oxSWCNHs was confirmed by activation of the glucocorticoid response element-driven transcription in mouse bone marrow stromal ST2 cells, and the induction of alkaline phosphatase in mouse osteoblastic MC3T3-E1 cells. Moreover, no cytotoxicity was observed in the presence of oxSWCNHs.

The well-known anticancer agent cisplatin (CDDP) has also been encapsulated into oxSWCNHs using nanoprecipitation [67]. In order to apply CDDP@oxSWCNHs to a DDS, the release rate of CDDP must be slow in order to prevent drug dissipation before reaching the tumor. In fact, the release rate of CDDP from oxSWCNHs in PBS and a culture medium was much slower than the dissolution rate of free CDDP powder; that is, a 72 h period was required for CDDP release from CDDP@oxSWCNHs to saturate the PBS, but only 24 h for the free CDDP. The CDDP released from CDDP@oxSWCNHs was effective in terminating the growth of human lung-cancer cells, but the oxSWCNHs themselves had no such effect. These results suggested that oxSWCNHs would be potentially useful as a drug carrier.

When the release rate of CDDP@oxSWCNHs was compared to that of CDDP@hSWCNHs [68], the extent of CDDP release from CDDP@oxSWCNHs in PBS was only 15%, but was 70% from CDDP@hSWCNHs. This variation was considered due to the –COONa and –ONa groups at the hole edges of the oxSWCNHs decreasing the hole diameters and hindering the release of CDDP from the oxSWCNHs. The hole size enlargement and control of the number of functional groups at the hole edges of oxSWCNHs were also shown to be effective in producing the necessary CDDP, releasing quantities at a slow rate. Such requirements were satisfied by applying a special hole-opening method [69] where, to change the diameters of the oxSWCNH holes, the holes were opened up by combusting the as-grown SWCNHs in dry air, heating them to a target temperature at a rate of $1\,°C\,min^{-1}$. With regards to the least amount of CDDP found outside of the oxSWCNHs, the optimum target temperature was 500 °C. The CDDP@oxSWCNHs obtained were shown to release the CDDP slowly and abundantly (ca. 80%) over 50 h, which favored their use as CDDP carriers *in vivo*. When comparing the amount of CDDP released in PBS from CDDP@oxSWCNHs (quick), the amount released from CDDP@oxSWCNHs (slow) was much greater.

When various potential erythropoietin (EPO) carriers were investigated, a high adsorption on CNTs was revealed, whereas that on SWCNHs was low [70]. The

SWCNHs, which are considered to possess a greater available surface area for adsorption than CNTs, did not elicit a higher serum EPO level. Hence, not only is the adsorption space important to serve as an effective drug delivery tool, but also the structure of the adsorbent. In another study, the CDDP@oxSWCNHs were shown to have a higher antitumor efficiency than intact CDDP, both *in vitro* and *in vivo* [71]. *In vitro*, the enhancement of anticancer efficiency of CDDP@oxSWCNHs was caused by the adherence of oxSWCNHs to the cells. This would increase the local concentration of CDDP released from CDDP@oxSWCNHs, and result in effective cell killing. The high total released quantity and slowly release of CDDP could maintain a high local concentration over a long period. *In vivo*, the CDDP@oxSWCNHs remained in the tumor tissues for a considerable period, and this contributed to the higher anticancer efficiency of CDDP@oxSWCNHs when compared to intact CDDP. In the meantime, the oxSWCNHs were also shown to have an *in vivo* anticancer effect.

In an aqueous environment, such as the blood, hydrophobic nanomaterials tend to form large agglomerates, which might block capillary vessels and lead to serious consequences. Although SWCNHs have shown potential as carriers in DDSs, their inherent hydrophobicity poses an impediment to their clinical application. Notably, their insolubility in water and great tendency to agglomerate (micrometer-size) restrict their extensive use in biological applications. Consequently, the ability to endow SWCNHs with a molecular dispersibility under aqueous conditions represents a major challenge in this field. An example of this was the improvement in the solubility of SWCNHs by amination [72]. Upon treatment with $NaNH_2$, the carbon nanohorn aggregate (NHA) agglomerates could be converted, in their entirety, into water-soluble amino-NHAs (a-NHAs) that were taken up by mammalian cells but did not show significant cytotoxicity. The noncovalent functionalization of CNTs with various polymers has been studied extensively to improve the dispersibility of the molecules, although some of these methods have employed carcinogenic or indiscriminately cytolytic molecules such as pyrene or sodium dodecylsulfate (SDS). Elsewhere, the noncovalent modification of oxSWCNHs using an amide-linked poly(ethylene glycol)–doxorubicin (PEG–DXR) was reported and yielded well-dispersed PEG–DXR–oxSWCNH complexes having diameters of approximately 160 nm – a size that would be expected to produce an enhanced permeation and retention (EPR) effect [73]. The properties confined by PEG included thermal stability, resistance to proteolysis, increased water solubility, decreased antigenicity and immunogenicity, and a slower clearance from the plasma. The cytotoxicity of the PEG–DXR–oxSWCNHs was evaluated by subjecting NCI-H640 human non-small cell lung cancer cells to terminal deoxynucleotidyl transferase dUTP nick end labeling (TUNEL) staining. The complexes would not be rapidly cleared from the blood by the liver and spleen, and would exert an EPR effect in any tumors which showed neovascularization. Moreover, the complexes exhibited a DXR-dependent apoptotic activity against cancer cells. Because PEG is nontoxic, is stable and has a low immunogenicity, a conjugate comprised of PEG and a peptide aptamer (NHBP-1) was used for the noncovalent modification of SWCNHs [74]. This modification endowed the hydrophobic oxSWCNHs with a good dispersibility under aqueous conditions. Moreover, the CDDP could be loaded onto the inner spaces of surface-modified oxSWCNHs to produce well-dispersed, CDDP-loaded CNHs (CDDP@oxSWCNHs/20PEG–NHBP), the cytotoxicity of which was estimated by measuring the metabolic activity of viable cells using the WST-1 assay (Roche Diagnostics GmbH, Mannheim, Germany). Consequently, CDDP@oxSWCNHs/20PEG-NHBP was shown to exert a dose-dependent cytotoxic effect on NCI-H640 cells. Moreover, the oxSWCNHs/20PEG–NHBP complexes themselves had no effect on cell proliferation, which indicated that the cytotoxic effect was due to the released CDDP. In another report, CNH agglomerates were successfully isolated by attaching gum arabic (GA) to achieve a stable suspension through steric stabilization [75]. In this case, GA was used as a typical copolymer because it was biocompatible, and its protein moieties could be conjugated with biological cargoes for intracellular delivery. The modified CNHs showed excellent biocompatibility because of their high purity and the surface modification of GA. The cytotoxicity of CNHs was investigated by incubating modified CNHs with human cervical cancer cells (HeLa cells) in a DMEM culture containing serum. The experimental results showed the modified CNHs to be nontoxic, and most likely to enter the cells through via endocytic pathways. Notably, the CNHs showed promise as a vehicle for intracellular deliveries.

The results of preliminary studies with SWCNH–DDSs have indicated problems involving the pharmacokinetics of SWCNHs, notably with regard to the excretion of these compounds from the body. Evidence has been provided that these compounds are trapped by the liver, lung, or spleen, so that their complete excretion is difficult. In an attempt to overcome these problems, the number of reactive sites on SWCNHs was increased through extensive chemical modification, by using biocompatible molecules. For example, light-assisted H_2O_2 oxidation SWCNHs (LAOx–SWCNHs) were modified with bovine serum albumin (BSA) through diimide-activated amidation; this resulted in a product with an enhanced hydrophilicity that would disperse effectively in aqueous solution [76]. The cellular uptake of the BSA-modified LAOx–SWCNHs conjugates was investigated, and their incorporation into cultured mammalian cells via an endocytosis pathway observed. When H460 cancer cells were incubated with the dispersion of LAOx–SWCNHs–BSA, cellular uptake was observed, using confocal microscopy. The observations showed that 91% of the H460 cells internalized the LAOx–SWCNHs–BSA complexes, but located them mainly around the cell membranes rather than in the cell interior; hence, it appeared that the cell internalization was energy-dependent and the uptake mechanism was considered to be endocytosis. Recently, SWCNHs were initially functionalized by the direct addition of amine functions, to render them dispersible in aqueous buffers, and subsequently labeled with a fluorescent dye for imaging. The biocompatibility of the functionalized SWCNHs (f-SWCNHs) towards primary murine macrophages was explored in detail [77]. As the macrophages survived well *in vitro* for a few days without stimulation, cell viability after incubation with f-SWCNHs was examined. Specific cell death was not observed in the presence of f-SWCNHs, in comparison with untreated cells; hence, it was confirmed that the f-SWCNHs did not exert any lethal effect after their penetration into the cells. The resultant f-SWCNHs were biocompatible and evaluated on their capacity to be integrated into a cellular system. The f-SWCNHs were also shown to be effectively phagocytosed by primary murine macrophages, without affecting cell survival. A rapid uptake of f-SWCNHs might be particularly advantageous if a strategy targeting macrophages were envisaged. Recently, a double photodynamic therapy (PDT) and photohyperthermia (PHT) cancer phototherapy system, in which zinc phthalocyanine (ZnPc) was the PDT agent, oxSWCNHs was the PHT agent, and protein (BSA)-enhanced biocompatibility was fabricated [78]. For this, ZnPc was loaded onto oxSWCNHs, and BSA then attached to the carboxyl groups of the oxSWCNHs. When ZnPc–oxSWCNHs–BSA was injected into tumors that had been transplanted subcutaneously into mice, subsequent laser irradiation caused the tumors to disappear. Thus, the SWCNHs not only had PHT effects by themselves, but also provided a platform for the attachment/loading of PDT agents.

Currently, during clinical therapy, patients must be treated by repetitive drug administration, because the drugs employed are rapidly metabolized and eliminated from the body. Unfortunately, this often results in a high dosage uptake, a low efficacy, and adverse side effects. In order to improve such therapy, therefore, the controlled release of drugs – including the period of release and the site – are important factors. The key here is that DDS is a nontoxic material; moreover, it is able to incorporate drugs and to release them at the desired site(s) and rate(s) over a prolonged period, so that the drug concentration is maintained within the therapeutic range. Previously, vancomycin (VCM) hydrochloride was successfully incorporated into oxSWCNHs by using a simple solvent evaporation method that took advantage of the interaction between VCM and oxSWCNHs [79]. The oxSWCNHs were initially modified with phospholipid (PL)–PEG to improve their dispersion in aqueous systems. In VCM–oxSWCNHs–PEG, the release of VCM was mainly controlled by the interaction between VCM and oxSWCNHs, such that VCM could be released in a slow and stable manner. Consequently, oxSWCNHs was proposed to act as a potential VCM carrier, due to such sustained drug release. In another report, the *in vitro* and *in vivo* antitumor activities and lymphatic migration of water-dispersed and anticancer drug-bound oxSWCNHs (PEG–DXR–oxSWCNHs) was described [80]. When the PEG–DXR–oxSWCNHs were administered intratumorally to human non-small-cell lung cancer-cell NCI-H460-bearing mice, they were shown to be effective in reducing tumor growth, which was associated with a prolonged DXR retention in the tumor. These findings suggested that water-dispersed SWCNHs might serve as a potential drug carrier for local chemotherapy. Recently, a report was made of the precise ultrastructural localization and quantification of SWCNHs, to clearly identify the sites of uptake [81]. For this, Gd_2O_3 nanoparticles were embedded within SWCNH aggregates (Gd_2O_3@SWCNHaq) to facilitate their detection and quantification. Following intravenous injection of the as-prepared Gd_2O_3@SWCNHaq into mice, the largest amount (70–80% of that injected) was identified in the liver, 12% was found in the spleen, and 1–2% in the stomach/intestine or the lungs. Within the liver, the Gd_2O_3@SWCNHaq was not found in the hepatocytes, but rather was

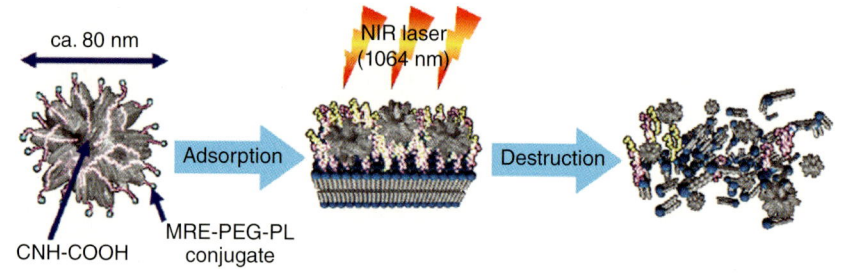

Figure 4.3 Method for the elimination of microorganisms using MRE–CNH complexes and NIR laser irradiation (1064 nm). The MRE selectively targets the microbe, the PEG chains affect the water dispersibility of the CNH–COOH molecules, and the hydrophobic carbon chains of the phospholipid (PL) bind noncovalently to the surfaces of the CNH–COOH molecules via hydrophobic interactions. Reproduced from Ref. [82].

localized in the Kupffer cells, where they were located primarily in the phagosomes and occasionally in phagolysosome-like vacuoles.

The application of CNHs as potent laser therapeutic agents for the highly selective elimination of microorganisms has also been reported. In one study, a molecular recognition element (MRE)- PEG-PL-carboxylic-functionalized SWCNH (SWCNH-COOH) complex was used [82]. The choice of an appropriate MRE in the near-infrared (NIR) laser-driven SWCNH system allowed the selective elimination of harmful multidrug-resistant microorganisms, such as methicillin-resistant *Staphylococcus aureus* (MRSA) (Figure 4.3). In addition, small molecules (<2 nm) can easily infiltrate the interior space of SWCNHs through the nanopores in the walls of carboxylic-functionalized SWCNHs, and this made possible the controlled release of these small molecules. Thus, MRE-PEG-PL-SWCNH-COOH has the potential to become not only a highly selective NIR laser exothermic material, but also a nanocarrier for various drugs. The construction of such multifunctional nanomaterials will, undoubtedly, become a "hot topic" in the field of nanomedicine. Subsequently, a novel approach for the use of NIR laser-driven SWCNHs as an antiviral agent was reported [83], whereby a T7 tag antibody (T7 tag Ab)–SWCNH complex, which would effectively eliminate the T7 bacteriophage (a model virus) upon NIR laser irradiation, was successfully synthesized. The functionalized SWCNH nanomaterials also exhibited the photo-exothermic elimination of harmful viruses, such as human immunodeficiency virus (HIV), severe acute respiratory syndrome (SARS), and avian influenza virus.

4.4.3
SWCNHs Used in Magnetic Resonance Analysis

Further research into SWCNH applications in the living body, and the assessment of any associated toxicological hazards, will require non-anatomic *in vivo* observations of the motion and tissue accumulation of SWCNHs. For this purpose, magnetic resonance imaging (MRI) might be used, if an MRI contrast agent were to be attached to the SWCNHs. The potential of Gd_2O_3-deposited oxSWCNHs as a positive MRI contrast agent has been demonstrated previously [84]. In this case, the T_1 relaxation rate (T_1^{-1}) of water protons was enhanced significantly following the addition of Gd_2O_3–oxSWCNHs to the agarose gel, and seen to increase in line with the Gd concentration. The T_1^{-1} was more than 10-fold higher than that for the water protons in the oxSWCNHs/PBS/agarose system with a Gd concentration of 25 mM, suggesting that ultrafine Gd_2O_3 nanoparticles on oxSWCNHs would function as a positive MRI contrast agent. As a sharp bright image was observed at a low concentration of Gd_2O_3–oxSWCNHs, it was relatively straightforward to identify the oxSWCNHs locations in the T_1-weighted MR image. In order to apply SWCNHs as a drug-delivery carrier, the drug-carrying SWCNHs should be accumulated in the living body so as to maximize the efficacy of chemotherapeutic treatments. Hence, it is feasible that the Gd_2O_3–oxSWCNHs could serve as a labeled drug-delivery carrier. In addition to the Gd^{3+} complexes, superparamagnetic magnetite (i.e., magnetite Fe_3O_4) nanoparticles with diameters of less than about 25 nm may serve as another effective contrast agent for MRI. In this case, $Fe(OAC)_2$ was first deposited onto oxSWCNHs and then heated to 400 °C; this resulted in a superparamagnetic magnetite (Fe_3O_4, 6 nm diameter) that was strongly attached to the oxSWCNHs (MAGoxSWCNHs), and which made possible the use of MRI with MAGoxSWCNHs in living mice [85]. Prior to using MAGoxSWCNHs for *in vivo* MRI, the single-dose acute toxicity of MAGoxSWCNHs in mice was investigated to determine the optimum dosage. For this, a suspension of MAGoxSWCNHs in PBS was administered intravenously at doses of 2 or 8 mg kg^{-1} body weight. All mice survived the entire test period, and no clinical signs of abnormalities were observed. The preliminary *in vivo* toxicity test indicated that a MAGoxSWCNHs dose below 8 mg kg^{-1} was likely to be nontoxic for mice. When the accumulative behavior of MAGoxSWCNHs in an anesthetized mouse was studied using *in vivo* MRI, the spleen was seen to darken with time following MAGoxSWCNHs administration. Subsequent histopathological examinations showed that the MAGoxSWCNHs had accumulated in the spleen within 120 min of intravenous administration, and this resulted in the time-dependent darkening of the spleen observed in the MR images (Figure 4.4). A time-dependent darkening of the kidneys in the MR images was also observed, which suggested that MAGoxSWCNHs had accumulated in these organs. None of the mice used for the MRI studies showed any abnormalities after recovery from anesthesia. The significant MRI contrast observed in these studies confirmed the feasibility of using magnetite-nanoparticle labeling for the *in vivo* visualization of oxSWCNHs.

4.4.4
Biosensing Applications of SWCNHs

The biosensing applications of SWCNHs were also reported. SWCNHs are essentially metal-free and can be used directly for biosensing studies. Initially, SWCNHs

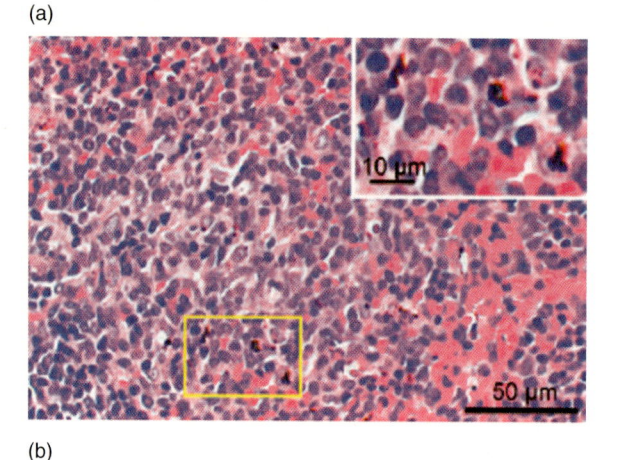

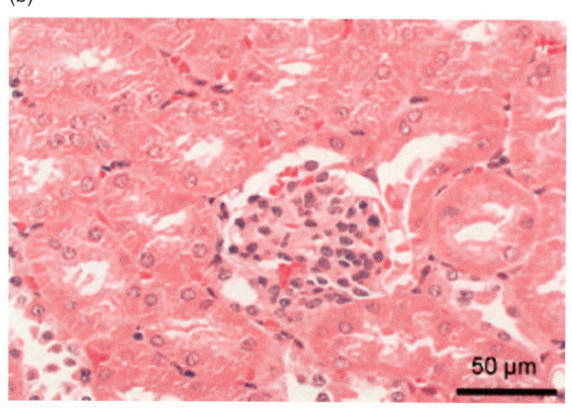

Figure 4.4 Optical microscopy images of mouse tissues recorded at 120 min after the intravenous injection of 6 mg kg^{-1} body weight of MAGoxSWCNHs. The tissues were stained with hematoxylin and eosin. (a) Spleen tissue. The inset shows a magnified image of a selected area; (b) Kidney tissue. The absence of MAGoxSWCNHs in the kidneys suggested that ultrafine aggregates of the MAGoxSWCNHs were distributed at a relatively low concentration in these organs. However, black particles of agglomerated MAGoxSWCNHs were observed in the spleen. Reproduced from Ref. [85].

were used to fabricate a glucose biosensor [86] which was constructed by encapsulating glucose oxidase in the Nafion–SWCNHs composite. The mediated glucose biosensor showed a linear range from 0 to 0.6 mM, together with a high sensitivity (1.06 μA mM^{-1}) and stability, and was able to avoid the interference which commonly coexisted. Later, SWCNHs were used as a biocompatible matrix for fabricating hydrogen peroxide biosensor [21, 87]. The direct immobilization of acid-stable and thermostable soybean peroxidase (SBP) on an SWCNH-modified electrode surface can realize the direct electrochemistry of SBP. Based on the direct electrochemistry of SBP, a mediator-free H_2O_2 biosensor was constructed which showed good thermostability. *Myoglobin*, an oxygen-transport protein found in mammalian muscle, has a molecular weight of about 17 000 Da, and its structure is well known. The activity of myoglobin can be greatly enhanced by suitable modification; moreover, it is much cheaper than peroxidases and is very suitable for large-scale production using well-known recombinant means. The above-described features make myoglobin a promising component for hydrogen peroxide biosensors. As myoglobin has an isoelectric point (pI) of about 7.2, and can easily be adsorbed to negatively charged poly(sodium styrene sulfonate) (PSS) under weakly acidic solutions, the SWCNHs which had been noncovalently functionalized with PSS were used to assemble myoglobin and to facilitate the electrochemical communication between myoglobin and composite-modified electrode. The as-prepared myoglobin-based biosensor demonstrates favorable properties for the determination of H_2O_2, and the noncovalent functionalization of SWCNHs will facilitate their application in areas of biosensors and electrochemistry [87]. Recently, the use of a SWCNH-modified glassy carbon electrode was also investigated for the simultaneous determination of uric acid (UA), dopamine (DA), and ascorbic acid (AA) [88]. In this case, the voltammetric peaks of UA, DA, and AA were each well-resolved at the SWCNH-modified glassy carbon electrode, with UA in particular showing a reversible redox peak at the SWCNH-modified glassy carbon electrode that enabled its detection with an improved sensitivity of more than one order of magnitude.

4.5 Conclusions

Due to their unique structures and properties, SWCNHs have been successfully used in the field of biomedicine yet, despite such ongoing effort and investigation, the topic of SWCNHs is still in its infancy. However, major opportunities are foreseen for the application of SWCNHs in the biomedical field, assuming that the fundamental issues concerning their manipulation and fabrication can be overcome, and the solubility of functionalized SWCNHs achieved. Indeed, in the future, water-dispersed SWCNHs are expected to serve as novel drug carriers for local chemotherapy against tumors, with the antitumor activity of these complexes augmented by various refinements and/or small drug depositions (e.g., cisplatin) within their interior spaces. Consequently, a variety of functionalized SWCNHs should be explored in order to develop their biomedical applications. At the same time, smaller oxSWCNHs – which would be expected to be excreted more readily from the body – should be prepared for biomedical applications. The smaller oxSWCNHs would shed light on the systemic administration of water-dispersed oxSWCNHs. Moreover, in order to achieve complex excretion, the SWCNHs could be chemically modified by the addition of various biocompatible molecules, so as to increase the number of reactive sites. These new and innovative CNH-based hybrid materials should also be prepared and studied for biomedical applications.

Acknowledgments

The authors are very grateful to Professor S. Iijima (Solution Oriented Research for Science and Technology in Japan Science and Technology Agency) for a generous offer of SWCNHs. These studies were kindly supported by the National Natural Science Foundation of China (No. 20875086 & 20505016), the Ministry of Science and Technology of the People's Republic of China (No.2006BAE03B08), and the Department of Sciences & Technology of Jilin Province (20070108 and 20082104).

References

1 Schüth, F. and Schmidt, W. (2002) Microporous and mesoporous materials. *Advanced Materials*, **14**, 629–38.
2 Xia, Y., Yang, P., Sun, Y., Wu, Y., Mayers, B., Cates, B., Yin, Y., Kim, F. and Yan, H. (2003) One-dimensional nanostructures: synthesis, characterization, and applications. *Advanced Materials*, **15**, 353–89.
3 Stein, A. (2003) Advances in microporous and mesoporous solids – highlights of recent progress. *Advanced Materials*, **15**, 763–75.
4 Hamley, I.W. (2003) Nanotechnology with soft materials. *Angewandte Chemie International Edition*, **42**, 1692–712.
5 Kroto, H.W., Heath, J.R., O'Brien, S.C., Curl, R.F. and Smalley, R.E. (1985) C_{60}: buckminsterfullerene. *Nature*, **318**, 162–3.
6 Iijima, S. (1999) Helical microtubules of graphitic carbon. *Nature*, **354**, 56–8.
7 Iijima, S. and Ichihashi, T. (1993) Single-shell carbon nanotubes of 1-nm diameter. *Nature*, **363**, 603–5.
8 Ugart, D. (1992) Curling and closure of graphitic networks under electron-beam irradiation. *Nature*, **359**, 707–9.
9 Krishnan, A., Dujardin, E., Treacy, M.M.J., Hugdahl, J., Lynum, S. and Ebbesen, T.W. (1997) Graphitic cones and the nucleation of curved carbon surfaces. *Nature*, **388**, 451–4.
10 Iijima, S., Wakabayashi, T. and Achiba, Y. (1996) Structures of carbon soot prepared by laser ablation. *Journal of Physical Chemistry*, **100**, 5839–43.
11 Iijima, S., Yudasaka, M., Yamada, R., Bandow, S., Suenaga, K., Kokai, F. and Takahashi, K. (1999) Nano-aggregates of single-walled graphitic carbon nanohorns. *Chemical Physics Letters*, **309**, 165–70.
12 Kasuya, D., Yudasaka, M., Takahashi, K., Kokai, F. and Iijima, S. (2002) Selective production of single-wall carbon nanohorn aggregates and their formation mechanism. *Journal of Physical Chemistry B*, **106**, 4947–751.
13 Azami, T., Kasuya, D., Yoshitake, T., Kubo, Y., Yudasaka, M., Ichihashi, T. and Iijima, S. (2007) Production of small single-wall carbon nanohorns by CO_2 laser ablation of graphite in Ne-gas atmosphere. *Carbon*, **45**, 1364–9.
14 Azami, T., Kasuya, D., Yuge, R., Yudasaka, M., Iijima, S., Yoshitake, T. and Kubo, Y. (2008) Large-Scale production of single-wall carbon nanohorns with high purity. *Journal of Physical Chemistry C*, **112**, 1330–4.
15 Ikeda, M., Takikawa, H., Tahara, T., Fujimura, Y., Kato, M., Tanaka, K., Itoh, S. and Sakakibara, T. (2002) Preparation of carbon nanohorn aggregates by cavity arc jet in open air. *Japanese Journal of Applied Physics, Part 2*, **41** (7B), L852–4.
16 Takikawa, H., Ikeda, M., Hirahara, K., Hibi, Y., Tao, Y., Ruiz Jr, P.A., Sakakibara, T., Itoch, S. and Iijima, S. (2002) Fabrication of single-walled carbon nanotubes and nanohorns by means of a torch arc in open air. *Physica B*, **323**, 277–9.
17 Yamaguchi, T., Bandow, S. and Iijima, S. (2004) Synthesis of carbon nanohorn particles by simple pulsed arc discharge

ignited between pre-heated carbon rods. *Chemical Physics Letters*, **389**, 181–5.

18 Sano, N., Nakano, J. and Kanki, T. (2004) Synthesis of single-walled carbon nanotubes with nanohorns by arc in liquid nitrogen. *Carbon*, **42**, 686–8.

19 Sano, N. (2004) Low-cost synthesis of single-walled carbon nanohorns using the arc in water method with gas injection. *Journal of Physics D: Applied Physics*, **37**, L17–20.

20 Sano, N., Kimura, Y. and Suzuki, T. (2008) Synthesis of carbon nanohorns by gas-injected arc-in-water method and application to catalyst-support for polymer electrolyte fuel cell electrodes. *Journal of Materials Chemistry*, **18**, 1555–60.

21 Shi, L.H., Liu, X.Q., Niu, W.X., Li, H.J., Han, S., Chen, J.A. and Xu, G.B. (2009) Hydrogen peroxide biosensor based on direct electrochemistry of soybean peroxidase immobilized on single-walled carbon nanohorn modified electrode. *Biosensors & Bioelectronics*, **24**, 1159–63.

22 Utsumi, S., Miyawaki, J., Tanaka, H., Hattori, Y., Itoi, T., Ichikuni, N., Kanob, H., Yudasaka, M., Iijima, S. and Kaneko, K. (2005) Opening mechanism of internal nanoporosity of single-wall carbon nanohorn. *Journal of Physical Chemistry B*, **109**, 14319–24.

23 Bandow, S., Rao, A.M., Sumanasekera, G.U., Eklund, P.C., Kokai, F., Takahashi, K., Yudasaka, M. and Iijima, S. (2000) Evidence for anomalously small charge transfer in doped single-wall carbon nanohorn aggregates with Li, K and Br. *Applied Physics A*, **71**, 561–4.

24 Bandow, S., Kokai, F., Takahashi, K., Yudasaka, M., Qin, L.C. and Iijima, S. (2000) Interlayer spacing anomaly of single-wall carbon nanohorn aggregate. *Chemical Physics Letters*, **321**, 514–19.

25 Murata, K., Kaneko, K., Kokai, F., Takahashi, K., Yudasaka, M. and Iijima, S. (2000) Pore structure of single-wall carbon nanohorn aggregates. *Chemical Physics Letters*, **331**, 14–20.

26 Murata, K., Kaneko, K., Steele, W.A., Kokai, F., Takahashi, K., Kasuya, D., Hibarahara, K., Yudasaka, M. and Iijima, S. (2001) Molecular potential structures of heat-treated single-wall carbon nanohorn assemblies. *Journal of Physical Chemistry B*, **105**, 10210–16.

27 Murata, K., Kaneko, K., Steele, W.A., Kokai, F., Takahashi, K., Kasuya, D., Yudasaka, M. and Iijima, S. (2001) Porosity evaluation of intrinsic intraparticle nanopores of single-wall carbon nanohorn. *Nano Letters*, **1**, 197–9.

28 Murata, K., Hirahara, K., Yudasaka, M., Iijima, S., Kasuya, D. and Kaneko, K. (2002) Nanowindow-induced molecular sieving effect in a single-wall carbon nanohorn. *Journal of Physical Chemistry B*, **106**, 12668–9.

29 Fan, J., Yudasaka, M., Miyawaki, J., Ajima, K., Murata, K. and Iijima, S. (2006) Control of hole opening in single-wall carbon nanotubes and single-wall carbon nanohorns using oxygen. *Journal of Physical Chemistry B*, **110**, 1587–91.

30 Bekyarova, E., Hanzawa, Y., Kaneko, K., Silvestre-Albero, J., Sepulveda-Escribano, A., Rodriguez-Reinoso, F., Kasuya, D., Yudasaka, M. and Iijima, S. (2002) Cluster-mediated filling of water vapor in intratube and interstitial nanospaces of single-wall carbon nanohorns. *Chemical Physics Letters*, **366**, 463–8.

31 Miyawaki, J., Yudasaka, M. and Iijima, S. (2004) Solvent effects on hole-edge structure for single-wall carbon nanotubes and single-wall carbon nanohorns. *Journal of Physical Chemistry B*, **108**, 10732–5.

32 Bekyarova, E., Kaneko, K., Yudasaka, M., Murata, K., Kasuya, D. and Iijima, S. (2002) Micropore development and structure rearrangement of single-wall carbon nanohorn assemblies by compression. *Advanced Materials*, **14**, 973–5.

33 Yang, C.M., Kasuya, D., Yudasaka, M., Iijima, S. and Kaneko, K. (2004) Microporosity development of single-wall carbon nanohorn with chemically induced coalescence of the assembly structure. *Journal of Physical Chemistry B*, **108**, 17775–82.

34 Yang, C.M., Noguchi, H., Murata, K., Yudasaka, M., Hashimoto, A., Iijima, S. and Kaneko, K. (2005) Highly ultramicroporous single-walled carbon nanohorn assemblies. *Advanced Materials*, **17**, 866–70.

35 Yuge, R., Ichihashi, T., Miyawaki, J., Yoshitake, T., Iijima, S. and Yudasaka, M. (2009) Hidden caves in an aggregate of single-wall carbon nanohorns founded by using Gd_2O_3 probes. *Journal of Physical Chemistry C*, **113**, 2741–4.

36 Miyawaki, J., Yuge, R., Kawai, T., Yudasaka, M. and Iijima, S. (2007) Evidence of thermal closing of atomic-vacancy holes in single-wall carbon nanohorns. *Journal of Physical Chemistry C*, **111**, 1553–5.

37 Fan, J., Yuge, R., Miyawaki, J..R., Kawai, T., Iijima, S. and Yudasaka, M. (2008) Close-open-close evolution of holes at the tips of conical graphenes of single-wall carbon nanohorns. *Journal of Physical Chemistry C*, **112**, 8600–3.

38 Tagmatarchis, N., Maigne, A., Yudasaka, M. and Iijima, S. (2006) Functionalization of carbon nanohorns with azomethine ylides: towards solubility enhancement and electron-transfer processes. *Small*, **2**, 490–4.

39 Pagona, G., Rotas, G., Petsalakis, I.D., Theodorakopoulos, G., Fan, J., Maigne, A., Yudasaka, M., Iijima, S. and Tagmatarchis, N. (2007) Soluble functionalized carbon nanohorns. *Journal of Nanoscience and Nanotechnology*, **7**, 3468–72.

40 Cioffi, C., Campidelli, S., Brunetti, F.G., Meneghetti, M. and Prato, M. (2006) Functionalisation of carbon nanohorns. *Chemical Communications*, 2129–31.

41 Petsalakis, I.D., Pagona, G., Theodorakopoulos, G., Tagmatarchis, N., Yudasaka, M. and Iijima, S. (2006) Unbalanced strain-directed functionalization of carbon nanohorns: a theoretical investigation based on complementary methods. *Chemical Physics Letters*, **429**, 194–8.

42 Cioffi, C., Campidelli, S., Sooambar, C., Marcaccio, M., Marcolongo, G., Meneghetti, M., Paolucci, D., Paolucci, F., Ehli, C., Rahman, G.M.A., Sgobba, V., Guldi, D. and Prato, M. (2007) Synthesis, characterization, and photoinduced electron transfer in functionalized single wall carbon nanohorns. *Journal of the American Chemical Society*, **129**, 3938–45.

43 Mountrichis, G., Pispas, S. and Tagmatarchis, N. (2007) Grafting living polymers onto carbon nanohorns. *Chemistry – A European Journal*, **13**, 7595–9.

44 Mountrichis, G., Tagmatarchis, N. and Pispas, S. (2009) Functionalization of carbon nanohorns with polyethylene oxide: synthesis and incorporation in a polymer matrix. *Journal of Nanoscience and Nanotechnology*, **9**, 3775–9.

45 Pagona, G., Karousis, N. and Tagmatarchis, N. (2008) Aryl diazonium functionalization of carbon nanohorns. *Carbon*, **46**, 604–10.

46 Rubio, N., Herrero, A., Meneghetti, M., Diaz-Ortiz, A., Schiavon, M., Prato, M. and Vazquez, E. (2009) Efficient functionalization of carbon nanohorns via microwave irradiation. *Journal of Materials Chemistry*, **19**, 4407–13.

47 Pagona, G., Tagmatarchis, N., Fan, J., Yudasaka, M. and Iijima, S. (2006) Cone-end functionalization of carbon nanohorns. *Chemistry of Materials*, **18**, 3918–20.

48 Pagona, G., Sandanayaka, A.S.D., Araki, Y., Fan, J., Tagmatarchis, N., Charalambidis, G., Coutsolelos, A.G., Boitrel, B., Yudasaka, M., Iijima, S. and Ito, O. (2007) Covalent functionalization of carbon nanohorns with porphyrins: nanohybrid formation and photoinduced electron and energy transfer. *Advanced Functional and Materials*, **17**, 1705–11.

49 Pagona, G., Sandanayaka, A.S.D., Hasobe, T., Charalambidis, G., Coutsolelos, A.G., Yudasaka, M., Iijima, S. and Tagmatarchis, N. (2008) Characterization and photoelectrochemical properties of nanostructured thin film composed of carbon nanohorns covalently functionalized with porphyrins. *Journal of Physical Chemistry C*, **112**, 15735–41.

50 Sandanayaka, A.S.D., Pagona, G., Fan, J., Tagmatarchis, N., Yudasaka, M., Iijima, S., Araki, Y. and Ito, O. (2007) Photoinduced electron-transfer processes of carbon nanohorns with covalently linked pyrene chromophores: charge-separation and electro-migration systems. *Journal of Materials Chemistry*, **17**, 2540–6.

51 Zhu, J., Yudasaka, M., Zhang, M.F., Kasuya, D. and Iijima, S. (2003) A surface modification approach to the patterned

assembly of single-walled carbon nanomaterials. *Nano Letters*, **3**, 1239–43.

52 Zhu, J., Kase, D., Shiba, K., Kasuya, D., Yudasaka, M. and Iijima, S. (2003) Binary nanomaterials based on nanocarbons: a case for probing carbon nanohorns' biorecognition properties. *Nano Letters*, **3**, 1033–6.

53 Pagona, G., Araki, Y., Fan, J., Tagmatarchis, N., Yudasaka, M., Iijima, S. and Ito, O. (2006) Electronic interplay on illuminated aqueous carbon nanohorn-porphyrin ensembles. *Journal of Physical Chemistry B*, **110**, 20729–32.

54 Pagona, G., Fan, J., Maigne, A., Yudasaka, M., Iijima, S. and Tagmatarchis, N. (2007) Aqueous carbon nanohorn-pyrene-porphyrin nanoensembles: controlling charge-transfer interactions. *Diamond and Related Materials*, **16**, 1150–3.

55 Pagona, G., Sandanayaka, A.S.D., Maigne, A., Fan, J., Papavassiliou, G.C., Petsalakis, I.D., Steele, B.R., Yudasaka, M., Iijima, S., Tagmatarchis, N. and Ito, O. (2007) Photoinduced electron transfer in aqueous carbon nanohorn-pyrene-tetrathiafulvalent architectures. *Chemistry–A European Journal*, **13**, 7600–7.

56 Moutrichas, G., Ichihashi, T., Pispas, S., Yudasaka, M., Iijima, S. and Tagmatarchis, N. (2009) Solubilization of carbon nanohorns by block polyelectrolyte wrapping and templated formation of gold nanoparticles. *Journal of Physical Chemistry C*, **113**, 5444–9.

57 Murata, K., Kaneko, K., Kanoh, H., Kasuya, D., Takahashi, K., Kokai, F., Yudasaka, M. and Iijima, S. (2002) Adsorption mechanism of supercritical hydrogen in internal and interstitial nanospaces of single-wall carbon nanohorn assembly. *Journal of Physical Chemistry C*, **106**, 11132–8.

58 Bekyarova, E., Murata, K., Yudasaka, M., Kasuya, D., Iijima, S., Tanaka, H., Kahoh, H. and Kaneko, K. (2003) Single-wall nanostructured carbon for methane storage. *Journal of Physical Chemistry B*, **107**, 4681–4.

59 Bekyarova, E., Hashimoto, A., Yudasaka, M., Hattori, Y., Murata, K., Kanoh, H., Kasuya, D., Ijima, S. and Kaneko, K. (2005) Palladium nanoclusters deposited on single-walled carbon nanohorns. *Journal of Physical Chemistry B*, **109**, 3711–14.

60 Sano, N. and Ukita, S. (2006) One-step synthesis of Pt-loaded carbon nanohorns for fuel cell electrode by arc plasma in liquid nitrogen. *Materials Chemistry and Physics*, **99**, 447–50.

61 Yang, C.M., Kim, Y.J., Endo, M., Kanoh, H., Yudasaka, M., Iijima, S. and Kaneko, K. (2007) Nanowindow-regulated specific capacitance of supercapacitor electrodes single-wall carbon nanohorns. *Journal of the American Chemical Society*, **129**, 20–1.

62 Sano, N., Kinugasa, M., Otsuki, F. and Suehiro, J. (2007) Gas sensor using single-wall carbon nanohorns. *Advanced Powder Technology*, **18**, 455–66.

63 Zhu, S.Y., Fan, L.S., Liu, X.Q., Shi, L.H., Li, H.J., Han, S. and Xu, G.B. (2008) Determination of concentrated hydrogen peroxide at single-walled carbon nanohorn paste electrode. *Electrochemistry Communications*, **10**, 695–8.

64 Lynch, R.M., Voy, B.H., Glass, D.F., Mahurin, S.M., Zhao, B., Hu, H., Saxton, A.M., Donnell, R.L. and Cheng, M.D. (2007) Assessing the pulmonary toxicity of single-walled carbon nanohorns. *Nanotoxicology*, **1**, 157–66.

65 Miyawaki, J., Yudasaka, M., Azami, T., Kubo, Y. and Iijima, S. (2008) Toxicity of single-walled carbon nanohorns. *ACS Nano*, **2**, 213–26.

66 Murakami, T., Ajima, K., Miyawaki, J., Yudasaka, M., Iijima, S. and Shiba, K. (2004) Drug-loaded carbon nanohorns: adsorption and release of dexamethasone in vitro. *Molecular Pharmaceutics*, **1**, 399–405.

67 Ajima, K., Yudasaka, M., Murakami, T., Maigne, A., Shiba, K. and Iijima, S. (2005) Carbon nanohorns as anticancer drug carriers. *Molecular Pharmaceutics*, **2**, 475–80.

68 Ajima, K., Yudasaka, M., Maigne, A., Miyawaki, J. and Iijima, S. (2006) Effect of functional groups at hole edges on cisplatin release from inside single-wall carbon nanohorns. *Journal of Physical Chemistry B*, **110**, 5773 -8.

69 Ajima, K., Maigne, A., Yudasaka, M. and Iijima, S. (2006) Optimum hole-opening condition for cisplatin incorporation in

single-wall carbon nanohorns and its release. *Journal of Physical Chemistry B*, **110**, 19097–9.

70 Venkatesan, N., Yoshimitsu, J., Ito, Y., Shibata, N. and Takada, K. (2005) Liquid filled nanoparticles as a drug delivery tool for protein therapeutics. *Biomaterials*, **26**, 7154–63.

71 Ajima, K., Murakami, T., Mizaoguchi, Y., Tsuchida, K., Ichihashi, T., Iijima, S. and Yudasaka, M. (2008) Enhancement of *in vivo* anticancer effects of cisplatin by incorporation inside single-wall carbon nanohorns. *ASC Nano*, **2**, 2057–64.

72 Isobe, H., Tanaka, T., Maeda, R., Noiri, E., Solin, N., Yudasaka, M., Iijima, S. and Nakamura, E. (2006) Preparation, purification, characterization, and cytotoxicity assessment of water-soluble, transition-metal-free carbon nanotube aggregates. *Angewandte Chemie International Edition*, **45**, 6676–80.

73 Murakami, T., Fan, J., Yudasaka, M., Iijima, S. and Shiba, K. (2006) Solubilization of single-wall carbon nanohorns using a PEG-doxorubicin conjugate. *Molecular Pharmaceutics*, **3**, 407–14.

74 Matsumura, S., Ajima, K., Yudasaka, M., Iijima, S. and Shiba, K. (2007) Dispersion of cisplatin-loaded carbon nanohorns with a conjugate comprised of an artificial peptide aptamer and polyethylene glycol. *Molecular Pharmaceutics*, **4**, 723–9.

75 Fan, X.B., Tan, J., Zhang, G.L. and Zhang, F.B. (2007) Isolation of carbon nanohorn assemblies and their potential for intracellular delivery. *Nanotechnology*, **18**, 1951031–6.

76 Zhang, M.F., Yudasaka, M., Ajima, K., Miyawaki, J. and Iijima, S. (2007) Light-assisted oxidation of single-wall carbon nanohorns for abundant creation of oxygenated groups that enable chemical modifications with proteins to enhance biocompatibility. *ACS Nano*, **1**, 265–72.

77 Lacotte, S., Garcia, A., Decossa, M., Al-Jamal, W., Li, S.P., Kostatelos, K., Muller, S., Prato, M., Dumortier, H. and Bianco, A. (2008) Interfacing functionalized carbon nanohorns with primary phagocytic cells. *Advanced Materials*, **20**, 2421–6.

78 Zhang, M.F., Murakami, T., Ajima, K., Tsuchida, K., Sandanayaka, A.S.D., Ito, O., Iijima, S. and Yudasaka, M. (2008) Fabrication of ZnPc/protein nanohorns for double photodynamic and hyperthermic cancer phototherapy. *Proceedings of the National Academy of Sciences of the United States of America*, **105**, 14773–8.

79 Xu, J.X., Yudasaka, M., Kouraba, S., Sekido, M., Yamamoto, Y. and Iijima, S. (2008) Single wall carbon nanohorn as a drug carrier for controlled release. *Chemical Physics Letters*, **461**, 189–92.

80 Murakami, T., Sawada, H., Tamura, G., Yudasaka, M., Iijima, S. and Tsuchida, K. (2008) Water-dispersed single-wall carbon nanohorns as drug carriers for local cancer chemotherapy. *Nanomedicine*, **3**, 453–63.

81 Miyawaki, J., Matsumura, S., Yuge, R., Murakami, T., Sato, S., Tomida, A., Tsuruo, T., Ichihashi, T., Fujinmi, T., Irie, H., Tsuchida, K., Iijima, S., Shiba, K. and Yudasaka, M. (2009) Biodistribution and ultrastructural localization of single-walled carbon nanohorns determined *in vivo* with embedded Gd_2O_3 labels. *ACS Nano*, **3**, 1399–406.

82 Miyako, E., Nagata, H., Hirano, K., Makita, Y., Nakayama, K. and Hirotsu, T. (2007) Near-infrared laser-triggered carbon nanohorns for selective elimination of microbes. *Nanotechnology*, **18**, 4751031–7.

83 Miyako, E., Nagata, H., Hirano, K., Sakamoto, K., Makita, Y., Nakayama, K. and Hirotsu, T. (2008) Photoinduced antiviral carbon nanohorns. *Nanotechnology*, **19**, 0751061–6.

84 Miyawaki, J., Yudasaka, M., Imai, H., Yorimitsu, H., Isobe, K., Nakamura, E. and Iijima, S. (2006) Synthesis of ultrafine Gd_2O_3 nanoparticles inside single-wall carbon nanohorns. *Journal of Physical Chemistry B*, **110**, 5179–81.

85 Miyawaki, J., Yudasaka, M., Imai, H., Yorimitsu, H., Isobe, H., Nakamura, E. and Iijima, S. (2006) *In vivo* magnetic resonance imaging of single-walled carbon nanohorns by labeling with magnetite nanoparticles. *Advanced Materials*, **18**, 1010–14.

86 Liu, X.Q., Shi, L.H., Niu, W.X., Li, H.J. and Xu, G.B. (2008) Amperometric glucose biosensor based on single-walled carbon nanohorns. *Biosensors & Bioelectronics*, **23**, 1887–90.

87 Liu, X.Q., Li, H.J., Wang, F.A., Zhu, S.Y., Wang, Y.L. and Xu, G.B. (2010) Functionalized single-walled carbon nanohorns for electrochemical biosensing. *Biosensors & Bioelectronics*, **25**, 2194–9.

88 Zhu, S.Y., Li, H.J., Niu, W.X. and Xu, G.B. (2009) Simultaneous electrochemical determination of uric acid, dopamine, and ascorbic acid at single-walled carbon nanohorn modified glassy carbon electrode. *Biosensors & Bioelectronics*, **25**, 940–3.

5
Bio-Inspired Magnetic Carbon Materials

Elby Titus, José Gracio, Duncan P. Fagg, Manoj K. Singh and Antonio C. M. Sousa

5.1
Introduction

In recent years, carbon nanomaterials have received considerable attention for applications in life sciences. The various allotropic forms of carbon (diamond, graphite and fullerenes) have been suggested as unique new substances for these applications, due to their outstanding properties such as biocompatibility, excellent stability, and easy surface functionalizability. Furthermore, the coupling of magnetism with carbon science opens many promising research directions where magnetic carbon is created for specific applications in a systematic and controllable way. Recent investigations of magnetism in certain forms of carbon have demonstrated this tailorability, identifying applications in nanotechnology, medicine, and telecommunications. Magnetic nanocarbons also find application in high-density memory devices and in quantum computers. As carbon materials are generally compatible with living tissue, these magnetic nanostructures could also be useful in medical applications, such as magnetic resonance imaging (MRI) and the targeted delivery of drugs to specific parts of the body. As such, ferromagnetism in carbon is of real theoretical and experimental scientific interest, with potential added value impact to the community. In the following discussions, the aim is to highlight the many advantages that magnets made from carbon may offer over metal magnets.

Although magnetism – one of the most curious phenomena observed in solids – has been studied for centuries, there is still no comprehensive theory that provides a full explanation. Several models and mechanisms have been proposed that partially explain the phenomenon, while some aspects require further study and might even be termed "unsolved mysteries" [1]. Magnetism is observed in elements which have a net magnetic moment in the solid state due to unpaired electrons in their atoms; these elements include iron, nickel, and cobalt. One of the last places that ferromagnetism would be expected to occur is in carbon, because its electrons pair up preferentially to form covalent bonds. However, despite the covalent nature of carbon, defects and irregularities in pure carbon materials can give rise to electrons that are unpaired with other electrons. As each unpaired electron can then produce a magnetic field by its spinning, the material itself can then become magnetic when all of these spins are aligned. Carbon is one of the most interesting elements in the biosphere, and is a central part of life. Carbon is also lightweight, very stable, simple to process, and cheap to produce – all factors that lead to great possibilities for carbon's application.

In fact, several recent experimental and theoretical reports [2–8] have appeared which deal with the magnetic properties of carbon systems, including diamond, graphite, and fullerenes [C_{60}/carbon nanotubes (CNTs)]. The experimentalists [9] have identified different types of carbon magnets, and classified them as: (i) chains containing interacting radicals; (ii) carbon materials with a mixture of sp2 and sp3 coordinated atoms; (iii) amorphous carbon structures containing trivalent elements (e.g., phosphorus, nitrogen, boron); (iv) the nano form of allotropic carbon; and (v) fullerenes.

The proposed mechanisms of magnetism in various carbon-based materials have recently been reviewed [10]. Most of the reports on intrinsic magnetism in carbon materials have been supported through defect mechanisms. Typically, the structure of sp^3 and sp^2 bonding has a major influence on the formation of magnetism in carbon material, and various synthetic methods have been described for the fabrication of magnetic carbon by induced magnetism. *Ion implantation* is an established technique used to induce magnetism in carbon materials. For example, trivalent impurities such as nitrogen, phosphorus and boron can be incorporated into the carbon matrix, and this has been shown to lead to ferromagnetism [11]. Physical methods (e.g., laser ablation, plasma, ion implantation) to incorporate foreign bodies into carbon materials offer advantages when compared to chemical or wet methods. For example, recent reports have been made of laser-ablated glassy carbon being prepared under an argon atmosphere and exhibiting ferromagnetic behavior up to 90 K, along with high saturation magnetization [12].

The aim of this chapter is to explore the magnetic properties of carbon materials (diamond, graphite and CNTs), and to highlight their importance in the coming era of the life sciences. Thus, a general review of the magnetism in carbon materials is provided, together with an explanation for the allotropic forms of carbon and their magnetic behavior, and the physics underlying the unique properties of these structures and devices. The chapter will provide a broad overview of the latest developments in this emerging and important field, with emphasis placed on nanostructured materials, nanoscale characterization and new techniques for the fabrication of nanomagnetic material, as well as their chemical and physical properties. Finally, the applications of these magnetic carbons in biomedicine are highlighted.

5.2
Allotropic Forms of Carbon

Carbon appears in Group 14 of the Periodic Table, and has the electronic structure $1s^2\ 2s^2\ 2p^2$. Among the crystalline allotropic forms of carbon, the first allotropic form – diamond – is three-dimensional (3-D) and is a semi-conductor; the second form – graphite, is two-dimensional (2-D) and is a conductor; and finally the third form – fullerenes – are one-dimensional (1-D) in CNTs, zero-dimensional in "bucky balls," and can be either conductors or semi-conductors. The bucky balls belonging to the family of fullerenes have unusual properties. Recently, *graphene* has been identified as another newly discovered material, and is part of the family of graphite. There is a wide range of electronic properties among the allotropic forms of carbon, which vary from superconductivity to ferromagnetism. Indeed, it is this versatility that has attracted considerable attention in science, and has already been widely used for commercial applications.

5.3
Magnetism in Diamond

The intrinsic magnetism rarely observed in diamond can be explained in terms of defects, vacancies, and impurities [13]. One of the novel ways of manipulating magnetism in diamond is through defects and vacancies. For example, when a magnetic field is applied to diamond, it acts as a large insulator with a high electric resistivity, and no free electrons exist to induce any magnetic response [14].

Magnetism can also be induced in diamond by a variety of techniques, with ion implantation being one of the most successful. For example, Talapatra *et al.* [11] identified ferromagnetism in nanosized diamond particles that had been implanted with nitrogen and carbon ions, and explained its presence by defect generation. In the study of Talapatra *et al.* [11], the highest magnetization was achieved for the nitrogen-doped samples; however, no information could be provided regarding the impurity concentration in the samples in relation to the measured values of the magnetic moments. Diamond can be easily doped by using boron, and it has a high potential in electronic devices [15]. Notably, the electronic behavior of the diamond was sensitive to the doping doses; typically, a lightly boron-doped diamond showed p-type character [16], while a heavily boron-doped diamond exhibited metallic conduction [17]. The electrical conductivity and magnetic properties of heavily boron-doped diamond films have been reported [18]. In particular, boron doping in diamond changes its conducting behavior, and a metal to insulator transition in boron-implanted diamond was reported recently [19]. Elsewhere [20], the superconducting behavior of boron-doped diamond bulk samples (ca. 2.8 ± 0.5 atom% boron) has also been noted.

5.3.1
Biomedical Applications of Magnetic Diamond

Gem quality diamond is regarded as a precious ornament, and is valued in terms of its weight (karats), gem quality, color, and appearance. The inertness of diamond is advantageous as it does not react with the body; diamond also has a prime importance in medical astrology. The scope of diamond in medicine has been enhanced enormously by the creation of nanodiamond (ND), which is nanoscale in dimension but with fewer karats compared to the gem diamond. Nanodiamond is also biocompatible, and has shown promise in playing a significant role in improving cancer treatment by limiting the uncontrolled exposure of toxic drugs to the body. In a recent study [21], the claim was made that aggregated clusters of nanodiamond were ideal for carrying a chemotherapeutic agent, to shield that agent from normal cells so as not to kill them, and then to release the agent slowly only after it had reached the cellular target. Another merit of the nanodiamond material is its inertness, which leads to an avoidance of the inflammation that is a major concern when dealing with cancer; inflammation may not only retard the activity of anticancer drugs but also enhance the growth of cancer cells. The success of such drug delivery lies in the attainment of the correct tissue; moreover, the carrier should have no reaction with the body after drug delivery, and nanodiamond satisfies each of these conditions due to its inertness and biocompatibility. The solubility of nanodiamond is also an interesting factor for consideration in drug delivery.

The manipulation of nanodiamond with regards to its magnetic properties is of considerable interest, not only for application in devices such as magnetoresistance (MR) sensors, but also for basic research in biological and medical systems. The doping of diamond and its magnetic properties has been discussed earlier. When Wang *et al.* [22] studied the MR effect of boron-doped heteroepitaxial and polycrystalline diamond films, they included different device structures with magnetic fields ranging from 0 to 5 T. The MR variation under magnetic field was more notable, and the effect in heteroepitaxial diamond films was greater than that in polycrystalline films. The MR of the films at 500 K can still be observed at 5 T, indicating the possibility of a high-temperature-operating Hall device. Fei *et al.* [23] investigated the electrical properties and MR of boron-doped polycrystalline diamond films grown on p-typed Si (100), using the hot filament chemical vapor deposition (HFCVD) method, and observed variations in the quality of diamonds according to the atomic concentration and grain size, which further affected the carrier mobility and longitudinal MR. For a magnetic field (B) of 20 T and temperature 300 K, the longitudinal resistance change rate $\Delta \rho_{//}/\rho_0$ is up to 20%. Meanwhile, $\Delta \rho_{//}/\rho_0$ is proportional to $\mu^2 B^2$ in a low field, and proportional to $\mu^{1.5} B$ in a high field. The advantage of this study is that the authors could demonstrate the property in a high field. Choy *et al.* explained negative MR in ultracrystalline diamond by strong or weak localization [24], while Russel *et al.* [25] measured MR in p-type semiconducting diamond for magnetic fields up to 170 kG. Longitudinal effects were measured in the <$\bar{1}11$> and <$1\bar{1}3$> directions, and transverse effects were measured for current I in the <$1\bar{1}3$> direction and magnetic field H in the <110> direction. By using a theory which assumes an isotropic isothermal solid in which conduction is by holes from three uncoupled valance bands associated with spherical energy surfaces, and in which mixed scattering by lattice vibrations and ionized impurities is applicable, lattice mobilities of different values were calculated for different bands. Recent studies on ND biological applications revealed these materials to be much more biocompatible than most other carbon nanomaterials, including fullerenes and CNTs. In addition, they are nontoxic, small in size, and have large specific surface area. Moreover, they can be easily functionalized with biomolecules, making nanodiamonds attractive for a variety of biomedical applications, both *in vitro* and *in vivo*. One very promising approach is the use of magnetic nanodiamonds as markers and carriers for biomolecules. The magnetic particles to be used as markers can be detected by highly sensitive MR, and this results in a purely electronic signal. In addition, magnetic NDs used as carriers can be manipulated on-chip via currents. This combination of sensing and manipulating magnetic diamond particles represents a promising choice for future integrated "lab-on-a-chip" systems. Recently, Willems et al. [26] reported the observation of a negative MR in boron-doped ND films at low temperatures. The presence of superconducting regions were identified inside the insulating films as being the cause of the negative MR. The experimental observations were also explained by modeling the systems, which consisted of a distribution of superconducting granules, the global properties of which can be tuned by the intergrain distance.

5.4
Magnetism in Graphite

Graphite is composed of stacks of 2-D nets of hexagons, where every carbon atom has three nearest neighbors; this means that three out of four electrons form sp2 bonds in the graphite plane, while the remaining π orbital is perpendicular to the surface and forms metallic π bands across the surface. The interaction between layers is of a weak, van der Waals type. The graphite in its pure form is the most diamagnetic material due to the lack of d- and f- electrons, which are generally assumed to be necessary for the occurrence of ferromagnetism at relatively high temperatures ($T > 15$ K). However, when a magnetic field is applied to graphite, a semi-metal (3.5×10^{-5} m Ω), the π-electrons are free to circulate along the hexagonal carbon rings and are responsible for inducing a magnetic field that opposes the external field [14]. The findings of this study are supported by Raman's concept of molecular ring currents. The difference in magnetic susceptibility of graphite and diamond prompted Raman to postulate the flow of currents around the ring system of graphite in response to an applied magnetic field. The same authors [14] also noted the magnetic susceptibility behavior between diamond and graphite as (-5.0×10^{-7} emu g^{-1}) and (-7.3×10^{-6} emu g^{-1}), respectively.

According to a theoretical study [27], the magnetism in graphite is due to various defects in its honeycomb structure. Magnetic properties induced by defects on graphite structures, such as pores, the edges of planes and topological defects, have also been theoretically predicted. The possible coexistence of sp3 and sp2 bonds has been also used to justify the occurrence of these properties [28]. Some reports have proved the existence of weak ferromagnetic magnetization loops in highly-oriented pyrolytic graphite (HOPG) [29].

A simple and inexpensive chemical route, based on a vapor-phase reaction, is reported to obtain bulk ferromagnetic graphite [30]. This magnetic graphite was produced by the reaction of pristine graphite with controlled amounts of oxygen released from the decomposition of CuO at high temperature. This chemical attack created pores and stacking structures, and also increased the exposed edges of the graphene planes, to produce a foamy-like graphite. However, the use of powder graphite as a reagent seems to be crucial because of the high reactivity to chemical attacks, due to the very large exposed surface and the previous existence of defects. Xia et al. [31] demonstrated that room-temperature magnetism can occur in HOPG, by using low-energy carbon ion implantation. Lehtinen et al. [32] also confirmed the presence of ferromagnetism in HOPG, using density functional theory (DFT) calculations.

5.4.1
Biomedical Applications of Magnetic Graphite

Graphite with ferromagnetic properties has been actively pursued for potential medical applications, including integrated imaging, diagnosis, and therapy. Transition metals (TMs) and their alloys produce powerful magnetic signals that could be useful when detecting tumors by using MRI. The magnetic properties of transition-metal oxides have been investigated for many years, but they are easily oxidized, which makes their application very difficult from a practical standpoint. Yet, despite this problem, a large number of reports have been focused on TM oxide nanoparticles. However, the saturation magnetization and Curie temperature of these oxide materials are reduced when compared to pure TMs. Moreover, naked TMs have been shown to be chemically active materials, which makes them unsuitable for use in the body. Consequently, much effort is currently being expended to fabricate core–shell structures where TM nanoparticles are encapsulated inside an outer shell [33].

A wide range of physical and chemical properties can be realized in core–shell structures, depending on the selection of the two components (i.e., the core and the shell). An enhanced band luminescence has been identified in core–shell structures with extreme band gap differences, while a core–shell with metal-semiconductor components shows enhanced optical properties. It has been reported that a more efficient energy transfer between the semiconducting shell and the metal core is possible, by matching the plasmon resonance excitation energy of the metal core to the band gap energy of the semi-conducting shell [34]. Metal-semiconducting core–shell structures are generally prepared using wet chemical methods [35]. In this case, the synthesis method includes the growth of the shell around the core by harvesting the seed (core); the thickness of the shell can then be adjusted by feeding the correct amount of precursor. Methane is the most commonly used precursor for the coating of graphitic shells [36]. Graphite is of particular interest as it is chemically and thermally stable, and is also biocompatible. Seo et al. [37] have created and encapsulated a carbon-coated iron–cobalt alloy as a MRI contrast agent. For this, the alloy was formed by heating encapsulated iron and nickel in silica at high temperature (800 °C), after which the graphite coating was prepared under a methane atmosphere. Finally, a polymer coating was applied to provide stability in water. By using scanning electron microscopy (SEM), it was possible to characterize the resulting particles as having a core–shell structure. The magnetic and optical properties of these materials were demonstrated, and the cultured cell which absorbed the nanoparticle could easily be imaged by using MRI. Subsequent toxicology studies revealed no problems associated with the repeated injection of the nanomaterial, which suggested a potential advantage of this ferromagnetic nanomaterial-related MRI contrast agent when compared to conventional gadolinium-based contrast agents. In addition, useful images could be obtained by using only 10% of the metal dose required when employing gadolinium-based contrast agents. When the material was injected into rabbits, the image could be maintained for 20 min. Moreover, the investigators predicted that they would be able to target these ferromagnetic nanomaterials to tumors, thus providing an ability not only to image cancer cells using MRI, but also to treat cancer using near-infrared light-activated thermal therapy that was capable of travelling several centimeters through biological tissues.

Recently, the clear imaging by an Fe/Co core carbon graphitic shell contrast agent has been reported [37, 38]. Both, *in vitro* and *in vivo* imaging provide excellent images of high quality, enabling the effective study of infectious tissues. The MR of graphite has also found application in biosensors, with boronated graphite specimens with negative MR properties having been reported. This material was prepared from a grafoil heat-treated at 3000 °C for 30 min [39]. The transport properties, such as electrical resistivity, Hall coefficient and transverse MR, and total magnetic susceptibility, were monitored for this specimen and compared with the data for pristine graphite. For the specimen with a high boron concentration, a negative transverse MR at liquid nitrogen temperature and a weakly temperature-dependent resistivity were observed [40].

5.5
Magnetism in Carbon Nanotubes/Fullerenes

To date, all discussions have focused on the use of allotropic forms of carbon, diamond, and graphite as bulk materials. Unlike these bulk forms of carbon, 1-D carbon materials (e.g., CNTs/fullerenes, graphene) offer better chances of magnetism due to the easy restructuring of C–C bonds; moreover, these particles have sizes in the nanometer range.

Interest in research on nanoscale magnetic materials is, therefore, steadily increasing: nanostructured magnetic materials exhibit new and interesting physical properties, which cannot be found in bulk. The nanosized magnetic material has great potential for technical applications, such as MR sensors, biosensors, and drug delivery. In the following subsection the properties of magnetic CNTs/fullerenes are addressed in detail, especially with regards to their fabrication and characterization, and the physics underlying the unique properties of these structures. The presence of ferromagnetism in CNTs has been verified both experimentally [41] and theoretically [42]. However, the specific source of the magnetism observed in these experiments remains the subject of debate, with several possible alternatives.

The chances of ferromagnetism occurring in CNTs are high due to their extraordinary structure. As a result of their nanoscaled size and cylindrical symmetry, CNTs have stimulated several studies to investigate their electronic and magnetic properties [43]. A CNT can be regarded as one large carbon molecule which is formed by folding single-layer graphite into a cylinder; typically, the diameter of a CNT can be measured in nanometers, while its length can reach macroscopic dimensions [44]. The high aspect ratio in CNTs determines their exceptionally high Young's modulus [45], whereas the CNT's electrical conductivity depends heavily on the diameter and the helicity, which is the angle between the most highly packed chains of atoms and the axis of the cylinder [46]. Two types of CNT exist: (i) those with walls, which contain a single layer of carbon atoms, termed single-walled CNTs (SWNTs) [47]; and (ii) nanotubes with walls consisting of several concentric cylindrical graphite layers, termed multi-walled CNTs (MWNTs) [48]. The hexagonal lattice structure of CNTs gives rise to three types of SWNT, the diameters of which vary between 0.4 and 4 nm. From theoretical predictions, the curvature effect and the misorientation of $2p_z$ orbitals on a cylindrical surface play an important role in the electronic structures [49]. Today, scanning tunneling microscopy (STM) is an important tool for the study of energy band gaps in materials. Indeed, the effects of curvature effect and of the critical magnetic field on the magnetization of CNTs have recently been studied in great detail [50]. Another possibility for magnetism in CNTs is generated by the presence of defects. Although CNTs are generally treated as defect-free due to the highly stable bonds in carbon formation, experimental observations [51] have revealed the presence of defects in CNTs. Such defects, whether introduced during the synthetic process or stress-induced, can be divided into three categories [52], namely: (i) Stone–Wales (SW) defects; (ii) defects induced by change of a C–C bond; and (iii) vacancy defects formed from dangling bonds. According to a recent study conducted by Ajayan et al. [53], an ideal single vacancy with three dangling bonds is unstable. It is also possible to form a heptagonal structure by a combination of two from three dangling bonds.

Magnetic ordering due to defects in fullerenes is confirmed by inducing pressure and irradiating with protons [54]. However, experimental evidence of defect-induced magnetism in CNT samples is still lacking due to the presence of metal catalysts. Ferromagnetism associated with dangling bonds is shown by calculations on open-ended zigzag nanotubes; ferromagnetic ordering associated with dangling bonds at zigzag edges [55] has been observed. In this case, first-principle calculations were used for the study of vacancies [21], and ferromagnetism associated with CNTs having a single vacancy in the range 4–8 was confirmed. Theoretical studies [56] have also shown that the intrinsic ferromagnetism of CNTs could be induced by single-walled defects. The open zigzag boron nitride (BN) nanotubes (NTs) also exhibited a ferromagnetism which might be due to stable unpaired electrons at the openings [57]. These BN NTs [58, 59] are similar in structure to CNTs, where B and N are substituted for the entire carbon atoms in the CNTs.

This type of structure leads to the possibility of fabricating a nanoscale, spin-polarized field emitter or a STM probe. Recent calculations have shown that the defects created by carbon atoms on CNTs have a magnetic moment [60, 61], and this is supported by experiments which suggest the presence of localized spin moments in CNTs [62], most likely related to the dangling bonds on the surface. In order to understand how the adatom–vacancy defect pair can contribute globally to the magnetism of carbon systems, it is necessary to consider the properties of vacancies in graphite and CNTs. The magnitude of the adatom magnetic moment is dependent on the radii and chiralities of the CNTs [63].

Magnetism can be induced in CNTs/fullerenes by using a variety of methods, including doping with non-magnetic atoms, hydrogen bonding, transition metal contacting/filling, ion implantation, or by physically damaging the CNTs (e.g., with acid treatment or ultrasonication) so as to create defects and vacancies [64–66]. Magnetism in CNTs, created by hydrogen doping, has been reported by Yang et al. [67]. Such magnetism can also be induced in BN NTs by doping with carbon atoms into the lattice during synthesis [59], and extra charges would be injected synchronously into the BN system. In contrast, previous experimental and theoretical studies have shown that the localized sharp resonant valence band states, as induced by topological defects at the tube tips, play the primary role in the field electron emission of CNTs [68], with electrons being emitted from the tips of open CNTs [69]. It is to be expected that the introduction of C atoms into the mouths of open armchair (BN) NTs would induce their local ferromagnetism, or cause them to be magnetized. This effect is due to a structural transformation from the original closed-shell system [i.e., open armchair (BN) NTs] to the open-shell system (i.e., C-doped NTs). Magnetoresistance [70] is a significant property developed by the magnetization of CNTs/fullerenes. The main advantage of CNTs and fullerenes is their generation of giant (G) MR or tunnel (T) MR, due to the 1-D structure.

Notably, MR has a major role in biomedicine (see Section 5.5.1 for a detailed description of applications). Recently, a group of investigators [71] described the spin field-effect transistor in CNT systems, whereby they could control the spin-polarized current through a CNT by using a third gate electrode that was capacitively coupled to the CNT. The main experimental findings here were that:

- the MR oscillated as a function of the gate voltage, just as with linear conductance;
- the MR was low near the conductance peak, but high in the conductance valleys;
- the MR in some samples may even have become negative near the conductance peak.

These findings can be explained by spin-polarized resonant tunneling through the discrete energy levels formed in a finite CNT.

In recent study conducted by the present authors, MR oscillations were demonstrated in nickel-coated CNTs, and observed to be a function of the gate voltage [72]. The existence of both coulomb blockade and a coulomb staircase at room temperature in MWNT-based heterostructures has also been reported (Figure 5.1).

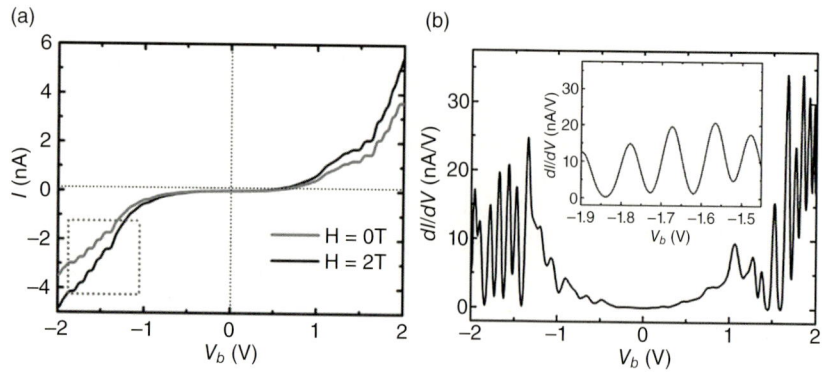

Figure 5.1 (a) Coulomb blockade and coulomb staircase displayed in the nickel-coated MWNTs; (b) dI/dV versus V_b plot of the nickel-coated MWNT [72].

5.5.1
Biomedical Applications of Magnetic Carbon Nanotubes/Fullerenes

There are several potential applications of CNTs that could be of major interest in the field of the life sciences; these include gas sensors, biosensors, targeted drug delivery, and tissue engineering. Magnetic CNTs, however, mainly find application in MRI, drug delivery, and biosensors [73]. The CNTs can be rendered magnetic in many ways for these applications, including either coating/contacting or filling with ferromagnetic material. The present authors have successfully filled CNTs with nickel and cobalt [74, 75], such that the enhanced ferromagnetic property obtained can be utilized in medical imaging. The main advantage of these CNTs is the thermal and structural stability of the ferromagnetic material due to the protective coating of carbon; such materials may find application, for instance, in MRI.

The ferromagnetic-coated/contacted CNTs represent other interesting structures that can be applied as biosensors. The high percentage of MR, along with specific MR features, can be exploited in the development of biosensors [76]. For example, tunnel magnetoresistance (TMR) is an MR effect that occurs in magnetic tunnel junctions (MTJs). These devices normally consist of two ferromagnetic materials separated by a non-ferromagnetic material, and the phenomenon occurs as electrons tunnel from one ferromagnetic material to other, through the non-ferromagnetic material. This process cannot be predicted by classical physics – it is strictly a quantum phenomenon. It is also preferable to have very thin non-ferromagnetic layer. The effect varies from material to material, with the choice of a perfect material being highly important to achieve a maximum degree of TMR. Although the TMR effect was originally discovered several decades ago, in 1975, by Jullière in Fe/Ge-O/Co junctions at 4.2 K [77], the percentage TMR obtained at room temperature was less than 1%. However, a variety of materials was later introduced with enhanced TMR values. Half-metallic ferromagnets, which are metallic for the majority-spin state and semiconducting with an energy gap for the minority-spin state, leading to 100% spin polarization at the Fermi level, represent one of the key materials for obtaining huge TMR. Indeed, today TMR effects of up to 6000 at room temperature are frequently observed in junctions of the Fe/MgO/Fe system [78].

CNTs are known for their one-dimensional nature, and many reports have been made which prove single-electron tunneling (SET) through CNTs. Although SET would be expected to give a high TMR% in CNTs, the difficulty with a CNT is the entangled nature that develops during its growth, unless it has been specially treated for the growth of isolated CNTs. The present authors have used a novel technique [79] – termed "double plasma-enhanced hot filament chemical vapor deposition" for the deposition of isolated and vertically aligned CNTs on a desired substrate, and have recently reported the MR property in nickel-coated, vertically aligned CNTs. The MR property of the sample was confirmed by low field measurements which were performed from −100 to 100 Oe and back to −100 Oe. The resistance peak appeared as the field moved through zero, and showed a change in resistance from the parallel to anti-parallel alignment of the magnetizations.

Information on the TMR% can also be obtained from the current–voltage (I–V) characteristics of the sample. A coulomb blockade (CB) region was observed between −1 V to 1 V, and a stepwise increment of the current (CS) was observed beyond the CB region. The (I–V_b) curves exhibited similar features at zero and applied magnetic field; the similarity in (I–V_b) at zero and applied field has been reported [80]. The oscillation of MR as a function of bias voltage has been observed. Typically, the sample without laser chemical vapor deposition (LCVD) also shows no CS. The approximately null current in the CB region, and the CS beyond the CB region, is a clear indication of the SET phenomenon which, together with the associated MR, occur only if the transport of electrons from one electrode to another is inhibited due to the extremely high electrostatic energy ($e^2/2C$, where e is the charge of the electron and C is the capacitance) of a single electron compared to the thermal energy, k_BT. When the bias voltage increases and exceeds the threshold ($V^{th} = e/2C$), the current begins to increase. If the resistance of two junctions is similar ($R1 \approx R2$), then the current is increased smoothly with a bias voltage. The MR effect can be understood in terms of Julliére's model [77], which has been used recently to explain the MR in the ferromagnetic tunnel junction (FTJ) [81, 82]. The ferromagnetic nanomaterial, when coated onto CNTs, has been applied to the development of magnetic force microscopy (MFM) tips [83]. One area of interest would be to use this tip as a highly sensitive, single molecule for studying weak biomolecular forces, including those of DNA [76]. In addition to biosensor applications, ferromagnetic nanomaterial-coated CNTs have found application in drug-delivery systems, as they can be conjugated with specific antibodies and then used for the labeling and sorting of cells. The possibility of applying ferromagnetic nanoparticles with diameters of 10–100 nm for potential drug- or gene-delivery purposes has also been reported recently [84]. In this case, the larger the magnetic particle, the greater was the magnetic moment in the magnetic field, and the stronger the force that could be exerted against the blood flow when delivering the pharmaceutical tag. Notably, magnetic nanoparticles designed for drug delivery – a process termed "magnetofection" – must also be nontoxic and completely biocompatible. Magnetofection offers great potential for transporting drugs directly to appropriate cells, along a magnetic field gradient. Moreover, the technique also provides a rapid and highly efficient system of drug delivery.

Ferromagnetism may also be induced in CNTs by filling the hollow CNTs with ferromagnetic material, with such CNTs applications range from individual filled tubes in sensors for magnetic scanning probe microscopy, to contrast agents in MRI. The identification of synthesis routes to produce CNTs filled with magnetic materials, and their consequent extraordinary magnetism, has opened a new link between magnetism and CNTs. The main advantage of ferromagnetic encapsulation is that the CNTs effectively protect the ferromagnetic material against oxidation, such that their long-term stability is enhanced. Since Ajayan *et al.* [85] first succeeded in opening and filling CNTs with molten Pb, several attempts have been made to fill CNTs with a variety of metals or oxides. Nanocapillarity methods for filling CNTs with molten silver nitrate [86] and with sulfur, selenium, and cesium, have also been demonstrated. To date, two main methods have been used to prepare transition metal-filled CNTs: (i) an *in situ* synthesis during CNT growth [87]; and (ii) template formation based on mesoporous alumina [88].

The *in situ* process generally involves the pyrolysis of a transition metal powder, with or without an additional carbon source, such as acetylene or C_{60} [89]. Unfortunately, it is difficult to exert a precise control on the sublimation rate of the ferrocene powder by simply adjusting the preheating temperature in the cold zone, and consequently this technique normally results in a low Fe filling rate and thick carbon coatings – that is, a high C:Fe ratio.

The template process requires the preparation of CNTs in porous alumina by catalytic pyrolysis, prior to the transition metal being filled into the CNT cavities by other means [90]. Because this technique is complicated, and is unsuited to continuous filling, a simple and controllable process is required for the production of transition metal-filled CNTs with a high and continuous filling rate and a low C:Fe atomic ratio. CNTs filled with transition metals (Fe, Co, and Ni) have been synthesized experimentally using chemical vapor deposition (CVD), with methane as a precursor [74, 75]. Recent experiments have also shown that the positions of SWNTs can be controlled by CVD, using Ar-ion laser irradiation as a source of heat [91]. SWNTs filled with Fe wires have also been synthesized experimentally, based on the wet chemistry process [92].

The filling of ferromagnetic materials into CNTs has been demonstrated using arc discharge [93], capillary infiltration, the electrolysis of molten salt, and CVD [74]. The latter approach represents an efficient method for the *in situ* growth of ferromagnetic-filled CNTs, with the critical parameters for growth including control of the precursor gas (methane, hydrogen) and the growth temperature.

The present authors have demonstrated nickel and cobalt filling of CNTs, using the CVD technique.

An increased coercivity of the encapsulated transition metal nanowires was also observed [74], in agreement with the results of Jang et al. [94], with the first iron-filled SWNTs being obtained by Cheng et al. [95]. A ferromagnetic behavior at room temperature, with a hysteresis loop and a large magnetic anisotropy, was observed. The physical properties obtained by the introduction of metals into MWNTs were shown to depend on the character of the interatomic bonds between the carbon and the metal, the type of metal crystal structure which can be imposed by compression forces, and the symmetry of the carbon.

One major difficulty in filling SWNTs is very small diameter of the tube, which is close to 1 nm. However, the problem is more easily overcome in the case of MWNTs, where the tube diameter may range from 10 to 30 nm. As an example, ferrite nanowires have been synthesized by the vigorous stirring of an aqueous solution, containing cobalt and iron nitrates together with MWNTs [96], while a chemical method of filling CNTs with different magnetic materials has been described by Seifu et al. [97].

CNTs filled with iron atoms and nanowires are of particular interest, because of the possible modification of electronic and magnetic properties, and their application to biomedicine [98]. Notably, iron nanowires are protected from oxidation, and the structure is more stable when placed inside CNTs. The results or previous studies have shown that, among the transition metals (iron, cobalt, nickel) placed on Cu(111), Fe exhibited the largest magnetic moment [99]. The stable structures and magnetic properties of Fe-filled SWNTs on Ni(1 1 1) and Cu(1 1 1) were investigated based on first-principles calculations [100]. The SWNT was transformed into an arch-like structure when the iron atom was placed inside, close to the Ni surface; a similar situation occurred when the iron-filled SWNT was placed on the Cu surface. To explain this, the authors have proposed that, when the Fe-filled (3,3) chiral SWNT was adsorbed onto the Ni surface, the C–C bonds of the CNT were broken by the Fe wire and the metal surface, leading to the arch-like structure. Iron nanowires have also been encapsulated in CNTs using pyrolysis [101]; in this case, when the Fe nanowire was placed inside the (3,3) chiral SWNT, the original metallic characteristics of the SWNT were changed into those of a semiconductor. The stable structure and magnetic moment of an Fe-filled SWNT, when placed on Ni(111), were seen to depend heavily on the position of the Fe atom in the SWNT. The magnetic and electronic properties of iron-filled CNTs were also found to differ with the varying CNT diameters [102].

The present authors have noted differences between the ferromagnetic properties of magnetic material-filled and magnetic material-contacted CNTs. Hence, the coercivity of a nickel nanorod [74] was shown to be less than that of nickel particles (Figure 5.2) [72]. It has been reported that iron nanoparticles encapsulated within the carbon structure can induce constraints on the use of nanotubes for future device applications. Examples of this include CNT-doped conductive polymer nanocomposites [103], biosensors, actuators, and molecular electronics [104].

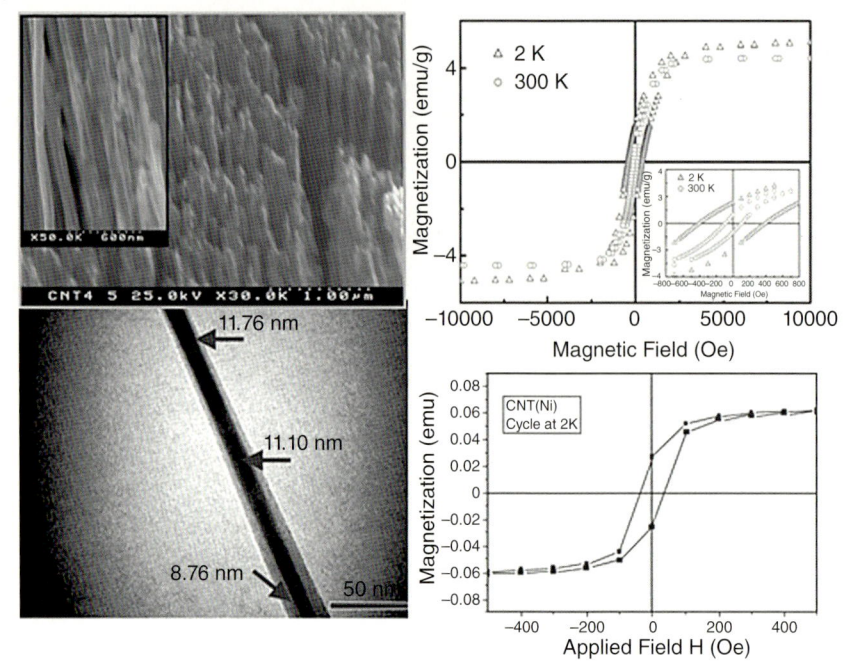

Figure 5.2 Superconducting quantum interference device (SQUID) analysis (magnetic properties) of nickel-contacted (top) and nickel-filled (bottom) CNTs [72, 74].

5.6
Magnetism in Graphene

The magnetic properties of vacancies in graphene have been described in detail by Ma et al. [105]. As with CNTs, the defects, vacancies and zigzag edges in graphene may be responsible for an intrinsic magnetism. According to Hjort et al. [106], by using the Hückel method it can be shown that the three symmetric atoms which neighbor the vacancy on a graphite surface can contribute an extra π electron to the system, and this will give rise to an unpaired spin; in other words, the vacancy would be magnetic.

Other types of defect, such as holes and cracks, can also lead to magnetism. In each of these cases the disruption of graphene π-bonds will lead to the appearance of a density of states (DOS) at the Fermi level. Electron–electron exchange interactions would lead to the polarization of these states, thus inducing an itinerant magnetism. More recently, DFT calculations have been considered the best approach to studying the vacancy in graphene, and due to an enhanced interest in graphene and other nanomaterials, many quantum chemical analyses of graphene have been conducted, mostly using the DFT approach [107]. An intrinsic magnetism in graphene has been reported in several different situations. In most cases, the magnetism was observed in graphene with zigzag edges, whether as nanodots [108] or nanoribbons [109]. Subsequently, Zheng et al. [110] reported the influence of atom/molecular attachment (hydrogen, fluorine, and oxygen atoms) to the graphene zigzag edges. The magnetic moment, spin density and energy gap were all greatly influenced by the quantity of molecules attached to the zigzag edges of graphene. Notably, these types of structure might have potential application in sensors, as the spin of the electrons can be manipulated conveniently in 2-D structures so as to generate magnetism. The π-bonds can be disrupted by the addition of dopants, or by ion implantation [111]. Substrate-induced magnetism in epitaxial graphene buffer layers has been reported by Ramasubramanyam et al. [112]. However, unlike other studies on exfoliated graphene sheets [113], attention here was focused on graphene deposited on a silicon carbide substrate (after annealing 4H- or 6H-SiC) by epitaxial growth. In this case, the adatoms were chosen to be in the thermodynamically favored T4 site [114]. Whilst other adatom reconstructions (e.g., Si tetramers [115]) might indeed occur at the interface, several general features of magnetism in epitaxial graphene can be inferred from the T4 adatom model, as well as the bulk-terminated model.

5.6.1
Biomedical Applications of Magnetic Graphene

The possible existence of room-temperature ferromagnetism in graphene, which is attributed to the edge defects and functional groups attached to graphene [109], has aroused immense interest in biomedicine. The 2-D and honeycomb lattice of sp^2-bonded carbon atoms are favorable in graphene for the manipulation of magnetism. Recently, it has been shown that single-layer graphene can be cut lithographically, and that the edges so-formed will generate magnetism [115]. It has also been shown how the spin is aligned along the edges of graphene. Graphene has also been converted into graphone by hydrogen coverage, such that the new material is ferromagnetic [116].

Graphene has emerged as a versatile material with outstanding ferromagnetic properties that could prove useful in biomedicine. A very large value of magnetic resonance has been recently predicted in graphene [117]. Notably, it is has been suggested that graphene can utilize spin in addition to charges; moreover, the potential for long spin relaxation times would be advantageous when fabricating magnetic resonance devices for application in medical imaging.

Graphene is also an excellent material for use as a biosensor. Indeed, a recent report described the ability of DNA to turn a fluorescent light switch on and off when close to graphene that was being used to create a biosensor. Other possible applications for a DNA–graphene biosensor include the diagnosis of diseases such as cancer, the detection of toxins in tainted food and the detection of pathogens from biological weapons. The use of graphene–DNA structures has also been proposed for drug delivery in gene therapy, due to the high stability of the complex.

As the properties of graphene are greatly influenced by its single-layer structure, the method used to prepare graphene will be critical in order to control the graphene layers. In addition, an in-depth structural analysis of graphene sheets is highly desirable if this material is to be exploited in the field of nanoelectronics.

5.7
Conclusion

Nanostructured magnetic carbon materials have the potential to significantly impact the biomedical area. Although magnetism is one of the oldest fields of science, today it is at the forefront of the nanotechnology revolution. The past decade has witnessed great interest in the magnetism of structures at nanoscale dimensionality, and investigators worldwide have begun to study the ferromagnetism of nanomaterials by inducing magnetism into allotropic forms of carbon. Research into ferromagnetic nanomaterials is not only important for studying fundamental magnetic phenomena, but also is essential for next-generation biomedicine. In this respect, the most important areas include the quest for magnetic resonance devices, drug-delivery systems, and biosensors.

References

1 Miller, J.S. (2001) Unsolved mysteries in molecule-based magnets – a personal view. *Polyhedron*, **20**, 1723–5.
2 Takai, K., Oga, M., Enoki, T. and Taomoto, A. (2004) Effect of heat-treatment on magnetic properties of non-graphitic disordered carbon. *Diamond and Related Materials*, **13**, 1469–73.
3 Lehtinen, P.O., Foster, A.S., Ayuela, A., Vehvilaèinen, T.T. and Nieminen, R.M. (2004) Structure and magnetic properties of adatoms on carbon nanotubes. *Physical Review B*, **69**, 155422–7.
4 Ma, Y., Lehtinen, P.O., Foster, A.S. and Nieminen, R.M. (2004) Magnetic properties of vacancies in graphene and single-walled carbon nanotubes. *New Journal of Physics*, **68**, 78675–84.
5 Shtogun, Y.V. and Woods, L.M. (2009) Electronic and magnetic properties of deformed and defective single wall carbon nanotubes. *Carbon*, **47**, 3252–62.
6 Rocha, A.R., Padilha, J.E., Fazzio, A. and da Silva, A.J.R. (2008) Transport properties of single vacancies in nanotubes. *Physical Review B*, **77**, 153406–10.
7 Shan, B., Lakatos, G.W., Peng, S. and Cho, K. (2005) First-principles study of band-gap change in deformed nanotubes. *Applied Physics Letters*, **87**, 73109–173112.
8 Tsetseris, L. and Pantelides, S.T. (2009) Adatom complexes and self-healing mechanisms on graphene and single-wall carbon nanotubes. *Carbon*, **47**, 901–8.
9 Latham, C.D., Heggie, M.I., Gamez, J.A., Suarez-Martinez, I., Ewels, C.P. and Briddon, P.R. (2008) The di-interstitial in graphite. *Journal of Physics Condensed Matter*, **20**, 395220–8.
10 Hod, O. and Scuseria, G.E. (2008) Half-metallic zigzag carbon nanotube dots. *ACS Nano*, **2**, 2243–9.
11 Talapatra, S., Ganesan, P.G., Kim, T., Vajtai, R., Huang, M., Shima, M., Ramanath, G., Srivastava, D., Deevi, S.C. and Ajayan, P.M. (2005) Irradiation-induced magnetism in carbon nanostructures. *Physical Review Letters*, **95**, 097201.
12 Rode, A.V., Elliman, R.G., Gamaly, E.G. and Veinger, A.I. (2002) Electronic and magnetic properties of carbon nanofoam

produced by high-repetition-rate laser ablation. *Applied Surface Science*, **197–198**, 644–9.
13. Goolaup, S., Adeyeye, A.O. and Singh, N. (2005) Magnetic properties of diamond-shaped $Ni_{80}Fe_{20}$ nanomagnets. *Journal of Physics D: Applied Physics*, **38**, 2749–54.
14. Haddon, R.C. (1995) Magnetism of the carbon allotropes. *Nature*, **378**, 249–55.
15. Butler, J.E., Geis, M.W., Krohn, K.E., Lawless, J., Jr, Deneault, S., Lyszczarz, T.M., Flechtner, D. and Wright, R. (2003) Exceptionally high voltage Schottky diamond diodes and low boron doping. *Semiconductor Science and Technology*, **18**, S67–S71.
16. Mizuochia, N., Ogurab, M., Watanabeb, H., Isoyaa, J., Okushib, H. and Yamasaki, S. (2004) EPR study of hydrogen-related defects in boron-doped p-type CVD homoepitaxial diamond films. *Diamond and Related Materials*, **13**, 2096–9.
17. Yokoya, T., Nakamura, T., Matsushita, T., Muro, T., Takano, Y., Nagao, M., Tanouchi, T., Kawarada, H. and Oguchi, T. (2005) Origin of the metallic properties of heavily boron-doped superconducting diamond. *Nature*, **438**, 647–50.
18. Manivannan, A., Underwood, S., Morales, E.H. and Seehra, M.S. (2003) Magnetic and electrical characterization of heavily boron-doped diamond. *Materials Characterization*, **51**, 329–33.
19. Tshepe, T., Kasl, C., Prins, J.F. and Hoch, M.J.R. (2004) Metal-insulator transition in boron-ion-implanted diamond. *Physical Review B*, **70**, 245107–14.
20. Ekimov, E.A., Sidorov, V.A., Bauer, E.D., Mel'nik, N.N., Curro, N.J. and Thompson, J.D. (2004) Superconductivity in diamond. *Nature*, **428**, 542–5.
21. Belavin, V.V., Bulusheva, L.G. and Okotrub, A.V. (2004) Modifications to the electronic structure of carbon nanotubes with symmetric and random vacancies. *International Journal of Quantum Chemistry*, **96**, 239–46.
22. Wang, W.L., Liāo, K.J. and Wang, B.B. (2000) Magnetoresistance effect of p-type diamond films in various doping levels at different temperatures. *Diamond and Related Materials*, **9**, 1612–16.
23. Fei, Y.J., Yang, D., Wang, X., Menga, Q.B., Wang, X., Xiong, Y.Y., Nie, Y.X. and Feng, K.A. (2002) Magnetoresistance of boron-doped chemical vapor deposition polycrystalline diamond films. *Diamond and Related Materials*, **11**, 49–52.
24. Choy, T.C., Stoneham, A.M., Ortuñoa, M. and Somoza, A.M. (1972) Negative magnetoresistance in ultrananocrystalline diamond: strong or weak localization. *Applied Physics Letters*, **92**, 012120–3.
25. Russel, K.J. and Leivo, W.J. (1972) Hi-field magnetoresistance of semiconducting diamond. *Journal of Applied Physics*, **6**, 4588–94.
26. Willems, B.L., Zhang, G., Vanacken, J., Moshchalkov, V.V., Janssens, S.D., Williams, O.A., Haenen, K. and Wagner, P. (2009) Negative magnetoresistance in boron-doped nanocrystalline diamond films. *Journal of Applied Physics*, **106**, 033711–16.
27. Yang, X., Xia, H., Qinc, X., Lia, W., Dai, Y., Liu, X., Zhao, M., Xia, Y., Yan, S. and Wang, B. (2009) Correlation between the vacancy defects and ferromagnetism in graphite. *Carbon*, **47**, 1399–406.
28. Esquinazi, P., Spemann, D., Hohne, R., Setzer, A., Han, K.-H. and Butz, T. (2003) Induced magnetic ordering by proton irradiation in graphite. *Physical Review Letters*, **91**, 227201–5.
29. Spemann, D., Han, K.H., Esquinazi, P., Hohne, R. and Butz, T. (2004) Ferromagnetic microstructures in highly oriented pyrolytic graphite created by high energy proton irradiation. *Nuclear Instruments and Methods in Physics Research B*, **219–220**, 886–90.
30. Pardo, H., Faccio, R., Araujo-Moreira, F.M., de Lima, O.F. and Mombru, A.W. (2006) Synthesis and characterization of stable room temperature bulk ferromagnetic graphite. *Carbon*, **44**, 565–9.
31. Xia, H., Li, W., Song, Y., Yang, X., Liu, X., Zhao, M., Xia, Y., Song, C., Wang, T.-W., Zhu, D., Gong, J. and Zhu, Z. (2008) Tunable magnetism in carbon-ion-implanted highly oriented

pyrolytic graphite. *Advanced Materials*, **20**, 4679–83.

32 Lehtinen, P.O., Foster, A.S., Ma, Y., Krasheninnikov, A.V. and Nieminen, R.M. (2004) Irradiation-induced magnetism in graphite: a density functional study. *Physical Review Letters*, **93**, 187202–6.

33 Luo, W., Xie, Y., Wu, C. and Zheng, F. (2008) Spherical CoS_2@carbon core–shell nanoparticles: one-pot synthesis and Li storage property. *Nanotechnology*, **19**, 075602–8.

34 Costa-Fernandez, J.M., Pereiro, R. and Sanz-Medel, A. (2006) The use of luminescent quantum dots for optical sensing. *Trends in Analytical Chemistry*, **25**, 207–18.

35 Bönnemann, H. and Richards, R.M. (2001) Nanoscopic metal particles. 2 Synthetic methods and potential applications. *European Journal of Inorganic Chemistry*, **1434**, 2455–80.

36 He, C.N., Du, X.W., Ding, J., Shi, C.S., Li, J.J., Zhao, N.Q. and Cui, L. (2006) Low-temperature CVD synthesis of carbon-encapsulated magnetic Ni nanoparticles with a narrow distribution of diameters. *Carbon*, **44**, 2330–56.

37 Seo, W.S., Lee, J.H., Sun, X., Suzuki, Y., Mann, D., Liu, Z., Terashima, M., Yang, P.C., McConnell, M.V., Nishimura, D.G. and Dai, H. (2006) FeCo/graphitic-shell nanocrystals as advanced magnetic-resonance-imaging and near-infrared agents. *Nature Materials*, **5**, 971–6.

38 Pol, S.V., Frydman, A., Churilov, G.N. and Gedanken, A. (2005) Fabrication and magnetic properties of Ni nanospheres encapsulated in a fullerene-like carbon. *Journal of Physical Chemistry B*, **109**, 9495–8.

39 Sugihara, K., Hishiyama, Y. and Kaburagi, Y. (2000) Electronic and transport properties of boronated graphite – 3D-weak localization effect. *Molecular Crystals and Liquid Crystals Science and Technology*, **340**, 367–71.

40 Hishiyama, Y., Kaburagi, Y. and Sugihara, K. (2000) Negative magnetoresistance and magnetic susceptibility of boronated graphite. *Molecular Crystals and Liquid Crystals Science and Technology*, **340**, 337–42.

41 Vo, T., Wu, Y.D., Car, R. and Robert, M. (2008) Structures, interactions, and ferromagnetism of Fe-carbon nanotube systems. *Journal of Physical Chemistry C*, **112**, 8400–7.

42 Mattis, D.C. (2005) Theory of ferromagnetism in carbon foam. *Physical Review B*, **71**, 144424–9.

43 Kane, A.A., Sheps, T., Branigan, E.T., Apkarian, V.A., Cheng, H.M., Hemminger, J.C., Hunt, S.R. and Collins, P.G. (2009) Graphitic electrical contacts to metallic single-walled carbon nanotubes using Pt electrodes. *Nano Letters*, **9**, 3586–91.

44 Zheng, L.X., O'Connell, M.J., Doorn, S.K., Liao, X.Z., Zhao, Y.H., Akhadov, E.A., Hoffbauer, M.A., Roop, B.J., Jia, Q.X., Dye, R.C., Peterson, D.E., Huang, S.M., Liu, J. and Zhu, Y.T. (2004) Ultralong single-wall carbon nanotubes. *Nature Materials*, **3**, 673–6.

45 Treacy, M.M.J., Ebbesen, T.W. and Gibson, J.M. (1996) Exceptionally high Young's modulus observed for individual carbon nanotubes. *Nature*, **381**, 678–80.

46 Qin, L.C., Iijima, S., Kataura, H., Maniwa, Y., Suzuki, S. and Achiba, Y. (1997) Helicity and packing of single-walled carbon nanotubes studied by electron nanodiffraction. *Chemical Physics Letters*, **268**, 101–6.

47 Harutyunyan, A.R., Chen, G., Paronyan, T.M., Pigos, E.M., Kuznetsov, O.A., Hewaparakrama, K., Kim, S.M., Zakharov, D., Stach, E.A. and Sumanasekera, G.U. (2009) Preferential growth of single-walled carbon nanotubes with metallic conductivity. *Science*, **326**, 116–20.

48 Crespi, V.H. (1999) Local temperature during the growth of multiwalled carbon nanotubes. *Physical Review Letters*, **82**, 2908–10.

49 Gulseren, O., Yildirim, T. and Ciraci, S. (2002) Systematic ab initio study of curvature effects in carbon nanotubes. *Physical Review B*, **65**, 153405–9.

50 Tsai, C.C., Chen, S.C., Shyu, F.L., Chang, C.P. and Lin, M.F. (2005) Magnetization of carbon nanotubes. *Physica E*, **30**, 86–92.

51. Osváth, Z., Vértesy, G., Tapasztó, L., Wéber, F., Horváth, Z.E., Gyulai, J. and Biró, L.P. (2005) Atomically resolved STM images of carbon nanotube defects produced by Ar+ irradiation. *Physical Review B*, **72**, 045429–35.

52. Zhou, L.G. and Shi, S.Q. (2003) Adsorption of foreign atoms on Stone-Wales defects in carbon nanotube. *Carbon*, **41**, 579–625.

53. Zhang, Y., Talapatra, S., Kar, S., Vajtai, R., Nayak, S.K. and Ajayan, P.M. (2007) First-principles study of defect-induced magnetism in carbon. *Physical Review Letters*, **99**, 107201–5.

54. Esquinazi, P., Hohne, R., Han, K.H., Setzer, A., Spemann, D. and Butz, T. (2004) Magnetic carbon: explicit evidence of ferromagnetism induced by proton irradiation. *Carbon*, **42**, 1213–18.

55. Kim, Y.H., Choi, J. and Chang, K.J. (2003) Defective fullerenes and nanotubes as molecular magnets: an ab initio study. *Physical Review B*, **68**, 125420–4.

56. Carpio, A., Bonilla, L.L., de Juan, F. and Vozmediano, M.A.H. (2008) Dislocations in graphene. *New Journal of Physics*, **10**, 053021–34.

57. Moradian, R. and Azadi, S. (2008) Magnetism in defected single-walled boron nitride nanotubes. *Europhysics Letters*, **83**, 17007–13.

58. Golberg, D., Bando, Y., Tang, C. and Zhi, C. (2007) Boron nitride nanotubes. *Advanced Materials*, **19**, 2413–32.

59. Zhou, G. and Duan, W. (2007) Spin-polarized electron current from carbon-doped open armchair boron nitride nanotubes: implication for nano-spintronic devices. *Chemical Physics Letters*, **437**, 83–6.

60. Krasheninnikov, A.V., Nordlund, K., Lehtinen, P.O., Foster, A.S., Ayuela, A. and Nieminen, R.M. (2004) Adsorption and migration of carbon adatoms on zigzag carbon nanotubes. *Carbon*, **42**, 1021–5.

61. Liu, H. (2006) Induced magnetic moment in defected single-walled carbon nanotubes. *Journal of Physics: Conference Series*, **29**, 194–7.

62. Berber, S. and Oshiyama, A. (2006) Reconstruction of mono-vacancies in carbon nanotubes: atomic relaxation vs. spin polarization. *Physica B*, **376–377**, 272–5.

63. Liu, C.P. and Xu, N. (2008) Magnetic response of chiral carbon nanotori: the dependence of torus radius. *Physica B*, **403**, 2884–7.

64. Ma, Y., Lehtinen, P.O., Foster, A.S. and Nieminen, R.M. (2005) Hydrogen-induced magnetism in carbon nanotubes. *Physical Review B*, **72**, 085451–7.

65. Curiale, J., Sánchez, R.D., Troiani, H.E., Ramos, C.A., Pastoriza, H., Leyva, A.G. and Levy, P. (2007) Magnetism of manganite nanotubes constituted by assembled nanoparticles. *Physical Review B*, **75**, 224410–19.

66. Krusin-Elbaum, L., Newns, D.M., Zeng, H., Derycke, V., Sun, J.Z. and Sandstrom, R. (2004) Room-temperature ferromagnetic nanotubes controlled by electron or hole doping. *Nature*, **431**, 672–6.

67. Yang, X. and Wu, G. (2009) Itinerant flat-band magnetism in hydrogenated carbon nanotubes. *ACS Nano*, **7**, 1646–50.

68. Carroll, D.L., Redlich, P. and Ajayan, P.M. (1997) Electronic structure and localized states at carbon nanotube tips. *Physical Review Letters*, **78**, 2811–14.

69. Hao, S., Zhou, G., Duan, W., Wu, J. and Gu, B.L. (2006) Tremendous spin-splitting effects in open boron nitride nanotubes: application to nanoscale spintronic devices. *Journal of American Chemical Society*, **128**, 8453–8.

70. Roche, S. and Saito, R. (2001) Magnetoresistance of carbon nanotubes: from molecular to mesoscopic fingerprints. *Physical Review Letters*, **87**, 246803–7.

71. Son, Y.W., Cohen, M.L. and Louie, S.G. (2007) Electric field effects on spin transport in defective metallic carbon nanotubes. *Nano Letters*, **7**, 3518–22.

72. Titus, E., Singh, M.K., Cabral, G., Paserin, V., Ramesh, B.P., Blau, W.J., Ventura, J., Araújo, J.P. and Grácio, J. (2009) Fabrication of vertically aligned carbon nanotubes for spintronic device applications. *Journal of Materials Chemistry*, **19**, 7216–21.

73 Ananta, J.S., Matson, M.L., Tang, A.M., Mandal, T., Lin, S., Wong, K., Wong, S.T. and Wilson, L.J. (2009) Single-walled carbon nanotube materials as T_2-weighted MRI contrast agents. *Journal of Physical Chemistry C*, **113**, 19369–72.

74 Tyagi, P.K., Singh, M.K., Misra, A., Palnitkar, U., Misra, D.S., Titus, E., Ali, N., Cabral, G., Gracio, J., Roy, M. and Kulshreshtha, S.K. (2004) Preparation of Ni-filled carbon nanotubes for key potential applications in nanotechnology. *Thin Solid Films*, **469–470**, 127–30.

75 Ramesh, B.P., Blau, W.J., Tyagi, P.K., Misra, D.S., Ali, N., Gracio, J., Cabral, G. and Titus, E. (2006) Thermogravimetric analysis of cobalt-filled carbon nanotubes deposited by chemical vapour deposition. *Thin Solid Films*, **494**, 128–32.

76 Baselt, D.R., Lee, G.U., Natesan, M., Metzger, S.W., Sheehan, P.E. and Colton, R.J. (1998) A biosensor based on magnetoresistance technology. *Biosensors & Bioelectronics*, **13**, 731–9.

77 Julliere, M. (1975) Tunneling between ferromagnetic films. *Physics Letters A*, **54**, 225–6.

78 Miyazaki, T. and Tezuka, N. (1995) Giant magnetic tunneling effect in Fe/Al_2O_3/Fe junction. *Journal of Magnetism and Magnetic Materials*, **139**, L231–4.

79 Cabral, G., Titus, E., Misra, D.S. and Gracio, J. (2008) Selective growth of vertically aligned carbon nanotubes by double plasma chemical vapour deposition technique. *Journal of Nanoscience and Nanotechnology*, **8**, 4029–32.

80 Yakushiji, K., Mitani, S., Ernult, F., Takanashi, K. and Fujimori, H. (2007) Spin-dependent tunnelling and coulomb blockade in ferromagnetic nanoparticles. *Physics Report*, **451**, 1–35.

81 Liu, R.S., Pettersson, H., Michalk, L., Canali, C.M. and Samuelson, L. (2007) Probing spin accumulation in Ni/Au/Ni single-electron transistors with efficient spin injection and detection electrodes. *Nano Letters*, **7**, 81–5.

82 Wiesendanger, R. (1994) *Scanning Probe Microscopy and Spectroscopy*, Cambridge University Press, New York.

83 Deng, Z., Yenilmez, E., Leu, J., Hoffman, J.E., Straver, E.W.J., Dai, H. and Moler, K.A. (2004) Metal-coated carbon nanotube tips for magnetic force microscopy. *Applied Physics Letters*, **85**, 623–6.

84 Takeda, S., Mishima, F., Terazono, B., Izumi, Y. and Nishijima, S. (2006) Development of magnetic force-assisted new gene transfer system using biopolymer-coated ferromagnetic nanoparticles. *IEEE Transactions on Applied Superconductivity*, **16**, 61–1890.

85 Ajayan, P.M. and Iijima, S. (1993) Capillarity-induced filling of carbon nanotubes. *Nature*, **361**, 333–4.

86 Sloan, J., Wright, D.M., Woo, H.G., Bailey, S., Brown, G., York, A.P.E., Coleman, K.S., Hutchison, J.L. and Greena, L.H. (1999) Capillarity and silver nanowire formation observed in single walled carbon nanotubes. *Chemical Communications*, 699–700.

87 Liu, S. and Wehmschulte, R.J. (2005) A novel hybrid of carbon nanotubes/iron nanoparticles: iron-filled nodule-containing carbon nanotubes. *Carbon*, **43**, 1550–5.

88 Ajayan, P.M., Stephan, O. and Redlich, P.H. (1995) Carbon nanotubes as removable templates for metal oxide nanocomposites and nanostructures. *Nature*, **375**, 564–8.

89 Wang, W., Wang, K., Ruitao, L.V., Wei, J., Zhang, X., Kang, F., Chang, J., Shu, Q., Wang, Y. and Wu, D. (2007) Synthesis of Fe-filled thin-walled carbon nanotubes with high filling ratio by using dichlorobenzene as precursor. *Carbon*, **45**, 1105–36.

90 Sadasivan, V., Richter, C.P., Menon, L. and Williams, P.F. (2005) Electrochemical self-assembly of porous alumina templates. *AIChE Journal*, **51**, 649–56.

91 Fujiwara, Y., Maehashi, K., Ohno, Y., Inoue, K. and Matsumoto, K. (2005) Position-controlled growth of single-walled carbon nanotubes by laser-irradiated chemical vapor deposition. *Japanese Journal of Applied Physics*, **44**, 1581–4.

92 Borowiak-Palen, E. (2008) Iron filled carbon nanotubes for bio-applications. *Materials Science-Poland*, **26**, 413–19.

93 Demoncy, N., Stephan, O., Brun, N., Colliex, C., Loiseau, A. and Pascard, H. (1998) Filling carbon nanotubes with metals by the arc-discharge method: the key role of sulfur. *European Physical Journal B*, **4**, 147–57.

94 Jang, J.W., Lee, K.W., Oh, I.H., Lee, E.M., Kim, I.M., Lee, C.E. and Lee, C.J. (2008) Magnetic Fe catalyst particles in the vapor-phase-grown multi-walled carbon nanotubes. *Solid State Communications*, **145**, 561–4.

95 Cheng, J., Zou, X.P., Zhu, G., Wang, M.F., Su, Y., Yang, G.Q. and Lü, X.M. (2009) Synthesis of iron-filled carbon nanotubes with a great excess of ferrocene and their magnetic properties. *Solid State Communications*, **149**, 1619–22.

96 Pham-Huu, C., Keller, N., Estournès, C., Ehret, G. and Ledoux, M.J. (2002) Synthesis of $CoFe_2O_4$ nanowire in carbon nanotubes: a new use of the confinement effect. *Chemical Communications (Cambridge)*, **17**, 1882–3.

97 Seifu, D., Hijji, Y., Hirsch, G. and Karna, S.P. (2008) Chemical method of filling carbon nanotubes with magnetic material. *Journal of Magnetism and Magnetic Materials*, **320**, 312–15.

98 Geng, F. and Cong, H. (2006) Fe-filled carbon nanotube array with high coercivity. *Physica B*, **382**, 300–4.

99 Kishi, T., Kasai, H., Nakanishi, H., Dino, W.A. and Komori, F. (2004) Electronic structure of Fe, Co, Ni nanowires on Cu (111). *Surface Science*, **566–568**, 1052–6.

100 David, M., Kishi, T., Kisaku, M., Dino, W.A., Nakanishi, H. and Kasai, H. (2007) First principles investigation on Fe-filled single-walled carbon nanotubes on Ni (111) and Cu (111). *Journal of Magnetism and Magnetic Materials*, **310**, 748–50.

101 Jin-Phillipp, N.-Y. and Rühle, M. (2004) Carbon nanotube/metal interface studied by cross-sectional transmission electron microscopy. *Physical Review B*, **70**, 245421–5.

102 Kisaku, M., Mahmudur, R.M.D., Kishi, T., Matsunaka, D., Matsunaka, R.T.A., Wilson, D.A., Nakanishi, H. and Kasai, H. (2005) Diameter dependent magnetic and electronic properties of single-walled carbon nanotubes with Fe nanowires. *Japanese Journal of Applied Physics*, **44**, 882–8.

103 Yılmaz, F. and Kucukyavuz, Z. (2009) Conducting polymer composites of multiwalled carbon nanotube filled doped polyaniline. *Journal of Applied Polymer Science*, **111**, 680–4.

104 Guo, M., Chen, J., Li, J., Nie, L. and Yao, S.Z. (2004) Carbon nanotubes-based amperometric cholesterol biosensor fabricated through layer-by-layer technique. *Electroanalysis*, **16**, 1992–8.

105 Ma, Y., Lehtinen, P.O., Foster, A.S. and Nieminen, R.M. (2004) Magnetic properties of vacancies in graphene and single-walled carbon nanotubes. *New Journal of Physics*, **6**, 68–83.

106 Hjort, M. and Stafstrom, S. (2000) Modeling vacancies in graphite via the Hueckel method. *Physical Review B*, **61**, 14089–94.

107 Banerjee, S. and Bhattacharyya, D. (2008) Electronic properties of nano-graphene sheets calculated using quantum chemical DFT. *Computational Materials Science*, **44**, 41–5.

108 Hod, O., Barone, V. and Scuseria, G.E. (2008) Half-metallic graphene nanodots: a comprehensive first-principles theoretical study. *Physical Review B*, **77**, 035411–16.

109 Brey, L. and Fertig, H.A. (2006) Electronic states of graphene nanoribbons studied with the Dirac equation. *Physical Review B*, **73**, 235411–15.

110 Zheng, H. and Duley, W. (2008) First-principles study of edge chemical modifications in graphene nanodots. *Physical Review B*, **78**, 045421–5.

111 Choi, S., Jeong, B.W., Kim, S. and Kim, G. (2008) Monovacancy-induced magnetism in graphene bilayers. *Journal of Physics Condensed Matter*, **20**, 235220–4.

112 Ramasubramaniam, A., Medhekar, N.V. and Shenoy, V.B. (2009) Substrate-induced magnetism in epitaxial graphene buffer layers. *Nanotechnology*, **20**, 275705–11.

113 Northrup, J.E. and Neugebauer, J. (1995) Theory of the adatom–induced

reconstruction of the SiC(0001) surface. *Physical Review B*, **52**, R17001–5.

114 Rutter, G.M., Guisinger, N.P., Crain, J.N., Jarvis, E.A.A., Stiles, M.D., Li, T., First, P.N. and Stroscio, J.A. (2007) Imaging the interface of epitaxial graphene with silicon carbide via scanning tunneling microscopy. *Physical Review B*, **76**, 235416–21.

115 Cong, C.X., Yu, T., Ni, Z.H., Liu, L., Shen, Z.X. and Huang, W. (2009) Fabrication of graphene nanodisk arrays using nanosphere lithography. *Journal Physical Chemistry C*, **113**, 6529–32.

116 Zhou, J., Wang, Q., Sun, Q., Chen, X.S., Kawazoe, Y. and Jena, P. (2009) Ferromagnetism in semihydrogenated graphene sheet. *Nano Letters*, **9**, 3867–70.

117 Soodchomshom, B., Ming Tang, I. and Hoonsawat, R. (2009) Dirac tunneling magnetoresistance in a double ferromagnetic graphene barrier structure. *Physica E*, **41**, 1310–14.

6
Multi-Walled Carbon Nanotubes for Drug Delivery

Nicole Levi-Polyachenko

6.1
Introduction

Why should nanosized carbon particles be helpful in drug delivery? Nanomaterials of the same elemental composition as their bulk counterparts have unique characteristics due to quantum confinement of the electrons, phonons, and plasmons in one-dimensional (1-D) materials such as carbon nanotubes (CNTs) [1]. In addition, their aspect ratio (length/diameter) and chemistry influences their intracellular and *in vivo* response. Multi-walled nanotubes (MWNTs) consist of many nested tubes, creating multiple concentric sidewalls, and have diameters of a few nanometers up to hundreds of nanometers, and lengths in the micron range, depending on the growth conditions utilized. MWNTs can be employed as a drug-delivery vehicle, by external functionalization and/or intertube filling with the modality of choice, or by inducing hyperthermia to aid in drug delivery; therefore, their ability to deliver therapeutic agents is an aspect that is explored in this chapter.

Many therapeutics cannot be delivered orally, for reasons of dosing restrictions and poor bioavailability. Following intravenous (IV) delivery, therapeutic agents are less likely to be degraded but will reach their target areas more rapidly and at higher doses. Nonetheless, problems persist with IV administration, including protein aggregation on the therapeutic agent, immune clearance, aggregation of the therapeutic agent leading to drug inactivation, and poor localization of the drug, such as delivery to the peritoneal cavity or traversal of the blood–brain barrier. Thus, the MWNTs may serve as vehicles to minimize these negative aspects and aid in more efficient drug delivery.

One factor to be considered when using MWNTs for drug delivery is their interaction with the cell membranes, including receptors on the membrane surface and subcellular structures. Cheng and Porter showed, by using high-resolution transmission electron microscopy (TEM), that MWNTs may penetrate not only the cytoplasm of mammalian cells but also on occasion the nucleus, leading to a 20–25% decrease in cell viability [2]. MWNTs can also enhance cell permeabilization for the transfection and delivery of therapeutic agents, and thus would be of major benefit for processes such as gene therapy, vaccination, and electrochemotherapy. One mechanism for MWNT cell interaction was demonstrated by Hirano *et al.*, where it was shown that MWNTs can disrupt the plasma membrane, and adhere to the plasma membrane by reacting with MARCO (macrophage receptor with collagenous structure) [3], which specifically recognizes environmental particles. The authors suggested that exposure to the MWNTs would induce MARCO upregulation, and that once MARCO was activated, the plasma membrane would attempt to extend around the MWNTs. In doing this, the membrane would become injured (disrupted).

MWNTs are excellent candidates for drug delivery as they have larger inner diameters and greater surface areas than single-walled nanotubes (SWNTs) for molecule functionalization. One effective method would be to functionalize a targeting modality to the nanotube exterior, and to fill the interior of the tube with a drug by using capillary action. Because the number of molecules that actually are conjugated onto the nanotubes can be quite low, it may be very difficult to fabricate sufficient conjugated material for clinical use. Although the aspect ratio of the CNTs can enhance their ability to carry more therapeutic molecules, and also influences their interaction with incident radiation, very high-aspect-ratio materials can cause damage to cell membranes, as well as difficulties in immune system recognition and clearance. Thus, the aspect ratio – and especially the length of the CNT – should be carefully considered to minimize biopersistence. In order to better enhance drug delivery, the nanotubes can be sugar-coated to aid their dispersion in aqueous media; for this, gum arabic, amylose, amylopectin, cyclodextrin, β-galactosidase, glucose and dextran [4–8] have been used. Although proteins bind to nanotubes nonspecifically, a common technique to minimize protein absorption is to functionalize the nanotubes with poly(ethylene glycol) (PEG) [9, 10]. Other molecules that have been functionalized onto the nanotubes to aid their dispersion in aqueous media include bovine serum albumin (BSA), histidine, and tryptophan, all of which were shown to bind very strongly to the nanotubes [11, 12].

Following growth, nanotubes often must be cleaned to remove any graphitic carbon and/or excess catalyst. Cleaning can be carried out using oxidative treatments, either by sonicating the nanotubes in acids or annealing them in a pressurized, high-temperature, oxygen environment. Cleaning can also remove the caps at the ends of the tubes, create holes in the sidewalls, and shorten the tubes by chemically breaking them. Consequently, the aspect ratios can be controlled by using cleaning procedures, and holes or open-ended tubes made available for intratubular filling. Oxidative cleaning also enhances the solubility of the CNTs by the addition of carboxylic acid or hydroxyl groups to the sidewalls [13]. Furthermore, MWNTs with dopant atoms introduced into their sidewalls are currently being investigated for medicinal applications, as they may have a reduced toxicological potential; an example of this is nitrogen-doped MWNTs [14].

Both, viscosity and surface tension are important for the capillary filling of CNTs. Fluids with a surface tension close to that of water are optimal, as is the use of a supersaturated solvent so that, on evaporation of the solvent the drug molecules will aggregate along the interior walls of the nanotubes [15]. Ideally, for the release of the drug at a specific site, the drug would be encapsulated within the tube with minimal drug along the exterior. However, it is difficult to remove drugs that are adhered to the outside of the tube without first removing drug at the interior, unless the nanotube ends are capped.

The template synthesis of CNTs can result in nanotubes with large diameters. For example, Kim et al. used template synthesis to prepare CNTs with 500 nm diameters, and then dissociated and filled the nanotubes by capillary action with 50 nm polystyrene beads [16]. Filling of the tube was achieved via capillary action, whereby evaporation of the filling liquid leaves a void, so as to produce a flow of suspension from the droplet in the pipette into the CNTs (see Figure 6.1).

An alternative method for loading chemical vapor deposited, template-grown CNTs with diameters between 200 and 300 nm was reported by Bazilevsky et al., who loaded the CNTs with polystyrene spheres [17]. The technique used was slightly modified from that of Kim et al., in that they applied the CNTs to a substrate, in this case a TEM grid. After allowing the nanotube solvent to evaporate, there remained a thin film of nanotube. The polystyrene spheres were then applied, suspended in a solvent that would rapidly evaporate and cause the bead solution to be sucked into the nanotubes by capillary action. As Bazilevsky et al. were working with a dry substrate, they also attempted to apply a polymer, namely polycaprolactone (PCL), to the nanotubes; subsequently, by using TEM, a filling of the tubes with PCL was observed. The self-sustained diffusion led to a continual pulling of the fluids into the CNTs, while depositing material from the supersaturated solution onto the interior surface of the nanotubes, due to evaporation of the solvent. In another report, Bazilevsky et al. used CNTs of 50–100 nm diameter, and investigated whether there was any difference between low-molecular-weight and higher-molecular-weight polymers in terms of whether they would fill the CNTs, or not [18]. Subsequently, polyethylene oxide (PEO) or PCL with molecular weights in excess of 100 kDa were shown to be incapable of filling the tubes; rather, the polymer solutions needed to be quite dilute for filling to occur. One notable result of these studies was that the CNTs did not fill completely, but instead became filled with a foam-type structure due to condensation of the polymer, Brownian motion, and intermolecular forces.

A final interesting method for filling large-diameter CNTs (ca. 500 nm diameter) was developed by Nadarajan et al., who employed centrifugal forces to fill the nanotubes [19]. This method of nanotube filling is not dependent on the viscosity or the surface tension of the fluid, as a viscous sodium alginate solution containing quantum dots (QDs) and polystyrene nanospheres was used. Having applied the large-diameter CNTs to the top of the alginate gel, a centrifugal force was then applied such that the CNTs were pulled through the viscous material; this caused the QDs and polystyrene particles to be encapsulated into the tubes (see Figure 6.2). A rapid capping method was then used to close and seal the ends of the tubes; this involved the application of calcium chloride to the alginate tubes, to effectively crosslink the alginate. Any excess alginate was removed from the outside of the

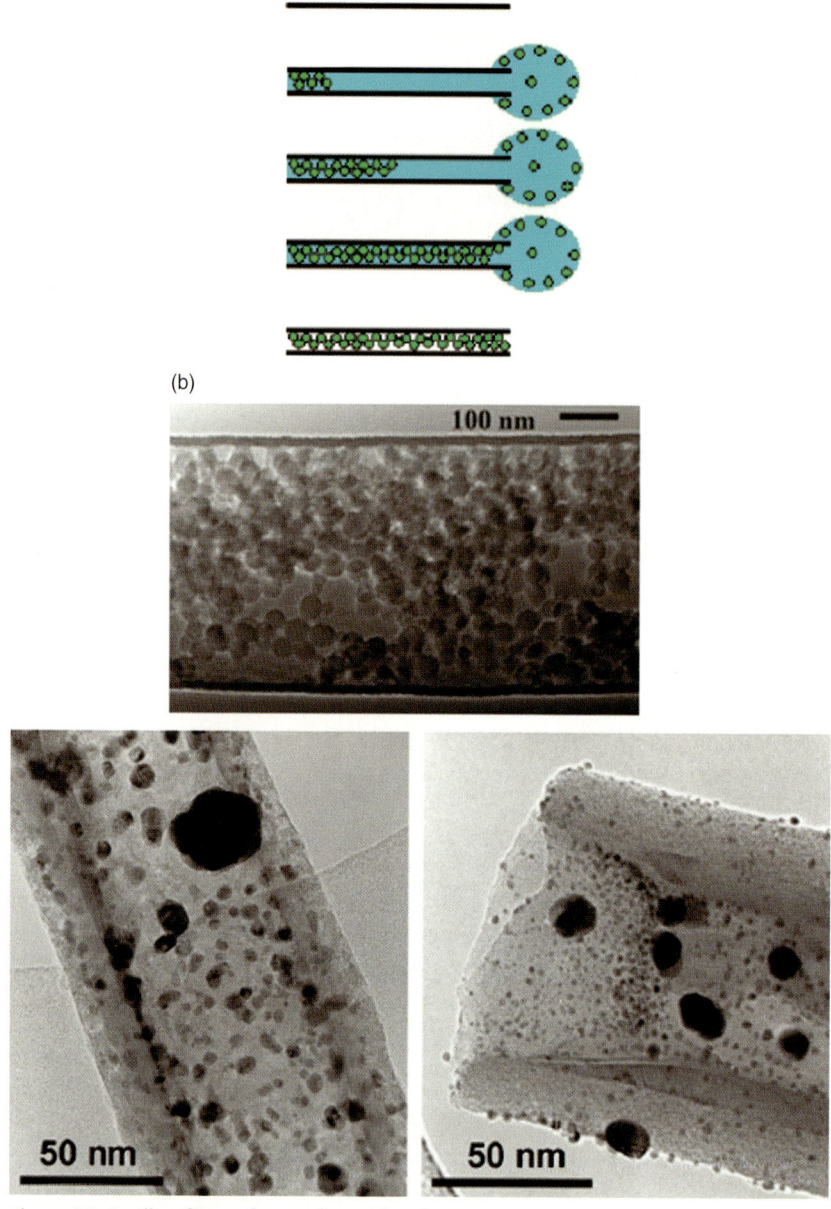

Figure 6.1 Capillary filling of nanotubes with polystyrene spheres and palladium nanocrystals involves aggregation of the agent along the interior of the nanotubes due evaporation of the solvent to leave particles inside the tube. Reproduced from Ref. [16].

(a) (b) (c) (d)

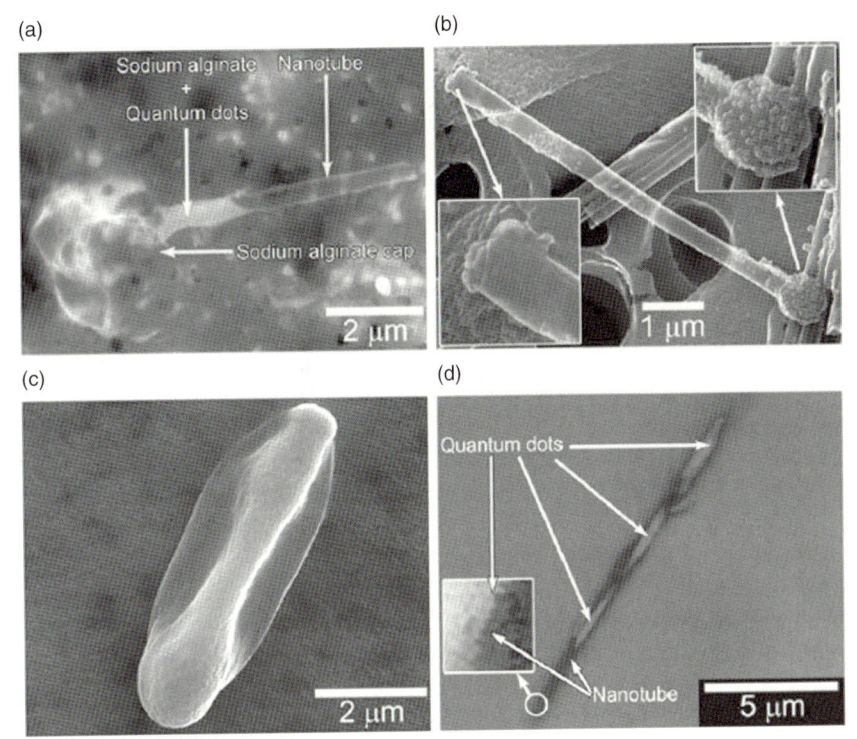

Figure 6.2 Carbon nanotubes filled with quantum dots in a sodium alginate carrier solvent. Reproduced from Ref. [19].

nanotubes by repeated saline washing, which left the caps at the tube ends intact. The loading efficiency was found to be 40–50%, while the internalized material could be removed by using ultrasound. It was also found that, for nanotubes of 500 nm diameter, the maximum size of nanoparticle loading was 40 nm. Above this diameter, larger particles caused sedimentation and a decrease in the loading efficiency. Clearly, the use of alginate for nanotube filling represents an excellent choice since, depending on the molecular weight of the alginate, it will degrade within specific time intervals to release the drugs over time.

Although, currently, many reports and resources have been devoted to the study of SWNTs for medical applications, MWNTs are considered to be equally interesting and may offer new methods for the development of drug-delivery mechanisms. While MWNTs may exhibit specific interactions with cell receptors, they also have the unique ability to retain the conformational structure of any attached molecules. Interactions between the sidewalls can influence the thermal and phonon energy states of MWNTs, such that their larger interior diameters and exterior size offers the ability to carry more therapeutic molecules, both inside and/or outside, than can SWNTs. In this chapter, attention is focused specifically on the use of MWNTs across multiple needs for drug delivery, including the ability of MWNTs to deliver agents capable of modifying genetic function, transdermal delivery, antibiotic agents, and the delivery of chemotherapeutic agents – either directly via intracellular transport, or indirectly via hyperthermia to induce intracellular transport. Within this chapter will be highlighted the broad range of drug-delivery applications that MWNTs can facilitate, in addition to a review of some initial data relating to the use of MWNTs as drug-delivery mediators. On the basis that the reader is now more familiar with CNTs, the aim of the chapter is to subdivide therapeutic delivery into areas of gene therapy (Section 6.2), antibacterial (Section 6.3), wound-healing (Section 6.4), chemotherapeutic delivery via intracellular transport (Section 6.5.1), and hyperthermia (Section 6.5.2).

6.2
Gene Therapy

Gene therapy consists of a series of vectors delivered to cells in order to mediate gene regulation in the cells and mitigate diseases such as cancer and hereditary conditions. Since some types of vector that are used in gene therapy include viral delivery and DNA delivery, conjugated nanotubes represent an ideal therapeutic delivery mechanism for DNA, peptides, small interfering RNA (siRNA), and antisense oligonucleotides. Typically, DNA can be attached to CNTs by either covalent or noncovalent methods; the covalent methods include an initial linking onto the CNT using COOH and OH groups that are developed during CNT oxidation (cleaning/shortening). DNA can be attached in either a single-stranded form or short double-stranded form noncovalently, by the DNA simply wrapping around the nanotube [20]. In order to purify DNA-wrapped CNTs, separation can be achieved using gel electrophoresis [21].

One reason for using CNTs for plasmid DNA delivery is that it is very difficult for plasmid DNA to enter the cell nucleus, with previous methods that have included the use of viral vectors having led to immunological problems, inflammation, or even cancer [22]. Other cationic molecules that have been used to deliver plasmid DNA include lipids or dendrimers; these cause DNA to be condensed into a (mostly) spherical shape, with molecular interactions between the carrier molecule and the DNA determining gene expression. Hence, spherical molecules may be less ideal than the rod-shaped CNTs [22]. Transfection can also be achieved by attaching a plasmid DNA to a CNT. For this, Cai et al. achieved significant success when using "nanospearing" for the delivery of plasmid DNA [23]. For this, the CNTs were first grown with a magnetic nickel core, after which the DNA was loaded onto the CNTs and the nanotubes with DNA stirred magnetically in the presence of a monolayer of cells; this caused the cells to be "speared" by the DNA-loaded nanotubes. This method produced 76% of MCF breast cancer cells that had CNTs embedded, with 90% cell viability. The goal here was to deliver the mammalian expression vector (pEGFP-c1) and, on spearing, the Bal17 cells had a 100% enhanced green fluorescent protein (EGFP) expression. CNTs without a nickel core were used as a control, and these failed to cause any transduction, as no spearing would occur when the magnetic core was absent. In addition, when the technique was tested against B cells and primary neurons (which can be difficult to transfect), 100% and 80% transfection was obtained, respectively. Notably, although spearing led to a perturbation of the cell membranes, the membranes recovered. Consequently, nanospearing was considered to represent a highly efficient means of DNA transfection.

Anti-sense methods are nucleic acid-based therapies that can be used to modulate gene expression and regulate protein synthesis. Unfortunately, however, anti-sense oligodeoxynucleotides have a poor intracellular uptake and are easily degraded. Jia et al. used QDs to fluorescently label MWNTs that had been modified with polyethylenimine (PEI) and anti-sense molecules [24]. This resulted in the anti-sense oligodeoxynucleotides entering the nucleus following MWNTs entry into the cytoplasm, and led to a 20% cellular apoptosis of HeLa cervical cancer cells; by comparison, only 3% apoptosis was achieved with unlabelled MWNTs which served as a control.

When Singh et al. placed the plasmid DNA, pCMW, βgal-expressing galactosidase onto functionalized SWNTs and MWNTs, an increased gene expression was achieved that was five- to tenfold higher than for the delivery of plasmid DNA alone [22]. Subsequently, Singh et al. prepared three formulations of cationic SWNTs or MWNTs in order to condense plasmid DNA; these included amine-functionalized SWNTs and MWNTs, and lysine-functionalized SWNTs. The surface area and charge density proved to be critical parameters for electrostatic complex formation between the functionalized CNTs and DNA. Singh et al. also showed that an ammonium-functionalized SWNT, when complexed with plasmid DNA, produced a tenfold increase in gene expression levels compared to the naked DNA. In this case, the DNA was found to bind more tightly to MWNTs than to SWNTs, and that the MWNTs could bind and condense with DNA more tightly than could SWNTs. A tighter binding of the DNA to MWNTs may lead to an inhibition of intracellular release, while modifications for a more efficient release of DNA from the MWNTs once inside the cell would benefit the use of MWNTs for plasmid DNA delivery.

Microglial cells are one type of cell that behave as immune effector cells (capable of phagocytosis and antigen presentation) in the central nervous system, and are capable of interacting with MWNTs. Kateb et al. loaded zeocin plasmid DNA or siRNA onto MWNTS and delivered the complexes to BV2 microglial cells, and also to the GL2A1 glioma cell line [25]. The results not only showed DNA to have a higher loading capacity than siRNA on the MWNTs, but also that the BV2 cells could more efficiently internalize MWNTs than could the glioma cells. Thus, MWNTS could potentially selectively deliver an immunogenetic therapy directly to the microglial cells.

A second mechanism for immunotherapy was developed by Yander et al., in which they attached the malarial peptide AMA1 to MWNTS, and then injected the complex intraperitoneally into mice, so as to develop antibodies [26]. The correct conformation of the peptide was found to be retained when it was attached to the CNT. The fact that the conformation of the malarial peptide was retained indicated that the attachment of such molecules to MWNTs might be beneficial to retain the structure of the molecules and, in turn, to enhance drug delivery for immunotherapeutic purposes. Possible mechanisms for gene therapy using MWNTs include: (i) retention of the proper molecular structure of molecules when attached to the CNTs; and (ii) an ability of the CNTs to cross the cell membrane to enhance the presentation of molecules functionalized to the sidewalls of MWNTs, compared to delivery without a MWNT carrier.

6.3
Antibacterial Therapy

Although not necessarily acting as a drug-delivery agent, CNTs can act as antibacterial agents by damaging bacterial cell walls. For example, when Kang et al. examined the cytotoxicity of SWNTs and MWNTs towards *Escherichia coli* [27], two methods were used: (i) a suspension was prepared of the CNTs nanotubes and *E. coli*; or (ii) a thin film of the CNTs was prepared, on top of which the *E. coli* were deposited. Consequently, the SWNTs were found to be much more toxic than the MWNTs. In the case of the deposited film, the SWNTs caused an 87% decrease in cell viability, compared to 31% for the MWNTs. For the aqueous suspensions, the SWNTs produced an 80% decrease in cell viability, and the MWNTs a 24% decrease. The authors surmised that the SWNTs were much more toxic due to their mechanisms for affecting the bacteria's cell membrane integrity, metabolic activity and morphology. It was suggested that the smaller diameter of the SWNTs would facilitate their partial penetration into the *E. coli* cells, by disrupting the cell membrane. Liu et al. also examined the toxicity of SWNTs, but did so by comparing dispersed to aggregated nanotubes, and found the dispersed state to be more toxic [28]. For this, two Gram-negative bacteria (*E. coli* and *Pseudomonas aeruginosa*) and two Gram-positive bacteria (*Staphylococcus aureus* and *Bacillus subtilis*) were compared. The SWNTs were found not to inhibit cell growth or to cause oxidative stress; consequently, it was hypothesized that the toxicity observed was caused by a nanospearing-type phenomenon. By using atomic force microscopy (AFM), the softer Gram-positive cells were found to be more vulnerable to piercing by the SWNTs. When Kang et al. compared Gram-negative *E. coli* and *P. aeruginosa* and Gram-positive *Staphylococcus epidermidis* and *B. subtilis* [29], SWNT resulted in an 80% cell inactivation compared to MWNTs, resulting in a 30% cell inactivation in the case of *E. coli*, whereas for *P. aeruginosa* the SWNTs caused over 75% cell inactivation compared to only 20% by the MWNTs. The data for *S. epidermidis* showed that SWNTs caused a 65% decrease in cell inactivation, whereas MWNTs caused only a 45% decrease. Finally, for *B. subtilis* a 15% decrease in cell inactivation by SWNT was compared to a 5% decrease by MWNTs. These results were in compliance with data acquired by others, that the SWNT appeared to be more toxic towards bacteria than did the MWNTs. However, in Kang's case the Gram-negative bacteria were found to be more susceptible to the nanotubes, and showed a greater reduction in cell viability than did the Gram-positive bacteria, despite the mechanisms by which the nanotubes were applied to the bacteria being quite different in these two studies.

In another exposure of E. coli MG1655 and an additional bacterium, *Cupriavidus metallidurans* CH34, Simon-Deckers et al. showed that MWNTs, when sugar-coated with gum arabic, caused a 50–60% reduction in the *E. coli* population, at a concentration of 100 mg l^{-1} [30]. The second bacterial cell line, *C. metallidurans*, is known to be resistant to environmental toxins, and was unaffected by MWNTs. One exciting point to note regarding this study was that the MWNTs were adsorbed onto the cell surface, but not internalized (see Figure 6.3). This contrasted with the case of SWNTs, which actually penetrate the bacterial cells, potentially causing cell death.

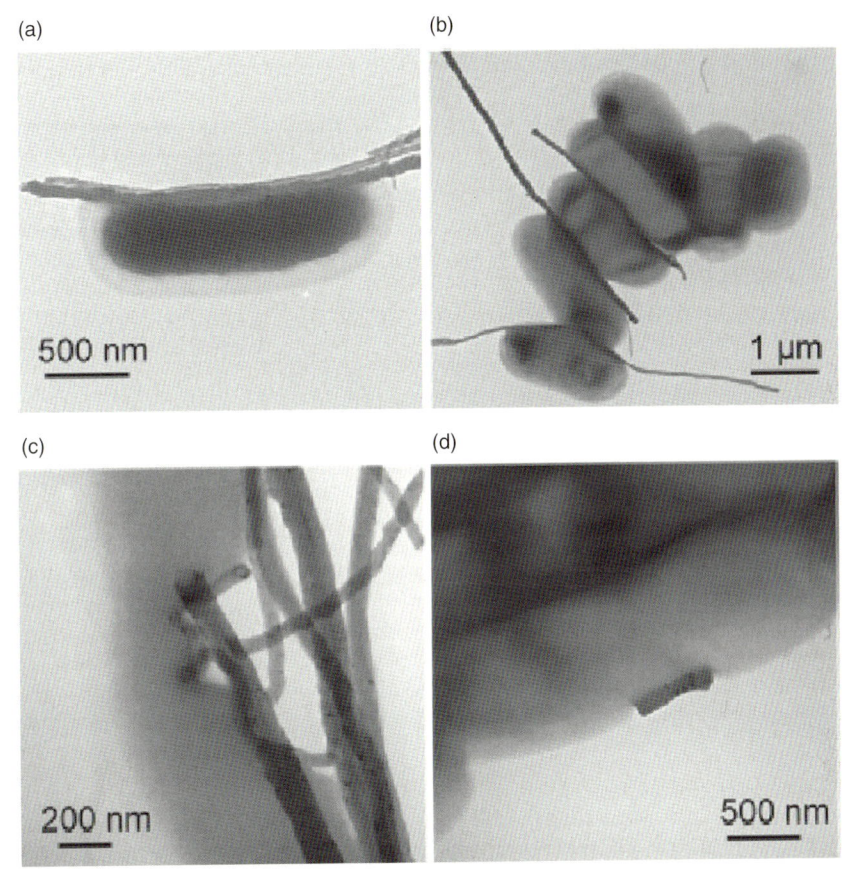

Figure 6.3 TEM images of MWNTs exposed to *E. coli* (b) and *C. metallidurans* (a,c,d). The images show that MWNTs are associated with the bacterial cell wall and may disrupt it, but do not penetrate into the interior of the bacterium. Reproduced from Ref. [30].

As an alternative to cell spearing of bacteria as an antibacterial modality, much interest has been expressed in the delivery of antibiotics by attachment to CNTs. Amphotericin B, an antibiotic often used to treat fungal infections, is highly toxic towards mammalian cells, mainly because of its aggregation in aqueous media. Consequently, the proposal was made that CNTs could be used to minimize such aggregation, and hence the toxicity of Amphotericin B. When Wu et al. conjugated amphotericin B onto MWNTs, this led to a reduction in toxicity of more than 30% in a Jurkat lymphoma cell line [31]. As with other studies, the MWNT-amphotericin B complex was shown to locate found inside the cells, close to the nuclear membranes but not within the nucleus. The length and loading of Amphotericin B was also shown to depend on the acid oxidation of the CNTs, as has been shown by others for SWNTs. After performing an evaluation in lymphoma cells, and confirming the reduction in toxicity when amphotericin B was attached to the CNTs, Wu et al. evaluated the complex against three different types of fungus, namely *Candida parapsilosis*, *Cryptococcus neoformans*, and *Candida albicans*. A significant reduction was found in the minimum inhibitory concentration (MIC) for amphotericin B when bound to MWNTs. This was believed due to the higher degree of solubility of amphotericin B when attached to MWNTs, and that many amphotericin B molecules were attached to each CNT. Whilst the SWNT- and MWNT- amphotericin B complexes showed similar MIC-values for *C. parapsilosis* (1.6 μg ml^{-1}) and *C. neoformans* (0.8 μg ml^{-1}), both complexes showed significantly lower MICs than for amphotericin B alone (20 μg ml^{-1} for *C. parapsilosis*, 5 μg ml^{-1} for *C. neoformans*). For the inactivation of *C. albicans*, MWNTs complexed to amphotericin B has proved to be a more effective antifungal delivery agent than SWNTs (6.4 μg ml^{-1} for MWNTs versus 3.8 μg ml^{-1} for SWNTq), perhaps because a greater quantity of amphotericin B can be complexed onto the larger nanotubes. Although, at present, few investigations are being conducted into the delivery of antibiotic agents attached to CNTs for the development of efficient antibacterial agents, when based on the results obtained to date, this avenue of study would appear to show much promise. Indeed, for maximal antibiotic delivery, the MWNTs may prove to be ideal based on their large surface area for functionalization, which allows them to carry more molecules per nanotube than SWNTs but without causing membrane damage.

6.4
Wound Healing

In non-medical applications, CNTs are often combined with polymers to enhance the polymer's characteristics, such as thermal and electrical transport [32]. Biologically compatible polymers that are capable of releasing a drug and degrading over time are ideal for use as transdermal and *in vivo* patch materials for delivery. For example, CNTs can be incorporated into biopolymers and aid drug delivery by modulating the release of a drug by thermal, electrical, or diffusion mechanisms. As nanotubes can be functionalized or wrapped with polymers to aid solubility, composites of nanotubes and biopolymers might be used to encapsulate drugs for localized delivery.

Hydrogels are polymers that absorb water readily and can release therapeutic agents slowly, through passive diffusion. Typical hydrogel materials that have been examined for medicinal applications include xanthan gum, guar gum, polyethylene oxide, polyvinyl alcohol, chitosan/polyethylene glycol blends, dextran, sodium alginate, polyhydroxymethyl methacrylate, and gelatin [33–36]. In fact, nanotubes have been shown to stabilize gelatin hydrogels for biological applications [33]. In the case of hydrogels, swelling – as well as bound and unbound water – may serve as a very valuable technique for the delivery of drugs. For example, Haider *et al.* showed that levels of both bound and unbound water in gelatin–MWNT complexes was decreased due to the hydrophobic nature of the nanotubes [37]. It was also shown that higher concentrations of MWNTs (>1%) reduced gelatin swelling by inhibiting the water transport between gelatin molecules. Likewise, the gelatin film could be flexed by controlling the osmotic pressure difference between the cathode and anode sides of the film. This was achieved by applying a voltage of 6–9 V to a doped film in an ionic solution. When added to the gelatin film, the CNTs – being electrically conductive materials – caused an increase in bending such that, the higher the nanotube loading, the greater the degree of bending. The bending was shown to be reversible because it was simply due to an influx of water on the cationic or anionic sides of the hydrogel. The MWNTs could specifically modify the amount of water, and hence also the mechanical forces and ion regulation through the film, leading to a more efficient drug delivery from this type of film.

One problem encountered with the transdermal patch drug delivery systems in current clinical use is that, although they can deliver an initial "burst" quickly, the amounts of drug delivered over the remaining patch lifetime decrease rapidly. One way of modulating drug release would be to devise an electrically stimulated patch, and indeed CNT may be very effective for this purpose. The biopolymer chitosan has a backbone chain that allows for electrical conductivity, but its conductivity is much too low for use in electrically stimulated drug release. However, it is possible easily to blend nanotubes with chitosan chains, either by wrapping the chains around individual CNTs or, for low concentrations of CNT compared to chitosan, to develop flexible films of chitosan that are doped with CNT, so as to enhance the thermal, electrical, or diffusive properties of chitosan.

Chitosan, as a cationic polysaccharide, and hyaluronic acid (HA), a glycosaminoglycan, are both frequently used in wound-healing therapies, to control the hydration states of the tissues. Typically, chitosan films are beneficial for increased healing and reducing the bacterial contamination of burns, while HA is often added to emollient creams and cosmetics to increase the hydration state of the skin. Formulations of chitosan, HA or salmon sperm DNA with 0.5% SWNTs were prepared by solution casting, and the amount of neurotrophin-3 (NT-3) release was measured with and without electrical stimulation [38]. The chitosan–SWNT films were seen to swell dramatically (by 700%) compared to normal, due to the SWNTs acting as crosslinkers between the chitosan molecules and thus increasing the area between chitosan molecules, allowing more water to be retained. To release the NT-3 growth factor, a ± 1 mA, 100 ms square wave at 250 Hz was used as a stimulus. On electrical stimulation, NT3 release from the SWNT–chitosan film was increased 6.5-fold, while the SWNT–HA film showed a twofold increase, and SWNT + DNA showed no significant increase. As chitosan has a lower impedance than HA or DNA – and thus, with the addition of SWNT, an even lower impedance – an additional NT-3 released occurred on electrical stimulation. The fact that each film released only about 30% of the NT-3 passively indicated a tight binding of the growth factor to SWNT, regardless of the biopolymer used. This represents an example of a wound-healing patch for therapeutic delivery. Although SWNTs were used, an extension of similar materials using MWNTs would also be advantageous, depending on the need for polymer chain interaction and electrical stimulation. In order to prompt the formation of an electrical percolation pathway in a biopolymer film for therapeutic drug release, MWNTs might be more influential than SWNTs, due to their typically longer lengths.

The development and growth of aligned nanotube arrays can allow for specific drug transport. For example, Majumder et al. set up a CNT array that would allow only specific molecules to pass down the 7 nm-diameter tube and into the skin, under the effect of an applied voltage [39]. In this case, a "gate-keeper" type molecule (a long-chain molecule that can interact electrostatically with an applied bias to cover the pores of the nanotubes) was developed and added to the top of the CNT array. When a negative bias was applied (up to −200 mV), the pores were opened by the electrostatic displacement of the gate-keeper molecule; this allowed fluid transport, but a positive bias (up to 7 mV) caused the pores to close. This type of array was used to create a significant decrease in the rate of diffusion of nicotine through a patch and into the test skin, as shown in Figure 6.4. The fluid flow was

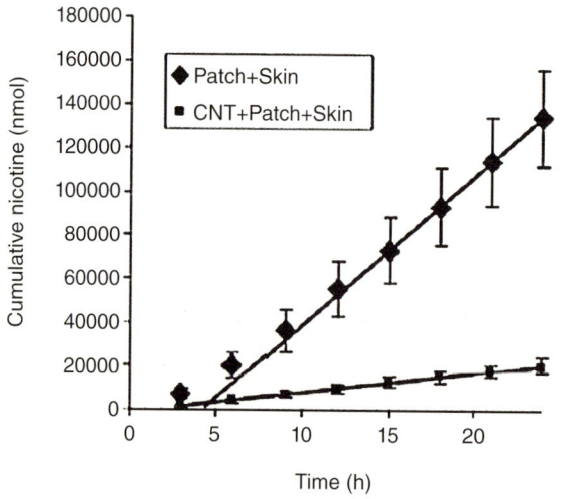

Figure 6.4 Nicotine flux from a transdermal patch through skin, with and without nanotubes embedded in the patch. Reproduced from Ref. [38].

fivefold than down conventional membranes, due to the way that the liquids flowed down the nanometer-sized CNTs.

One option for MWNTs, that might be either beneficial or serve as a hindrance in wound healing, is that MWNTs have been shown to decrease the transepithelial electrical resistance (TEER), specifically in human airway epithelial cells. In contrast, SWNTs caused a much smaller change in the TEER. This effect was monitored by Rotoli et al., who showed that MWNTs appear to interfere with the tight junctions in epithelial cell complexes [40]. Consequently, Rotoli and coworkers prepared an epithelial monolayer, applied the MWNTs, and showed there to be an eightfold increase in mannitol permeability over the control. Decreases in the electrical resistance of the cell membrane of certain cell types might be beneficial for the further enhancement of drug delivery. However, as shown by these studies, there may be a quite negative effect if permeability is not required in a specific area.

Im et al. prepared a PEO and pentaerythritol triacrylate composite with MWNTs and ketoprofen, by electrospinning [41]. In order to evaluate the release of ketoprofen, a portion of the electrospun materials was applied to a piece of mouse skin in a standard Franz diffusion-type cell, and a voltage applied to the skin, which was suspended in phosphate-buffered saline (PBS). The voltages ranged from 1 to 15 V, and were applied for up to 600 min, which represents a long time for drug delivery by electrical stimulation. With no stimulation and no nanotubes, a 33% release of the drug occurred, and this was increased to 46.7% with electrical stimulation. With nanotubes and electrical stimulation, the release of ketoprofen from the fibers was assessed at 85.6%. In addition, a decrease was seen in the burst rate of the drug when MWNTs were added into the electrospun material. It was suggested by the authors that the PEO had dissolved in water, and hence released the drug as the polymer dissolves while the ketoprofen in direct contact with the polymer is also released. The MWNTs may aid in this process by modifying the amount of bound and unbound water in the composite. Encapsulated materials, such as nanotubes and other high-aspect-ratio particles are known to align with the long direction of electrospun fibers. The MWNTs were also seen to locate in the middle of the electrospun polymer fibers. The authors used a MWNTs loading of 10% of the PEO, which was a high loading for the nanotubes of a composite. Nonetheless, it might be possible to reduce the percentage of CNTs by furthering enhancing dispersion of the tubes within the electrospun fibers.

6.5 Chemotherapy

Currently, cancer is the second-highest cause of death in the United States, with more than one million new cases being discovered annually [42]. Current treatments include surgical removal, chemotherapy, radiation, and cryotherapy or ablation therapies.

6.5.1
Hyperthermic Drug Delivery Using CNTs

Nanotubes filled with therapeutic agents, and hyperthermic delivery using CNTs, represent two methods that will offer significant contributions to localized drug delivery techniques. As shown in Figure 6.5, the infrared spectrum between 700–1100 nm is the region with absorption minima for hemoglobin and water, which makes it ideal for medical applications [43, 44]. Nanotubes are metallic or semiconducting in nature, and thus behave as a dipole antenna and will strongly absorb infrared radiation [45]. As nanotube bundling can decrease the infrared coupling efficiency and nanotube heating, a good dispersion of the tubes is important. CNTs absorb electromagnetic radiation throughout the infrared spectrum due to transitions between the first and second van Hove singularities, which induce phonon

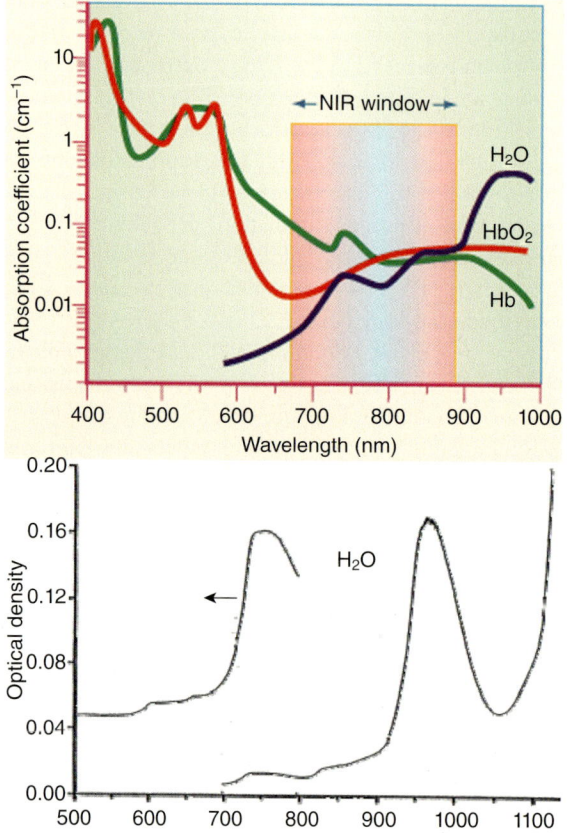

Figure 6.5 Absorption minima of water and hemoglobin from 700–900 nm (upper panel) and absorption minima of water to show the range from 1000–1100 nm (lower panel). Reproduced from Refs [42, 43].

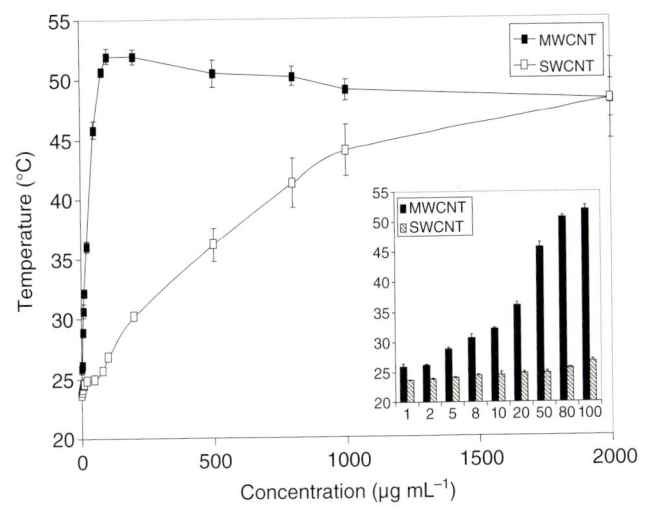

Figure 6.6 Well-dispersed MWNTs generate a greater temperature than well-dispersed SWNTs for all concentrations up to 2 mg ml^{-1}. Reproduced from Ref. [53].

and plasmon resonances along the tube length, thus generating heat [46–50]. Some lasers have emissions close to 800 nm (diode lasers) and 1064 nm (Nd:YAG lasers); these are used clinically, especially for applications in dermatology and plastic (cosmetic) surgery. MWNTs might be more efficient than SWNTs due to their additional mass and interaction between concentric tubes, whereas metallic tubes should be more efficient than semi-conducting tubes for heat generation upon infrared stimulation, as shown in Figure 6.6 [51–53].

Hyperthermic therapies used for increased drug delivery have often been found to be most effective if elevated temperatures are maintained for a few hours [54]. In clinical hyperthermic chemoperfusion, the bulk tumors are first removed, after which a heated chemotherapeutic is circulated at the tumor site for 2 h to minimize any micrometastases that cannot be easily resected [54]. Mild hyperthermia increases cellular metabolism and cell membrane permeability for increased drug uptake by cells [55]. As MWNTs can generate temperature increases rapidly, they can be used to raise the temperature locally and enhance therapeutic uptake by malignant cells.

Certain primary cancers tend to metastasize to the organs and walls of the peritoneal cavity; specifically, these include colorectal, gastric, and breast cancers, in addition to primary intraperitoneal cancers of the pancreas and liver. Although, peritoneal micrometastasis may not be detected and even removed during gross tumor debulking, they could be managed by the direct delivery of chemotherapeutic agents. Although IV chemotherapy delivery is beneficial for many tumor types, systemically delivered agents often do not reach the peritoneum in sufficiently high doses to be very effective, and therefore a regional dose delivery would be

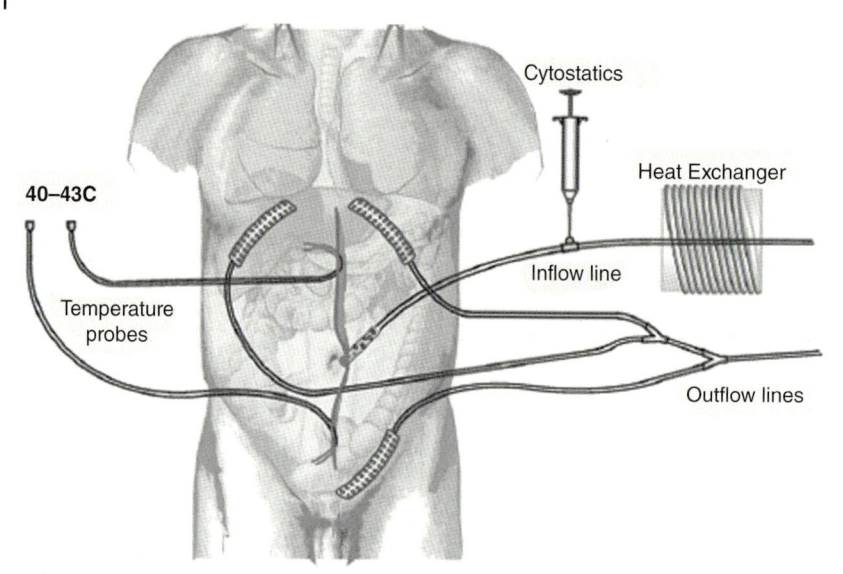

Figure 6.7 Circuit diagram for intraperitoneal hyperthermic chemoperfusion. Reproduced from Ref. [58].

preferred [56]. The technique of intraperitoneal hyperthermic chemoperfusion (IPHC) delivers drugs, at temperatures between 40–43 °C, up to 3 mm into the peritoneal tissue in order to target micrometastatic lesions [57]. A schematic of the perfusion set-up is shown in Figure 6.7. Although survival rates have been improved dramatically when using IPHC in combination with surgical resection, the drawbacks to the procedure include an increased time that the patient must be anesthetized, and large quantities of drug perfusate (up to 2 liters) [57]. The delivery of certain chemotherapeutic agents can be optimized with hyperthermia, and these include: doxorubicin, melphalan, mitomycin C (MMC), mitoxantrone, gemcitabine, etoposide, and the platinum-based agents carboplatin, cisplatin and oxaliplatin, the cellular efflux of which is minimized by hyperthermia [58, 59]. Clinically established data are available as to which chemotherapeutic agents interact well with hyperthermia, and at what temperatures they are most effective [59, 60].

As shown by Levi-Polyachenko *et al.* (Figures 6.8 and 6.9), MWNTs can be used to generate rapid hyperthermia (to 42 °C), and can also increase the cellular uptake of the drugs MMC and oxaliplatin by colorectal cancer cells, leading to a reduction in the cell population similar to the standard 2-h hyperthermia treatment at 42 °C [61]. Laser stimulation without MWNT led to similar results as with control cells treated at 37 °C.

Klingeler *et al.*, developed the CNT thermometer by filling MWNTs with copper halides to measure the temperature of nanotubes by using magnetic resonance, specifically the proton relaxation rate [62]. This nanotube thermometer could be

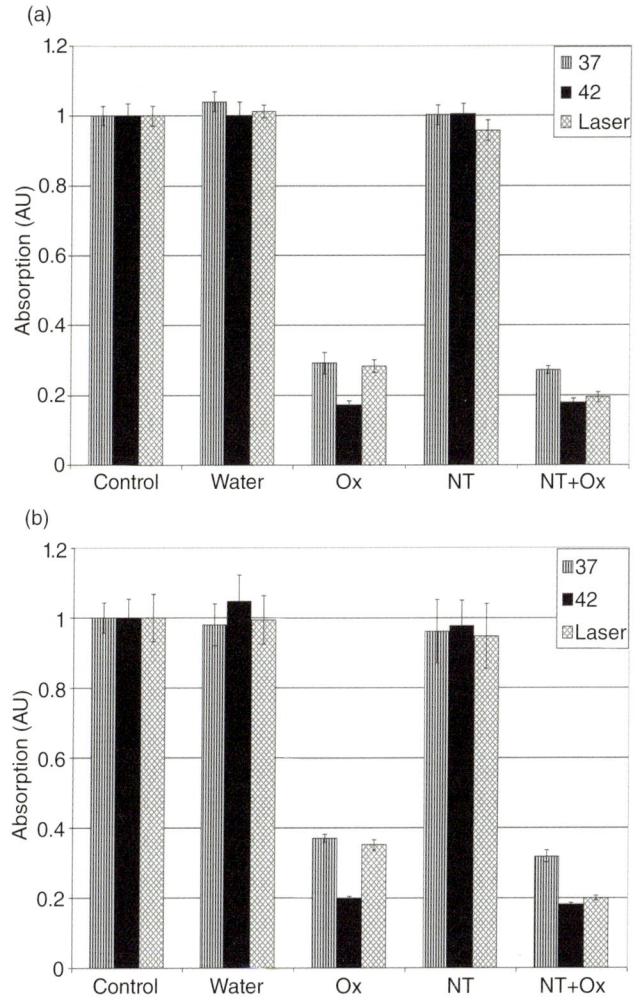

Figure 6.8 The response of (a) HCT 116 and (b) RKO colorectal cancer cell lines to nanotubes (NT), laser application and oxaliplatin (Ox). Reproduced from Ref. [61].

used to generate local hyperthermia which could then be measured using magnetic resonance imaging (MRI) thermometry. This is the first step in the development of nanotubes, the interior of which contains a chemotherapeutic agent that can be released under hyperthermic conditions.

Nanotube-induced hyperthermia does not only have to be light-induced; rather, it can also be induced by the application of a radiofrequency field (see Figure 6.10). However, the radiofrequency experiments reported to date have used large quantities of SWNTs (500 mg ml^{-1}), whereas others have used 100 mg ml^{-1} MWNTs, and Kam *et al.* used 5 mg ml^{-1} folic acid ligand-targeted SWNTs to achieve sufficient

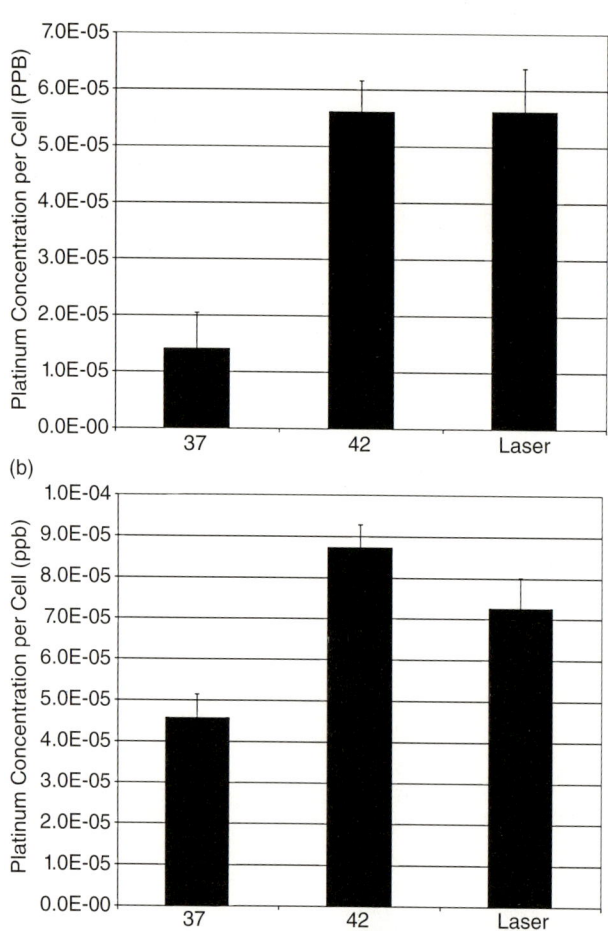

Figure 6.9 The amount of platinum per RKO (a) or HCT 116 (b) colorectal cancer cell treated rapidly at 37, 42, or 42 °C by the infrared laser stimulation of nanotubes. Reproduced from Ref. [61].

heating [53, 63–65]. To date, MWNTs have not been used with radiofrequency to generate hyperthermia, although it is expected that a similar trend would be observed – and possibly even enhanced – by a more efficient heat transfer from MWNTs.

6.5.2
Drug Transport Using CNTs

Currently, there is a clinical limitation for the application of electric fields through tissues. Electric fields are commonly applied by using a needle or plate electrodes

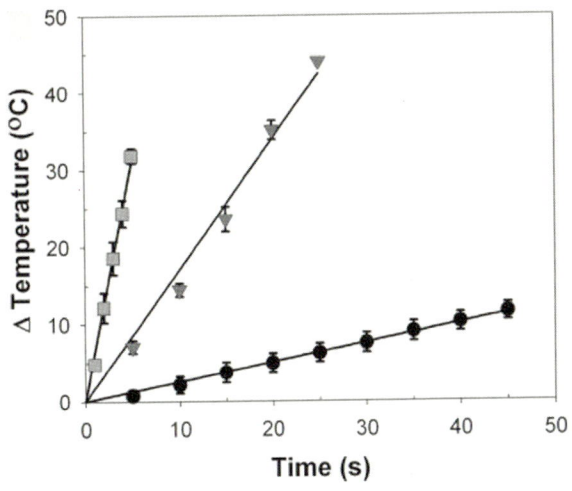

Figure 6.10 Hyperthermia generated by nanotubes stimulated with radiofrequency (600 W). The circles indicate no SWNT, triangles indicate SWNT at a concentration of 50 mg l^{-1}, and squares indicate SWNT at a concentration of 250 mg l^{-1}. Reproduced from Ref. [63].

to the tissue; typically, the plate electrodes when used clinically will apply electric fields of between 0.5 and 100 kV cm^{-1} to the tissue, which is quite high [66]. The MWNTs can be used to generate very high electric fields, specifically at the ends of the CNT, while MWNTs have a longitudinal metallic response and a transverse mode insulator response. Based on the principles that MWNTs behave as metals in a longitudinal field, and as insulators in a transverse field, MWNTs have been used to generate electric fields. The retention of intracellular trypan blue (a dye that easily permeates dead or compromised cell membranes) was investigated by Raffa *et al.*, who showed that two types of wave are required in order to generate an electropermeabilization of the MWNTs [66]. An alignment wave was found to be critical for aligning the nanotubes with respect to the transverse mode of the electric field, in order to effectively generate electropermeabilization and the uptake of the trypan blue in cells. Hence, an electric field was applied to cells in a MWNT-containing solution on an electrode chip. The concentration of the MWNT solution was 10 μg ml^{-1}, and the field was applied at 35 V cm^{-1} for the longitudinal mode, for 40 ms at 15 V cm^{-1} for the alignment wave, and for 20 μs with an applied frequency of 1 Hz and 500 pulses to induce the uptake of trypan blue. Raffa *et al.* also investigated cells in an MWNT-containing suspension in a cuvette, whereby 55 V cm^{-1} was applied as the maximum electric field voltage. An 80% efficiency of open porosity was found in the cell membranes, and this enabled trypan blue to be transported effectively into the cells. Consequently, intracellular trypan blue was observed within 15 min after electropermeabilization.

Yu *et al.* attached a gonadotropin-releasing hormone (GnRH) onto MWNTs by using an amide linkage [67], and subsequently examined the delivery mechanism of the MWNT–GnRH complexes in DU145 prostate and HeLa ovarian cells. Although CNTs were visualized inside the cells, they were not present in the nucleus. Delivery using the MWNT–GnRH complex resulted in an 82% death rate of DU145 cells, but of only 35% for HeLa cells (this difference occurred specifically because HeLa cells do not express GnRH receptors). Notably, the separate delivery of GnRH and MWNT proved nontoxic to the cells. Pastorin *et al.* identified a rapid kinetic uptake (within 1 h) of MWNTs that were noncovalently linked with a fluorescent probe and the chemotherapeutic agent, methotrexate. The MWNTs were seen to be localized around the nuclear membrane, but were unable to penetrate the nucleus [68].

Very interesting results have been reported by Iijima and coworkers for the delivery of cisplatin and other platinum compounds, when using single-wall carbon nanohorns (SWNHs) of 2–5 nm diameter, with a length of approximately 40–50 nm. Thousands of SWNHs are able to aggregate into a larger particle with a diameter of approximately 80–100 nm. In these studies, the SWNHs were first oxidized in order to form holes in the sidewalls, and then gently mixed with a cisplatin compound in a solvent that could be quickly evaporated, leaving only the cisplatin behind (see Figure 6.11). The result was a 12% encapsulation of cisplatin within the nanohorns, and an 80% release of cisplatin from the nanohorns over a 50 h period [69]. The major finding, however, was that a very slow oxidation of the nanohorns, at high temperature in air, produced larger holes in the nanohorns; this, in turn, led to a greater uptake of cisplatin, as well as a greater release of cisplatin over time.

In a further report, Ajima *et al.* showed that oxidized SWNHs had COOH groups and OH groups that formed specifically at the edges of the holes in the oxidized horns [70]. Unfortunately, this proved to be problematic, as there was a strong interaction between these groups and the sodium in the PBS that was used for drug release. The release of cisplatin was slowed threefold from these oxidized tubes, with only a 15% release despite an original overall filling of the nanotubes of 70%. It was found that the release rate could be controlled by oxidation of the SWNHs; that is, the longer the oxidation time the larger the holes and the less plugging would be required due to sodium interaction with the COOH and OH groups on the horns. In 2007, Matsamura reported an improvement in Ajima's mechanism for the cisplatin loading of carbon nanohorns. One problem with the loading of carbon nanohorns is their aggregation, leading to an inefficient release rate. Consequently, the nanohorns were noncovalently functionalized with PEG and the peptide aptamer NHBP1, prior to the addition of cisplatin [71]. Subsequently, the average size hole in the nanohorn was found to be 2–3 nm, while the major axis of the cisplatin molecule was 0.5 nm.

One major problem of aggregation is caused by the salts present in ionic solutions or cell culture media, and used to disperse the carbon nanohorns. The authors required the aptamer because PEG alone was unable to achieve adequate nanohorn dispersion, and the addition of the PEG–NHBP1 failed to inhibit cisplatin release. The most significant studies conducted to date for filling carbon nanomaterials with a chemotherapeutic agent has been the filling of SWNHs with

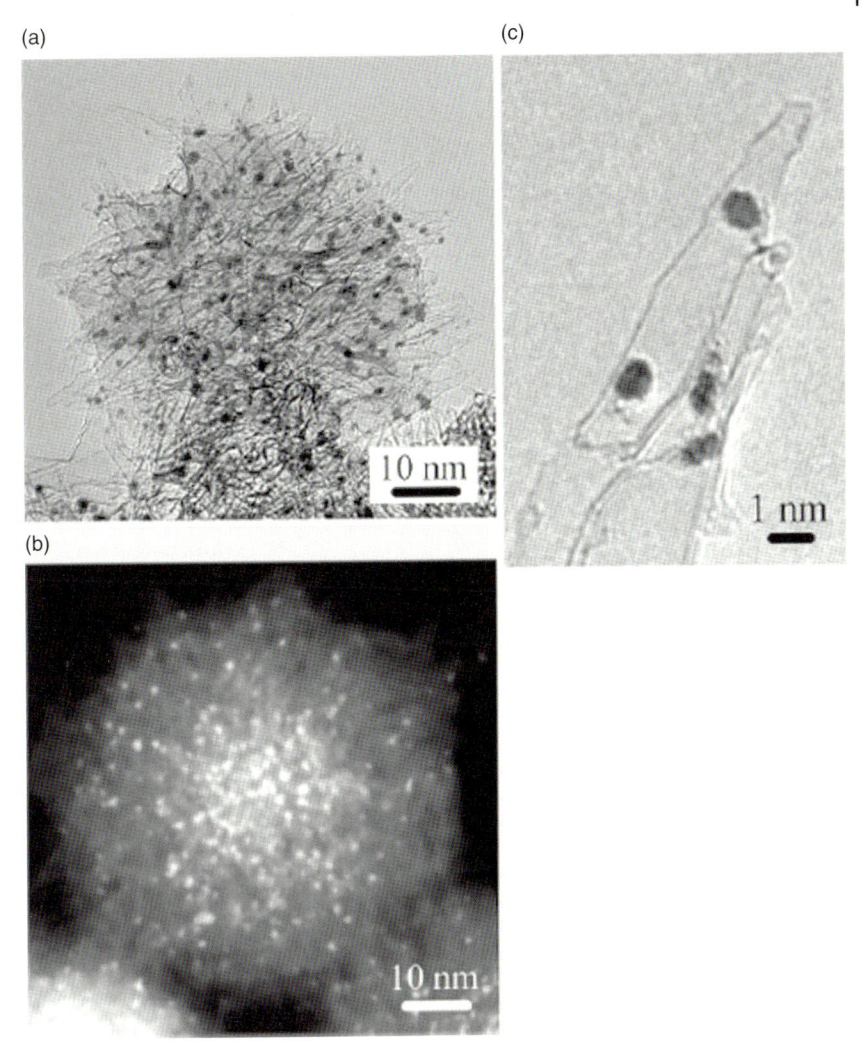

Figure 6.11 (a) TEM image and (b, c) inverse contrast TEM images of cisplatin loaded into carbon nanohorns. Reproduced from Ref. [70].

cisplatin. Although cisplatin is an ideal drug for filling, due to its compact size, palladium crystals are readily formed on the interior surfaces of the MWNT (see Figure 6.1). Therefore, the filling of a MWNT with cisplatin in a manner similar to the filling of carbon nanohorns is a distinct possibility. One advantage of using MWNTs is the amount of chemotherapeutic agent that one tube can carry; the main challenge, however, would be to contain the drug within the nanotube and to release it optimally, at a specific location.

The major mechanism for cancer metastasis is the migration of micrometastases from the primary tumor lesion into the lymph nodes; consequently, the targeting of lymph nodes might be exceptionally important when targeting micrometastases. In order to more efficiently localize the drug, gemcitabine, to the lymph nodes of a mouse, Yang *et al.* attached magnetite (Fe_3O_4) nanoparticles by precipitation onto MWNTs [72]. They also grafted poly(acrylic acid) (PAA) onto the MWNTs in order to aid the dispersion of the nanotubes. A magnet-driven approach was then used to distribute the nanotubes containing the drug into the lymph nodes. Initially, when the complex was injected subcutaneously into the left foot pad of rats, neither toxic effects nor skin problems were observes at the injection site. However, after 3 h the lymph node closest to the injection site was examined, and found to contain MWNTs, although no MWNTs were seen in the other organs. It was suggested that a subcutaneous delivery of the complex might result in an accumulation in the lymph nodes, due to the enhanced permeability and retention (EPR) effect which is quite common with many nanoparticles. By employing the magnetic properties of the magnetite that had been precipitated onto the MWNTs, it was possible to suture a magnet of 1800 Gauss onto the lymph node closest to the injection site. Notably, compared to controls, after 24 h the highest concentration of gemcitabine was present in the targeted lymph node, while the lowest concentration was present in the blood plasma.

6.6
Summary and Future Perspectives

The larger aspect ratio and larger diameters of the MWNTs allow them to carry more molecules, either outside or inside, so as to minimize the aggregation of therapeutic agents and/or retain the conformational structure of the attached agent, leading to an enhanced performance of the agent. Whilst the filling of MWNTs has been accomplished by capillary action or centrifugation, one challenge remains regarding the loading of sufficient material onto or into a MWNT as to be valuable for clinical use. Current mechanisms of drug release can include enzymatic degradation, or capillary and hyperthermic emptying of the interior tubular contents, using light or sound. Previously, MWNTs have been used for the efficient delivery of plasmid DNA to regulate the cell cycle and apoptosis, and also to deliver peptides to mediate antibody synthesis and, consequently, immunity. As an antibacterial agent, MWNTs have been shown to be less toxic than SWNTs, apparently based on their larger diameter, and they also lack any significant negative consequences on membrane integrity. Although MWNTs appear not to penetrate bacterial cell walls as readily as SWNTs, they do interact strongly with the cell walls for the delivery of cytotoxic agents. Typically, MWNTs have been shown to interact with the plasma membrane of eukaryotic cells, which can be used for either positive consequences (by disrupting the plasma membrane for enhanced drug delivery) or negative consequences (by disrupting the plasma membrane of cells, the membranes of which should not be manipulated). Perhaps the interaction of the membrane could be modified by altering the aspect ratio: shorter MWNTs may disrupt cell membranes to a lesser degree, thus imparting less damage while still being able to localize and deliver therapeutic agents. MWNTs can also aid in wound healing, by significantly influencing the release of therapeutic agents for polymer membranes by diffusion, or electrical stimulation. Certainly, MWNTs can be as equally beneficial as SWNTs in the delivery of chemotherapeutic agents by intracellular transport or hyperthermia-induced methods. In fact, MWNTs are superior to SWNTs when creating rapid increases in temperature in response to infrared stimulation. Finally, it is to be expected that MWNTs may be beneficial for the delivery of chemotherapeutic platinum compounds, by mechanisms similar to those used by the SWNHs.

The future use of MWNTs in medical applications is just beginning to emerge. Indeed, many of the medical applications that are commonly evaluated using SWNTs may function even better when using MWNTs, although some of the fundamental mechanisms as to why MWNTs *are* better must first be understood. The development of polymer–MWNT composites for the delivery of therapeutic agents holds significant potential as a solid first step for using MWNTs to optimize transdermal drug delivery. The benefits of using MWNTs for intracellular transport, either directly or indirectly via hyperthermia, may extend to many areas of medicine that include wound healing, the control of bacterial infections, and the control of cancer by therapeutic agents or genetic regulation. Clearly, the unique aspects of using MWNTs for drug delivery are on the cusp of development of medical applications and, if successful, could be transferred rapidly into clinical use.

References

1 Dresselhaus, M.S., Dresselhaus, G., Charlier, J.C. and Hernandez, E. (2004) Electronic, thermal and mechanical properties of carbon nanotubes. *Philosophical Transactions of the Royal Society of London Series A-Mathematical Physical and Engineering Sciences*, **362**, 2065–98.

2 Cheng, C., Muller, K.H., Koziol, K.K.K., Skepper, J.N., Midgley, P.A., Welland, M.E. and Porter, A.E. (2009) Toxicity and imaging of multi-walled carbon nanotubes in human macrophage cells. *Biomaterials*, **30**, 4152–60.

3 Hirano, S., Kanno, S. and Furuyama, A. (2008) Multi-walled carbon nanotubes injure the plasma membrane of macrophages. *Toxicology and Applied Pharmacology*, **232**, 244–51.

4 Star, A., Steuerman, D.W., Heath, J.R. and Stoddart, J.F. (2002) Starched carbon nanotubes. *Angewandte Chemie - International Edition*, **41**, 2508–12.

5 Bandyopadhyaya, R., Nativ-Roth, E., Regev, O. and Yerushalmi-Rozen, R. (2002) Stabilization of individual carbon nanotubes in aqueous solutions. *Nano Letters*, **2**, 25–8.

6 Dodziuk, H., Ejchart, A., Anczewski, W., Ueda, H., Krinichnaya, E., Dolgonos, G. and Kutner, W. (2003) Water solubilization, determination of the number of different types of single-wall carbon nanotubes and their partial separation with respect to diameters by complexation with eta-cyclodextrin. *Chemical Communications*, 986–7.

7 Matsuura, K., Hayashi, K. and Kimizuka, N. (2003) Lectin-mediated supramolecular

junctions of galactose-derivatized single-walled carbon nanotubes. *Chemistry Letters*, **32**, 212–13.

8 Pompeo, F. and Resasco, D.E. (2002) Water solubilization of single-walled carbon nanotubes by functionalization with glucosamine. *Nano Letters*, **2**, 369–73.

9 Lin, Y., Allard, L.F. and Sun, Y.P. (2004) Protein-affinity of single-walled carbon nanotubes in water. *Journal of Physical Chemistry B*, **108**, 3760–4.

10 Menna, E., Scorrano, G., Maggini, M., Cavallaro, M., Della Negra, F., Battagliarin, M., Bozio, R., Fantinel, F. and Meneghetti, M. (2003) Shortened single-walled nanotubes functionalized with poly(ethylene glycol): preparation and properties. *ARKIVOC*, **12**, 64–73.

11 Wang, S.Q., Humphreys, E.S., Chung, S.Y., Delduco, D.F., Lustig, S.R., Wang, H., Parker, K.N., Rizzo, N.W., Subramoney, S., Chiang, Y.M. and Jagota, A. (2003) Peptides with selective affinity for carbon nanotubes. *Nature Materials*, **2**, 196–200.

12 Huang, W.J., Taylor, S., Fu, K.F., Lin, Y., Zhang, D.H., Hanks, T.W., Rao, A.M. and Sun, Y.P. (2002) Attaching proteins to carbon nanotubes via diimide-activated amidation. *Nano Letters*, **2**, 311–14.

13 Kagan, V.E., Bayir, H. and Shvedova, A.A. (2005) Nanomedicine and nanotoxicology: two sides of the same coin. *Nanomedicine: Nanotechnology, Biology and Medicine*, **1**, 313–16.

14 Carrero-Sanchez, J.C., Elias, A.L., Mancilla, R., Arrellin, G., Terrones, H., Laclette, J.P. and Terrones, M. (2006) Biocompatibility and toxicological studies of carbon nanotubes doped with nitrogen. *Nano Letters*, **6**, 1609–16.

15 Monthioux, M. (2002) Filling single-wall carbon nanotubes. *Carbon*, **40**, 1809–23.

16 Kim, B.M., Qian, S. and Bau, H.H. (2005) Filling carbon nanotubes with particles. *Nano Letters*, **5**, 873–8.

17 Bazilevsky, A.V., Sun, K., Yarin, A.L. and Megaridis, C.M. (2008) Room-temperature, open-air, wet intercalation of liquids, surfactants, polymers and nanoparticles within nanotubes and microchannels. *Journal of Materials Chemistry*, **18**, 696–702.

18 Bazilevsky, A.V., Sun, K.X., Yarin, A.L. and Megaridis, C.M. (2007) Selective intercalation of polymers in carbon nanotubes. *Langmuir*, **23**, 7451–5.

19 Nadarajan, S.B., Katsikis, P.D. and Papazoglou, E.S. (2007) Loading carbon nanotubes with viscous fluids and nanoparticles – a simpler approach. *Applied Physics A - Materials Science and Processing*, **89**, 437–42.

20 Zheng, M., Jagota, A., Semke, E.D., Diner, B.A., Mclean, R.S., Lustig, S.R., Richardson, R.E. and Tassi, N.G. (2003) DNA-assisted dispersion and separation of carbon nanotubes. *Nature Materials*, **2**, 338–42.

21 Wang, J.X., Li, M.X., Shi, Z.J., Li, N.Q. and Gu, Z.N. (2004) Electrochemistry of DNA at single-wall carbon nanotubes. *Electroanalysis*, **16**, 140–4.

22 Singh, R., Pantarotto, D., Mccarthy, D., Chaloin, O., Hoebeke, J., Partidos, C.D., Briand, J.P., Prato, M., Bianco, A. and Kostarelos, K. (2005) Binding and condensation of plasmid DNA onto functionalized carbon nanotubes: toward the construction of nanotube-based gene delivery vectors. *Journal of the American Chemical Society*, **127**, 4388–96.

23 Cai, D., Mataraza, J.M., Qin, Z.H., Huang, Z.P., Huang, J.Y., Chiles, T.C., Carnahan, D., Kempa, K. and Ren, Z.F. (2005) Highly efficient molecular delivery into mammalian cells using carbon nanotube spearing. *Nature Methods*, **2**, 449–54.

24 Jia, N.Q., Lian, Q., Shen, H.B., Wang, C., Li, X.Y. and Yang, Z.N. (2007) Intracellular delivery of quantum dots tagged antisense oligodeoxynucleotides by functionalized multiwalled carbon nanotubes. *Nano Letters*, **7**, 2976–80.

25 Kateb, B., Van Handel, M., Zhang, L., Bronikowski, M.J., Manohara, H. and Badie, B. (2007) Internalization of MWNTs by microglia: possible application in immunotherapy of brain tumors. *NeuroImage*, **37**, S9–S17.

26 Yandar, N., Pastorin, G., Prato, M., Bianco, A., Patarroyo, M.E. and Manuel Lozano, J. (2008) Immunological profile of a *Plasmodium vivax* AMA-1 N-terminus peptide-carbon nanotube conjugate in an infected *Plasmodium*

berghei mouse model. *Vaccine*, 26, 5864–73.

27 Kang, S., Herzberg, M., Rodrigues, D.F. and Elimelech, M. (2008) Antibacterial effects of carbon nanotubes: size does matter. *Langmuir*, 24, 6409–13.

28 Liu, S.B., Wei, L., Hao, L., Fang, N., Chang, M.W., Xu, R., Yang, Y.H. and Chen, Y. (2009) Sharper and faster 'nano darts' kill more bacteria: a study of antibacterial activity of individually dispersed pristine single-walled carbon nanotube. *ACS Nano*, 3, 3891–902.

29 Kang, S., Mauter, M.S. and Elimelech, M. (2009) Microbial cytotoxicity of carbon-based nanomaterials: implications for river water and wastewater effluent. *Environmental Science and Technology*, 43, 2648–53.

30 Simon-Deckers, A., Loo, S., Mayne-L'hermite, M., Herlin-Boime, N., Menguy, N., Reynaud, C., Gouget, B. and Carriere, M. (2009) Size-, composition- and shape-dependent toxicological impact of metal oxide nanoparticles and carbon nanotubes toward bacteria. *Environmental Science and Technology*, 43, 8423–9.

31 Wu, W., Wieckowski, S., Pastorin, G., Benincasa, M., Klumpp, C., Briand, J.P., Gennaro, R., Prato, M. and Bianco, A. (2005) Targeted delivery of amphotericin B to cells by using functionalized carbon nanotubes. *Angewandte Chemie - International Edition*, 44, 6358–62.

32 Levi, N., Czerw, R., Xing, S.Y., Iyer, P. and Carroll, D.L. (2004) Properties of polyvinylidene difluoride-carbon nanotube blends. *Nano Letters*, 4, 1267–71.

33 Li, H., Wang, D.Q., Liu, B.L. and Gao, L.Z. (2004) Synthesis of a novel gelatin-carbon nanotubes hybrid hydrogel. *Colloids and Surfaces B - Biointerfaces*, 33, 85–8.

34 El Sherbiny, I.M., Lins, R.J., Abdel-Bary, E.M. and Harding, D.R.K. (2005) Preparation, characterization, swelling and *in vitro* drug release behaviour of poly N-acryloylglycine-chitosan interpolymeric pH and thermally-responsive hydrogels. *European Polymer Journal*, 41, 2584–91.

35 Jain, A., Kim, Y.T., Mckeon, R.J. and Bellamkonda, R.V. (2006) In situ gelling hydrogels for conformal repair of spinal cord defects, and local delivery of BDNF after spinal cord injury. *Biomaterials*, 27, 497–504.

36 Wu, J., Wei, W., Wang, L.Y., Su, Z.G. and Ma, G.H. (2007) A thermosensitive hydrogel based on quaternized chitosan and poly(ethylene glycol) for nasal drug delivery system. *Biomaterials*, 28, 2220–32.

37 Haider, S., Park, S.Y., Saeed, K. and Farmer, B.L. (2007) Swelling and electroresponsive characteristics of gelatin immobilized onto multi-walled carbon nanotubes. *Sensors and Actuators B - Chemical*, 124, 517–28.

38 Thompson, B.C., Moulton, S.E., Gilmore, K.J., Higgins, M.J., Whitten, P.G. and Wallace, G.G. (2009) Carbon nanotube biogels. *Carbon*, 47, 1282–91.

39 Majumder, M., Stinchcomb, A. and Hinds, B.J. (2010) Towards mimicking natural protein channels with aligned carbon nanotube membranes for active drug delivery. *Life Sciences*, 86, 563–8.

40 Rotoli, B.M., Bussolati, O., Bianchi, M.G., Barilli, A., Balasubramanian, C., Bellucci, S. and Bergamaschi, E. (2008) Non-functionalized multi-walled carbon nanotubes alter the paracellular permeability of human airway epithelial cells. *Toxicology Letters*, 178, 95–102.

41 Im, J.S., Bai, B.C. and Lee, Y.S. (2010) The effect of carbon nanotubes on drug delivery in an electro-sensitive transdermal drug delivery system. *Biomaterials*, 31, 1414–19.

42 American Cancer Society (2004) Cancer Facts and Statistics.

43 Weissleder, R. (2001) *Nature Biotechnology*, 19, 319.

44 Braun, C. and Smirnov, S. (1993) Why is water blue. *Journal of Chemical Education*, 70, 612.

45 Wang, Y., Kempa, K., Kimball, B., Carlson, J.B., Benham, G., Li, W.Z., Kempa, T., Rybczynski, J., Herczynski, A. and Ren, Z.F. (2004) Receiving and transmitting light-like radio waves: antenna effect in arrays of aligned carbon nanotubes. *Applied Physics Letters*, 85, 2607–9.

46 Kouklin, N., Tzolov, M., Straus, D., Yin, A. and Xu, J.M. (2004) Infrared absorption properties of carbon nanotubes synthesized by chemical vapor

deposition. *Applied Physics Letters*, **85**, 4463–5.

47 Bachilo, S.M., Strano, M.S., Kittrell, C., Hauge, R.H., Smalley, R.E. and Weisman, R.B. (2002) Structure-assigned optical spectra of single-walled carbon nanotubes. *Science*, **298**, 2361–6.

48 Dresselhaus, M.S. and Eklund, P.C. (2000) Phonons in carbon nanotubes. *Advances in Physics*, **49**, 705–814.

49 Maksimenko, S.A., Slepyan, G.Y., Nemilentsau, A.M. and Shuba, M.V. (2008) Carbon nanotube antenna: far-field, near-field and thermal-noise properties. *Physica E: Low-Dimensional Systems and Nanostructures*, **40**, 2360–4.

50 Hepplestone, S.P. and Srivastava, G.P. (2006) The intrinsic lifetime of low-frequency zone-centre phonon modes in silicon nanowires and carbon nanotubes. *Applied Surface Science*, **252**, 7726–9.

51 Launay, S., Fedorov, A.G., Joshi, Y., Cao, A. and Ajayan, P.M. (2006) Hybrid micro-nano structured thermal interfaces for pool boiling heat transfer enhancement. *Microelectronics Journal*, **37**, 1158–64.

52 Park, K.J. and Jung, D. (2007) Enhancement of nucleate boiling heat transfer using carbon nanotubes. *International Journal of Heat and Mass Transfer*, **50**, 4499–502.

53 Burke, A., Ding, X.F., Singh, R. *et al.* (2009) Long-term survival following a single treatment of kidney tumors with multiwalled carbon nanotubes and near-infrared radiation. *Proceedings of the National Academy of Sciences of the United States of America*, **106**, 12897–902.

54 Loggie, B.W., Fleming, R.A. and Geisinger, K.R. (1996) Cytologic assessment before and after intraperitoneal hyperthermic chemotherapy for peritoneal carcinomatosis. *Acta Cytologica*, **40**, 1154–8.

55 Shen, P., Hawksworth, J., Lovato, J., Loggie, B.W., Geisinger, K.R., Fleming, R.A. and Levine, E.A. (2004) Cytoreductive surgery and intraperitoneal hyperthermic chemotherapy with mitomycin C for peritoneal carcinomatosis from nonappendiceal colorectal carcinoma. *Annals of Surgical Oncology*, **11**, 178–86.

56 Levine, E.A., Stewart, J.H., Russell, G.B., Geisinger, K.R., Loggie, B.L. and Shen, P. (2007) Cytoreductive surgery and intraperitoneal hyperthermic chemotherapy for peritoneal surface malignancy: experience with 501 procedures. *Journal of the American College of Surgeons*, **204**, 943–53.

57 Sugarbaker, P.H., Mora, J.T., Carmignani, P., Stuart, O.A. and Yoo, D. (2005) Update on chemotherapeutic agents utilized for perioperative intraperitoneal chemotherapy. *Oncologist*, **10**, 112–22.

58 Reingruber, B., Boettcher, M.I., Klein, P., Hohenberger, W. and Pelz, J.O.W. (2007) Hyperthermic intraperitoneal chemoperfusion is an option for treatment of peritoneal carcinomatosis in children. *Journal of Pediatric Surgery*, **42**, e17–21.

59 Mohamed, F., Marchettini, P., Stuart, O.A., Urano, M. and Sugarbaker, P.H. (2003) Thermal enhancement of new chemotherapeutic agents at moderate hyperthermia. *Annals of Surgical Oncology*, **10**, 463–8.

60 Hildebrandt, B., Wust, P., Ahlers, O., Dieing, A., Sreenivasa, G., Kerner, T., Felix, R. and Riess, H. (2002) The cellular and molecular basis of hyperthermia. *Critical Reviews in Oncology Hematology*, **43**, 33–56.

61 Levi-Polyachenk, N.H., Merkel, E.J., Jones, B.T., Carroll, D.L. and Stewart, J.H. (2009) Rapid photothermal intracellular drug delivery using multiwalled carbon nanotubes. *Molecular Pharmaceutics*, **6**, 1092–9.

62 Klingeler, R., Hampel, S. and Buchner, B. (2008) Carbon nanotube based biomedical agents for heating, temperature sensing and drug delivery. *International Journal of Hyperthermia*, **24**, 496–505.

63 Gannon, C.J., Cherukuri, P., Yakobson, B.I., Cognet, L., Kanzius, J.S., Kittrell, C., Weisman, R.B., Pasquali, M., Schmidt, H.K., Smalley, R.E. and Curley, S.A. (2007) Carbon nanotube-enhanced thermal destruction of cancer cells in a noninvasive radiofrequency field. *Cancer*, **110**, 2654–65.

64 Torti, S., Byrne, F., Whelan, O., Levi, N., Ucer, B., Schmid, M., Torti, F., Akman, S., Liu, J., Ajayan, P., Nalamasu, O. and

Carroll, D. (2007) Thermal ablation therapeutics based on CN_x multi-walled nanotubes. *International Journal of Nanomedicine*, **2**, 707–14.

65 Kam, N.W.S., O'Connell, M., Wisdom, J.A. and Dai, H.J. (2005) Carbon nanotubes as multifunctional biological transporters and near-infrared agents for selective cancer cell destruction. *Proceedings of the National Academy of Sciences of the United States of America*, **102**, 11600–5.

66 Raffa, V., Ciofani, G., Vittorio, O., Pensabene, V. and Cuschieri, A. (2010) Carbon nanotube-enhanced cell electropermeabilisation. *Bioelectrochemistry*, **79**, 136–41.

67 Yu, B.Z., Yang, J.S. and Li, W.X. (2007) In vitro capability of multi-walled carbon nanotubes modified with gonadotrophin releasing hormone on killing cancer cells. *Carbon*, **45**, 1921–7.

68 Pastorin, G., Wu, W., Wieckowski, S., Briand, J.P., Kostarelos, K., Prato, M. and Bianco, A. (2006) Double functionalisation of carbon nanotubes for multimodal drug delivery. *Chemical Communications*, 1182–4.

69 Ajima, K., Maigne, A., Yudasaka, M. and Iijima, S. (2006) Optimum hole-opening condition for cisplatin incorporation in single-wall carbon nanohorns and its release. *Journal of Physical Chemistry B*, **110**, 19097–9.

70 Ajima, K., Yudasaka, M., Maigne, A., Miyawaki, J. and Iijima, S. (2006) Effect of functional groups at hole edges on cisplatin release from inside single-wall carbon nanohorns. *Journal of Physical Chemistry B*, **110**, 5773–8.

71 Matsumura, S., Ajima, K., Yudasaka, M., Iijima, S. and Shiba, K. (2007) Dispersion of cisplatin-loaded carbon nanohorns with a conjugate comprised of an artificial peptide aptamer and polyethylene glycol. *Molecular Pharmaceutics*, **4**, 723–9.

72 Yang, D., Yang, F., Hu, J.H., Long, J., Wang, C.C., Fu, D.L. and Ni, Q.X. (2009) Hydrophilic multi-walled carbon nanotubes decorated with magnetite nanoparticles as lymphatic targeted drug delivery vehicles. *Chemical Communications*, 4447–9.

7
Carbon Nanotube-Based Three-Dimensional Matrices for Tissue Engineering

Izabela Firkowska and Michael Giersig

7.1
Introduction

Today, nanotechnology poses new frontiers in science and technology, the essence of this highly multi- and interdisciplinary field being not only the ability to work at atomic and molecular levels, but also to create structures or devices with a fundamentally new molecular organization. One of the many fields to have benefited from the rapid evolution of nanotechnology – and, in particular, from the discovery of nanoscaled materials – is that of tissue engineering. The aim of this interdisciplinary field is to develop biological substitutes (such as artificial extracellular matrices, also termed "scaffolds") that are capable of repairing or regenerating the functions of a damaged tissue. In order to engineer materials capable of supporting the structure and function of human tissue, a deep understanding is required not only of the cell itself but also of the extracellular matrix (ECM) interactions that take place within the tissues. For several decades, the artificial scaffolds that have been designed to support cell and tissue regeneration have traditionally been focused on a macroscopic level. The main aim was to match the properties similar to those of natural tissues, but without reconstructing the nanoscale features observed in native tissues.

All tissues of the human body contain differentiated cells living in an ECM that demonstrates a hierarchical organization from nano- to macroscopic length scale; consequently, it is clear that cells are naturally accustomed to interact with nanometer length-scale elements [1]. It is for this reason that nanoscale surface features are considered to be key determinants of the cellular response. Indeed, it is commonly believed that the topography, engineered with nanoscale structural elements analogous to dimensions present in the ECM, is critical for the correct function of each specialized tissue. These new discoveries in nanotechnology, combined with new abilities of engineering, have enabled materials scientists and engineers to design and fabricate novel scaffolds by incorporating nanoscale features, thus imitating the characteristics of the natural ECM. However, in order

Nanomaterials for the Life Sciences Vol.9: Carbon Nanomaterials. Edited by Challa S. S. R. Kumar
Copyright © 2011 WILEY-VCH Verlag GmbH & Co. KGaA, Weinheim
ISBN: 978-3-527-32169-8

to accomplish the construction of more biomimetic cellular environments, the fundamental design principles that determine how cells and tissues form and function as hierarchical assemblies of nanoscale and micrometer-scale components must first be understood. Accordingly, much attention has been focused on cell-material–surface interactions, with particular interest on those where the properties try to mimic the dimensions present in natural tissues. During the past few years, many investigations have demonstrated a powerful influence of nanoscale topography on cellular behavior, ranging from changes in cell adhesion, spreading, and/or cytoskeletal organization to the regulation of gene expression [2–4]. Moreover, it has been shown that nanoscaled topography may induce various responses of the same cell type, independent of the underlying material chemistry [5, 6].

The nanoscale structure of the ECM provides a natural net of complex nanofibers, which support and guide the cell behavior. Each fiber conceals clues that pave the way for the cell to form tissues as complex as bone, heart, and liver [7]. Until recently, an ECM-mimicking fiber with a nanoscale diameter has been the missing constituent in cell scaffold design [8]. One of the most promising nanometer-sized cylinders that could imitate nanofibers present in native ECM, are carbon nanotubes (CNTs). Zhao et al. [9], for example, have demonstrated the potential of nanotubes to mimic the role of collagen, the major component of the ECM protein in the human body, and to serve as a scaffold for the growth of hydroxyapatite. Apart from nanoscale dimensions, CNTs possess numerous physical, chemical, and mechanical properties, which make them distinct from other nanofibrous materials used for biological applications [10]. In particular, their extremely high strength, lightness, and electrical conductivity enable the creation of biomimetic constructs with highly predictable physical properties. Moreover, the latest expansion and availability of methods for the chemical modification and biofunctionalization of CNTs [11] has made possible the creation of CNT-based scaffolds that possess bioactive surfaces and thus are capable of interacting positively with cells, by means of enhanced cell adhesion, proliferation, migration, and differentiated function. Although, in contrast to nanofibrous polymers such as polyglycolic acid (PGA), poly(L-lactic acid) (PLLA) and poly(lactic-co-glycolic acid) (PLGA), which are frequently used for scaffold fabrications [12], CNTs do not exhibit a biodegradable nature, the above-mentioned advantages more than counterbalance this drawback.

In this chapter, the suitability of CNTs as biomaterials for tissue engineering, and the fabrication, properties and performance of three-dimensional (3-D) CNT-based matrices will be discussed.

7.2
Carbon Nanotubes

Following the initial discovery of CNTs by Iijima during the early 1990s [13], carbon-based nanotechnology has developed rapidly as a platform technology for

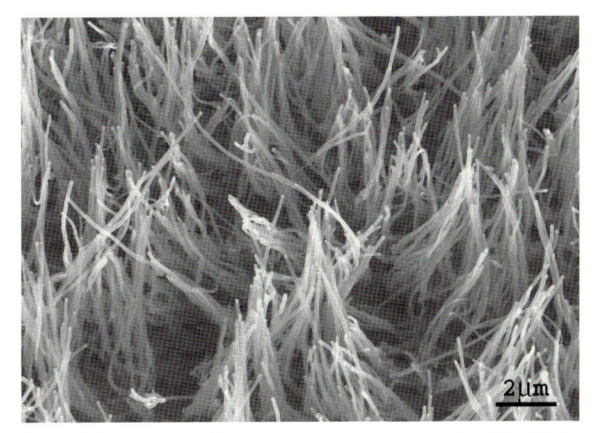

Figure 7.1 Scanning electron microscopy image of multiwalled carbon nanotubes.

a variety of uses, including biomedical applications. Currently, CNTs are considered to consist of a rolled-up graphene sheet that forms a concentric cylinder possessing a nanometer-scale diameter, but with a much greater length that results in a very large aspect ratio (Figure 7.1). The unique properties of CNTs are defined by their structure (chirality), diameter, and length. Two main types of CNT have been identified, namely single-walled carbon nanotubes (SWNTs), which are composed of single tubes, and multi-walled carbon nanotubes (MWNTs), which are composed of groups of SWNTs.

It is possible to synthesize CNTs by using either of three methods, namely chemical vapor deposition (CVD) [14], arc-discharge [15], and laser ablation [16]. Of the methods used for CNT production, CVD is the most frequently used commercially. The technique involves the decomposition of hydrocarbon gases on the substrate in the presence of metal catalyst particles, with the synthesis of CNTs often being either thermally or plasma-enhanced. Depending on the growth conditions, CNTs of various lengths (from nanometers to millimeters) and widths (1 to 100 nm) can be produced.

The small radius, large specific surface, and σ-π rehybridization make CNTs very attractive for chemical and biological applications, because of their strong sensitivity to both chemical and/or environmental interactions [17]. The chemical functionalization of CNTs represents a very promising target, as this can lead to improvements in CNT solubility and processability. It also allows the exceptional properties of CNTs to be combined with those of other types of material. To date, several methods have been developed for the functionalization of CNTs, including the covalent functionalization of sidewalls, noncovalent exohedral functionalization (e.g., with surfactants and polymers), endohedral functionalization, and defect functionalization. Chemical groups present on the CNTs may also serve as anchor groups for further functionalization; typical examples include biological and bioactive species such as proteins or nucleic acids [18, 19]. This

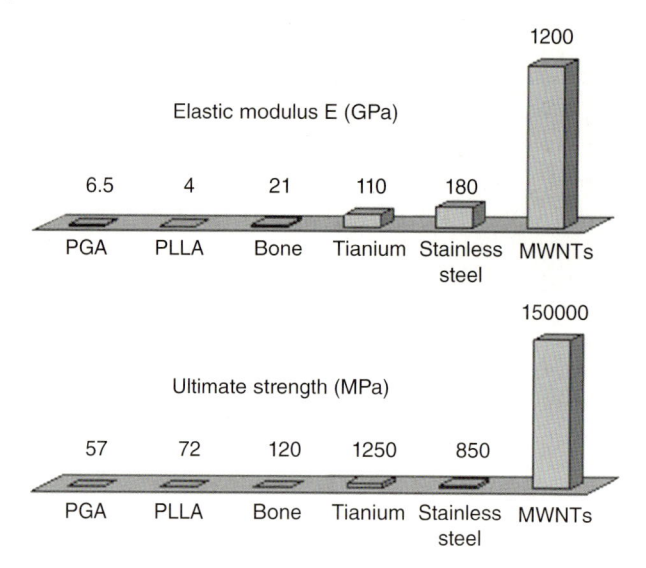

Figure 7.2 Illustration of the elastic modulus and strength of carbon nanotubes [17] and common tissue engineering materials: polyglycolic acid (PGA), poly(L-lactic acid) (PLLA) [20], bone, titanium, and stainless steel [21].

type of bioconjugation is especially attractive for the biomedical applications of CNTs.

Mechanically, CNTs are currently the strongest fibers known to mankind, with the extraordinary high tensile strength (150 GPa) and excellent flexibility (reversible bending up to angles of 120°) of CNTs providing them with superiority over popular materials used in tissue engineering (such as PGA, PLLA, titanium, and steel), and making them ideal candidates for the production of lightweight, high-strength bone materials. For comparative purposes, Figure 7.2 shows the elastic modulus and strength of CNTs, bone, and several other common materials used in bone-tissue engineering.

To summarize, it is believed that this unique combination of properties may be especially useful in the creation of matrices for tissue engineering, the main benefits being their highly predictable biological and physical properties.

7.3
Carbon Nanotubes for Matrix Enhancement

The artificial matrix plays a very important role in tissue engineering, since it is responsible not only for defining the space that the engineered tissue occupies but also for aiding the process of tissue development [22]. One of the few requirements that the matrix must fulfill is that of mechanical stability, as this is crucial for maintaining a predesigned tissue structure. Mechanical stability depends mainly on the selection of a biomaterial and the architectural design of the matrix. Although polymers, such as PLLA or PLGA are the primary materials for matrices in various tissue engineering applications, they lack the necessary mechanical strength (see Figure 7.2), and neither can these materials be functionalized as easily as CNTs. The extraordinary mechanical properties of CNTs renders them very attractive as reinforcing fillers for the production of a new generation of tissue matrices. The functional groups, which can be readily introduced onto the materials' surfaces, greatly enhance the CNTs' connections with a wide variety of polymeric matrices, thus improving the mechanical properties of the nanocomposites. In fact, current data have revealed that, when dispersed in a polymer, CNTs can provide significant improvements in the mechanical properties of a nanocomposite [23–25]. To date, CNTs have been incorporated into a host of synthetic polymers and biopolymers. Typically, when CNTs are merged with chitosan they significantly enhance the mechanical strength of the nanocomposite [26]; for example, the incorporation of only 0.8 wt% of CNTs into a chitosan matrix led to improvements of 93% and 99%, respectively, in the Young's modulus and tensile strength of the nanomaterial.

Apart from polymer enhancement, CNTs have also been used to reinforce ceramic matrices. For example, Gao *et al.* [27] have successfully fabricated CNT/$BaTiO_3$ composites, where the addition of 1 wt% CNTs increased the fracture toughness by about 240%. Likewise, by using plasma-sprayed techniques, CNTs have been uniformly distributed in a brittle hydroxyapatite (HA) bioceramic coating, improving the fracture toughness of the nanocomposite by 56% [28].

The results of the above-mentioned studies have demonstrated that the mechanical properties of matrices can be significantly improved with CNTs. Moreover, the fact that the addition of very small amounts of CNTs is sufficient for matrix enhancement may counterbalance their nondegradable nature.

CNTs have also been used to create electrically conductive polymers and tissue matrices, with the capacity to provide controlled electrical stimulation. Indeed, it has been reported that current-conducting CNT/polymer composites can promote various osteoblast cell functions. For example, the application of an alternating current to these nanocomposites led to an increase in osteoblast proliferation by 46%, and in calcium deposition of 307% [29]; taken together, these results suggested that CNT-based composites might be used to stimulate bone formation. Other studies have been directed toward exploiting the electrical properties of CNTs for the purpose of healing neurological and brain-related injuries. As an example, Gheith *et al.* [30] have used electrically conductive layer-by-layer (LbL)-assembled, modified SWNT films to stimulate the neurophysiological activity of neural cells. The electrophysiological measurements showed that SWNT films could be used to electrically stimulate significant ion conductance in neuronal cells. This indicated a good electrical coupling between the LbL film and the neuronal cells in the lateral electrical configuration.

7.4
Cellular Responses to CNT-Based Matrices

Whereas, mechanical reinforcement was the initial motivation for the use of CNTs, there is evidence that CNTs can both accelerate and direct cell growth. Several *in vitro* studies have been conducted to investigate the interaction between CNTs or nanocomposites and mammalian cells, and it was shown subsequently that a collagen matrix with embedded SWNTs could sustain high cell viability in the case of smooth muscle cells [31]. In studies conducted by Zanello *et al.* [32], the proliferation and function of osteoblast cells seeded onto five differently functionalized CNTs were examined. Typically, the bone cells were shown to prefer electrically neutral CNTs, and this sustained both osteoblast growth and bone-forming functions. A subsequent follow-up study was conducted to investigate the adhesion properties of osteoblasts, fibroblasts, neuronal cells and astrocytes on polycarbonate urethane/carbon nanotube (PU/CNT) nanocomposites [33]. The study results revealed that cell functions which contributed to glial scar-tissue formation (astrocytes) and fibrous-tissue encapsulation (fibroblasts) had decreased. The possibility of using nanotubes as substrates for nerve cell growth, and also as probes for neural functions at the nanometer scale, has been reported by Mattson *et al.* [34]. These authors showed that neurons grown on CNTs which had been functionalized with a bioactive molecule, 4-hydroxynonenal, developed multiple neurites and extensive branching. Moreover, it also became possible to control the characteristics of neurite outgrowth by manipulating the charge carried by the functionalized CNTs. As shown by Hu *et al.* [35], neurons plated onto positively charged CNTs exhibited more numerous growth cones, a longer neurite outgrowth, and a greater degree of neurite branching when compared to neurons grown on negatively charged nanotubes.

7.5
CNT Engineering into Three-Dimensional Matrices

Although the above-mentioned studies mostly involved investigations into cell adhesion and proliferation on randomly oriented nanotubes and CNT/polymers, the CNTs can also be arranged into 3-D matrices with well-defined periodic architectures.

7.5.1
Vertically Aligned CNT-Based Matrices

Carbon nanotubes, particularly those of a multiwall type, can be grown as forests on flat substrates by using the hot-filament plasma-enhanced chemical vapor deposition (PECVD) technique, the details of which have been extensively reported [36]. Typically, growth takes 5–10 min, depending on the length requirement for the CNTs. Scanning electron microscopy (SEM) images of the typical forest, at low and medium magnification, are shown in Figure 7.3a. The CNT arrays produced are seen to exhibit a perfect vertical alignment, with a very good separation between the individual nanotubes.

Kempa et al. [36] have shown also that periodic arrays of well-aligned CNTs can be fabricated inexpensively on Ni dots, produced via the process of self-assembly nanosphere lithography (NSL). In this process, a suspension of polystyrene nanospheres is applied onto the water surface, where self-assembly into a hexagonally close-packed monolayer (Figure 7.3b) takes place. The nanosphere monolayer is subsequently used as a mask for the deposition of a catalyst (Ni, Co, or Fe), using electron beam evaporation. Following deposition of the Ni dots, the polystyrene particles are removed chemically so as to reveal a honeycomb pattern of Ni dots (of quasi-triangular shape) (Figure 7.3b, right) that can be used directly to prepare growth-aligned CNT arrays, using a hot-filament PECVD technique.

Representative SEM images of the periodically aligned nanotubes, where the position of the nanotubes corresponds to the honeycomb pattern of the nickel catalysts, are shown in Figure 7.3c. On most of the periodic islands, only single nanotubes were grown, although double growth may also take place (see the enlarged SEM image in Figure 7.3c, right). The reason for this may relate to the growth conditions themselves, or to imperfections in the nickel structure. However, it is most likely that the final step in the substrate preparation – the annealing phase – influences the creation of additional smaller catalytic centers surrounding the main growth core, and this subsequently contributes to the growth of shorter and thinner CNTs. This imperfection can be eliminated by replacing the quasi-triangular catalyst islands with round catalyst dots [36].

This new type of dimensional arrangement of CNTs was investigated for the purpose of studying cell adhesion, growth, and migration. Whereas, many studies which have focused on cell interaction and migration on patterned surfaces have shown the morphology and orientation of the cells to change in response to microscale topographies [3, 37, 38], relatively few investigations have been made of the response of the cells to nanoscale topographies and, in particular, to carbon nanopillars. Rovensky et al. [39], for example, studied the behavior of mouse fibroblasts on random arrays of silicon whiskers. Other studies have examined the response of human fibroblasts and endothelial cells to silicone elastomeric pillars [40, 41]. Based on the results of these studies, Giannona et al. [42] investigated the behavior of human osteoblast cells adhered to arrays of vertically aligned CNTs (VACNTs), mainly by monitoring the effect of the periodicity of the VACNTs on cell attachment and morphology, using SEM.

The morphology of osteoblasts growing on a CNT-carpet revealed the cells to be flattened, well spread, and attached to the tips of the individual CNTs (Figure 7.4a–c). In comparison to conventional plastic dishes, the cells clearly preferred to adhere to the CNT tips. The osteoblasts also seemed to exert forces on CNTs by becoming attached to the tips of the nanotubes. As shown in a higher-magnification image (Figure 7.4c), the nanotubes appeared to be stressed and deflected under cellular forces.

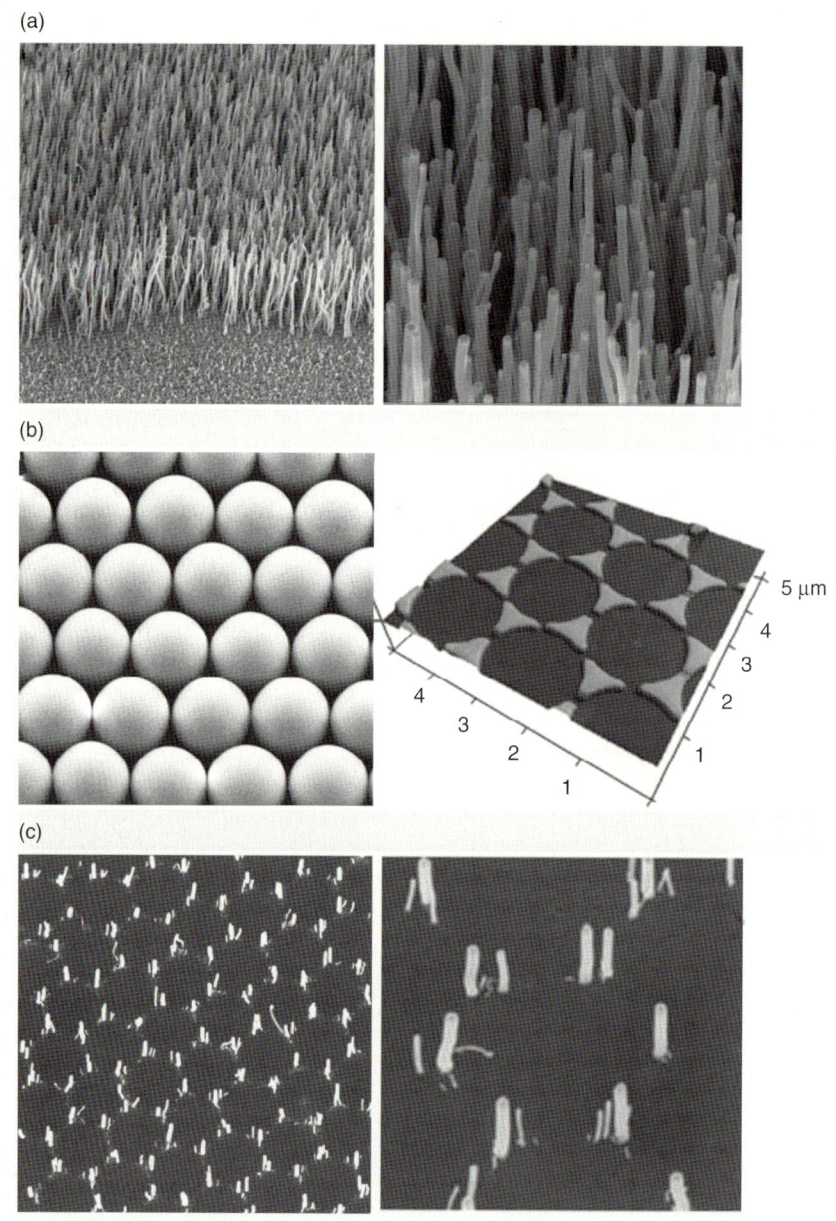

Figure 7.3 SEM images of vertically aligned carbon nanotubes (CNTs) produced using the plasma-enhanced CVD method. (a) Randomly distributed nanotubes; (b) Hexagonally aligned nanotubes were grown on substrates prepared by means of nanosphere lithography process; (c) Periodically aligned CNTs. The right-hand panels show higher magnifications.

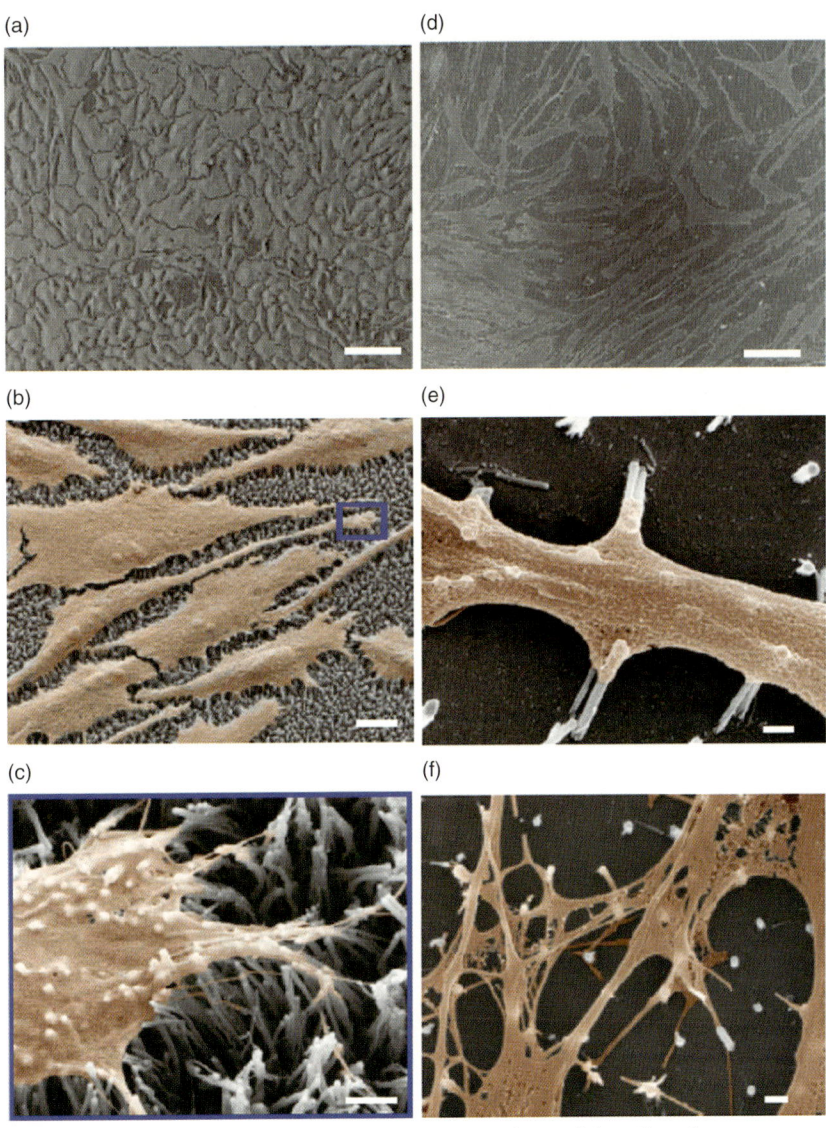

Figure 7.4 Color-enhanced SEM images depicting the influence of topography on osteoblast cell attachment and orientation. (a–c) Cells cultured on randomly vertically aligned carbon nanotubes (VACNTs) are flattened, with their visible nucleus located in the central part of the cell. Scale bars = 100 μm, 10 μm, and 1 μm, respectively; (d–f) Osteoblasts grown on periodically aligned nanotubes display changes in their morphology and orientation. Scale bar = 100 μm.

When the effect of the periodicity of VACNTs on cell attachment and morphology was assessed, clear differences were shown in cell morphology and behavior when the VACNTs were organized in a hexagonal pattern (Figure 7.4). Adaptation of the cells to these patterned CNTs appeared to influence their shape and orientation, with osteoblasts typically responding to the CNTs by changing their morphology. As in the case of the CNT-forest, the attachment of osteoblast cells occurred on the tips of the periodically aligned CNTs. Notably, the cells did not reach down to the surface between the CNTs, but rather remained fixed to their tips, suspended above the sapphire surface (see Figure 7.4e). The osteoblasts growing on this substrate also produced many cell extensions (Figure 7.4f). The presence of these extensions on periodic VACNTs was consistent with the observed alignment, as the cell extensions were responsible for cell movement and surface sensing.

The effect of surface topography on cell adhesion has also been assessed by an evaluation of the focal adhesion protein development after a 24 h period of cell culture. The use of immunofluorescence staining permitted study of both vinculin abundance and colocalization with the cytoskeleton. Vinculin is a plasma membrane-associated protein which is found in focal adhesions and is involved in the coupling of actin-based microfilaments to the adhesion plaque. Based on the fact that vinculin is one of the most prominent proteins of the focal adhesions, it has been used as an ideal marker protein to label the focal contacts formed in osteoblasts which have adhered to VACNTs [43].

The results presented in Figure 7.5 show that the vinculin (green) and actin (red) distributions differ depending on whether they are on a smooth surface (glass) or on a nanopatterned surface (VACNTs). In general, cells on glass reveal a random distribution of vinculin throughout their cell bodies, whereas osteoblasts growing on VACNTs generate well-organized vinculin clusters at the ends of many actin stress fibers. Taken together, these well-constituted and long vinculin clusters, together with the well-organized actin bundles, indicate good cell adhesion.

In summary, the results presented here reveal that topographical features have a significant influence on the attachment and growth of osteoblast cells. Surface topography in terms of the distribution of VACNTs plays an important role in cell shape alteration, and also influences the direction of cell movement. Despite the fact that both surfaces were favorable to cell attachment and proliferation, the osteoblast-like cells grew differently on substrates with randomly- and periodically-distributed VACNTs. It was observed that the alignment of osteoblast-like cells would be significantly influenced by the periodicity of the individual CNTs.

7.5.2
Three-Dimensional Cavity Network of Interconnected Nanotubes

The VACNTs matrices can be further modified to an interlocking resistive network of interconnected nanotubes, the main feature of which is a regular 3-D architecture. This transformation process, described in detail by Correa-Duarte et al. [44], is based on an acid treatment that induces capillary and tensile forces between the aligned tubes, and subsequently spreads the solution over the whole silicon sub-

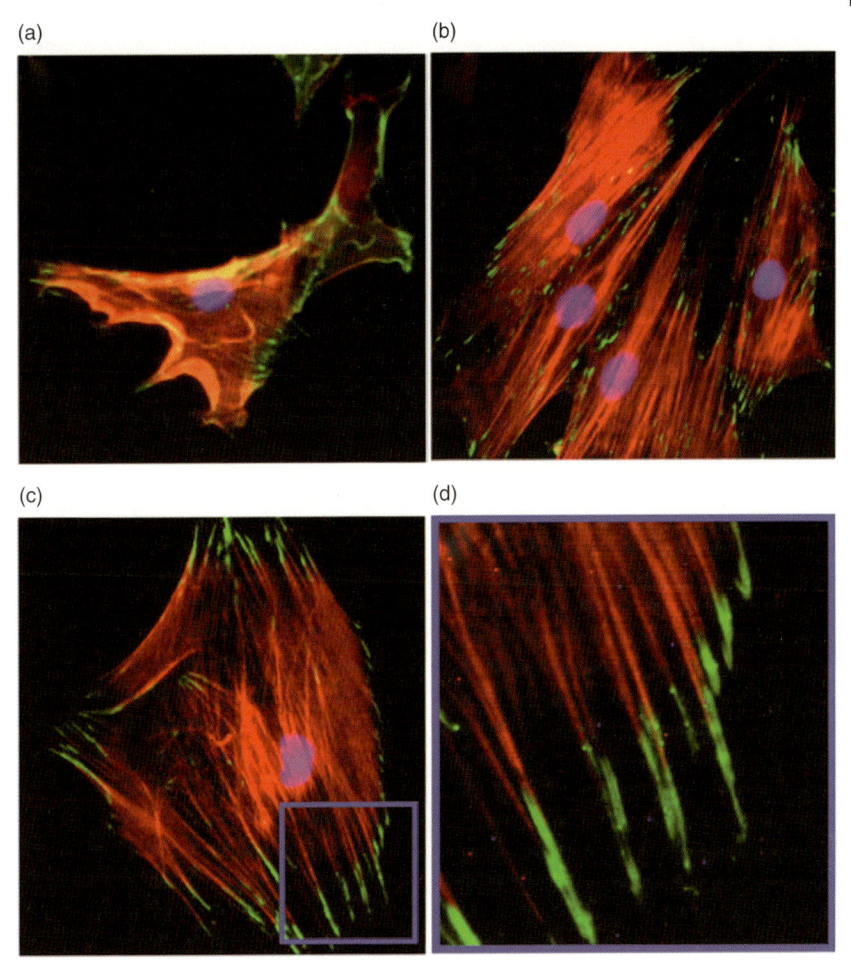

Figure 7.5 Immunofluorescence images of osteoblast cells stained for actin (red), vinculin (green), and nucleus (blue). (a) Cells adhering to glass; (b–d) Vertically aligned CNTs.

strate in such a way that all the interspaces among the aligned tubes become soaked. The latter process generates a hydrostatic dilation stress to the solution, which is larger at higher densities, leading to a flattening of the nanotubes. When nanotubes collapsing from opposite directions meet between two regions of lower density, a 3-D structure composed of honeycomb-like polygons is formed. The SEM images of VACNTs 50 µm in length, bonded to a silicon substrate so as to form a 3-D scaffold after functionalization treatment, are shown in Figure 7.6.

Depending on the length of the aligned MWNTs, it is possible to modulate their spatial distribution. Thus, this technique can be used to create pyramid-like structures (Figure 7.6b) or interconnected cavities with a volume sufficiently large to

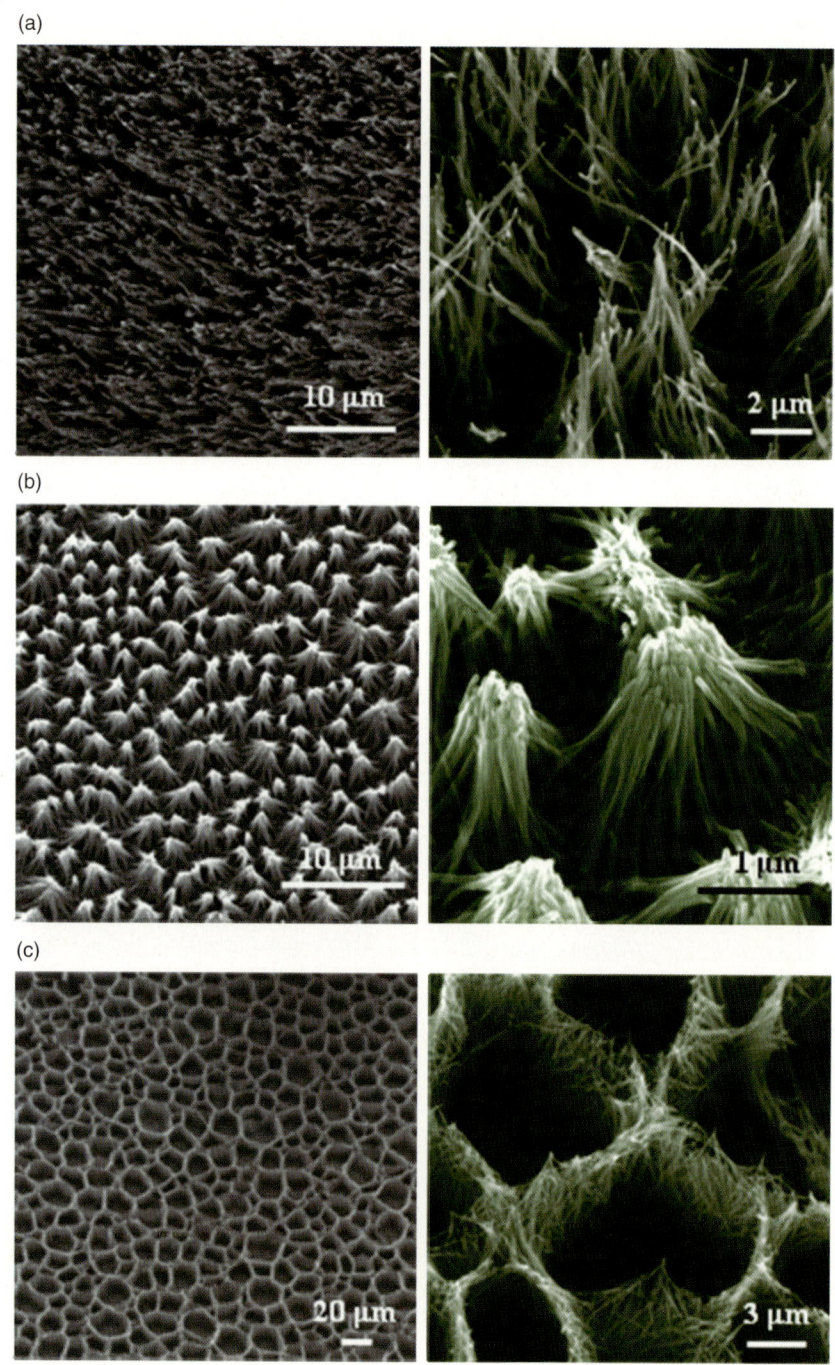

Figure 7.6 Transformation of VACNTs structure (a) into a pyramid-like structure with basal planes of ca. 3 μm (b), and network of crosslinked CNT walls, forming cavities (c). The right-hand panels show higher magnifications.

harbor a specific cell type under investigation (Figure 7.6c). Cavities varying between 5 and 60 µm and 5–15 µm diameter can be obtained by modeling MWNTs of 50 and 35 µm in length, respectively.

The high-resolution SEM image of the walls of the CNTs network (Figure 7.7a) shows the large surface area available for cell attachment (Figure 7.7b), and indicates the superior mechanical properties of the network due to crosslinking of the MWNTs.

As the main concern regarding the use of CNTs in biomedical applications is their possible cytotoxicity [45–47], the testing of CNT-based matrices on living organisms is essential for their successful utilization. Until now, very few reports have been made demonstrating the biocompatibility of CNT-based substrates with

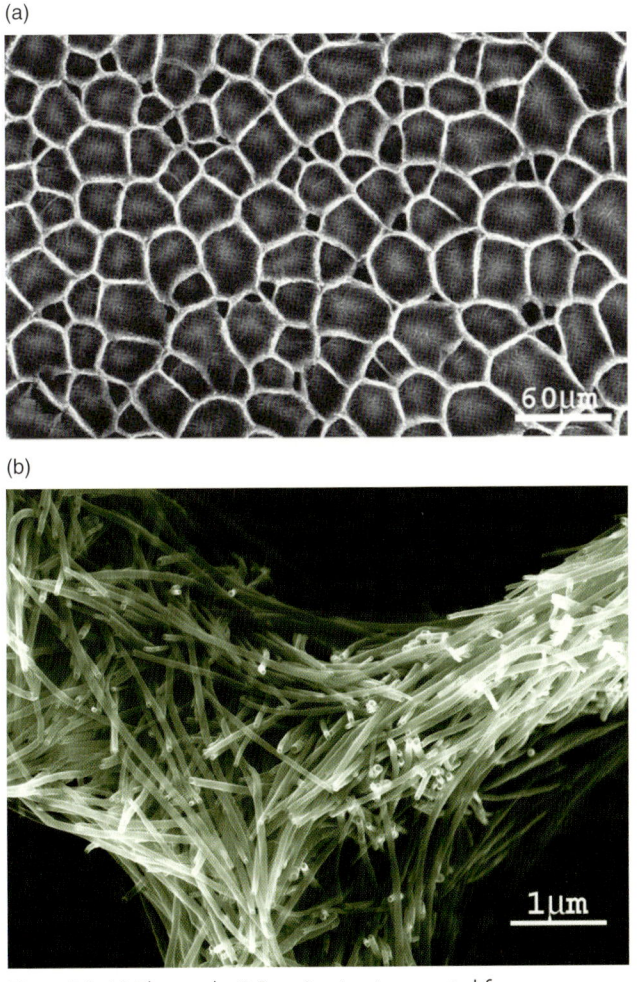

Figure 7.7 (a) The regular 3-D-cavity structure created from 50 µm length nanotubes; (b) High-magnification image of intercrossed tubes in the walls of the cavities.

various cell cultures. MacDonald *et al.* [31], for example, showed that collagen–CNT hybrids could sustain smooth muscle cell viability, while chemically modified CNTs have been demonstrated as a suitable scaffold material to grow neuron and osteoblast cells, without affecting cell activity [30, 35, 48]. Taking into account these contradictory opinions regarding the toxicity of CNTs, the matrices produced were tested for their biological compatibility with a mouse fibroblast cells culture, L 929.

In this case, cell growth onto the 3-D-cavity network and cell morphology were determined using SEM. As depicted in Figure 7.8, isolated fibroblasts were found to have spread onto the nanotube network, with elongated cytoplasm projections attaching to the walls of the cavities. After 7 days of growth, the fibroblasts had formed a confluent layer covering the surface of the network (Figure 7.8b); that

(a)

(b)

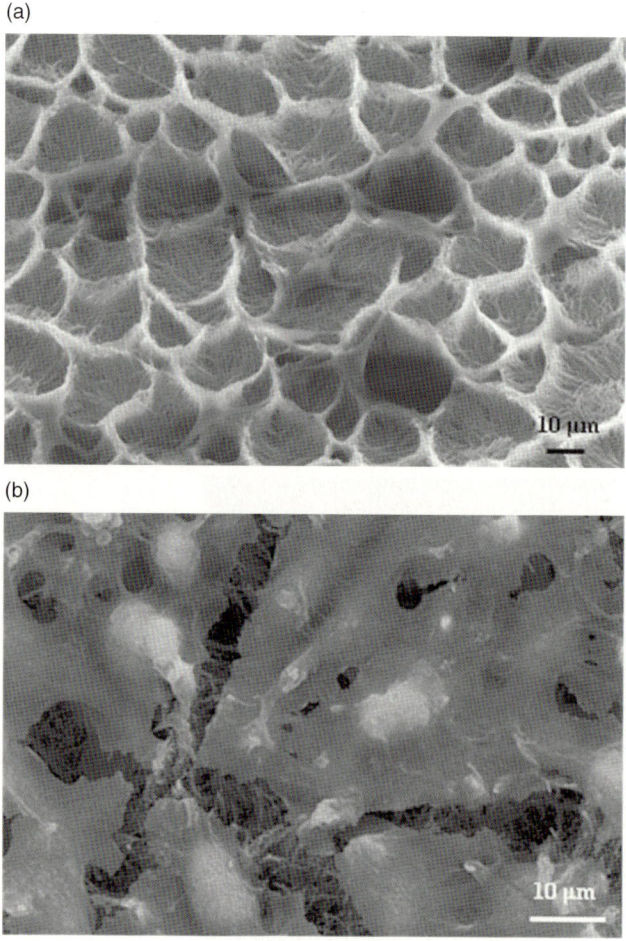

Figure 7.8 Scanning electron microscopy image of mouse fibroblasts growing on MWNT-based matrices after 1 day (a), and after 7 days (b).

7.5 CNT Engineering into Three-Dimensional Matrices

is, most of the surface of the MWNT-based network was hidden beneath the cell layer.

Based on the fibroblast cell activity after incubation on CNTs matrices, the authors excluded any toxic impact of CNTs on cell viability, thereby indicating that a CNT-based network with regular 3-D structure is highly biocompatible and provides a good basis to stimulate robust tissue formation.

7.5.3
Freestanding MWNT-Based Matrix

One of the common shortcomings of the MWNT-based matrices discussed above is the lack of the freestanding form; that is, both VACNTs and 3-D-cavity networks are firmly attached to the silicon substrate, which excludes the possibility of their *in vivo* implantation. In order to overcome this problem, Firkowska *et al.* [49] proposed the fabrication process which combines the conventional NSL and LbL techniques.

The fabrication method involves the preparation of hexagonally ordered polystyrene microspheres (Figure 7.9), and the subsequent deposition of CNTs by means of the LbL technique. The LbL approach is based on the sequential dipping of a substrate into solutions (dispersions) of oppositely charged components [50], so as to produce multilayer assemblies that are held together by the combination of

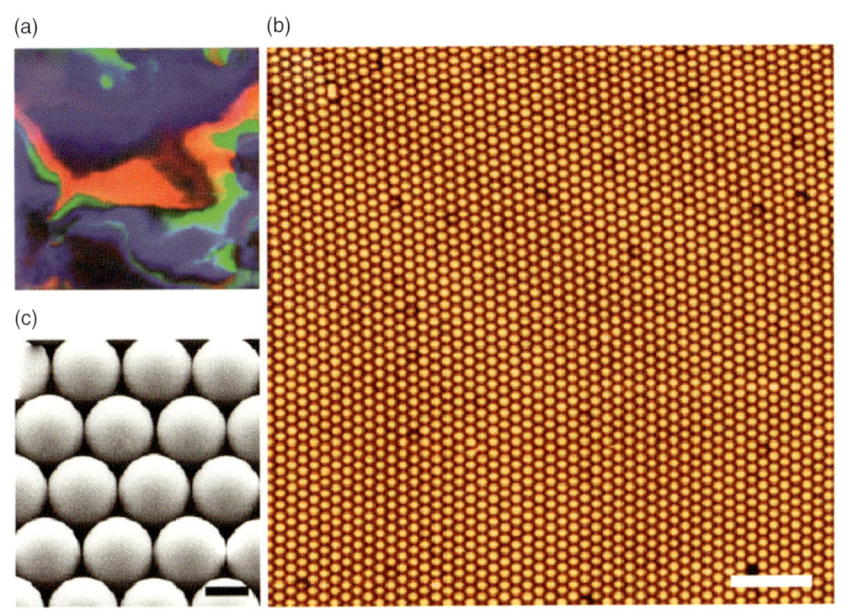

Figure 7.9 (a) Digital image, (b) SEM image and (c) atomic force microscopy image of a typical polystyrene mask deposited on a silicon substrate and subsequently used for CNT assembly. Scale bars = 10 μm (panel b) and 1 μm (panel c).

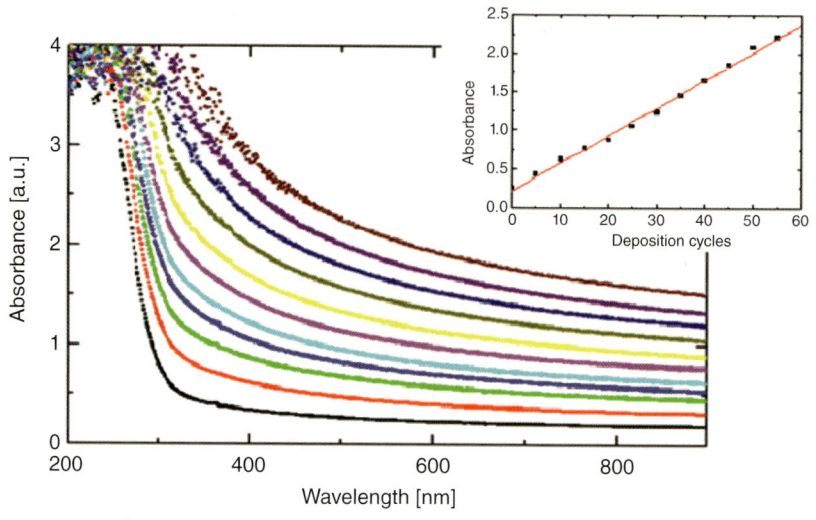

Figure 7.10 Absorption spectra of MWNT-PE layer-by-layer composite. The insert shows absorbance at 550 nm versus the number of deposited MWNT-PE bi-layers [54].

attractive electrostatic and dispersive forces [51]. The LbL films created in this way demonstrate a very homogeneous distribution of the components, and a high concentration of CNT contents, which may reach 50% [52].

The LbL pattern of deposition can be demonstrated by monitoring the optical density of the film as it grows. As shown in Figure 7.10 for the assembly of MWNTs and polyelectrolyte (PE) [53], the absorbance increases very regularly and with a constant increment, which indicates the deposition of essentially the same amount of nanotubes in each deposition cycle. The growth of the LbL film can be also monitored using SEM, as shown in Figure 7.11b.

Among some advantageous features of the LbL and NSL techniques, the tolerance to the substrate shapes, control over the film thickness, surface topography, and matrix chemical composition, matrix–substrate link should be mentioned. Unlike methods used for VACNTs and 3-D-cavity network fabrication, the techniques discussed in this section allow for a simple separation of the produced matrix from the supporting substrate. The freestanding form of a CNT-based matrix can be obtained using a chemical delamination process, which involves treatment of the polystyrene mask with tetrahydrofuran (THF). As depicted in Figure 7.11b, after THF treatment the polystyrene particles are completely dissolved to leave the free-standing nanotube-based matrix with a thin PE membrane on the top. This polymer residuum, which results from the infiltration of the PEs between the microspheres, is subsequently removed using a reactive-ion etching (RIE) process. Such a prepared matrix demonstrates an exceptional architecture consisting of crosslinked nanotubes self-assembled into an ordered, cavity-like structure (Figure 7.11c,d).

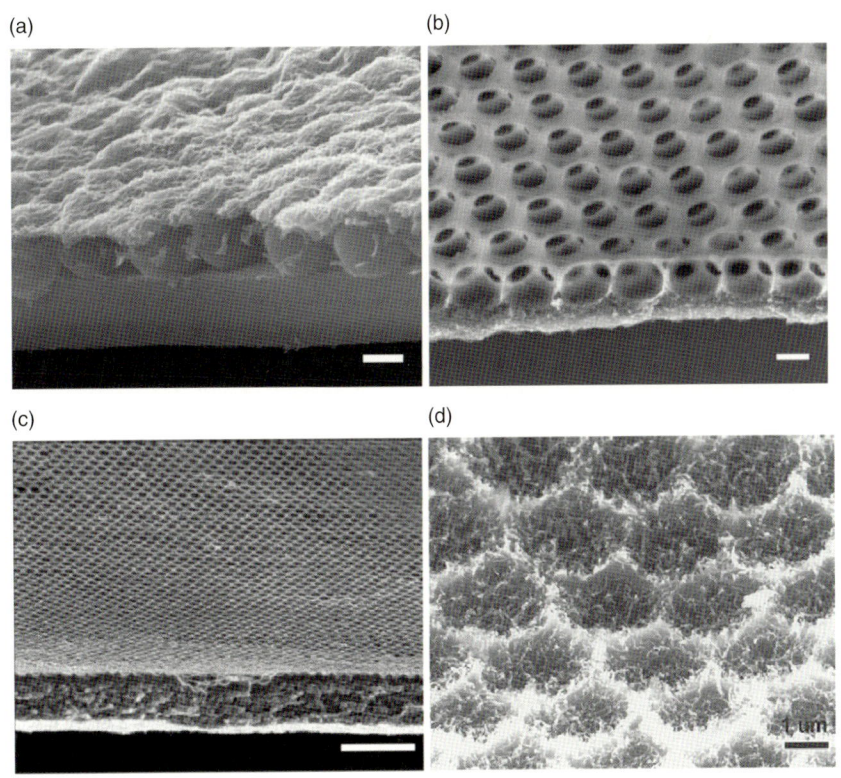

Figure 7.11 SEM images depicting steps in the fabrication of a MWNT-based matrix. (a) CNTs deposition on a polystyrene mask; (b) Free-standing LbL film after chemical delamination of the microspheres. Scale bar = 1 µm; (c,d) Cavity-like topography after reactive-ion etching. Scale bars = 10 µm.

In order to determine whether this freestanding matrix, which is rich in micro- and nanotopographical cues, is capable of providing a favorable platform for cell adhesion and proliferation, osteoblast cell-growth on the above-mentioned matrix has been visualized by means of SEM. The SEM images of osteoblasts grown for 7 days on the matrix are shown in Figure 7.12. The cells were seen to be flat, with a visible cell nucleus protruding in the center (Figure 7.12b), and to be well-spread and to cover almost the entire surface. The typical cell diameters for flat osteoblasts were of the order of 40 µm, and similar to those of osteoblasts found on the surface of natural bone [32]. Subsequent SEM studies revealed that the MWNT-based matrices produced allowed a flattening and spreading of the osteoblasts, and showed an adequate cell shape for proliferation.

Cell adhesion to a substrate influences cell morphology as well as cell proliferation and differentiation [55]. Thus, it could be said that the adhesion of cells to the surface of a matrix is one of the major factors responsible for its biocompatibility. SEM investigations, carried out at high magnification, have revealed the

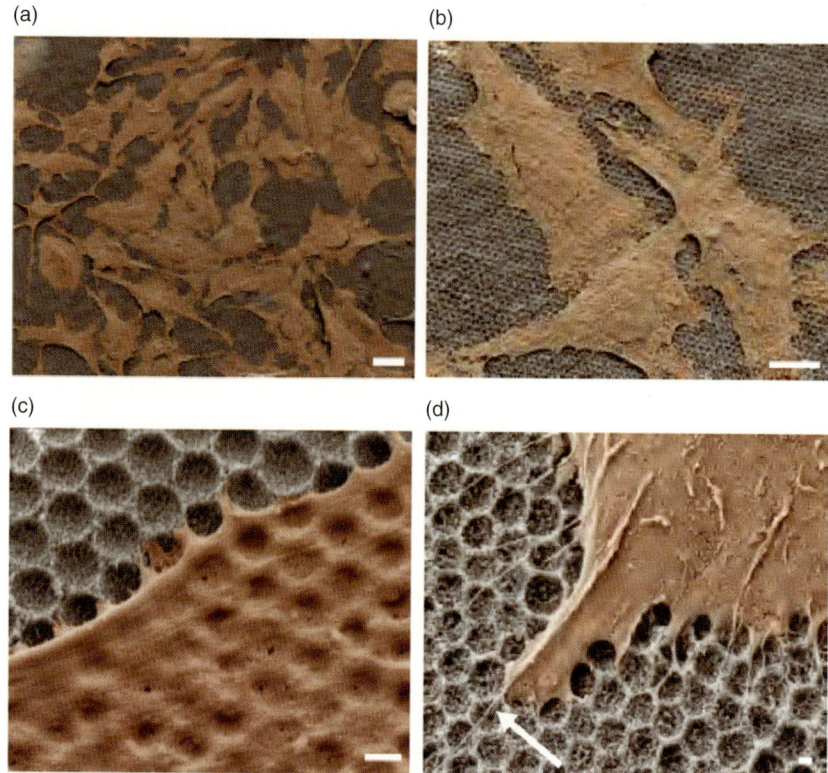

Figure 7.12 SEM images showing the morphology of osteoblast-like cells and their physical contact with the CNT structure. (a) The flat cell bodies extend over almost the entire area of observation; (b) The cell nucleus protrudes in the center. Scale bar = 10 μm; (c,d) A cell with a long thread-like cytoplasmic prolongation (arrow). Scale bar = 1 μm. For better cell visualization, the gray-scale was pseudo-colored with graphic software.

morphology of physical contacts between the cells and the CNT matrix. As shown in Figure 7.12c, the cavity-like structure can be seen beneath the part of the cell body, which strongly suggests the presence of tight junctions and adhesion mechanisms in the matrix. As osteoblasts are anchorage-dependent cells, their adhesion is a precondition for subsequent cell function [55]. Various SEM investigations have also revealed the presence of thin, threadlike cytoplasmic prolongations (Figure 7.12d); these pseudopods have a diameter in the range of 10 to 20 nm, close to the size of the MWNT diameter.

7.5.3.1 Modification of the MWNT-Based Matrix Surface with Bioactive Calcium Phosphate Nanoparticles

It is well known that, apart from the surface topography, cells are also sensitive to differences in the chemical properties of materials. It has been reported on many occasions that variations in the chemistry of the outermost functional groups of a surface evidently affect the cell response [34, 35, 56–58]. One of the great advantages of using CNTs to engineer matrices for cell growth is that the CNTs offer not only long-term durability, together with nanoscale topography, but also functionalization. It has been shown that CNTs can be chemically modified in a defined manner to control neurite outgrowth and branching [35]. Adequately functionalized CNTs have been also used successfully as scaffolds for the growth of artificial bone mineral (hydroxyapatite mineralization) [9].

In this section, a brief description is provided of a novel approach for the bioactivation of MWNT-based matrices with calcium phosphate nanoparticles (CP-NPs). Calcium phosphate is one of the most widely investigated materials for bone tissue engineering, based on the fact that the skeletal system contains about 70% inorganic material composed of calcium phosphate [59]. Due to its close chemical and crystal resemblance to bone mineral, calcium phosphate has an excellent biocompatibility. Indeed, numerous *in vivo* and *in vitro* assessments have indicated that calcium phosphate can support the attachment, differentiation, and proliferation of osteoblasts [60, 61]. Based on these results, CNT matrices with bioactive nanoparticles would be expected to provide a more desirable growth environment for the human hipbone osteoblast (HOB) cells, compared to CNT matrices without attached nanoparticles.

The CP-NPs used to modify the MWNT-based matrices were synthesized from aqueous calcium nitrate and phosphoric acid in the presence of 2-carboxyethylphosphonic acid (CEPA). The thus-produced negatively charged nanoparticles had a discoidal shape with a diameter of 30–80 nm and a height of less than 5 nm [59]. The CP-NPs could be covalently attached to the CNT surfaces through interaction between the amine functionalities on the nanotube surface and carboxylic acid groups on nanoparticles stabilized with CEPA. The morphologies of the CNT matrices before and after functionalization with nanoparticles are presented in Figure 7.13. The SEM images obtained showed that although the matrix did not change its cavity-like topography, the CNTs which previously were easily visible on the surface were now mostly covered with CP-NPs. The findings of these SEM studies, together with energy-dispersive X-ray spectroscopy (EDX) measurements, indicated the successful deposition of the CP-NPs.

To verify whether the modification of CNTs with CP-NPs would enhance cell–surface interactions, osteoblast cells were seeded onto CNTs with and without bioactive nanoparticles (Figure 7.14). After the first day, there was no difference in cell density between the cells growing on bioactivated CNT-based matrix (Figure 7.14a) and those on the control (Figure 7.14b). However, on the third day, the average cell area was greater on the matrix with CP-NPs, reflecting the higher surface coverage by the cells. The enhanced osteoblast proliferation on the MWNT-based matrices modified with CP-NPs was also confirmed by proliferation studies, the results of which are shown in Figure 7.15. As expected, the number of osteoblasts was greater on the matrices enriched with CP-NPs.

Taken together, the results presented here demonstrate that the LbL-assembled composites from MWNTs can be successfully modified with CP-NPs, which

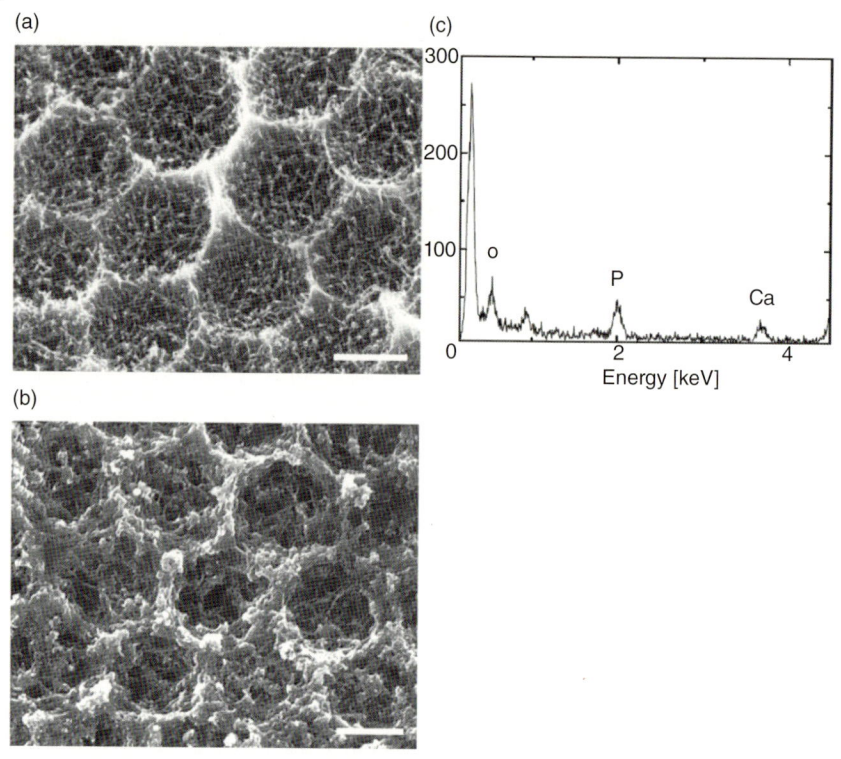

Figure 7.13 SEM images of a CNT-based matrix before (a) and after (b) incubation in the CP-NPs solution. Scale bar = 1 µm; (c) Energy-dispersive X-ray spectrum of CNTs coated with CP-NPs.

in turn leads to an effective increase of their biocompatibility with osteoblast cells.

7.6
Summary

The new discoveries in nanotechnology, combined with the ability of tissue engineering, have enabled the design and fabrication of matrices with nanoscale futures that imitate the characteristics of a natural ECM. In particular, CNTs with structures that mimic the nanoscale of native ECM, and have a high mechanical strength, excellent flexibility and low density, appear well suited as a biomaterial for the next generation of scaffolds. Although CNTs are not biodegradable, the above-mentioned advantages counterbalance this drawback. In addition, the potential cytotoxic effects related to CNTs may be diminished by chemical functionalization. Until now, CNTs have been shown capable of the successful

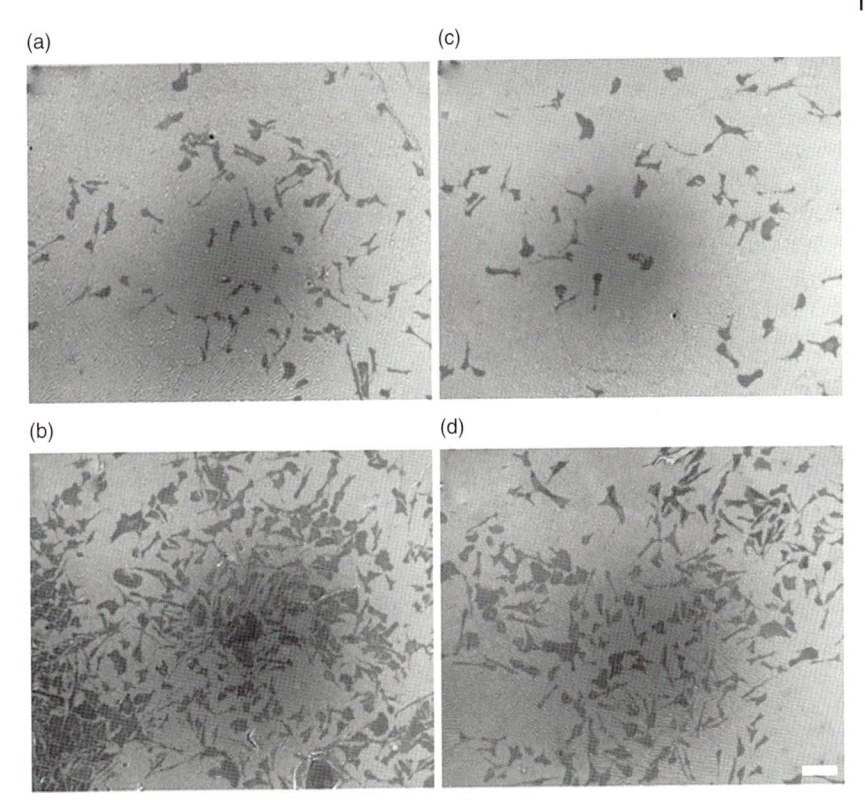

Figure 7.14 Representative SEM images of osteoblast cells cultured on MWNT-based matrices functionalized with (a,b) and without (c,d) CP-NPs after one day and three days, respectively. Scale bar = 100 mm.

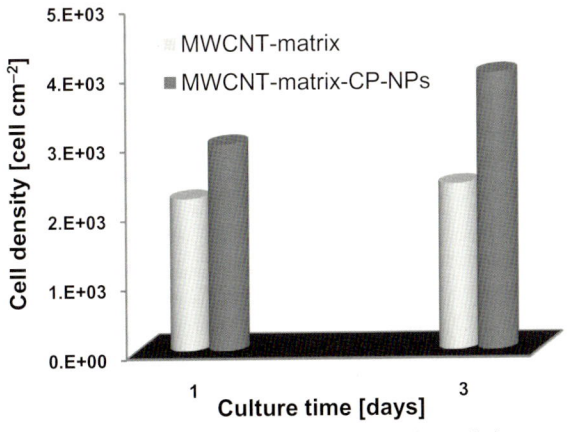

Figure 7.15 Graphical representation of osteoblast cell density on MWNT-based matrices with and without CP-NPs.

fabrication of matrices which, in addition to micro- and nano-features, possess the structural integrity and stability to retain their shape *in vivo*, with superior mechanical strength to support developing tissues and to withstand *in vivo* forces. The biocompatibility studies described in this chapter confirm the capability of nanotubes to support the long-term survival of both osteoblast and fibroblast cells.

References

1 Weiner, S. and Wagner, H.D. (1998) The material. Bone: structure-mechanical function. Relations. *Annual Review of Materials Science*, **28**, 271.

2 Dalby, M.J., Riehle, M.O., Sutherland, D.S., Agheli, H. and Curtis, A.S.G. (2004) Fibroblast response to a controlled nanoenvironment produced by colloidal lithography. *Journal of Biomedical Materials Research*, **69A**, 314.

3 Flemming, R.G., Murphy, C.J., Abrams, G.A., Goodman, S.L. and Nealey, P.F. (1999) Effects of synthetic micro- and nano-structured surfaces on cell behavior. *Biomaterials*, **20**, 573.

4 Dalby, M.J., Riehle, M.O., Yarwood, S.J., Wilkinson, C.D.W. and Curtis, A.S.G. (2003) Nucleus alignment and cell signaling in fibroblasts: response to a micro-grooved topography. *Experimental Cell Research*, **284**, 274.

5 Andersson, A.-S., Blasckhed, F., Eulervon, A., Richter-Dahlfors, A., Sutherland, D. and Kasemo, B. (2003) Nanoscale features influence epithelial cell morphology and cytokine production. *Biomaterials*, **24**, 3427.

6 Yim, E.K., Reano, R.M., Pang, S.W., Yee, A.F., Chen, C.S. and Leong, K.W. (2005) Nanopattern-induced changes in morphology and motility of smooth muscle cells. *Biomaterials*, **26**, 5405.

7 Stevens, M.M. and George, J.H. (2005) Exploring and engineering the cell surface interface. *Science*, **310**, 1135.

8 Bowlin, G.L. (2004) A new spin of scaffolds. *Materials Today*, **7** (5), 64.

9 Zhao, B., Hu, H., Mandal, S.K. and Haddon, R.C. (2005) A bone mimic based on the self-assembly of hydroxyapatite on chemically functionalized single-walled carbon nanotubes. *Chemistry of Materials*, **17**, 3235.

10 Harrison, B.S. and Atala, A. (2007) Carbon nanotube applications for tissue engineering. *Biomaterials*, **28**, 344.

11 Yang, W., Thordarson, P., Gooding, J.J., Ringer, S.P. and Braet, F. (2007) Carbon nanotubes for biological and biomedical applications. *Nanotechnology*, **41**, 412002.

12 Ma, P.X. (2008) Biomimetic materials for tissue engineering. *Advanced Drug Delivery Reviews*, **60** (2), 184.

13 Iijima, S. (1991) Helical microtubules of graphitic carbon. *Nature*, **354**, 56.

14 Ren, Z.F., Huang, Z.P., Xu, J.W., Wang, J.H., Bush, P., Siegel, M.P. and Provencio, P.N. (1998) Synthesis of large arrays of well-aligned carbon nanotubes on glass. *Science*, **282**, 1105.

15 Guo, T., Nikolaev, A., Thess, A., Colbert, D.T. and Smalley, R.E. (1995) Catalytic growth of single walled nanotubes by laser vaporization. *Chemical Physics Letters*, **243**, 49.

16 Jose-Yacaman, M. (1993) Catalytic growth of carbon microtubules with fullerene structure. *Applied Physics Letters*, **62**, 202.

17 Meyyappan, M. (2005) *Carbon Nanotubes. Science and Application*, CRS Press, Boca Raton, FL.

18 Prakash, R., Superfine, R., Washbum, S. and Falvo, M.R. (2006) Functionalization of carbon nanotubes with proteins and quantum dots in aqueous buffer solution. *Applied Physics Letters*, **88**, 063102.

19 Singh, K.V., Pandey, R.R., Wang, X., Lake, R., Ozkan, C.S., Wang, K. and Ozkan, M. (2006) Covalent functionalization of single walled carbon nanotubes with peptide nucleic acid: nanocomponents for molecular level electronics. *Carbon*, **44**, 1730.

20 Daniels, A.U., Chang, M.K.O. and Andriano, K.P. (1990) Mechanical properties of biodegradable polymers and

composites proposed for internal fixation of bone. *Journal of Applied Biomaterials*, **1**, 57.
21 Yuehuei, H. and Draughn, R.A. (2000) *Mechanical Testing of Bone and the Bone-Implant Interface*, CRC Press, Boca Raton, FL.
22 Harrison, B. and Atala, A. (2007) Carbon nanotube application for tissue engineering. *Biomaterials*, **28**, 344–53.
23 Coleman, J.N., Cadek, M., Blake, R., Nicolosi, V., Ryan, K.P., Belton, C., Fonseca, A., Nagy, J.B., Gunk'ko, Y.K. and Blau, W.J. (2004) High-performance nanotube-reinforced plastics: understanding the mechanism of strength increase. *Advanced Functional Materials*, **14**, 791–8.
24 Liu, T., Phang, I.Y., Shen, L., Chow, S.Y. and Zhang, W.-D. (2004) Morphology and mechanical properties of multiwalled carbon nanotubes reinforced nylon-6 composites. *Macromolecules*, **37**, 7214.
25 Coleman, J.N., Khan, U.M. and Gun'ko, Y.K. (2006) Mechanical reinforcement of polymers using carbon nanotubes. *Advanced Materials*, **18**, 689.
26 Wang, S.-F., Shen, L., Zhang, W.-D. and Tong, Y.-J. (2005) Preparation and mechanical properties of chitosan/carbon nanotubes composites. *Biomacromolecules*, **6**, 3067.
27 Gao, L., Jiang, L. and Sun, J. (2006) Carbon nanotube-ceramic composites. *Journal of Electroceramics*, **17**, 51.
28 Balani, K., Anderson, R., Laha, T., Andara, M., Tercero, J., Crumpler, E. and Agarwal, A. (2007) Plasmasprayed carbon nanotube reinforced hydroxyapatite coatings and their interaction with human osteoblasts *in vitro*. *Biomaterials*, **28**, 618.
29 Supronowicz, P.R., Ajayan, P.M., Ullmann, K.R., Arulanandam, B.P., Metzger, D.W. and Bizios, R. (2002) Novel current-conducting composite substrates for exposing osteoblasts to alternating current stimulation. *Journal of Biomedical Materials Research*, **5**, 499.
30 Gheith, M.K., Pappas, T.C., Liopo, A., Sinani, V.A., Shim, B.S., Motamedi, M., Wicksted, J.P. and Kotov, N.A. (2006) Stimulation of neural cells by lateral current in conductive layer-by layer films of single-walled carbon nanotubes. *Advanced Materials*, **18**, 2975.
31 MacDonald, R.A., Laurenzi, B.F., Viswanathan, G., Ajayan, P.M. and Stegemann, J.P. (2005) Collagen carbon nanotube composite materials as scaffolds in tissue engineering. *Journal of Biomedical Materials Research*, **74A**, 489.
32 Zanello, L.P., Zhao, B., Hu, H. and Haddon, R.C. (2006) Bone cell proliferation on carbon nanotubes. *Nano Letters*, **6**, 562.
33 Webster, T.J., Waid, M.C., McKenzie, J.L., Price, R.L. and Ejiofor, J.U. (2004) Nano-biotechnology: carbon nanofibers as improved neural and orthopaedic implants. *Nanotechnology*, **15**, 48.
34 Mattson, M.P., Haddon, R.C. and Rao, A.M. (2000) Molecular functionalization of carbon nanotubes and use as substrate for neuronal growth. *Journal of Molecular Neuroscience*, **14**, 175.
35 Hu, H., Ni, Y., Montana, V., Haddon, R.C. and Parpura, V. (2004) Chemically functionalized carbon nanotubes as substrate for neuronal growth. *Nano Letters*, **4**, 507.
36 Kempa, K., Kimball, B., Rybczynski, J., Huang, Z.P., Wu, P.F., Steeves, D., Sennett, M., Giersig, M., Rao, D., Carnaham, D.L., Wand, D.Z., Lao, J.Y., Li, W.Z. and Ren, Z.F. (2003) Photonic crystals based on periodic arrays of aligned carbon nanotubes. *Nano Letters*, **3**, 13.
37 Charest, J.L., Bryant, L.E., Garcia, A.J. and King, W.P. (2004) Hot embossing for micropatterned cell substrates. *Biomaterials*, **19**, 4767.
38 Anselme, K., Bigerelle, M., Noel, B., Lost, A. and Hardouin, P. (2002) Effect of grooved titanium substratum on human osteoblastic cell growth. *Journal of Biomedical Materials Research Part A*, **60**, 529.
39 Rovensky, Y.A., Bershadsky, A.D., Givargizov, E.I., Obolenskaya, L.N. and Vasilev, J.M. (1997) Spreading of mouse fibroblast on the substrate with the multiple spikes. *Experimental Cell Research*, **197**, 107.
40 von Recum, A.F. and van Kooten, T.G. (1995) The influence of micrography on cellular response and the implications for

silicone implants. *Journal of Biomaterials Science Polymer Edition*, **7**, 181.

41 van Kooten, T.G. and von Recum, A.F. (1999) Cell adhesion to textured silicone surfaces: the influence of time of adhesion and texture on focal contact and fibronectin fibril formation. *Tissue Engineering*, **5**, 223.

42 Giannona, S., Firkowska, I., Rojas-Chapana, J. and Giersig, M. (2007) Vertically aligned carbon nanotubes as cytocompatible material for enhanced adhesion and proliferation of osteoblast-like cells. *Journal of Nanoscience and Nanotechnology*, **7**, 1679.

43 Zamir, E. and Geiger, B. (2001) Molecular complexity and dynamics of cell-matrix adhesions. *Journal of Cell Science*, **114**, 3583.

44 Correa-Duarte, M.A., Wagner, N., Rojas-Chapana, J., Morsczeck, C., Thie, M. and Giersig, M. (2004) Fabrication and biocompatibility of carbon nanotube-based 3D networks as scaffolds for cell seeding and growth. *Nano Letters*, **4**, 2233.

45 Cui, D., Tian, F., Ozkan, C.S., Wang, M. and Gao, W. (2005) Effect of single wall carbon nanotubes on human HEK293 cells. *Toxicology Letters*, **155**, 73.

46 Manna, S.K., Sarkar, S., Barr, J., Wise, K., Barrera, E.V., Jejelowo, O., Rice-Ficht, A.C. and Ramesh, G.T. (2005) Single-walled carbon nanotubes induces oxidative stress and activates nuclear transcription factor-kB in human keratinocytes. *Nano Letters*, **5**, 1676.

47 Lam, C.-W., James, J.T., McCluskey, R. and Hunter, R.L. (2004) Pulmonary toxicity of single-wall carbon nanotubes in mice 7 and 9 days after intratracheal instillation. *Toxicological Sciences*, **77**, 126.

48 Price, R.L., Haberstroh, K.M. and Webster, T.J. (2004) Improved osteoblast viability in the presence of smaller nanometer dimensioned carbon fibres. *Nanotechnology*, **15**, 892.

49 Firkowska, I., Olek, M., Pazos-Perez, N., Rojas-Chapana, J. and Giersig, M. (2006) Highly ordered MWNT-based matrixes: topography at the nanoscale conceived for tissue engineering. *Langmuir*, **22**, 5427.

50 Caruso, F., Lichtenfeld, H., Giersig, M. and Möhwald, H. (1998) Electrostatic self-assembly of silica nanoparticle–polyelectrolyte multilayers on polystyrene latex particles. *Journal of the American Chemical Society*, **120**, 8523.

51 Kotov, N.A. (1999) Layer-by-layer self-assembly: The contribution of hydrophobic interactions. *Nanostructured Materials*, **12**, 789.

52 Mamedov, A.A., Kotov, N.A., Prato, M., Guldi, D.M., Wicksted, J.P. and Hirsch, A. (2002) Molecular design of strong single-wall carbon nanotube/polyelectrolyte multilayer composites. *Nature Materials*, **1**, 190.

53 Rogach, A.L., Koktysch, D., Harrison, M. and Kotov, N.A. (2000) Layer-by-layer assembled films of HgTe nanocrystals with strong infrared emission. *Chemistry of Materials*, **12**, 1526.

54 Olek, M., Ostrander, J., Jurga, S., Mohwald, H., Kotov, N., Kempa, K. and Giersig, M. (2004) Layer-by-layer assembly composites from multi wall carbon nanotubes with different morphologies. *Nano Letters*, **4**, 1889.

55 Anselme, K. (2000) Osteoblast adhesion on biomaterials. *Biomaterials*, **21**, 667.

56 Schwartz, Z., Lohmann, S.H., Oefinger, J., Bonewald, L.F., Dean, D.D. and Boyan, B.D. (1999) Implant surface characteristics modulate differential behaviour of cells in the osteoblast lineage. *Advances in Dental Research*, **13**, 38.

57 Bagambisa, F.B., Kappert, H.F. and Schilli, W. (1994) Cellular and molecular biological events at the implant interface. *Journal of Cranio-Maxillo-Facial Surgery*, **22**, 12.

58 Sirivisoot, S., Yao, C., Xiao, X., Sheldon, B.W. and Webster, T.J. (2007) Greater osteoblast functions on multiwalled carbon nanotubes grown from anodized nanotubular titanium for orthopedic application. *Nanotechnology*, **18**, 365102.

59 Andres, C., Sinani, V., Lee, D., Gun'ko, Y. and Kotov, N. (2006) Anisotropic calcium phosphate nanoparticles coated with 2-carboxyethylphosphonic acid. *Journal of Materials Chemistry*, **16**, 3964.

60 Phan, P.V., Grzanna, M., Chu, J., Polotsky, A., El-Ghannam, A., Heerden, D.V., Hungerford, D.S. and Frondoza,

C.G. (2003) The effect of silicon-containing calcium-phosphate particles on human osteoblast *in vitro*. *Journal of Biomedical Materials Research Part A*, **67**, 1001.

61 Webster, T.J., Ergun, C., Doremus, R.H., Siegel, R.W. and Bizios, R. (2000) Enhanced functions of osteoblasts on nanophase ceramics. *Biomaterials*, **21**, 1803.

8
Electrochemical Biosensors Based on Carbon Nanotubes

Jonathan C. Claussen, Jin Shi, Alfred R. Diggs, D. Marshall Porterfield and Timothy S. Fisher

8.1
Introduction

Electrochemistry is applied across diverse scientific and engineering fields, ranging from biomedical diagnostics to the development of fuel cell technology. Especially in the life sciences, electroanalytical chemistry is important and is exploited for the quantification and real-time sensing of a variety of analytes ranging from simple ions, to metabolic gases, and even to neurotransmitters. Recent advances in genomics, metabolomics, and proteomics, coupled with improved nanofabrication techniques, have propelled the field of electrochemical biosensing [1, 2]. Electrochemical biosensors offer the rapid, real-time, and ultra-sensitive detection of clinically important biomarkers such as those associated with diabetes [3], cancer [4], and Alzheimer's disease [5]. Recently, carbon allotropes such as carbon nanotubes (CNTs) [6] have been utilized in electrochemical biosensors to increase the sensor surface area and charge transport properties. Hence, electrochemical biosensors based on CNTs present a promising solution for detecting a myriad of clinically important biomarkers with high sensitivity, at low concentration levels.

The report by Sumio Iijima in 1991 on observed "helical microtubules of graphitic carbon" is considered the advent of CNTs [7]. Typically, CNTs consist of a rolled single layer (single-walled carbon nanotube; SWNT) or rolled multiple layers (multi-walled carbon nanotube; MWNT) of graphene sheets that have strong carbon–carbon sp^2 covalent bonds and are arranged in a hexagonal pattern [8, 9]. The concentric graphene sheets of MWNTs are separated by 3.4 Å [8], a separation distance similar to that found between individual graphene layers in graphite. The typical diameters of SWNTs vary from 1.2 nm to 5 nm, while those of MWNTs can vary more widely (10–50 nm) [10]. The small dimensions of CNTs are compatible in size with the approximate length scales of the basic building blocks of life (single cell ~1 μm, virus ~100 nm, protein ~10–50 nm, DNA duplex ~1 nm) [11]. The unique mechanical, electrical, and chemical properties of CNTs provide for strong stability and enhanced heterogeneous charge transport in a variety of biosensing applications. Thus, CNT-based biosensors offers a unique high-performance

Nanomaterials for the Life Sciences Vol.9: Carbon Nanomaterials. Edited by Challa S. S. R. Kumar
Copyright © 2011 WILEY-VCH Verlag GmbH & Co. KGaA, Weinheim
ISBN: 978-3-527-32169-8

platform for interfacing with biological systems and detecting and sensing numerous biomolecular agents [12].

CNT-based electrochemical biosensors offer numerous advantages over other biosensor platforms. First, they are capable of interfacing with biological agents with *in vitro* and *in vivo* applications for biocompatibility and biomolecular recognition [13, 14]. Second, the electronic signal transduction detection of electrochemical biosensors offers a highly sensitive label-free sensing scheme that allows for facile detection strategies. Finally, their relative low power requirements make them ideal for mobile sensing applications [15] that promote general usability and potential benefit for those who live in remote locations with limited access to healthcare facilities.

In this chapter, the recent advances and fundamental principles associated with electrochemical biosensors based on CNTs will be discussed. First, the various properties of CNTs, including the mechanical, electrical, and chemical/electrochemical properties that make them enticing for biosensing will be discussed. This will be followed by a general overview of electrochemical sensors, with special emphasis on amperometric sensing. Subsequently, a detailed review of CNT-based biosensor fabrication strategies including "top-down" and "bottom-up" fabrication paradigms will be provided. Finally, a comprehensive review of nonenzymatic and enzymatic CNT-based electrochemical biosensor applications will emphasize the latest research in this field.

8.2
CNT Properties

8.2.1
Mechanical

The high strength and rigidity of CNTs have been widely reported; for example, the Young's modulus for a single CNT has been reported as 1 TPa [16], with a tensile strength of 30 GPA [17]. The thermal conductivity of individual SWNTs and MWNTs at room temperature have reached values up to $10\,000\,W\,m^{-1}\,K^{-1}$ [18] and $830\,W\,m^{-1}\,K^{-1}$ [19], indicating the ease at which phonons propagate along CNTs. The high strength and rigidity of CNTs increases the durability and lifetime of electrochemical biosensors. In particular, the mechanical strength of CNTs is well-suited for atomic force microscopy (AFM) applications, where a CNT is mounted on the tip of a scanning probe microscope, whereas the excellent thermal conductivity of CNTs is beneficial to sensors that require enhanced heat dissipation [20].

8.2.2
Electrical

Due to their nearly one-dimensional (1-D) electronic structure, both SWNTs and MWNTs display ballistic electron transport at room temperature, at length scales on the order of a hundred nanometers to microns [21–23]. Current densities of $10^9 \, \text{amp} \, \text{cm}^{-2}$ have been measured in SWNTs [24], and $10^7 \, \text{amp} \, \text{cm}^{-2}$ in MWNTs [21]. These electronic transport properties, combined with the covalently bonded structure of CNTs, militate against current-induced electromigration at metal connects, and thus acts as superb interconnects in biosensing field-effect transistors (FETs) where a single SWNT can act as a gate between a metal source and metal drain lead [25]. Such SWNT FETs have been coated with enzymes [26] and synthetic oligonucleotides [27] for the detection of glucose and DNA hybridization, respectively. Furthermore, the electrical properties of CNTs, combined with their biocompatibility, make them well-suited for facilitating charge transport in redox reactions during electrochemical biosensing [28].

8.2.3
Chemical/Electrochemical

In the past, CNTs have been used to template the synthesis of nanoparticles of various compositions [29–33]. The hexagonal sp^2-bonded carbon lattice that makes up the cylindrical CNT structure provides unique electrochemical characteristics for biosensing [34]. In particular, SWNTs are quite useful in electrochemical biosensing applications [35] because of their extreme sensitivity to surface perturbations [36] and their high electron-transfer rates [37]. The large surface area and current-carrying capacity of SWNTs also enable an improved signal-to-noise ratio (SNR) during amperometric testing [38]. These characteristics, along with their relatively simple synthesis, make SWNTs an important and advantageous biosensing material.

The CNTs also demonstrate great chemical stability, leading to an inherent property of biocompatibility for biofunctionalization with a wide variety of biomolecular agents [39, 40]. Furthermore, the rich chemistry of CNTs allows for a variety of biofunctionalization strategies, including covalent attachment (chemical bond formation), noncovalent attachment (physio-absorption), and a hybrid method (noncovalent attachment of an "anchor" molecule followed by covalent attachment of a particular biomolecule) [41]. The chemical stability of the CNTs, combined with the aforementioned biofunctionalization strategies, have allowed them to be conjugated with numerous biological species including nucleic acids, carbohydrates, and proteins for biosensing experimentation.

In addition to being well-suited for a variety of biofunctionalization strategies, the chemical and electrochemical properties of CNTs permit enhanced heterogeneous charge transport during electrochemical biosensing. Due to their large aspect ratios, CNTs greatly increase the electroactive surface area of conventional electrodes. Moreover, this enlarged biosensor surface area increases the amount of docking points for biomolecule recognition agents, and accordingly enhances the electrical and charge transport between the CNT/analyte interface [42, 43]. Electrochemical charge transport is presumed to be significantly enhanced where there are breaks in the strong C–C bonds, such as are found at defect sites along the length of the nanotubes [44, 45]. Furthermore, broken nanotube ends can behave like the highly electroactive edge plan of highly ordered pyrolytic graphite (HOPG) created during strong acid treatments [46]. These acid treatments can also introduce oxygenated species, such as carboxylic acids, alcohols, and quinines [47, 48] which have also shown to increase heterogeneous charge transport [49, 50]. The cited "electrocatalytic properties" of CNTs most likely originate from residual catalyst metal impurities within the nanotubes themselves [51].

8.3
Electrochemical Biosensing

Electrochemical biosensors are transducers that convert biological information into electrical information. They can provide both qualitative and quantitative information on the existence and concentration of compounds in analyte [52]. Three common types of electroanalytical sensing modalities (potentiometric, conductometric, and amperometric) are typically utilized in electrochemical biosensing.

Amperometric sensors detect the heterogeneous charge transfer during redox reactions, such that the concentration of compounds can be measured in real time, without hysteresis [52]. The incorporation of enzymes into amperometric sensors as the biorecognition element has greatly improved their selectivity, and has been widely investigated ever since the enzymatic sensing of glucose was initiated by Clark and Lyons in 1962 [53, 54]. As a result, amperometric sensors are more widely studied in physiological biosensing-related research. Furthermore, the majority of CNT-based electrochemical biosensors utilize the amperometric sensing modality.

In succinct terms, amperometric biosensors involve heterogeneous charge transport (i.e., the exchange of an electron or electrons between a species in solution and an electrode surface [55]). In amperometric sensing, a bias is applied between the working and auxiliary electrodes that subsequently promotes electron transfer at the working electrode/solution interface. This electron transfer is monitored as a current, which is directly proportional to the concentration of analyte being oxidized or reduced within a zone of operation referred to as the "linear sensing region." A three-electrode potentiostat, consisting of a working, reference, and auxiliary electrode, is often utilized in amperometric measurements. In a typical three-electrode electrochemical set-up, the biosensor serves as the working electrode, a platinum wire as the auxiliary electrode, and Ag/AgCl as the reference electrode.

8.4
CNT-Based Electrode Fabrication

8.4.1
Adsorption

A common approach to fabricating CNT-based biosensors is to coat electrodes with CNTs via adsorption techniques. A CNT adsorption technique known as *abrasive immobilization* is perhaps one the simplest CNT/electrode fabrication strategies. The technique utilizes van der Waals force interactions between CNTs and glassy carbon (GC) electrodes, causing the CNTs to adhere to the electrode surface. Wang *et al.* were the first to report an abrasive immobilization technique for CNT/electrode bonding, by grinding a dry graphite electrode on a weighing paper containing CNT powder to intercalate the CNTs via adsorption [56]. Subsequently, Salimi *et al.* followed a similar approach by gently rubbing a polished basal plane pyrolytic graphite (BPPG) electrode surface on a fine qualitative filter paper containing MWNTs [57–59]. This electrode experienced an increase in current response to H_2O_2 oxidation during cyclic voltammetry (CV) experiments, as compared to bare BPPG electrodes. Following the immobilization of glucose oxidase (GOx) on electrodes via sol–gels, a well-defined amperometric response to glucose was reported for CNT-modified electrodes, as compared to a negligible response from unmodified electrodes [57]. The results of these experiments indicated that CNT-enhanced biosensors would improve the redox currents and decrease the overvoltage potentials during electrochemical sensing [57]. Indeed, abrasive immobilization represents a simple fabrication technique that does not utilize any additional chemical reagents that often promote enzyme degradation.

In addition to abrasive CNT/electrode immobilization strategies, CNT-composite fabrication presents another facile strategy by which CNTs may be adsorbed onto electrode surfaces. The CNT-composite fabrication process packs CNTs directly into a composite electrode, while maintaining the electrochemical properties of the nanotubes [60, 61]. Britto *et al.* first reported biosensors based on CNT-composite electrodes by packing a paste of MWNTs with bromoform into a glass tube for dopamine detection. This CNT-composite electrode displayed an improved reversibility and current response with dopamine in CV experiments, as compared to other carbon electrodes [60]. In addition to bromoform, mineral oil [61–63], Teflon [64], and epoxy [65] have also been used as binder materials in CNT-composite electrodes. Valentini *et al.* reported that distinct SWNT weight percentages within the binder materials of CNT-composite electrodes can significantly alter the electrochemical properties of the electrode, for better or for worse. For example, the experiments conducted by Valentini and coworkers concluded that, in order to achieve good electrochemical performance and reversibility, the optimal ratio of SWNTs and mineral oil for CNT-composite electrodes should be 75% (w/w). On the other hand, Wang *et al.* developed a binderless CNT-composite electrode by directly packing CNTs and GOx into a 21-gage needle. This CNT-modified needle electrode experienced an increased sensitivity towards the amperometric detection of H_2O_2, the reduced cofactor nicotinamide adenine dinucleotide (NADH), and glucose after enzyme immobilization [66]. Likewise, Zhao and coworkers developed a binderless CNT-composite electrode that displayed an enhanced electrochemical response to the oxidation of cysteine [67]. This CNT/Pt hybrid microelectrode was fabricated by grinding the etched tip of a Pt microelectrode on a flat plate containing CNTs. The etching process was continued until the etched microcavity at the electrode tip was filled with CNTs. In summary, the introduction of CNTs into composite electrodes enhances biosensor performance in two ways: (i) the CNTs help to form a nanoweb on the surface of the biosensor which aids in enzyme adsorption and stability [66, 68]; and (ii) the CNTs increase electron transfer during electrochemical reactions [28].

8.4.2
Covalent Bonding

The covalent linking of CNTs to the surface of electrodes has attracted much interest in recent years. Covalent bonding techniques prevent the CNT and enzyme leeching that often plagues electrodes immobilized with adsorbed CNTs. Various chemical crosslinkers have been used to form covalent bonds between CNTs and electrode surfaces [46, 69]. For example, Zeng et al. used (3-mercaptopropyl) trimethoxysilane (MPS) to form a three-dimensional (3-D) matrix of thiol groups on a gold electrode surface for subsequent CNT binding [69]. The procedure included the dispersal of MWNTs in dimethylformamide (DMF) (an effective dispersant for CNTs [70]), and subsequently casting the MWNT/DMF mixture onto the functionalized gold surface, where the S–Au bonds covalently linked the MWNTs to the gold electrode surface. This MWNT/gold electrode displayed a decreased peak separation and increased peak current during CV experimentation in $K_3Fe(CN)_6$ as compared to unmodified bare electrodes. This improvement in electrochemical response can be attributed to the increase in the effective surface area of the electrode by the covalently bonded CNTs. Furthermore, Zeng et al. showed that the peak electrochemical current associated with the immobilized CNTs is directly related to the MWNT film thickness. For example, the electrochemical current output signal increases as the MWNT thin film thickness increases, but it eventually levels off and decreases as the thin film thickness increases to a level that is physically and chemically unstable. Mancuso et al. demonstrated how the amperometric response for a CNT/MPS biosensor is greatly enhanced compared to a bare electrode during the oxidation of indole acetic acid (IAA) [46]. Additionally, redox hydrogels comprised of PVP–Os [synthesized with poly(4-vinylpyridine), Os(bpy)2Cl+/2+, and 2-bromoethylamine] and poly(ethylene glycol)-diglycidyl ether (PEGDG) have also been used as crosslinking agents for CNTs and enzymes such as GOx (see Figure 8.3) [71]. The covalent and/or ionic bonding stabilizes the CNT/enzyme structures dispersed on the electrode surfaces. An increased amperometric response to glucose has been exhibited, indicating that the PVP–Os and PEGDG (PVP–Os and PEGDG) have desirable biocompatibility and stability for enzymatic biosensing. Nevertheless, CNT/MPS biosensors have demonstrated perhaps the longest stability even after weeks of daily operation, due in part to their strong covalent binding schemes that crosslink the enzyme to CNTs, and the CNTs to electrode surfaces.

8.4.3
Polymer Entrapment

The polymer entrapment of CNTs and/or enzymes at electrode surfaces offers yet another popular approach to enhancing conventional electrodes with CNTs. This polymer entrapment process typically entails the creation of a separate polymer–CNT mixture that is subsequently cast on the electrode to form a nanocomposite layer of CNTs and enzyme. Many polymers have been reported for such applications, including Nafion [72–75], chitosan [76–78], silicate sol–gels [79–81], and polypyrrole (PPy) [82–84]. Wang et al. developed a glucose biosensor based on a mixture of CNTs and Nafion [75] (see Figure 8.1a). Nafion is a highly biocompatible conductive sulfonated tetrafluorethylene copolymer that is negatively charged, and thus possesses the ability to repel negatively charged electrochemical interferents such as ascorbic acid [85]. Biosensors based on CNT/Nafion nanocomposites have demonstrated increased amperometric sensitivity towards detecting NADH

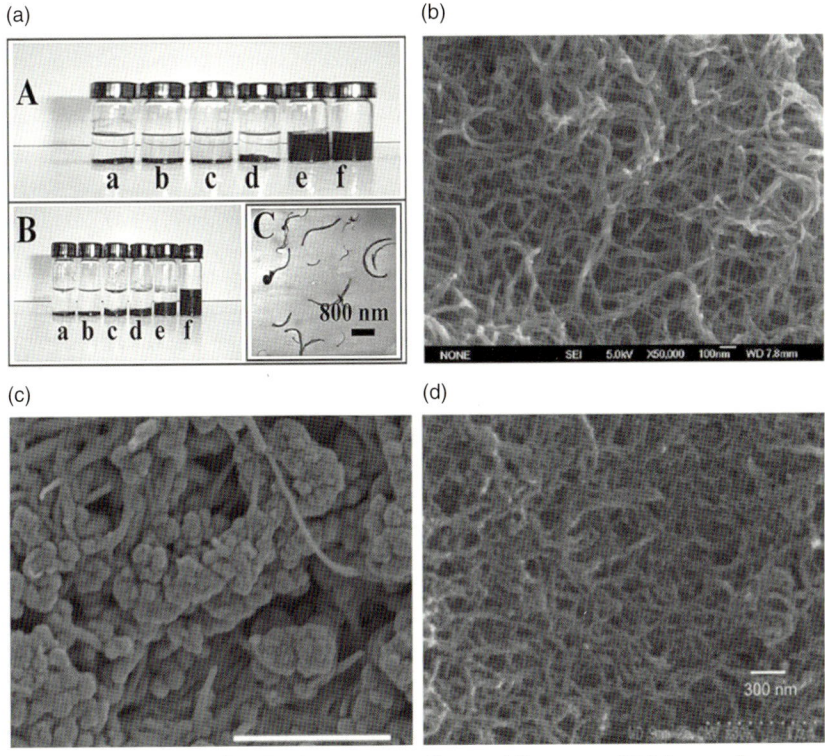

Figure 8.1 (a) Photographs of vials containing 0.5 mg ml^{-1} of SWNT (A) and MWNT (B) in different solutions: a = phosphate buffer (0.05 M, pH 7.4), b = 98% ethanol, c = 10% ethanol in phosphate buffer, d = 0.1% Nafion in phosphate buffer, e = 0.5% Nafion in phosphate buffer, f = 5% Nafion in ethanol. Also shown (C) is a transmission electron microscopy image of a 0.5% Nafion solution containing 0.3 mg ml^{-1} of MWNT. Reproduced with permission from Ref. [75]; © 2003, American Chemical Society; (b) Scanning Electron Microscopy, SEM, image of CNT-chitosan matrix. Reproduced with permission from Ref. [78]; © 2008, Elsevier; (c) SEM image of a MWNT sol–gel composite with 0.25 mg colloidal silica particles per mg MWNT. Reproduced with permission from Ref. [80]; © 2008, Elsevier; (d) SEM image of a PPy/MWNT–GOx film, prepared by using 1 mg ml^{-1} MWNT, 0.5 mg ml^{-1} GOx, and 0.5 M pyrrole at +0.7 V for 10 min. Reproduced with permission from Ref. [84]; © 2005, Elsevier.

and H_2O_2 [75], the electroactive products of enzymatic reactions with biomolecules such as glucose [72, 73, 75] and ethanol [74].

The conducting polymer chitosan has been incorporated into electrode design to assist in CNT and enzyme biofunctionalization. Chitosan is a film-forming linear polysaccharide that exhibits superb biocompatibility with biomolecules, and has an adhesive capability with chemically modified surfaces. CNT/chitosan layers can bind to glassy carbon electrodes via electrostatic forces (see Figure 8.1b) [78]. Biosensors based on CNT and chitosan have reported the oxidation of H_2O_2 and NADH at low working electrode potentials (< +300 mV) [76–78]; thus, interferents such as acetaminophen and ascorbic acid, that typically oxidize at higher working potentials (> +300 mV), should not generate significant background electrical noise. This reduction in background noise significantly increases the SNR of the biosensor during electrochemical sensing.

Silicate sol–gels, such as tetramethyl orthosilicate (TMOS) and tetraethyl orthosilicate (TEOS), have been used widely in electrochemical biosensing due to their ability to form porous 3-D matrix layers that can encapsulate enzyme molecules [86–89]. Additionally, sol–gel technology has been utilized to immobilize CNTs on biosensing electrodes [79–81]. For example, homogeneous suspensions of CNTs immobilized on electrode surfaces can be obtained by dispersing CNTs within sol–gels such as pretreated methyltrimethoxysilane (MTMOS) [79], propyltrimethoxysilane (PTMOS) [80], or methyltriethoxysilane (MTEOS) [81] solutions. When Gavalas *et al.* developed a CNT/sol–gel composite electrode, subsequent scanning electron microscopy (SEM) investigations revealed the porous nature of the sol–gel matrix, which is well-suited for CNT entrapment and analyte diffusion (see Figure 8.1c) [80]. Gong *et al.* reported that the electrochemical properties of a CNT/sol–gel electrode would vary, depending on the concentration of the MWNTs dispersed within the sol–gel matrix [81]. For example, when the CNT concentration is above 1.5 mg ml^{-1}, the electrode possesses electrochemical properties similar to that of a ceramic–CNT macroelectrode that displays a peak-shaped cyclic voltammogram in 1.0 mM $K_3Fe(CN)_6$, with a small peak separation that is indicative of a diffusion-limited redox process. However, when the CNT concentration is equal to or lower than 0.1 mg ml^{-1}, the electrode will exhibit the properties of a nanoelectrode ensemble (NEE), with a sigmoidal cyclic voltammogram that is indicative of enhanced mass transport by radial diffusion [90]. Thus, by altering the CNT concentrations in sol–gels, the mass transport and heterogeneous charge transport properties can be altered to tune the biosensor sensitivity. Furthermore, biosensors based on CNT/sol–gels have demonstrated an increased current response, a desirable reproducibility and stability, and decreased working potentials for amperometric H_2O_2 sensing compared to conventional sol–gel electrodes [79–81].

Finally, the conducting polymer polypyrrole (PPy) has been used extensively to immobilize CNTs onto electrodes for electrochemical biosensing. Chen *et al.* reported a method to use electro-polymerized PPy to physically entrap CNTs at the electrode surface [91], while Wang *et al.* noted that pretreated MWNTs modified with carboxyl groups and enzyme (GOx) can form, via electrostatic forces, a PPy/MWNT–GOx layer on an electrode surface (see Figure 8.1d) [84]. These results illustrated how a PPy matrix doped with CNTs could increase amperometric sensitivity towards glucose over a PPy matrix without CNTs. Moreover, the PPy/CNT/GOx nanocomposites displayed negligible electrochemical interference from electroactive species such as uric and ascorbic acid, at working potentials as high as +900 mV [84]. Thus, these PPy/MWNT–GOx electrodes experience enhanced sensitivity and selectivity during electrochemical biosensing. In addition to PPy, conductive polymers such as polyaniline (PAN) and the polyelectrolyte poly(diallyldimethylammonium chloride) (PDDA) have also been utilized in conjunction with CNTs to enhance the electrochemical biosensing performance [92, 93].

8.4.4
Aligned Arrays

Aligned-CNT electrodes offer numerous advantages in electrochemical biosensing. First, a new dimension of control over CNT orientation and spacing exists with aligned-CNT electrodes as compared to the previously discussed CNT-based electrode designs. Also, by controlling the spatial orientation of CNTs, functional groups such as oxygenated species can be developed at aligned-CNT ends. These functional groups can enhance charge transport and be used as docking points for numerous biofunctionalization strategies [28, 49]. Furthermore, this spatial control has led to a new class of electrodes, known as *nanoelectrodes*, that experience increased sensitivity due to enhanced mass transport of target species to the biosensor surface [90, 94].

The vast majority of aligned-CNT arrays are grown *in situ* from a metal catalyst deposited on the electrode surface. One of the earliest reported *in-situ* CNT fabrication techniques was a high-temperature pyrolysis technique where the CNTs were grown from a Fe/C catalyst. For example, Huang and coworkers utilized high-temperature pyrolysis to prepare an aligned-CNT thin film on a quartz plate that could be transferred to an electrode surface [95]. In this sensor design, $FeC_{32}N_8H_{16}$ (FePc), when located on a quartz plate substrate, acted as both the metal catalyst and carbon source required for the CNT growth process. Parallel-aligned CNTs were shown to grow from the substrate surface after the pyrolysis of FePc under Ar/H_2 at 800–1100 °C (see Figure 8.3). By dissolving this CNT substrate in strong acids such as HNO_3, the CNT layer could be separated and transferred to conventional carbon electrodes [95]. Gao et al. also developed a glucose biosensor based on this approach using a gold electrode [96]; a higher current density (current/area) was reported for the electrochemical response to glucose and a lower irreversible single oxidation peak, compared to a glassy carbon electrode without CNTs. Schulz et al. developed a CNT-modified needle biosensor for H_2O_2 sensing from a fabrication protocol, similar to Gao and coworkers. Although the CNT-modified needle biosensor experienced similar amperometric and cyclic voltammetric properties as the electrode reported by Gao et al. [97], the high-temperature growth conditions of these CNT electrodes was shown to hinder the manufacturability of CNT-based biosensors created via the high-temperature pyrolysis technique

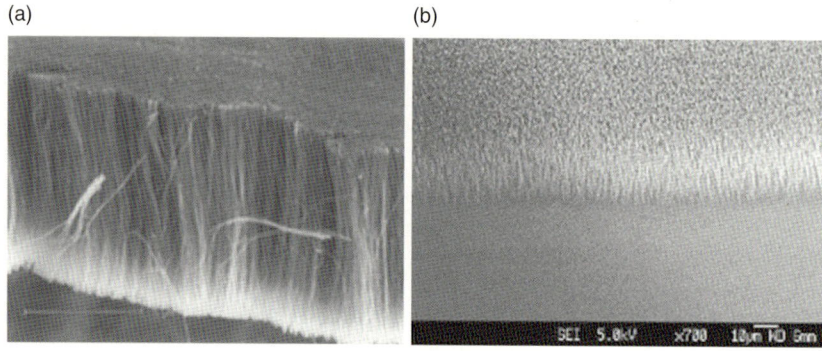

Figure 8.2 (a) Scanning Electron Microscopy, SEM, image of aligned CNTs based on a high-temperature pyrolysis growth procedure. Reproduced with permission from Ref. [95]; © 1999, American Chemical Society; (b) SEM image of aligned CNTs based on a microwave plasma chemical vapor deposition (MPCVD) CNT fabrication technique. Reprinted with permission from Ref. [98]; © 2003, Elsevier.

[95–97]. In contrast, the chemical vapor deposition (CVD) growth of aligned-CNT electrodes offered a lower temperature fabrication process for CNT synthesis.

CNT growth via a CVD protocol is arguably the most widely used method for developing aligned arrays of CNTs. For instance, Wang et al. described the growth of aligned CNTs using a microwave plasma-enhanced chemical vapor deposition (MPCVD) process that used nickel as a catalyst [68, 99]. The plasma formed by decomposition of the hydrocarbon gas caused an incorporation of carbon atoms into the nickel particles, and this led to the formation of the MWNTs. These MWNTs grew densely and vertically along the grain of the catalytic particles (see Figure 8.2). Wang et al. also reported that, by controlling the thickness of the nickel layer, the diameter of the aligned MWNTs could be controlled. Glucose sensing with these MWNT biosensors revealed a high stability of the immobilized enzyme, with 91% of the initial sensitivity retained after a three-month storage period. Unfortunately, one disadvantage with this approach is that nickel caps remained on the tube ends after the aligned MWNTs had been grown on the electrode surface. As discussed above, tube ends connected with oxygenated species will enhance heterogeneous charge transfer in electrochemical biosensing [28, 49]. In addition, oxygenated species located at the nanotube ends will provide docking points for covalent linking with enzymes. Therefore the metal-capped CNT ends diminish both the electroactivity and the biofunctionalization capability of the aligned-CNT arrays. Consequently, several CNT purification techniques have been developed to create aligned CNT-based electrodes that are more suitable for electrochemical biosensing.

Several CNT purification techniques have been developed to prepare aligned arrays of CNTs. For example, CNT purification technique was developed by Liu et al. that utilized a SWNT alignment method by self-assembly at room temperature [100]. After breaking up the SWNTs into shorter tubes via oxidation with sulfuric and nitric acids, the resultant open-ended tubes became chemically active. Subsequently, the chemically treated SWNTs were attached to the surface of gold electrodes via Au–S (thiol) bonds. Another advantage to the fabrication technique presented by Liu and coworkers is that the height of the aligned CNTs can be controlled by acid treatment [100]. Numerous other research groups have reported similar approaches to purifying and/or aligning CNT arrays on electrode surfaces for electrochemical biosensing [42, 101–104]. Additionally, aligned CNT electrode fabrication protocols have utilized metallic catalyst spacing techniques in order to grow spatially distanced CNT arrays through CVD processes. These spatially distanced CNT array electrodes are known as "nanoelectrodes."

8.4.4.1 Nanoelectrodes

Nanoelectrodes are electrodes with critical dimensions within the tens of nanometers [105]. Nanoelectrode arrays (NEAs) are more advantageous than macroelectrodes because of their enhanced mass transport [94], improved signal-to-noise ratios [106], and improved detection limits [107]. In addition, the reaction time and spatial resolution of electrochemical sensors scale inversely with the electrode radius [108]; therefore, the response time and detection limits of nanoelectrodes have continued to improve with decreasing size. These nanoscale electrodes are often comprised of nanoscale materials with inherent electrocatalytic properties such as CNTs and metallic nanoparticles. Thus, nanoelectrode arrays comprised of CNTs can be categorized as subgroups of vertically aligned CNT electrodes and CNT/metal hybrid electrodes.

One of the first NEAs was developed by Li *et al.* for DNA detection based on vertically aligned MWNTs embedded in SiO_2 [109]. In this case, the open ends of the MWNTs were subsequently functionalized with amine-terminated oligonucleotide probes for the highly sensitive detection of DNA target molecules (ca. 6 attomoles per experiment). One of the first nanoelectrode glucose biosensors was developed by Lin and coworkers (see Figure 8.3) [110]. For this, the vertically aligned CNT arrays were grown via CVD from Ni particles dispersed on a silicon based platform, while an epoxy passivation layer electrically isolated the distinct CNTs. An electrochemical treatment process introduced carboxylic acid groups onto the exposed tips of the CNTs, and carbodiimide chemistry was employed to functionalize the tips with the enzyme GOx. The ferricyanide cyclic voltammogram of the biosensor displayed a sigmoidal shape, which often is indicative of nanoelectrode behavior [94, 111]. The amperometric glucose sensing yielded a selective detection of glucose within a background of typical electroactive interferents found in blood (e.g., acetaminophen, ascorbic and uric acids), while the linear glucose sensing range was within typical blood glucose levels. Similarly, various other research groups have used CNT-based nanoelectrodes to increase sensitivity in electrochemical biosensing [38, 112, 113].

8.4.5
Hybrid (CNT/Metal Nanoparticle) Electrodes

CNTs decorated with metal nanoparticles represent another burgeoning category of CNT-based biosensors. The decoration of CNT-based biosensors with various transition metals (e.g., Pd [114], Pt [31], Au [115], Ag [116], and Cu [117]) increase

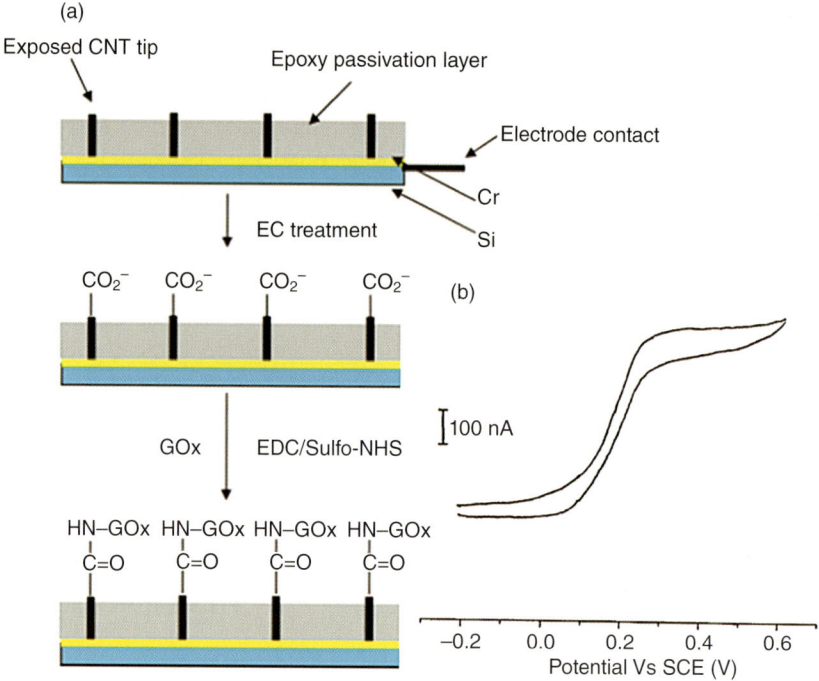

Figure 8.3 (a) Side view schematic diagram portraying CNT nanoelectrode ensembles. Electrochemical treatment of the CNT nanoelectrode ensembles leaves oxygenated species at the exposed tips of the CNTs, for subsequent biofunctionalization. The lower diagram shows coupling of the enzyme GOx to the electrochemically treated CNT tips; (b) Cyclic voltammogram recorded in a ferricyanide solution. The sigmoid shape of the scan is typical for nanoelectrode arrays. Reproduced with permission from Ref. [110]; © 2004, American Chemical Society.

the sensor surface area and improve the electrocatalytic nature of the biosensor towards the oxidation of chemical products such as H_2O_2 and the reduced cofactor NADH, that are produced during biological recognition events with oxidase and dehydrogenase enzymes, respectively. The size and morphology of these metallic nanoparticles plays a fundamental role in the electrochemical performance of the biosensor. For example, the effective surface area of the sensor is proportional to its capacitative current [118], and accordingly an increase in surface area should increase the amperometric signal output of the sensor during electrochemical testing. Additionally, the curvature of the nanoparticles has shown to affect the amount of biological agent that can be covalently immobilized to the metallic nanoparticles [119]. Furthermore, as with the case of vertically aligned CNT-based nanoelectrodes, spatially distance metallic nanoparticles on the biosensor surface can act as nanoelectrodes arrays, NEAs. NEAs can greatly improve the mass diffusion of electroactive species to the sensor surface itself [94], thus improving the biosensor sensitivity and response time.

Various methods have been developed to electrically link metal nanoparticles to CNT-based electrodes [31, 76, 77, 120, 121]. Numerous research groups have created Au/CNT-based glucose electrodes by decorating CNTs with Au nanoparticles via adsorption techniques [29, 122]. However, Hrapovic et al. linked Pt nanoparticles to SWNTs by charge interaction between Pt and Nafion-suspended SWNTs, where the negatively charged Nafion attracted the positively charged Pt [31]. Another somewhat contrasting (but commonly used) method for combining CNTs and metal nanoparticles is through polymer or sol–gel matrix immobilization. For example, Kang et al. described glucose biosensors based on uniform Pt–CNT–chitosan films where the amino group of chitosan facilitated the dissolvability of both the Pt- and –COOH-modified CNTs [77]. The electrodeposition of Pt and Au nanoparticles on CNT-modified electrodes using H_2PtCl_6 and $HAuCl_4$ salt solutions has been reported by various research groups [76, 121]. Perhaps one of the most successful glucose biosensors in terms of linear sensing range (0.01–50 mM) and detection limit (1.3 µM) was a SWNT/Au-coated Pd nanoparticle biosensor [123]. This biosensor created Pd nanocubes at SWNT defect sites, and subsequently coated them with Au via a facile electrodeposition strategy (see Figure 8.4). The templated MPCVD growth of SWNTs through a porous anodic alumina template incorporated with electrodeposited Au-coated Pd nanocubes, created an electrochemically active network that enhanced sensor sensitivity and produced a unique nanoenvironment that was well suited for enzyme immobilization.

8.5 Applications

8.5.1 Nonenzymatic Biosensing

CNTs incorporated into electrodes offer a tremendous platform for electrochemical biosensing, due to their unique electrical and chemical/electrochemical properties. As mentioned above, CNT-based biosensors offer enhanced heterogeneous charge transport during electrochemical sensing, due in part to their high electrically active surface areas and biocompatibility. Thus, CNT-based biosensors are well suited for the electrochemical detection of electroactive species at low concentrations. In this section, it will be shown how CNT-based biosensors have improved the sensitivity of biosensors towards the following four electroactive species: NADH; homocysteine; dopamine; and IAA.

8.5.1.1 Nicotinamide Adenine Dinucleotide (NADH)
In its oxidized form, NAD acts as a electron carrier for dehydrogenase enzymes [124]. The reduced form of nicotinamide adenine dinucleotide (NADH) is also the precursor to cyclic ADP-ribose, a stimulator of calcium signaling pathways [125], that donates its ADP-ribose moieties during the post-translational modification of

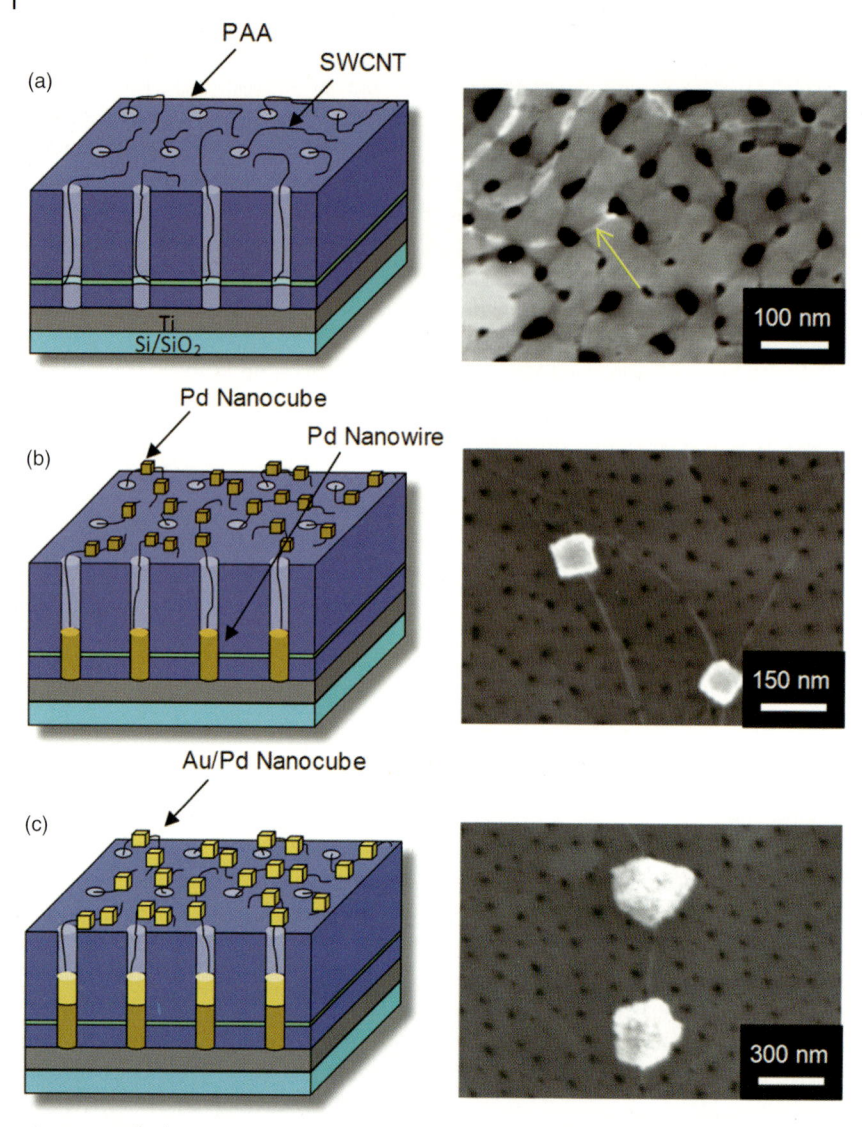

Figure 8.4 Tilted cross-sectional schematics with corresponding top-view field emission scanning electron microscopy (FESEM) images portraying the sequential fabrication process steps. (a) SWNTs grown from the pores of the porous anodic alumina (PAA) via MPCVD (the FESEM image shows a SWNT protruding from a pore and extending along the PAA surface); (b) Electrodeposition of Pd to form Pd nanowires in pores and Pd nanocubes on SWNTs (two such nanocubes are shown in the corresponding FESEM image); (c) Electrodeposition to coat the existing Pd nanocubes with a thin layer of Au. Reproduced with permission from Ref. [123]; © 2009, American Chemical Society.

proteins [126]. As several of the biological roles of NADH are known to depend on the inherent redox properties of the molecule itself, the redox nature of this compound has been widely investigated, using amperometry and CV methods [124, 127].

Although, in the early investigations, high working potentials (~1.0 V) were used to oxidize NADH on bare platinum electrodes [127], an unmodified glassy carbon electrode (GCE) was shown to be capable of oxidizing NADH at a lower working potential (~750 mV) [128]. Furthermore, Chen et al. lowered the overpotential required to oxidize NADH (to 645 mV), by using a GCE modified with ordered CNTs [129]. Subsequently, Musameh et al. reported that the addition of SWNTs and MWNTs to a GCE lowered the working potential necessary to oxidize NADH, to 60 mV [130]. In the biosensor fabrication protocol employed by Musameth et al., the CNTs were grown on an alumina template deposited onto a GCE, with Nafion acting as an adhesive layer. This low working potential is well suited for *in vivo* or *in vitro* experimentation, where high working potentials can cause tissues to be degraded and experiment results to be adversely affected. In addition, these CNT-modified electrodes were shown to be 70% more stable than the unmodified GCEs. Recently, Radoi et al. utilized SWNTs which were covalently linked to the redox mediator Varian Blue to detect NADH during flow injection analysis (FIA) [131]. In this case, a large pH and wide working potential range was established for this SWNT-based electrode.

8.5.1.2 Homocysteine

Cardiovascular disease and weak brittle bones have each been linked to increased blood levels of homocysteine (a sulfur-containing amino acid) [132–134]. Although the condition of *hyperhomocystienemia* (i.e., chronically high levels of homocysteine in the blood) is usually caused by genetic defects, milder forms of the condition can be attributed to prolonged fasting or to diets that are very deficient in vitamins [134, 135]. Depending on the severity of the condition, high blood levels of homocysteine can increase the risk of strokes, transient ischemic attack (TIA) and heart attack. Consequently, electroanalytical methods have been proposed as a rapid means of detecting homocysteine in the blood [136, 137].

The first type of electrode used to detect homocysteine was composed of gold and unmodified GCEs [138, 139], and required a high over-potential that was prone to electrochemical interference from electroactive species found endogenously in the test samples. In order to circumvent these high working potentials, Gong et al. utilized a GCE that had been modified with a composite layer of Nafion and MWNTs to oxidize homocysteine at a working potential of almost 0 V [136]. This low working potential greatly enhanced the selectivity of the biosensor towards the detection of homocysteine, as most of the interfering compounds were electrochemically inert under these conditions. By contrast, Lawrence et al. [140] operated a CNT paste electrode to detect homocysteine at a working potential of 700 mV. This large working potential was most likely necessary because of the high electrical resistance of the mineral oil used in the paste-making process, and was not an attribute of the CNTs themselves.

8.5.1.3 Dopamine

Dopamine is a neurotransmitter that activates five different receptors in the brain to perform various vital bodily functions [141] that range from raising the heart rate to supporting cognition [142, 143]. While low dopamine levels in the brain may affect cognition, chronically low levels may be a sign of Parkinsons's disease, that causes a slowing of motor functions and muscle tremors [144]. Thus, various CNT-based electrochemical biosensors have been developed in order to monitor the tissue levels of dopamine.

Various electrochemical methods, including CV and amperometry, have been used to detect dopamine, using biosensors. Unfortunately, however, *in vivo* electrochemical experiments are hampered by endogenous interferents (e.g., ascorbic acid) that exist within the brain tissue [145]. Thus, protective polymer layers (e.g., Nafion) are typically employed to increase selectivity towards dopamine [146]. For instance, Zhang *et al.* developed a biosensor composed of multilayers of MWNTs that were co-immobilized with diallydimethylammonium chloride that was able to detect dopamine selectively, even in the presence of ascorbic acid [145]. More recently, Wang *et al.* combined the redox mediator poly (3-methylthiophene) (P3MT) with SWNTs to create a dopamine sensor that demonstrated a detection limit of 0.5 nM [146]. This sensor employed differential pulse voltammetry to selectively detect dopamine amid ascorbic and uric acid interference. Moreover, this CNT-based electrode was found to be both highly stable and reproducible.

8.5.1.4 Indole Acetic Acid (IAA)

Indole acetic acid (IAA), a phytohormone also known as *auxin*, plays a key role in the development and growth of plants. Within the plant, IAA is mostly synthesized in the shoot apical meristem, from where it is transported basipetally from the shoot to the root tip, and then redistributed acropetally to the cortical and epidermal cells [147]. Young developing roots and leaves also synthesize and transport the hormone to the shoot apices, in a similar fashion. The polar gradient created from the shoot to root transport is essential for establishing vascular differentiation, proper organ development, various tropic effects, and apical dominance [148]. IAA is an electroactive compound that has been monitored using electrochemical biosensors [46, 149, 150].

Recently, CNTs have been incorporated into electrochemical biosensors by using a variety of methods, for the sensing of IAA. For instance, when Wu *et al.* [149] modified GCEs with MWNTs and a hydrophobic surfactant to sense IAA, the electrode designed had a detection limit of $0.02\,\mu M$ and a linear range of between 0.1 and $50\,\mu M$. Recently, CNTs have shown to increase the electrochemical signal received by the oxidation of endogenous IAA flux near the surface of Zea mays roots without the addition of exogenous IAA [154]. Moreover, Mancuso *et al.* [46] utilized microelectrodes modified with MWNTs to detect IAA flux using a noise and drift canceling technique called "self-referencing" [151–153]. Thus, Mancuso *et al.* were able to form a contour map of IAA influx in corn roots, where the bathing medium was doped with external auxin. It is likely that, one day, these CNT-based electrodes may pave the way for noninvasive IAA flux experiments close to the sensor–plant root interface.

8.5.2
Enzymatic Biosensing

CNT-based electrochemical biosensors offer a unique nanoenvironment for biocompatibility and biofunctionalization with proteins. This unique biosensor surface geometry provides conformational compatibility with proteins due to both steric and hydrophobic effects [11], allowing the proteins to wrap around and self-organize on the CNT surfaces [155, 156]. Thus, CNTs offer an advantageous platform for integration with enzymes.

Recently, various electrode surface modification approaches have incorporated the juxtaposition of CNTs and enzymes. One concomitant issue here is how to reconcile the immobilization of both the enzymes and the CNTs on the surface of electrodes, so that an optimal performance of the biosensors can be obtained. Existing methods include the encapsulation of enzymes and CNTs into one polymer layer [79, 157], the deposition of double or multiple layers containing CNTs and enzymes, respectively [158], and the direct absorption of enzymes onto the CNTs [159]. Gooding et al. have described a self-assembly attachment approach in which CNTs were incubated in a microperoxidase MP-11 solution in HEPES buffer. The enzymes were then shown to be attached to the ends of the tubes via covalent bonds, instead of being entrapped in gaps among the tubes themselves [160]. By using AFM to quantify the number of enzyme molecules and the number of CNTs present after linking, it was further shown that no enzymes were linked to the side walls of the CNTs. The AFM experiments showed the number of CNTs to be almost the same as the number of MP-11 enzymes [47, 160]. An enhanced electrode performance was produced when the peak current achieved for the MP-11/SWNT biosensor during CV experiments was more than threefold that of biosensors with no SWNTs [160]. Patolsky et al. reported similar approaches by linking the enzyme to the ends of CNTs [161]. Several groups have reported that the direct linking of enzymes to CNTs improves charge transport and enhances biosensor sensitivity during electrochemical biosensing [162–164]. These various methods for linking enzymes to CNTs demonstrate the biocompatibility of nanotubes with specific proteins such as enzymes. Some of the most popular enzyme/CNT immobilization schemes have included GOx, glutamate oxidase (GluOx) and alcohol oxidase (AOx) for the sensing of glucose, glutamate, and alcohol, respectively.

8.5.2.1 Glucose
Over the past few decades, glucose biosensors have attracted intense attention, mainly because diabetes has become a global health concern and, despite extensive efforts, no cure has yet been found [54]. In order to treat diabetes, the strict monitoring and control of blood glucose levels is imperative; consequently, accurate, rapid, and sensitive blood glucose biosensors would be very beneficial to perform this strict glycemic monitoring.

Enzymatic glucose biosensing was first conducted by Clark and Lyons at the Children Hospital in Cincinnati Ohio in 1962 [53]. In the biosensor developed for this purpose, the enzyme GOx was immobilized over an oxygen electrode to measure the amount of oxygen consumed in the GOx/glucose reaction [165]:

$$\text{D-glucose} + O_2 + H_2O \xrightarrow{\text{GOx}} \text{D-gluconic acid} + H_2O_2$$

Some years later, in 1973, Guilbault and Lubrano developed an amperometric glucose biosensor that oxidized the liberated hydrogen peroxide (H_2O_2) from the GOx/glucose reaction [166]:

$$H_2O_2 \rightarrow 2H^+ + O_2 + 2e^-$$

These two glucose-sensing experiments would lay the foundation for over three decades of progress in the amperometric sensing of glucose. Thus, the interaction between the GOx enzyme and glucose is quite possibly the most understood enzymatic reaction in all of electrochemistry. This understanding, combined with decades of research, have led to the amperometric sensing of glucose via GOx becoming the benchmark reaction for amperometric sensors, upon which the key characteristics of sensitivity, detection limit, and response time can be compared.

Numerous research groups have been able to functionalize SWNT arrays with GOx for the amperometric sensing of glucose. Lin et al. demonstrated how GOx could be immobilized to carboxyl groups located on the tips of SWNTs grown from a Si substrate [110]. This electrode was able to sense glucose concentrations of 15 mM, with a detection limit of 0.08 mM and a response time of 20–30 s. In addition, the electrode was also able to sense glucose at a negative overvoltage (−0.2 V) with no signal output from common blood interferents (e.g., uric acid, ascorbic acid, acetaminophen). A GCE electrode modified with SWNTs and immobilized with GOx, as developed by Yao and coworkers, was able to sense glucose with a linear sensing range of 0.1 mM to 5.5 mM, a detection limit of 0.05 mM, and a sensitivity of 62 $\mu A\,mM^{-1}\,cm^{-2}$ [35]. This SWNT electrode proved to be more sensitive than three MWNT-based electrodes of differing lengths which were compared in the same study. MWNTs arrays with a thick forest-like density were grown on a silicon substrate from a nickel catalyst by Wang et al.; subsequently, GOx was adsorbed onto the surface of the MWNT electrode and the biosensor was able to retain its stability (~86.7% of initial enzyme activity) for a four-month period [99]. Guan et al. developed a disposable glucose biosensor based on MWNTs immobilized with GOx on a screen-printed carbon electrode [167]. This relatively simplistic paste fabrication protocol, when combined with a glucose linear sensing range of up to approximately 20 mM, presented a potential alternative to current glucose test strip technologies.

Many research groups have combined metallic nanoparticles with CNT-based amperometric glucose biosensors to enhance sensor performance. Manso et al. reported colloidal gold nanoparticles dispersed on a CNT-Teflon substrate that displayed good reproducibility between five different electrodes over several different days, with a glucose detection limit of 10 mM [122]. A Au/MWNT-GOx electrode created by Wang et al. displayed a strong glucose response with high stability and a response time within 10–20 s [99]. A Pd–GOx–Nafion–CNT sensor showed a linear range up to 12 mM with a limit of detection of 0.15 mM, while selectively sensing glucose from uric and ascorbic acid blood interferents [168]. Yang et al. demonstrated how an electrode enhanced with CNTs and cobalt nanoparticles, when solubilized in an aqueous solution of chitosan, could sense glucose over a linear range of 0.01 mM to 10 mM, with a detection limit of 5×10^{-3} mM, and a response time of 10 s, while being capable of selectively sensing glucose in the background of common blood interferents [169]. One of the most impressive CNT/metal nanoparticle hybrid glucose sensor in terms of linear sensing range (0.01–50 mM) and detection limit (1.3 µM) consisted of Au/Pd nanocubes electrodeposited at SWNT defect sites [123]. In this case, the size, and morphology of the Pd nanocubes could be altered with a thin AU coating, thus demonstrating the capability to vary the performance characteristics of the biosensor.

In summary, glucose biosensors sensors have evolved dramatically since the initial experiments of Clark and Lyons in 1962. They now incorporate nanoscale materials to improve the sensitivity, response time, and detection limit. Indeed, some of the most impressive glucose-sensing results were achieved with biosensors based on CNTs and metallic nanoparticles. The vast amount of research conducted with CNT-based glucose biosensors demonstrates the biocompatibility of CNTs towards enzyme biofunctionalization. Thus, these glucose-sensing results have expanded the use of CNT-based biosensors across a broad spectrum of important biomolecule detection schemes, including those associated with glutamate and ethanol.

8.5.2.2 Glutamate

L-Glutamate is an important neurotransmitter in the central nervous system that is the primary mechanism for neuronal communication [170]. Abnormal levels of glutamate have been linked to diseases associated with schizophrenia, Parkinson's disease, and epilepsy [170, 171]. The real-time monitoring of neuronal glutamate levels could lead to therapeutic implant chips capable of monitoring and treating patients with neurological disorders [172]. Although the enzyme glutamate dehydrogenase (GluDH) has been used to measure glutamate via the amperometric sensing of NADH [173, 174], this method has certain drawbacks as the dehydrogenase enzymes require external cofactors that are typically not found endogenously in neuronal fluid. Moreover, the amperometric detection of NADH typically requires high electrode working potentials. In order to circumvent these challenges, the enzyme glutamate oxidase (GluOx) has been applied to the real-time monitoring of glutamate concentrations [175, 176]. The electroactive byproduct, H_2O_2, of the GluOx/glutamate reaction (see equations below) is oxidized at the sensor surface, and accordingly provides a measurable current signal during amperometric sensing:

$$\text{L-glutamate} + O_2 + H_2O \xrightarrow{\text{GluOx}} \alpha\text{-ketoglutarate} + NH_3 + H_2O_2$$

$$H_2O_2 \rightarrow 2H^+ + O_2 + 2e^-$$

Gerhardt's research group at the University of Kentucky has successfully immobilized GluOx on Pt microelectrodes coated with Nafion, which electrostatically repels anionic interferents, to monitor the release of L-glutamate in anesthetized rat brain [177]. In attempts to further improve sensor performance, including an enhancement of sensor sensivity, response time, and detection limits, CNTs have been employed in glutamate biosensor designs.

Huang et al. pasted a mixture of SWNTs modified with the mediator ferrocene and subsequently biofunctionalized the electrode with GluOx crosslinked to glutaraldehyde to produce a glutamate biosensor that was capable of sensing glutamate within the range of 1 to 7 μM [178]. A few years later, Meng et al. coated a GCE with SWNTs modified with the mediator thionine, which was subsequently biofunctionalized with GluOx crosslinked to glutaraldehyde [179]. The SWNT-based biosensor of Meng and coworkers amperometrically sensed glutamate with a linear sensing range of 0.5 to 400 μM and with a detection limit of 100 nM. Later, Tang et al. developed a mediator-free glutamate biosensor which consisted of GluDH dendrimer-encapsulated platinum nanoparticles and GluDH bonded covalently to MWNT films for glutamate sensing [180, 181]. Both MWNT/Pt–GluDH electrodes created by Tang and colleagues were able to sense glutamate amperometrically, with a detection limit of 10 nM and linear sensing ranges extending upwards to 60 and 250 μM. Perhaps the most creative electrochemical CNT-based glutamate biosensor was developed by Boo et al., who demonstrated how a single MWNT (30 nm diameter) could be mounted on a sharp-etched tungsten tip and isolated electrochemically by way of an ultraviolet-hardening process [182] (Figure 8.5a). This CNT nanoneedle biosensor was capable of sensing glutamate amperometrically within the 100 to 500 μM range, a detection limit capable of monitoring glutamine secretion inside a synaptic junction (Figure 8.5b).

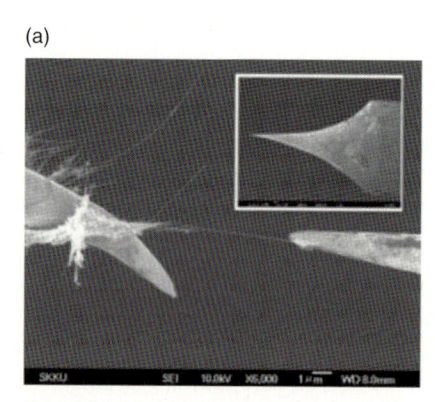

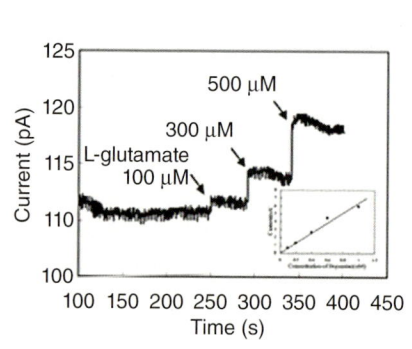

Figure 8.5 (a) Scanning Electron Microscopy SEM image of a MWNT nanoneedle electrode, showing the contact of a single MWNT with tungsten tip from the MWNT bundle. The inset shows the tungsten tip after etching; (b) Amperometric sensing of glutamate in HEPES buffer at a pH of 7.4 with the MWNT nanoneedle electrode. The inset shows the calibration curve. Reproduced with permission from Ref. [185]; © 2006, American Chemical Society.

The introduction of CNTs into electrochemical glutamate biosensors has significantly enhanced biosensor performance, including increases in the linear sensing range and a lowering of the detection limits for glutamate. Furthermore, it has been reported recently that CNTs can stimulate neuronal growth [183, 184], demonstrating their biocompatibility to neurons. These results demonstrate the potential capability that CNT-based electrodes have for possible implantation into the brain for *in vivo* neuronal glutamate sensing; consequently, CNT-based glutamate biosensors have great potential for assisting in the diagnosis and treatment of neurological diseases.

8.5.2.3 Ethanol

As ethanol is the product of anaerobic respiration for plants and bacteria, its electrochemical biosensing is important for monitoring bacterial metabolites produced during fermentation processes [185], food quality experiments [186], in new-generation bio-ethanol fuel technologies [187], in plant and bacteria physiology studies, and in the clinical analysis of blood, serum, saliva, urine, breath, and sweat.

Ethanol biosensors often utilize AOx [188, 189] to convert ethanol and O_2 into acetaldehyde and H_2O_2 (the electro-oxidative intermediate), according to the following chemical reaction:

$$\text{Ethanol} + O_2 \xrightarrow{\text{AOx}} CH_3CHO + H_2O_2$$

As with other oxidase-based electrochemical biosensors, an electric current is measured by oxidizing the resultant H_2O_2 at the electrode/solution interface accordingly [190].

Many ethanol electrochemical biosensors, however, also utilize alcohol dehydrogenase (ADH) for the amperometric detection of ethanol [191, 192]. The enzyme converts ethanol into acetaldehyde and NADH in the presence of the cofactor NAD^+, according to the following chemical equation; the NADH is subsequently electrochemically detected at the electrode/solution interface:

$$\text{Ethanol} + NAD^+ \xrightarrow{\text{ADH}} CH_3CHO + NADH + H^+$$

Both, AOx and ADH biosensors have distinct advantages and drawbacks. For example, as AOx ethanol biosensors require oxygen for their effective operation, oxygen depletion in the samples can lead to a significant degradation in biosensor performance. On the other hand, ADH ethanol biosensors require the presence of the cofactor NAD^+, which typically is not found endogenously in test samples and is thus prone to depletion. In terms of biosensor selectivity, both AOx- and ADH-based ethanol biosensors experience difficulties, as both enzymes are capable of reacting with either methanol or ethanol. Consequently, a laborious fractional distillation must often be carried out to eliminate any methanol from a test sample before ethanol sensing can begin. Nevertheless, ethanol biosensors have recently experienced enhanced sensitivity and selectivity due to the addition of CNTs to the electrode designs.

Clearly, the incorporation of CNTs into electrochemical ethanol biosensors has improved their performance [74, 188, 191, 192]. For example, Gouveia-Caridade *et al.* described an ethanol biosensor that used MWNTs and Nafion to enhance the sensitivity and selectivity of an ethanol biosensor [188]. Subsequently, the enzyme AOx was attached covalently to carboxyl groups at the MWNT tips, and crosslinked with bovine serum albumin and glutaraldehyde for stability. This ethanol biosensor was able to operate at a low working potential (−300 mV), while experiencing a sensitivity that was up to 20-fold greater than for conventional electrode materials. Later, Liaw *et al.* constructed an ethanol biosensor based on ADH co-immobilized with MWNTs and Nafion that experienced well-defined current responses during CV experimentation [74]. Santos and coworkers reported the details of an ethanol biosensor based on the co-immobilization of ADH and the mediator Meldola's Blue, that were both subsequently covalently linked to CNTs via glutaraldehyde [191]. This mixture of mediator, enzyme and CNTs was adhered to a graphite electrode by mineral oil to create the ethanol biosensor. The infusion of a mediator and CNTs allowed this biosensor to operate at a working potential of 0 mV, thus preventing interference from any endogenous electroactive species. The ethanol calibration plots provided by Santos and coworkers demonstrated the high amperometric sensitivity ($4.75\,\mu A\,mM^{-1}\,cm^{-2}$) and wide linear range ($0.05-10\,mmol\,l^{-1}$) of this CNT-based ethanol biosensor. Tsai *et al.* presented an ethanol biosensor which was based on ADH and poly(vinyl alcohol) (PVA) [192]; during fabrication, the PVA is pretreated with a freezing–thawing cycle and used as the immobilization media for both the CNT and ADH. As a result, significant improvements in sensor performance were reported, including an enhanced sensitivity ($196\,nA\,mM^{-1}$), a wide linear ethanol sensing range (up to 1.5 mM), and a fast response time (8 s) [193]. Moreover, Choi *et al.* developed a CNT-based ethanol biosensor with a response time of 2 s, an amperometric sensitivity of $51.6\,\mu A\,mM^{-1}\,cm^{-2}$, and a desirable long-term stability (75% of the original sensitivity was retained after four weeks) [194]. This biosensor utilized a titania–Nafion film with a large pore size and good biocompatibility to immobilize ADH and CNTs onto the electrode surface [194]. Another intriguing ethanol biosensor example was that developed by Umasankar *et al.*, and created by depositing MWNTs with Nafion onto a GCE [195]. Subsequently, the redox mediator poly(malachite green) (PMG) was electrodeposited onto the surface of the GCE modified with MWNTs. The electrochemical experimentation of this biosensor included a direct oxidation of ethanol using differential pulse voltammetry, demonstrating an example of an enzyme-less ethanol biosensor. In this case, well-defined voltammetric peaks were observed due to the enhanced electrocatalytic activity of the unique MWNT–NF–PMG film.

The incorporation of CNTs in ethanol biosensor construction has significantly enhanced electrochemical transduction, resulting in increased amperometric sensitivity [191], an extended linear sensing range [192] and shortened response times [194] for amperometric biosensors, as well as enhanced voltammetry responses for voltammetric biosensors [195]. Today, numerous industries – including those associated with alternative renewable energy, food quality control, and general clinical research – will continue to benefit from CNT-based electrochemical ethanol biosensors.

8.6 Conclusions

The recent advances and fundamental principles associated with CNT-based electrochemical biosensors continue to show much promise as research and applications advance. The unique mechanical, electrical, and chemical properties of CNTs and their applications to electroanalytical biosensing have been well established. Based on conventional and reliable CNT synthesis techniques, CNT-based biosensors have promise as mass-produced devices. A review of the literature for electrochemical biosensors, with special emphasis on amperometric sensing (i.e., the most ubiquitous electrochemical sensing modality), yields a dramatic new emphasis on CNTs and nanomaterial-enhanced strategies. This emphasis includes CNT-based biosensor fabrication strategies which utilize both "top-down" and "bottom-up" fabrication paradigms. The emergence of CNT-based nanoelectrodes arrays and their enhanced heterogeneous and mass transport charge properties further promises major advances in this field. CNT-based electroanalytical methods are profoundly advantageous for sensor approaches, and there are now numerous examples of nonenzymatic (e.g. NADH, homocysteine, dopamine, IAA) and enzymatic (e.g. glucose, glutamate, ethanol) CNT-based electrochemical biosensor applications in the recent research literature. Notably, these offer great promise for medical and biological research, in advancing the "toolkit" for fundamental research.

The juxtaposition of CNTs and electrochemical biosensing has created a cornucopia of research literature in the burgeoning field of nanotechnology, and only a portion of all sensing modalities and target analytes associated with CNT-based electrochemical biosensing have been covered in this chapter. However, it is believed that the greatest improvements in this field have originated from *in situ* CVD-grown CNT electrode designs, where control over CNT density and spacing is achieved through the precise placement of metal catalyst on the electrode surfaces. This interplay between the electrode surface area and CNT spacing in nanoelectrode array design has led to an enhanced mass transport of target analyte to the electrode surface by radial diffusion, and subsequently demonstrated marked improvements in key biosensor performance characteristics – including increased sensitivity, enlarged linear sensing regions, lower detection limits, and faster response times. In particular, the fusion of metal nanoparticles with CNTs continues to provide a unique nanoenvironment for numerous biofunctionalization modalities, while the properties of nanotubes and nanoparticles blend synergistically to enhance charge transport during redox reactions. Clearly, these fascinating properties of CNT-based biosensors will continue to propel the field of electrochemical biosensing for years to come.

References

1 Collins, F.S. and McKusick, V.A. (2001) *JAMA*, **285**, 540.
2 Sander, C. (2000) *Science*, **287**, 1977–8.
3 Claussen, J.C., Franklin, A.D., ul Haque, A., Porterfield, D.M. and Fisher, T.S. (2009) *ACS Nano*, **3**, 37–44. Claussen, J.C., Sungwon, K.S., ul Haque, A., Artiles, M.S., Porterfield, D.M. and Fisher, T.S. (2010) Electrochemical glucose biosensor of platinum nanospheres connected by carbon nanotubes. *Journal of Diabetes Science and Technology*, **4** (2), 312–19.
4 Mani, V., Chikkaveeraiah, B.V., Patel, V., Gutkind, J.S. and Rusling, J.F. (2009) Ultrasensitive immunosensor for cancer biomarker proteins using gold nanoparticle film electrodes and multienzyme-particle amplification. *ACS Nano*, **3** (3), 585–94.
5 Rahman, M.M., Umar, A. and Sawada, K. (2009) *The Journal of Physical Chemistry B*, **113** (5), 1511–16.
6 Wang, J. (2005) Carbon-nanotube-based electrochemical biosensors: A review. *Electroanalysis*, **17** (1), 7–14.
7 Iijima, S. (1991) *Nature*, **354**, 56–8.
8 Dai, H. (2002) *Frontiers in Surface and Interface Science*, **500**, 218–41.
9 Dresselhaus, M.S., Dresselhaus, G. and Avouris, P. (2001) *Carbon Nanotubes: Synthesis, Structure, Properties, and Applications*, Springer, Berlin.
10 Pumera, M. (2009) *Chemistry – A European Journal*, **15**, 4970–8.
11 Gruner, G. (2006) *Analytical and Bioanalytical Chemistry*, **384**, 322–35.
12 Kim, S.N., Rusling, J.F. and Papadimitrakopoulos, F. (2007) *Advanced Materials*, **19**, 3214.
13 Huang, W., Taylor, S., Fu, K., Lin, Y., Zhang, D., Hanks, T.W., Rao, A.M. and Sun, Y.P. (2002) *Nano Letters*, **2**, 311–14.
14 Shim, M., Shi Kam, N.W., Chen, R.J., Li, Y. and Dai, H. (2002) *Nano Letters*, **2**, 285–8.
15 Wang, J. (2005) *The Analyst*, **130**, 421–6.
16 Krishnan, A., Dujardin, E., Ebbesen, T.W., Yianilos, P.N. and Treacy, M.M.J. (1998) *Physical Review B*, **58**, 14013–19.
17 Yu, M.F., Files, B.S., Arepalli, S. and Ruoff, R.S. (2000) *Physical Review Letters*, **84**, 5552–5.
18 Yu, C., Shi, L., Yao, Z., Li, D. and Majumdar, A. (2005) *Nano Letters*, **5**, 1842–6.
19 Choi, T.Y., Poulikakos, D., Tharian, J. and Sennhauser, U. (2005) *Applied Physics Letters*, **87**, 013108.
20 Hafner, J.H., Cheung, C.L., Woolley, A.T. and Lieber, C.M. (2001) *Progress in Biophysics and Molecular Biology*, **77**, 73–110.
21 Frank, S., Poncharal, P., Wang, Z.L. and Heer, W.A. (1998) *Science*, **280**, 1744.
22 Kajiura, H., Huang, H. and Bezryadin, A. (2004) *Chemical Physics Letters*, **398**, 476–9.
23 Kong, J., Yenilmez, E., Tombler, T.W., Kim, W., Dai, H., Laughlin, R.B., Liu, L., Jayanthi, C.S. and Wu, S.Y. (2001) *Physical Review Letters*, **87**, 106801.
24 Javey, A., Qi, P., Wang, Q. and Dai, H. (2004) *Proceedings of the National Academy of Sciences of the United States of America*, **101**, 13408.
25 Baughman, R.H., Zakhidov, A.A. and De Heer, W.A. (2002) *Science*, **297**, 787.
26 Besteman, K., Lee, J.O., Wiertz, F.G.M., Heering, H.A. and Dekker, C. (2003) *Nano Letters*, **3**, 727–30.
27 Star, A., Tu, E., Niemann, J., Gabriel, J.C.P., Joiner, C.S. and Valcke, C. (2006) *Proceedings of the National Academy of Sciences of the United States of America*, **103**, 921.
28 Nugent, J.M., Santhanam, K.S.V., Rubio, A. and Ajayan, P.M. (2001) *Nano Letters*, **1**, 87–91.
29 Azamian, B.R., Coleman, K.S., Davis, J.J., Hanson, N. and Green, M.L.H. (2002) *Chemical Communications*, 366–7.
30 Franklin, A., Smith, J.T., Sands, T.D., Fisher, T., Choi, K.S. and Janes, D.B. (2007) *Journal of Chemical Physics*, **100**, 13756–62.
31 Hrapovic, S., Liu, Y., Male, K.B. and Luong, J.H.T. (2004) *Analytical Chemistry*, **76**, 1083–8.
32 Jiang, K., Eitan, A., Schadler, L.S., Ajayan, P.M., Siegel, R.W., Grobert, N.,

Mayne, M., Reyes-Reyes, M., Terrones, H. and Terrones, M. (2003) *Nano Letters*, **3**, 275–7.
33. Quinn, B.M., Dekker, C. and Lemay, S.G. (2005) *Journal of the American Chemical Society*, **127**, 6146–7.
34. Katz, E.E. and Willner, I.I. (2004) *ChemPhysChem*, **5**, 1084–104.
35. Yao, Y. and Shiu, K.K. (2007) *Analytical and Bioanalytical Chemistry*, **387**, 303–9.
36. Alivisatos, P. (2004) *Nature Biotechnology*, **22**, 47–52.
37. Hu, J., Odom, T.W. and Lieber, C.M. (1999) *Accounts of Chemical Research*, **32**, 435–45.
38. Heller, I., Kong, J., Heering, H.A., Williams, K.A., Lemay, S.G. and Dekker, C. (2005) *Nano Letters*, **5**, 137–42.
39. Sotiropoulou, S. and Chaniotakis, N.A. (2003) *Analytical and Bioanalytical Chemistry*, **375**, 103–5.
40. Zhou, W., Vavro, J., Guthy, C., Winey, K.I., Fischer, J.E., Ericson, L.M., Ramesh, S., Saini, R., Davis, V.A. and Kittrell, C. (2004) *Journal of Applied Physics*, **95**, 649.
41. Yang, W., Thordarson, P., Gooding, J.J., Ringer, S.P. and Braet, F. (2007) *Nanotechnology*, **18**, 412001.
42. Azamian, B.R., Davis, J.J., Coleman, K.S., Bagshaw, C.B. and Green, M.L.H. (2002) *Journal of the American Chemical Society*, **124**, 12664–5.
43. Hall, E.A.H., Gooding, J.J. and Hall, C.E. (1995) *Mikrochimica Acta*, **121**, 119–45.
44. Banks, C.E. and Compton, R.G. (2006) *The Analyst*, **131**, 15–21.
45. Moore, R.R., Banks, C.E. and Compton, R.G. (2004) *Analytical Chemistry*, **76**, 2677–82.
46. Mancuso, S., Marras, A.M., Magnus, V. and Balunka, F. (2005) *Analytical Biochemistry*, **341**, 344–51.
47. Gooding, J.J. (2005) *Electrochimica Acta*, **50**, 3049–60.
48. Koehne, J., Chen, H., Li, J., Cassell, A.M., Ye, Q., Ng, H.T., Han, J. and Meyyappan, M. (2003) *Nanotechnology*, **14**, 1239–45.
49. Chou, A., Bocking, T., Singh, N.K. and Gooding, J.J. (2005) Demonstration of the importance of oxygenated species at the ends of carbon nanotubes for their favourable electrochemical properties. *Chemical Communications*, 842–4.
50. Li, J. (2002) *Journal of Physical Chemistry B*, **106**, 9299–305.
51. Wang, J. (2007) *Chemical Reviews (Washington, DC, United States)*, **10**, 1021.
52. Wang, J. (1999) *Journal of Pharmaceutical and Biomedical Analysis*, **19**, 47–53.
53. Clark, L.C., Jr and Lyons, C. (1962) *Annals of the New York Academy of Sciences*, **102**, 29.
54. Wang, J. (2001) *Electroanalysis*, **13**, 983.
55. Diamond, D. (1998) *Principles of Chemical and Biological Sensors*, J. Wiley & Sons Inc., New York.
56. Wang, Z.-H., Liang, Q.-L., Wang, Y.-M. and Luo, G.-A. (2003) *Journal of Electroanalytical Chemistry*, **540**, 129–34.
57. Salimi, A., Compton, R.G. and Hallaj, R. (2004) *Analytical Biochemistry*, **333**, 49–56.
58. Salimi, A. and Hallaj, R. (2005) *Talanta*, **66**, 967–75.
59. Abdollah, S., Rahman, H. and Gholam-Reza, K. (2005) Amperometric detection of morphine at preheated glassy carbon electrode modified with multiwall carbon nanotubes. *Electroanalysis*, **17**, 873–9.
60. Britto, P.J., Santhanam, K.S.V. and Ajayan, P.M. (1996) *Bioelectrochemistry and Bioenergetics*, **41**, 121–5.
61. Rubianes, M.D. and Rivas, G.A. (2003) *Electrochemistry Communications*, **5**, 689–94.
62. Valentini, F., Amine, A., Orlanducci, S., Terranova, M.L. and Palleschi, G. (2003) *Analytical Chemistry*, **75**, 5413–21.
63. Roohollah Torabi, K., Wildgoose, G.G. and Compton, G.R. (2007) Room temperature ionic liquid carbon nanotube paste electrodes: overcoming large capacitive currents using rotating disk electrodes. *Electroanalysis*, **19**, 1483–9.
64. Wang, J. and Musameh, M. (2003) *Analytical Chemistry*, **75**, 2075–9.
65. Chen, G., Zhang, L. and Wang, J. (2004) *Talanta*, **64**, 1018–23.
66. Wang, J. and Musameh, M. (2003) Enzyme-dispersed carbon-nanotube electrodes: a needle microsensor for monitoring glucose. *The Analyst*, **128**, 1382–5.

67 Zhao, Y.-D., Zhang, W.-D., Chen, H. and Luo, Q.-M. (2003) *Sensors and Actuators B: Chemical*, **92**, 279–85.

68 Wang, S.G., Zhang, Q., Wang, R., Yoon, S.F., Ahn, J., Yang, D.J., Tian, J.Z., Li, J.Q. and Zhou, Q. (2003) *Electrochemistry Communications*, **5**, 800–3.

69 Zeng, B. and Huang, F. (2004) *Talanta*, **64**, 380–6.

70 Inam, F., Yan, H., Rees, M.J. and Pejis, T. (2008) Dimethylformamide: an effective dispersant for making ceramic–carbon nanotube composites. *Nanotechnology*, **19**, 195710.

71 Joshi, P.P., Merchant, S.A., Wang, Y. and Schmidtke, D.W. (2005) *Analytical Chemistry*, **77**, 3183–8.

72 Lin, Z., Chen, J. and Chen, G. (2008) *Electrochimica Acta*, **53**, 2396–401.

73 Tsai, Y.C., Li, S.C. and Chen, J.M. (2005) *Langmuir : ACS Journal of Surfaces and Colloids*, **21**, 3653–8.

74 Liaw, H.W., Chen, J.M. and Tsai, Y.C. (2006) *Journal of Nanoscience and Nanotechnology*, **6**, 2396–402.

75 Wang, J., Musameh, M. and Lin, Y.H. (2003) *Journal of the American Chemical Society*, **125**, 2408–9.

76 Kang, X., Mai, Z., Zou, X., Cai, P. and Mo, J. (2007) *Analytical Biochemistry*, **369**, 71–9.

77 Kang, X., Mai, Z., Zou, X., Cai, P. and Mo, J. (2008) *Talanta*, **74**, 879–86.

78 Wang, Y., Wei, W., Liu, X. and Zeng, X. (2009) *Materials Science and Engineering: C*, **29**, 50–4.

79 Chen, H. and Dong, S. (2007) *Biosensors and Bioelectronics*, **22**, 1811–15.

80 Gavalas, V.G., Law, S.A., Christopher Ball, J., Andrews, R. and Bachas, L.G. (2004) *Analytical Biochemistry*, **329**, 247.

81 Gong, K., Zhang, M., Yan, Y., Su, L., Mao, L., Xiong, S. and Chen, Y. (2004) *Analytical Chemistry (Washington DC)*, **76**, 6500–5.

82 Branzoi, V. and Pilan, L. (2008) *Molecular Crystals and Liquid Crystals*, **484**, 303–21.

83 Ekanayake, E.M.I.M., Preethichandra, D.M.G. and Kaneto, K. (2007) *Biosensors and Bioelectronics*, **23**, 107–13.

84 Wang, J. and Musameh, M. (2005) *Analytica Chimica Acta*, **539**, 209–13.

85 Dai, Z. and Möhwald, H. (2002) Highly stable and biocompatible nafion-based capsules with controlled permeability for low-molecular-weight species. *Chemistry – A European Journal*, **8** (20), 4751–5.

86 Bharathi, S. and Lev, O. (1998) *Analytical Communications- Royal Society of Chemistry*, **35**, 29–32.

87 Rickus, J.L., Dunn, B., Zink, J.I., Frances, S.L. and Chris, A.R.T. (2002) Optically based sol-gel biosensor materials, in *Optical Biosensors: Present and Future* (eds F. Ligler and C. Rowe-Taitt), Elsevier Science, Amsterdam, pp. 427–56.

88 Tsionsky, M., Gun, G., Glezer, V. and Lev, O. (1994) *Analytical Chemistry*, **66**, 1747–53.

89 Wang, J. and Pamidi, A.P.V. (1997) Sol-gel-derived gold composite electrodes. *Analytical Chemistry*, **69** (21), 4490–4.

90 Compton, R.G., Wildgoose, G.G., Rees, N.V., Streeter, I. and Baron, R. (2008) *Chemical Physics Letters*, **459**, 1–17.

91 Chen, G.Z., Shaffer, M.S.P., Coleby, D., Dixon, G., Zhou, W., Fray, D.J. and Windle, A.H. (2000) Carbon nanotube and polypyrrole composites: coating and doping. *Advanced Materials*, **12**, 522–6.

92 Mamedov, A.A., Kotov, N.A., Prato, M., Guldi, D.M., Wicksted, J.P. and Hirsch, A. (2002) *Nature Materials*, **1**, 190–4.

93 Rouse, J.H. and Lillehei, P.T. (2002) *Nano Letters*, **3**, 59–62.

94 Arrigan, D.W.M. (2004) *The Analyst*, **129**, 1157–65.

95 Huang, S., Dai, L. and Mau, A.W.H. (1999) *Journal of Physical Chemistry B*, **103**, 4223–7.

96 Gao, M., Dai, L. and Wallace, G.G. (2003) *Electroanalysis*, **15**, 1089–94.

97 Yun, Y.H., Bange, A., Shanov, V.N., Heineman, W.R., Halsall, H.B., Dong, Z., Jazieh, A., Yi Tue, D.W., Pixley, S., Behbehani, M. and Schulz, M.J. (2006) Fabrication and characterization of a multiwall carbon nanotube needle biosensor. *Proceedings, IEEE-NANO 2006, Sixth IEEE Conference*.

98 Wang, S.G., Zhang, Q., Yoon, S.F. and Ahn, J. (2003) *Scripta Materialia*, **48**, 409.

99. Wang, S.G., Zhang, Q., Wang, R. and Yoon, S.F. (2003) *Biochemical and Biophysical Research Communications*, **311**, 572–6.
100. Liu, Z., Shen, Z., Zhu, T., Hou, S. and Ying, L. (2000) *Langmuir*, **16**, 3569–73.
101. Chattopadhyay, D., Galeska, I. and Papadimitrakopoulos, F. (2001) *Journal of the American Chemical Society*, **123**, 9451–2.
102. Kim, B. and Sigmund, W.M. (2003) *Langmuir*, **19**, 4848–51.
103. Yu, X., Chattopadhyay, D., Galeska, I., Papadimitrakopoulos, F. and Rusling, J.F. (2003) *Electrochemistry Communications*, **408**, 408–11.
104. Yu, X.F., Mu, T., Huang, H.Z., Liu, Z.F. and Wu, N.Z. (2000) *Surface Science*, **461**, 199–207.
105. Zoski, C.G. (2002) *Electroanalysis*, **14**, 1041–51.
106. Weber, S.G. (1989) *Analytical Chemistry*, **61**, 295–302.
107. Menon, V.P. and Martin, C.R. (1995) *Analytical Chemistry*, **67**, 1920–8.
108. Penner, R.M., Heben, M.J., Longin, T.L. and Lewis, N.S. (1990) *Science*, **250**, 1118–21.
109. Li, J., Ng, H.T., Cassell, A., Fan, W., Chen, H., Ye, Q., Koehne, J., Han, J. and Meyyappan, M. (2003) *Nano Letters*, **3**, 597–602.
110. Lin, Y., Lu, F., Tu, Y. and Ren, Z. (2004) *Nano Letters*, **4**, 191–5.
111. Davies, T.J. and Compton, R.G. (2005) *Journal of Electroanalytical Chemistry*, **585**, 63–82.
112. Tu, Y., Lin, Y. and Ren, Z.F. (2003) *Nano Letters*, **3**, 107–9.
113. Tu, Y., Lin, Y., Yantasee, W. and Ren, Z. (2005) *Electroanalysis*, **17**, 79–84.
114. Lu, J., Do, I., Drzal, L.T., Worden, R.M. and Lee, I. (2008) *ACS Nano*, **2**, 1825–32.
115. Jena, B.K. and Raj, C.R. (2006) *Analytical Chemistry*, **78**, 6332–9.
116. Welch, C.M., Banks, C.E., Simm, A.O. and Compton, R.G. (2005) *Analytical and Bioanalytical Chemistry*, **382**, 12–21.
117. Pauliukaite, R. and Brett, C.M.A. (2005) *Electrochimica Acta*, **50**, 4973–80.
118. Bard, A.J. and Faulkner, L.R. (2001) *Electrochemical methods. Principles and applications*. 2nd edition. John Wiley & Sons, Inc., New York, pp. 250–2.
119. Cederquist, K.B. and Keating, C.D. (2009) Curvature effects in DNA: Au nanoparticle conjugates. *ACS Nano*, **3** (2), 256–60.
120. Evans, S.A.G., Elliott, J.M., Andrews, L.M., Barlett, P.N., Doyle, P.J. and Denuault, G. (2002) *Analytical Chemistry*, **74**, 1322–6.
121. Zou, Y., Xiang, C., Sun, L.X. and Xu, F. (2008) *Biosensors and Bioelectronics*, **23**, 1010–16.
122. Manso, J., Mena, M.L., Yánez-Sedeno, P. and Pingarrón, J. (2007) *Journal of Electroanalytical Chemistry*, **603**, 1–7.
123. Claussen, J.C., Franklin, A.D., ul Haque, A., Porterfield, D.M. and Fisher, T.S. (2009) *ACS Nano*, **3**, 37–44.
124. Valentini, F., Salis, A., Curulli, A. and Palleschi, G. (2004) *Analytical Chemistry*, **76**, 3244–8.
125. Ziegler, M. (2000) *European Journal of Biochemistry*, **267**, 1550–64.
126. Corda, D. and Di Girolamo, M. (2003) *EMBO Journal*, **22**, 1953–8.
127. Jaegfeldt, H. (1980) *Electroanalytical Chemistry*, **110**, 295–302.
128. Moiroux, J. and Elving, P.J. (1979) *Electroanalytical Chemistry*, **102**, 93–108.
129. Chen, J., Bao, J., Cai, C. and Lu, T. (2004) *Analytica Chimica Acta*, **516**, 29–34.
130. Musameh, M., Wang, J., Merkoci, A. and Lin, Y. (2002) *Electrochemistry Communications*, **4**, 743–6.
131. Radoi, A., Compagnone, D., Valcarcel, M.A., Placidi, P., Materazzi, S., Moscone, D. and Palleschi, G. (2008) *Electrochimica Acta*, **53**, 2161–9.
132. van Meurs, J.B.J., Dhonukshe-Rutten, R.A.M., Pluijm, S.M.F., van der Klift, M., de Jonge, R., Lindemans, J., de Groot, L.C.P.G.M., Hofman, A., Witteman, J.C.M., van Leeuwen, J.P.T.M., Breteler, M.M.B., Lips, P., Pols, H.A.P. and Uitterlinden, A.G. (2004) *New England Journal of Medicine*, **350**, 2033–41.
133. McLean, R.R., Jacques, P.F., Selhub, J., Tucker, K.L., Samelson, E.J., Broe, K.E., Hannan, M.T., Cupples, L.A. and Kiel, D.P. (2004) *New England Journal of Medicine*, **350**, 2042–9.

134 Selhub, J. (1999) *Annual Review of Nutrition*, **19**, 217–46.
135 Miller, J., Nadeau, M., Smith, D. and Selhub, J. (1994) *American Journal of Clinical Nutrition*, **59**, 1033–9.
136 Gong, K., Dong, Y., Xiong, S., Chen, Y. and Mao, L. (2004) *Biosensors and Bioelectronics*, **20**, 253–9.
137 Nekrassova, O., Lawrence, N.S. and Compton, R.G. (2003) *Talanta*, **60**, 1085–95.
138 Carvalho, F.D., Remião, F., Valet, P., Timbrell, J.A., Bastos, M.L. and Ferreira, M.A. (1994) *Biomedical Chromatography*, **8**, 134–6.
139 Vandeberg, P.J. and Johnson, D.C. (1994) *Analytica Chimica Acta*, **290**, 317–27.
140 Lawrence, N.S., Deo, R.P. and Wang, J. (2004) *Talanta*, **63**, 443–9.
141 Girault, J.-A. and Greengard, P. (2004) *Archives of Neurology*, **61**, 641–4.
142 Heijtz, R.D., Kolb, B. and Forssberg, H. (2007) *Physiology and Behavior*, **92**, 155–60.
143 Lokhandwala, M.F. and Jandhyala, B.S. (1979) *Journal of Pharmacology and Experimental Therapeutics*, **210**, 120–6.
144 Jankovic, J. (2008) *Journal of Neurology, Neurosurgery and Psychiatry*, **79**, 368–76.
145 Zhang, M., Gong, K., Zhang, H. and Mao, L. (2005) *Biosensors and Bioelectronics*, **20**, 1270–6.
146 Wang, H.-S., Li, T.-H., Jia, W.-L. and Xu, H.-Y. (2006) *Biosensors and Bioelectronics*, **22**, 664–9.
147 Taiz, L. and Zeiger, E. (2006) *Plant Physiology*, Sinauer Associates, Sunderland, Mass.
148 Blakeslee, J.J., Peer, W.A. and Murphy, A.S. (2005) *Current Opinion in Plant Biology*, **8**, 494–500.
149 Wu, K., Sun, Y. and Hu, S. (2003) *Sensors and Actuators B: Chemical*, **96**, 658–62.
150 Hernández, P., Galán, F., Nieto, O. and Hernández, L. (1994) *Electroanalysis*, **6**, 577–83.
151 Porterfield, D.M. (2002) *Journal of Plant Growth Regulation*, **21**, 177–90.
152 Porterfield, D.M. (2007) *Biosensors and Bioelectronics*, **22**, 1186–96.
153 Porterfield, D.M., Corkey, R.F., Sanger, R.H., Tornheim, K., Smith, P.J. and Corkey, B.E. (2000) *Diabetes*, **49**, 1511–16.
154 Mclamore, E.S., Diggs, A., Calvo Marzal, P., Shi, J., Blakeslee, J.J., Peer, W.A., Murphy, A.S. and Porterfield, D.M. (2010) *The Plant Journal*, **63**, 1004–16.
155 Balavoine, F., Schultz, P., Richard, C., Mallouh, V., Ebbesen, T.W. and Mioskowski, C. (1999) *Angewandte Chemie International Edition*, **38**, 1912–15.
156 Dieckmann, G.R., Dalton, A.B., Johnson, P.A., Razal, J., Chen, J., Giordano, G.M., Munoz, E., Musselman, I.H., Baughman, R.H. and Drapers, R.K. (2003) *Journal of the American Chemical Society*, **125**, 1770–7.
157 Choi, H.N., Han, J.H., Park, J.A., Lee, J.M. and Lee, W.-Y. (2007) *Electroanalysis*, **19**, 1757–63.
158 Lim, S., Wei, J., Lin, J., Li, Q. and You, J. (2005) *Biosensors and Bioelectronics*, **20**, 2341–6.
159 Jeykumari, D.R.S. and Sriman Narayanan, S. (2008) *Biosensors and Bioelectronics*, **23**, 1686–93.
160 Gooding, J.J., Wibowo, R., Liu, J.Q., Yang, W., Losic, D., Orbons, S., Mearns, F.J., Shapter, J.G. and Hibbert, D.B. (2003) *Journal of the American Chemical Society*, **125**, 9006–7.
161 Patolsky, F., Weizmann, Y. and Willner, I. (2004) *Angewandte Chemie International Edition*, **43**, 2113–17.
162 Cai, C. and Chen, J. (2004) *Analytical Biochemistry*, **332**, 75–83.
163 Guiseppi-Elie, A., Lei, C. and Baughman, R. (2002) Direct electron transfer of glucose oxidase on carbon nanotubes. *Nanotechnology*, **13**, 559.
164 Liu, Y., Wang, M., Zhao, F., Xu, Z. and Dong, S. (2005) *Biosensors and Bioelectronics*, **21**, 984–8.
165 Joseph, W. (2001) *Electroanalysis*, **13**, 983–8.
166 Guilbault, G.G. and Lubrano, G.J. (1973) *Analytica Chimica Acta*, **64**, 439–55.
167 Guan, W.-J., Li, Y., Chen, Y.-Q., Zhang, X.-B. and Hu, G.-Q. (2005) *Biosensors and Bioelectronics*, **21**, 508–12.
168 Lim, S.H., Wei, J., Lin, J., Li, Q. and KuaYou, J. (2005) *Biosensors and Bioelectronics*, **20**, 2341–6.

169 Yang, M., Jiang, J., Yang, Y., Chen, X., Shen, G. and Yu, R. (2006) *Biosensors and Bioelectronics*, **21**, 1791–7.

170 Nedergaard, M., Takano, T. and Hansen, A.J. (2002) *Nature Reviews Neuroscience*, **3**, 748–55.

171 Smythies, J. (1999) *European Journal of Pharmacology*, **370**, 1–7.

172 Jedlicka, S.S., Dadarlat, M., Hassell, T., Lin, Y., Young, A., Zhang, M., Irazoqui, P. and Rickus, J.L. (2009) *International Journal of Neural Systems*, **19**, 197–212.

173 Alvarez-Crespo, S.L., Lobo-Castañón, M.J., Miranda-Ordieres, A.J. and Tuñón-Blanco, P. (1997) *Biosensors and Bioelectronics*, **12**, 739–47.

174 Cui, Y., Barford, J.P. and Renneberg, R. (2007) *Enzyme and Microbial Technology*, **41**, 689–93.

175 Braeken, D., Rand, D.R., Andrei, A., Huys, R., Spira, M.E., Yitzchaik, S., Shappir, J., Borghs, G., Callewaert, G. and Bartic, C. (2009) *Biosensors and Bioelectronics*, **24**, 2384–9.

176 Varma, S., Yigzaw, Y. and Gorton, L. (2006) *Analytica Chimica Acta*, **556**, 319–25.

177 Day, B.K., Pomerleau, F., Burmeister, J.J., Huettl, P. and Gerhardt, G.A. (2006) *Journal of Neurochemistry*, **96**, 1626–35.

178 Huang, X.J., Im, H.S., Lee, D.H., Kim, H.S. and Choi, Y.K. (2007) *J Phys Chem C*, **111**, 1200–6.

179 Meng, L., Wu, P., Chen, G., Cai, C., Sun, Y. and Yuan, Z. (2009) *Biosens Bioelectron*, **24**, 1751–6.

180 Tang, L., Zhu, Y., Xu, L., Yang, X. and Li, C. (2007) *Talanta*, **73**, 438–43.

181 Tang, L., Zhu, Y., Yang, X. and Li, C. (2007) *Analytica Chimica Acta*, **597**, 145–50.

182 Boo, H., Jeong, R.A., Park, S., Kim, K.S., An, K.H., Lee, Y.H., Han, J.H., Kim, H.C. and Chung, T.D. (2006) *Analytical Chemistry*, **78**, 617–20.

183 Malarkey, E.B., Fisher, K.A., Bekyarova, E., Liu, W., Haddon, R.C. and Parpura, V. (2009) *Nano Letters*, **9**, 264–8.

184 Wang, K., Fishman, H.A., Dai, H. and Harris, J.S. (2006) *Nano Letters*, **6**, 2043–8.

185 Azevedo, A.M., Prazeres, D.M.F., Cabral, J.M.S. and Fonseca, L.P. (2005) *Biosensors and Bioelectronics*, **21**, 235–47.

186 Patel, N.G., Meier, S., Cammann, K. and Chemnitius, G.C. (2001) *Sensors and Actuators B: Chemical*, **75**, 101–10.

187 Balat, M. and Balat, H. (2009) *Applied Energy*, **86**, 2273–82.

188 Gouveia-Caridade, C., Pauliukaite, R. and Brett, C.M.A. (2008) *Electrochimica Acta*, **53**, 6732–9.

189 Yildiz, H.B. and Toppare, L. (2006) *Biosensors and Bioelectronics*, **21**, 2306–10.

190 Guilbault, G.G. and Lubrano, G.J. (1974) *Analytica Chimica Acta*, **69**, 189–94.

191 Santos, A.S., Pereira, A.C., Duran, N. and Kubota, L.T. (2006) *Electrochimica Acta*, **52**, 215–20.

192 Tsai, Y.-C., Huang, J.-D. and Chiu, C.-C. (2007) *Biosensors and Bioelectronics*, **22**, 3051–6.

193 Lee, C.-A. and Tsai, Y.-C. (2009) *Sensors and Actuators B: Chemical*, **138**, 518–23.

194 Choi, H.N., Young-Ku, L., Hoon, H.J. and Won-Yong, L. (2007) Amperometric ethanol biosensor based on carbon nanotubes dispersed in sol-gel-derived titania-Nafion composite film. *Electroanalysis*, **19**, 1524–30.

195 Umasankar, Y., Periasamy, A.P. and Chen, S.M. (2009) *Talanta*.

9
Single-Walled Carbon Nanotube Biosensors
Jeong-O Lee and Hye-Mi So

9.1
Introduction

In this chapter, attention is focused on the use of single-walled carbon nanotube field-effect transistors (SWNT-FETs) in bioapplications. In 1991, the discovery of carbon nanotubes (CNTs) had a major impact on science and technology, and their exceptional chemical and mechanical properties have since fascinated many scientists. A breakthrough in CNT research came about in 1998, when Tans and colleagues reported the transistor functionality of SWNTs at room temperature [1]. Since that time, extensive investigations have been conducted into the development of high-speed and high-performance electronic devices containing SWNTs. At the same time, the diversity of the applications of SWNT devices has increased, with the use of biosensors in molecular diagnostics being particularly popular, mainly because of the progress that has been made in molecular biology and genetics. Today, molecular biology has advanced to a level at which infinitesimal amounts of disease-related proteins, small molecules, and genes need to be identified, and it is vital that this task is carried out with high sensitivity and specificity. Hence, many research groups consider that nanotechnology can provide the solution to this problem. To this end, a wide variety of optical, electrical, thermal, and strain sensors based on nanotechnology have been developed. Electrical sensors are especially useful for point of care testing and miniaturization, and numerous biosensors that utilize CNTs or semiconductor nanowires have been developed during the past few decades.

Single-walled CNTs emerged as candidates for gas sensing in 2000 [2], although the first nanoscale FET biosensor used a Si nanowire FET as the active transducer [3]. Conventional FET sensors have limitations in both sensitivity and operating conditions (e.g., they cannot be operated in high-salt solutions), whereas nano-FET sensors are expected to be free of such limitations. The surface-to-volume ratio of one-dimensional (1-D) nanostructures (nanowires and nanotubes) is very large, which might improve sensor sensitivity, while the quantum nature of nanomaterials means that it is possible to develop sensors capable of detecting single

Nanomaterials for the Life Sciences Vol.9: Carbon Nanomaterials. Edited by Challa S. S. R. Kumar
Copyright © 2011 WILEY-VCH Verlag GmbH & Co. KGaA, Weinheim
ISBN: 978-3-527-32169-8

molecules. Indeed, the detection of a single virus with Si nanowire sensors was demonstrated by Patolsky and colleagues [4], and a single *Escherichia coli* cell has been detected using a SWNT-FET [5]. Yet, despite these benefits, nano-FET biosensors have not entered the market even after ten years' of extensive research. The reasons for this situation are manifold, but notably the cost to produce nano-FET sensors is very high, and their poor reproducibility and credibility also represent barriers to their market entry. Nonetheless, as the production costs continue to decrease rapidly, and the reproducibility of the nanomaterials gradually improves, it is possible that nano-FET-based sensors might enter the market during the next decade.

9.2
The Sensing Mechanisms of Nanotube Biosensors

In early CNT-based biosensors, the CNTs were used as electrodes to detect dopamine [6]. The majority of the disadvantages of conventional electrodes, such as glassy carbon electrodes (GCEs) or metal electrodes, can be overcome by using CNTs. Moreover, the sensitivity, stability, and response time of electrodes can also be improved by modifying them with CNTs. The electrochemical and electrical properties of CNTs mean that they are ideal materials for both electrodes and transducer components in biosensors. In this section, the sensing mechanisms of SWNT-FET biosensors, in which CNTs mainly serve as signal transducers, is considered.

The SWNT-FET sensors contain individual SWNTs or networks of SWNTs that act as the conducting channel, as well as metal source/drain electrodes and a Si substrate that serves as a back gate (see Figure 9.1). The binding of a target molecule onto receptors immobilized on the SWNT-FET can then be converted into an electrical signal.

The normal sensing mechanism of FET sensors is the so-called "electrostatic gating effect," in which the extra charges on the target molecule act as a "molecular gate." A direct charge transfer between the molecular adsorbates and SWNTs affects the channel conductance by changing the number of mobile charge carriers, so that the shift in the threshold voltage provides information about the analyte concentration. However, since SWNT-FETs operate as Schottky transistors [7], the contribution due to Schottky barrier modulation may be larger than

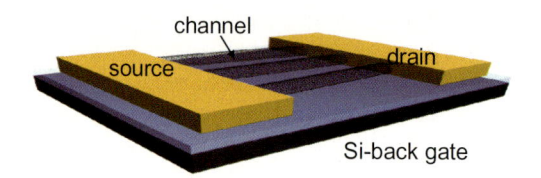

Figure 9.1 Schematic diagram of a SWNT-FET.

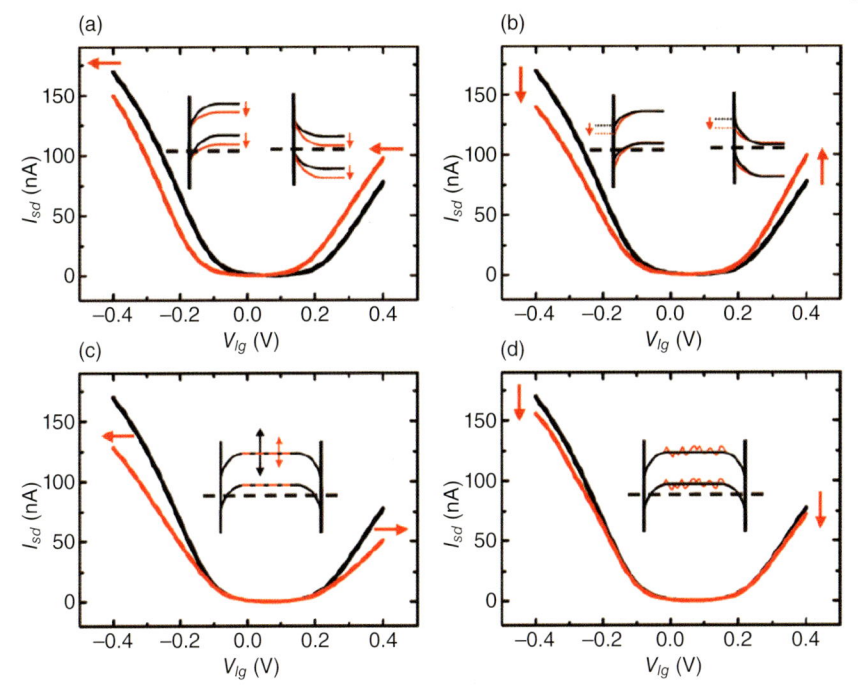

Figure 9.2 Possible sensing mechanisms of SWNT-FET biosensors. (a) Electrostatic gating; (b) Schottky barrier modulation; (c) Changes in the capacitance; (d) Mobility changes. Reproduced with permission from Ref. [9]; © 2008, American Chemical Society.

that of molecular doping, and consequently much debate has ensued regarding the sensing mechanism of SWNT-FET sensors [8]. In 2008, Heller and colleagues investigated the mechanism of SWNT-FET-based sensors, and showed that it is possible to identify the sensing mechanism by studying the electron and hole conduction regions [9]. Four routes can contribute to changes in the signal in SWNT-FET-based sensors, as shown in Figure 9.2, which highlights the evolution of the liquid gate transfer characteristics upon binding of the proteins. In Figure 9.2a, the electrostatic gating due to the adsorbed charged molecules results in a parallel shift of the $I-V_{lg}$ curve along the gate voltage axis. The majority of reported SWNT-FET sensors operate according to this mechanism, with the contact electrodes carefully screened from interactions. For example, Bradley et al. reported that the threshold voltage could be shifted towards the negative gate voltage by the donation of electrons from adsorbed proteins (e.g., streptavidin) to the CNTs [10].

In the case of a sensor operating with Schottky barriers, in which the adsorption of biomolecules at the contact barrier changes the metal work function, the conductance changes in the p- and n-branches are asymmetric (see Figure 9.2b). It should be noted that, as the p-channel current decreases, the n-channel current increases after the interaction. No shift in the gate threshold voltage is observed in this case, which is characteristic of Schottky barrier modulation. Since the first report in 2003, that the metal–SWNT contact barrier dominates changes in the conductance [8], various protein or DNA sensors with metal contacts as the active sensing area have been described. Further, the reduction in capacitance due to the adsorption of biomolecules might also contribute to changes in conductance (as shown in Figure 9.2c). Bestman et al. showed that an immobilization of the enzyme glucose oxidase (GOx) on an SWNT-FET caused a decrease in the conductance as a result of a change in the total capacitance of the tube [11]. However, a change in conductance due to a change in capacitance is only observed when the biomolecules are closely packed on the SWNT surfaces, as the quantum capacitance of SWNTs dominates the electrolyte interfacial capacitance. The charging of SWNTs under a given gate potential consists of an electrostatic component (C_e) and a quantum (C_q) component. As an object is shrunk to the nanoscale, its ability to store charges is no longer determined by the electrostatic force and the geometry of the object, but rather by the finite number of states available for the storage of electric charges. Therefore, in nanoscale objects such as nanotubes, the quantum capacitance for the storage of charge is determined by the density of states. In most back-gated SWNT-FETs, the electrostatic capacitance is smaller than the quantum capacitance; consequently, the former will dominate the capacitance of such devices because the electrostatic capacitance and quantum capacitance are connected in parallel, and thus the smaller capacitance dominates the total capacitance. In an electrolyte, however, the double-layer capacitance is larger than the quantum capacitance, and so the quantum capacitance can govern the device performance. If biomolecules adsorbed onto a SWNT form a closely packed protein layer that mobile ions are unable to penetrate, then their presence can dramatically change the electrostatic capacitance of the system.

Finally, a reduced carrier mobility represents a further possible mechanism for SWNT-FET biosensors (see Figure 9.2d). Since transport in SWNT-FETs is ballistic, Heller et al. showed that a reduced carrier mobility is not a dominant mechanism in SWNT-FET biosensors. However, as the length of the SWNT channel increases, then the reduced mobility which arises from the scattering might result in measurable effects on the electrical signal. Star et al. showed that biotin–streptavidin binding on SWNT-FETs could result in a change in the conductance [12], which was explained as follows: upon streptavidin–biotin binding, geometric deformations occur that lead to the formation of scattering sites on the nanotube, and thus to reduced conductance.

In conclusion, there are several possible mechanisms for the changes in conductance in SWNT-FET-based sensors due to biomolecular interactions, and consequently a careful device preparation is necessary to identify the mechanism of biosensing. Recent reports have shown that, although Schottky barrier-based sensing is the most sensitive technique, channel doping effects might provide more stable and reproducible signals [9].

9.3
The Immobilization of Biomolecules on SWNTs

In order to fabricate SWNT-FET biosensors, receptors such as antibodies or DNA molecules must first be immobilized on the active sensing area. In the case of sensors that use a Schottky barrier as the active sensing area, the immobilization of receptor molecules is more straightforward and is an established technique. However, with sensors that use channel doping, the receptor molecules should be immobilized on the sidewalls of the SWNTs, although this can affect the performance of a device, and therefore the performance of such biosensors. In this section, various approaches are described for the immobilization of receptor molecules on the sidewalls of SWNTs.

9.3.1
Covalent Binding

In the selective detection of target molecules, target-specific receptors are immobilized on the sidewalls of SWNTs by using either covalent or noncovalent binding. Covalent binding can be achieved through the formation of an amide bond with carboxyl groups, which can be effected by the acid treatment of SWNTs (see Figure 9.3). Although the binding obtained with carboxyl groups is strong, this may lead to significant changes in the excellent electrical properties of the SWNTs. In order to detect DNA hybridization, Wang et al. introduced carboxylic acid groups onto the surfaces or ends of nanotubes [13]. Subsequently, amino-terminated probe oligonucleotides were immobilized by the formation of covalent amide bonds between these carboxyl groups and the amino groups at the ends of the DNA oligonucleotides. The occurrence of hybridization between the probe and the

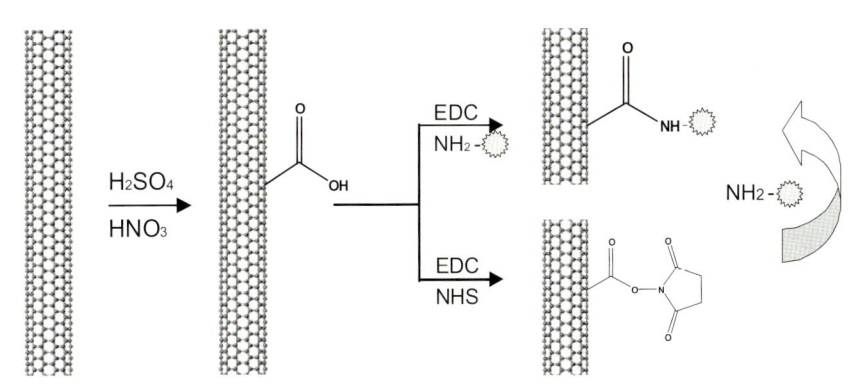

Figure 9.3 The chemical oxidation of SWNTs and the covalent binding of receptor molecules. EDC: 1-Ethyl-3-[3-dimethylaminopropyl]carbodiimide hydrochloride; NHS: N-hydroxysuccinimide.

target DNA oligonucleotides was confirmed by monitoring the changes in the voltammetric peak of methylene blue.

Zelada-Guillén et al. showed that aptamer-based sensors can be used to detect living microorganisms, by linking the aptamers to carboxylated SWNTs [14]. The aptamers used in these studies were modified with a five-carbon spacer and an amine group at the 3' end. Both, sensitivity and selectivity were increased by using SWNTs with covalently immobilized aptamers. This method can easily be applied for the attachment to the CNTs of any useful entity including amine groups, such as nucleic acids, polymers, dendrimers, and even inorganic nanoparticles.

9.3.2
Noncovalent Binding

Whilst the immobilization of receptor molecules through noncovalent binding with SWNTs is weaker than is achieved through covalent binding, it does have the main advantage that the excellent electrical properties of the CNTs are retained. In order to immobilize receptor molecules with a noncovalent interaction, linkers such as pyrene-based molecules or Tween 20 are required (Figure 9.4). In this case, the linker molecules are positioned through hydrophobic interactions such as π–π interactions with the sidewalls of the CNTs, and the receptor molecules are

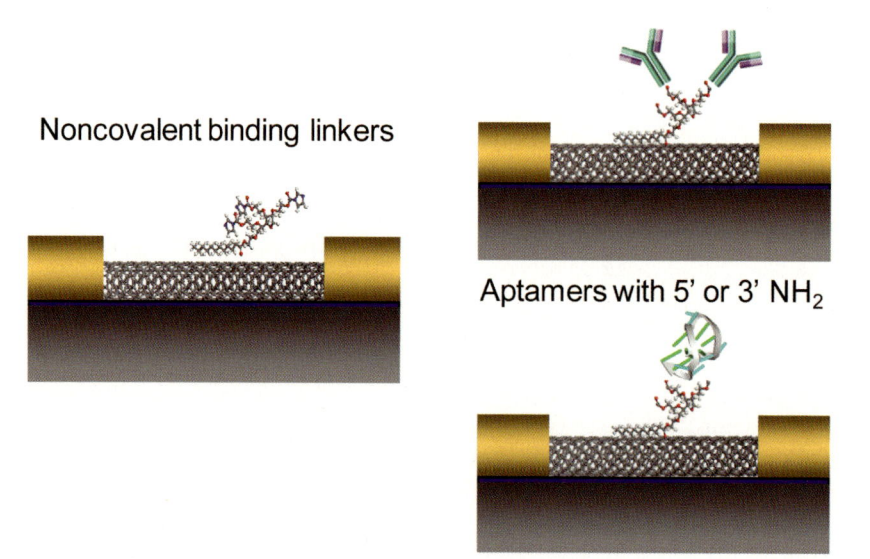

Figure 9.4 A schematic diagram of receptor immobilization with noncovalent binding linkers. In this example, CDI-Tween 20 was used as the linker molecule, and antibodies with lysine residues and aptamers labeled with NH_2 were immobilized on the sidewalls of the SWNTs.

immobilized through covalent binding with the linker molecules. In most studies, the receptor binding site is that of the *N*-hydroxysuccinimide (NHS) moiety of the linker molecule, which can react with the amine groups of receptor molecules.

For example, 1-pyrenebutanoic acid succinimidyl ester was irreversibly adsorbed onto the sidewall of a SWNT via π–π stacking of the pyrene group, and the active ester group was used to covalently bind the amino group of the receptor molecule [15]. To conjugate the biomolecules, Chen *et al.* functionalized the SWNTs with 1,1-carbonyldiimidazole (CDI)-activated Tween molecules, and then reacted these with biotin or staphylococcal protein A (SpA) [16]. The detection of target proteins with SWNT-FETs was then demonstrated by using noncovalently bound receptors; it was also shown that false signals due to the non-specific binding of biomolecules could be prevented by the immobilization of polyethylene oxide chains on the SWNTs.

9.3.3
Other Immobilization Methods (Metal Particles, etc.)

Although the covalent binding of receptors has the advantage of stable bond formation, the harsh oxidation process that creates carboxyl groups on SWNTs can degrade their excellent electronic properties. Further, the noncovalent binding method is problematic for certain interactions, because receptors can peel off the sensor surface upon binding, due to the weak bond between the linkers and the SWNTs.

Thus, instead of linking receptors directly to the sidewalls of SWNTs, it is possible to use supporters such as metallic nanoparticles. In this case, the active sensing area is larger than the bare SWNT, and a larger number of receptors can be immobilized on the sensor surface than in the direct linking method. Recently, Lo *et al.* showed that metal nanoparticles such as Au and Ni can be used to immobilize receptor molecules on the sidewalls of SWNTs [17]. Various techniques can be used to decorate metal nanoparticles onto the sidewalls of SWNTs. Among these, the electrochemical methods offer highly selective and uniform nanoparticle decoration [18]. Another advantage of using metal nanoparticles is that the oriented immobilization of receptor elements is made possible. In the case of Au nanoparticles, receptors with thiol (SH) groups can be covalently immobilized, and electrochemically grown Ni nanoparticles offer a base for biomolecules with hexahistidine moieties. In the study conducted by Lo and colleagues, single-chain variable fragment (scFv) antibodies modified with a hexahistidine tag were immobilized onto Ni nanoparticles decorating the sidewalls of SWNT-FETs; the change in conductance upon target binding was found to be much greater than for sensors with randomly oriented receptor molecules [17]. Although the sensing mechanism of metal-decorated sensors is not yet clearly understood, a change in the work function has been suggested [19]. In order to investigate the role of Ni nanoparticles in SWNT-FET sensors, low-temperature transport measurements were performed with Ni-decorated SWNT-FETs. At low temperatures, two types of periodic oscillation were observed, both of which were ascribed to the charging of the Ni nanoparticles. It was concluded that the electrochemically grown metal nanoparticles and SWNTs are resistively coupled; any extra charges generated by biomolecular interactions are possibly immediately transferred from the nanoparticles to the SWNTs.

9.4
Various Receptors for Nanotube Biosensors

To provide sensors with specificity, a recognition element that can specifically bind to the target should be chosen and immobilized on the sensor surface, as shown in Figure 9.5. In most cases, the components of immunoreactions are employed as the receptor–target (antibody–antigen) system, although enzymes can be used as the recognition elements for small molecules, and artificial antibodies such as aptamers and molecular imprints (MIPs) can be used instead of antibodies. Alternatively, instead of using whole antibodies, a type of fragment antibody has been developed, by using genetic engineering, that is much smaller than the whole antibodies. Various examples of the molecular recognition elements employed in nanotube FET sensors are described, and their advantages discussed, in the following subsections.

9.4.1
Aptamers

The first SWNT-FET sensors with aptamers as molecular recognition elements were demonstrated in 2005 [20]. *Aptamers* are single-strand nucleic acid oligomers that can bind a wide variety of targets, including proteins, small molecules, and cells. Aptamers are highly specific, can be mass produced once their sequence is known, are easily modifiable, and their production cost is much less than that of

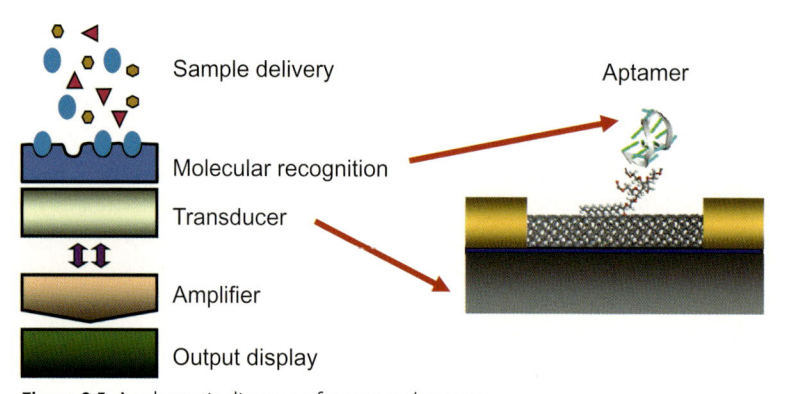

Figure 9.5 A schematic diagram of a sensor that uses aptamers as receptors and SWNT-FETs as the signal transducer.

antibodies. The major advantage of aptamers over antibodies in electronic sensing is that they are smaller; whereas, the average length of an antibody is ~10 nm, most aptamers are much shorter. Thus, aptamer–target binding can occur much closer to the sensor surface, which is advantageous for all surface-based sensors. The major challenge for FET sensors, however, is to increase the Debye screening length. When an electrode is immersed in an ionic solution, counterions in the solution rapidly screen the charges on the electrode. If there are many ions present (e.g., at high salt concentrations), then the neutralization of extra charge arises very close to the sensor surface, and consequently a sensor cannot detect the extra charges in a high-salt solution. As most body fluids contain high concentrations of salts, this effect is a serious problem for the practical use of sensors. One solution to the problem would be to combine microfluidic channels that can concentrate the target and exchange buffers within the sensing devices (though this approach is not discussed in detail here). Another option would be to decrease the distance between the target and the sensors, so that target binding could occur inside the electrical double layer. In this respect, aptamers are ideal candidates as they are much shorter than conventional antibodies [21].

The SWNT-FET sensors with aptamers as the molecular recognition elements were fabricated by using patterned chemical deposition and photolithography. In order to prevent the leakage of current from the devices, all electrodes except for the SWNT channels were insulated with SiO_x or a negative photoresist. The immobilization of the aptamers was achieved by using noncovalent binding linkers such as CDI-Tween 20 and pyrene-N-hydroxysuccinimide ester. First, the linkers were noncovalently adsorbed onto the sidewalls of the SWNTs via hydrophobic interactions, and the aptamers modified with NH_2 at the 5' (or 3') terminus were covalently linked to a succinimide ester or the active imidazole groups on the linkers. Any unreacted linkers were blocked with ethanolamine to prevent any nonspecific binding. Upon introduction of the target (thrombin), the conductance of the device was decreased instantly, and the signal tended to saturate after some time. The selectivity of the aptamer-immobilized SWNT-FET sensors was also tested. The SWNT-FET sensors with thrombin aptamers did not exhibit much change in conductance in the presence of elastase, a protein of the thrombin family which has a similar isoelectric point and molecular weight. This selectivity stems from the specificity of the aptamers, which possess a higher selectivity than antibodies. So and colleagues also showed that, due to the reversible conformations of aptamers, it is possible to fabricate recyclable sensors [20]. Aptamers have a specific 3-D conformation that can recognize their target, although this conformation is likely to change in high-salt solutions. Under high-salt conditions, the electrostatic interactions within the aptamer sequences are screened by ions in solution, such that the aptamers are unable to maintain their 3-D structure and the bound target is released into solution. So and colleagues washed the target-bound aptamer sensors with 6 M guanidine hydrochloride to regenerate the sensors, which could be used more than five times. Aptamers are compared with antibodies in Figure 9.6.

So and colleagues were the first to demonstrate the advantages of aptamers in SWNT-based sensors, while Maehashi *et al.* highlighted further advantages by

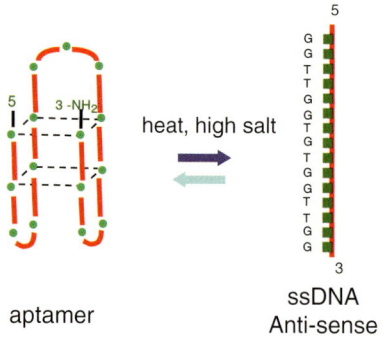

Figure 9.6 Aptamers versus antibodies. Aptamers lose their three-dimensional conformations when heated or in high-salt solutions, and release the bound targets.

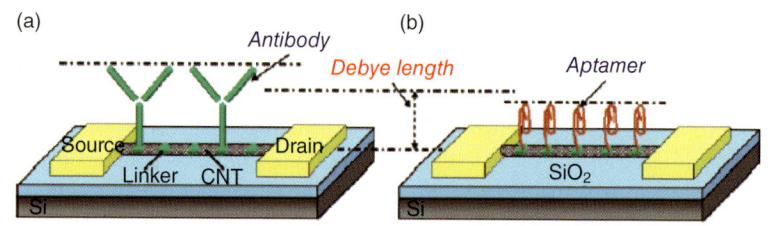

Figure 9.7 Aptamers versus antibodies. Since aptamers are much smaller than antibodies, target binding can occur inside the electrical double layer. Reproduced with permission from Ref. [22]; © 2007, American Chemical Society.

studying the effects of varying the ionic salt concentration [22]. As shown in Figure 9.7, aptamer-based sensors exhibit delicate changes at very low concentrations of the target molecules, whereas SWNT-FETs with antibodies as the molecular recognition element do not undergo significant changes, even with high concentrations of the target molecules.

In another example, aptamers specific to *E. coli* were used to detect the bacterium in water by using SWNT devices. So and colleagues used SWNT-FETs and statistical methods to determine *E. coli* [5], while Zelada-Guillén *et al.* employed an electrochemical technique to detect very low concentrations of *E. coli* [14]; the details of the detection of microorganisms are discussed in Section 9.5. Finally, Cella and colleagues developed single-stranded DNA (ssDNA) aptamers that were specific to the protective antigen component of the anthrax toxin, with the aim of fabricating SWNT-based chemiresistive sensors [23]. The sensor was found to display a wide dynamic range, from 1 to 800 nM, and high selectivity; moreover, the aptasensor could be reused six times by simply washing it with high-salt solutions.

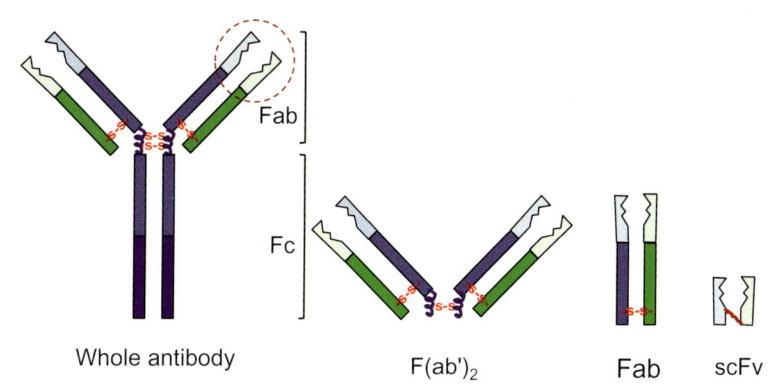

Figure 9.8 Schematic diagram of an antibody. For details, see the text.

9.4.2
Fragment Antibodies

Among the numerous advantages of aptamers over antibodies, in particular aptamer sensors may be much more sensitive than antibody sensors. Although a process for screening aptamers has already been established (SELEX, the systematic evolution of ligands by exponential enrichment), it is not trivial to find aptamers for certain targets. Thus, once the structure of an antibody is known, it might be advantageous to synthesize smaller antibodies by using genetic engineering.

The *immunoglobulins* are the basic units of antibodies. As shown in Figure 9.8, an immunoglobulin monomer consists of two different domains: (i) the Fc region, which refers to the stem in a Y-shaped molecule; this is found in most antibodies, and has a high structural similarity; and (ii) the "Fab" (fragment, antigen binding) region, which corresponds to an epitope for antigen binding. Since it is Fab that interacts directly with the target antigen, genetically engineered Fab can be used instead of whole antibodies. In the absence of the Fc region, Fab is much smaller than the whole antibody. Antibodies can be further fragmented for size reduction by removing the constant domain in Fab. For example, an scFv (single chain fragment antibody) can be prepared by linking variable regions of the immunoglobulin with short linkers. Both, Fab and scFv have been used successfully in SWNT sensors.

Kim and colleagues functionalized SWNT-FETs with: (i) whole immunoglobulin; (ii) F(ab')$_2$, and (iii) Fab, and then compared the responses of these SWNT-FETs after the introduction of human immunoglobulin G (IgG) at various concentrations [24]. All receptors were immobilized noncovalently with 1-pyrenebutanoic acid succinimidyl ester on the sidewalls of the SWNT-FETs, which were fabricated using a self-assembly method. The SWNT-FETs with Fab receptors exhibited an immediate decrease in conductance in the presence of 10 pg ml^{-1} target protein, and did not undergo any significant changes in the presence of phosphate-buffered saline (PBS), bovine serum albumin, fibrinogen, or streptavidin. However, SWNT-FETs containing whole antibody could not recognize target concentrations below 100 ng ml^{-1} human IgG. A decrease in conductance was observed only for highly concentrated target solutions (1000 ng ml^{-1} human IgG in PBS). Since the whole IgG antibody is much larger than the Debye screening length in PBS solution, most of the human IgG molecules were unable to interact with the SWNT-FET to produce a measurable signal. Antibody fragments represent a possible solution to this problem: F(ab')$_2$ and Fab are compared in Figure 9.8. F(ab')$_2$ consists of two Fab domains; the height of F(ab')$_2$ is approximately 5 nm, which is half that of a whole antibody. Therefore, charged target proteins might approach the sensor surface more easily for F(ab')$_2$: the sensitivity was found to be 100-fold higher with F(ab')$_2$. However, the angle between the two Fab domains was rather large, and the lateral dimensions of F(ab')$_2$ were too large to expect high sensitivity. In order to improve sensitivity further, Fab was tested as a molecular recognition element. Since Fab is much smaller (~3–5 nm) than the whole antibody or F(ab')$_2$, target binding might occur inside the electrical double layer, close to the SWNT surfaces. With Fab, the detection limit was lowered to 1 pg ml^{-1} human IgG.

It is also possible to separate antigen-binding motifs for use as smaller molecular receptors. A single-chain fragment antibody is shown schematically in Figure 9.8. Here, an scFv is a fusion of the variable regions of the heavy and light chains of immunoglobulins connected by a short linker, and is normally produced in bacterial cell cultures by using genetic engineering; indeed, it is possible to create special functionalities such as a hexahistidine tag [(his)$_6$] upon production. Recently, Lo *et al.* successfully employed a (his)$_6$-tagged scFv in a SWNT-FET to create sensors for carcinoembryonic antigen (CEA) [17]. In this experiment, (his)$_6$-tagged scFv was produced in an *E. coli* culture, and the oriented immobilization of scFv made possible through an interaction of the (his)$_6$ and Ni nanoparticles. To fabricate the sensors, Lo and colleagues produced SWNT-FETs by using a patterned growth technique, while Ni nanoparticles were produced using the electrochemical decoration technique. Normally, (his)$_6$ moieties are known to chelate with Ni^{2+} ions, and separation using (his)$_6$ is an established protocol in protein and antibody purifications. Both, Lo and colleagues and Park and colleagues [25] discovered that (his)$_6$ can also bind with electrochemically produced Ni nanostructures, and so used Ni nanoparticles and Ni nanohairs, respectively, to immobilize the (his)$_6$-tagged receptor molecules. The advantage of using (his)$_6$ was that the oriented immobilization of the receptor was possible. Most linkers used in SWNT sensors bind with the amine groups in antibodies or aptamers. In the case of antibodies, numerous amine-containing (lysine) residues are available, and the receptors are randomly oriented on the sensor surface. However, by introducing molecular glue at a specific position, it is possible to achieve an oriented immobilization of the receptors, which makes it possible to increase sensitivity. In the case of whole antibodies, as there are many lysine residues available in the Fc regions of antibodies, the orientation of antibodies is a much less significant problem. However, in the case of fragment antibodies such as

Fab or scFv, only a restricted number of lysine residues are available. Lo and colleagues compared the sensitivity of oriented scFv/SWNT-FET sensors that were immobilized with Ni nanoparticles and randomly oriented sensors that used 1-pyrene-N-hydroxysuccinimide ester. The Ni-decorated SWNT-FET sensors with $(his)_6$-tagged scFvs were found to be sensitive to delicate changes of $1\,ng\,ml^{-1}$ CEA, whereas sensors with randomly oriented scFvs did not undergo any significant changes, even at high ($>1\,\mu g\,ml^{-1}$) concentrations of CEA. When the crystal structure of the scFv was examined, there was found to be only one lysine residue in scFv available for immobilization to produce the sensor signal. Other lysine groups are involved in specific interactions, or are located near the antigen binding site. Since only one lysine residue is available out of the 10 lysine residues in scFv, the probability of CEA detection with a randomly oriented antibody is only 10%, whereas the antigen-binding pockets are fully available in oriented $(his)_6$-tagged scFv sensors.

The sensitivities of scFv sensors and SWNT-FET sensors with whole antibodies were also compared. Whole-antibody sensors were found to exhibit a change of ~20% in conductance with $100\,ng\,ml^{-1}$ CEA, whereas there was a 50–100% increase in conductance with $1\,ng\,ml^{-1}$ CEA in Ni-decorated, $(his)_6$-tagged scFv immobilized SWNT-FETs. This enhanced sensitivity could be explained in terms of the reduced length of the receptor and the increased sensing area due to Ni decoration.

In this section, the use of aptamers and antibody fragments has been described as "smaller" molecular receptors to enhance sensor sensitivity. Recently, Kim et al. showed that it is also possible to increase the sensitivity with a whole antibody instead of only with smaller receptors [26]. In their study, they prepared a mixed self-assembled monolayer (SAM) of 1-pyrenebutanoic acid succinimidyl ester and 1-pyrenebutanol. The 1-pyrenebutanoic acid succinimidyl ester acts as a linker that fuses the receptor antibodies, and 1-pyrenebutanol is used as a spacer. Sensors with linkers were found to exhibit only poor sensitivity, whereas SWNT-FET sensors in which the receptors were immobilized with the mixed SAM (the linker:spacer ratio was 1:3) was able to detect $1\,ng\,ml^{-1}$ target prostate-specific antigen. This enhanced sensitivity was attributed to the dilute receptor layers, which means that the target proteins can approach inside the electrical double layer with ease.

9.4.3
Enzymes and Proteins

The first SWNT-FET biosensor used the enzyme GOx as a molecular receptor, when Bestman et al. immobilized GOx on SWNT-FETs to produce glucose sensors [11]. GOx catalyzes the oxidation of β-D-glucose to D-glucono-1,5-lactone, with the production of hydrogen peroxide. Thus, when an increase was observed in the conductance of the GOx-immobilized SWNT-FET upon the introduction of 0.1 M glucose, it was attributed to a change in the conformation of GOx. In the same year, Star et al. immobilized biotin-N-hydroxysuccinimide esters on polymer [poly(ethylene imine) and poly(ethylene glycol)] -coated SWNT-FETs to monitor the binding of the protein streptavidin [12]. In this case, poly(ethylene imine) (PEI) was used to provide amine groups for the attachment of biotin molecules, and poly(ethylene glycol) (PEG) to prevent the nonspecific adsorption of proteins onto the SWNTs. A significant change in conductance was observed as a result of streptavidin binding, with charge transfer and enhanced scattering being suggested as possible mechanisms of the sensing process.

SWNT-FETs can also be employed for monitoring enzymatic interactions. For instance, Star and colleagues coated SWNT-FETs with starch, and monitored the changes in their electrical characteristics during the amyloglucosidase (AMG) reaction [27]. When AMG degrades the starch films on the SWNTs to glucose, the electrical characteristics are restored to values seen prior to starch coating. Several protein sensors based on immune reactions have also been reported. For example, both Park *et al.* and Li *et al.* reported the detection of tumor markers by using antibody-immobilized SWNT-FETs [28, 29]. Such antibody-immobilized SWNT-FET sensors may also be used to monitor small molecules [30].

9.4.4
Other Receptor Types

In addition to smaller biomolecules such as DNA or proteins, larger biomaterials such as cells and tissues can be used as receptors. Zhou *et al.* combined lipid bilayers on SWNT-FETs, and investigated the diffusion of lipid molecules and membrane-bound proteins [31]. They also showed that the specific binding of proteins to membrane-embedded receptors can be measured with SWNT-FETs. As a model system, a lipid bilayer with biotin was synthesized and immobilized on a SWNT-FET. The binding of streptavidin was found to result in a shift in the gate threshold voltage, and the charge density estimated from the shift was found to be comparable to the charge density of the biomolecules.

In another example, human olfactory receptors were employed as a recognition element in SWNT-FET sensors: the human olfactory receptors (hOR2AG1) were immobilized on a SWNT-FET, and the electronic responses to various odorant molecules measured [32]. The sensor was found to exhibit very high sensitivity (~100 fM), as well as superior selectivity. An olfactory receptor normally bears ionizable cysteine residues, and once a specific odorant molecule has bound to a receptor molecule, the latter shifts to a negatively charged state. Negatively charged molecules most likely affect the work function of the metal contact electrodes on the SWNT-FETs, which is then translated into a conductance change. Since olfactory receptors were used as the molecular recognition element, the detection of specific odorant molecules with single carbon atom resolution was possible. In particular, the sensor exhibited a clear decrease in conductance in the presence of amyl butyrate (AB), whereas no change in conductance was observed for butyl butyrate, a molecule which has one less carbon atom than AB.

9.5
The Application of Nanotube Biosensors to Pathogen Detection

Today, healthcare and medical diagnosis are the largest and the most important markets for biosensors. SWNT-based biosensors are expected to offer cheap, fast, and easy to use point-of-care testing devices. SWNT-FET DNA sensors capable of distinguishing single nucleotide polymorphisms have been demonstrated, as have SWNT-FET sensors that can detect small amounts of tumor markers. Although most tumor markers are not specific enough to confirm the diagnosis of cancer, they are widely used as indicators for cancer or metastasis and the recurrence of cancers. Glucose sensors based on SWNT-FETs have also been demonstrated.

Apart from biosensors for diagnostics and medicine, biosensors can also be used in environmental monitoring (air, water, and soil), chemical process control, and food industries. Further, tools for preventing bioterror or the outbreak of highly contagious diseases must be developed in the near future. In recent years, many people have suffered from swine flu, avian influenza, and severe acute respiratory syndrome (SARS), while food poisoning by pathogenic bacteria such as *E. coli* O157 is also a serious threat. There is also a risk in developing countries of waterborne epidemics. Thus, a sensor that could detect infectious diseases or pathogenic bacteria rapidly, and with high sensitivity, is required.

Currently, most infectious diseases and pathogenic bacteria are diagnosed with DNA-based molecular diagnostics, although strip sensors that use an immune interaction can be employed in primary screening. In many cases, the detection of DNA requires complicated sample preparation (cell lysis, purification, and polymerase chain reaction (PCR)) and trained personnel, but the sensitivity is very high owing to PCR amplification. The identification of infectious diseases could be performed via a different targeting route. The detection of DNA from viruses or bacteria might be one approach, while toxins released from pathogenic bacteria could also be used as "fingerprints." Surface-bound proteins or carbohydrates are widely used to identify pathogens, and sensors could target actual viruses or bacteria. In the latter case, antibodies or aptamers selected against the whole cell (virus) could be used as receptors. DNA detection, toxin detection, and whole-cell detection with SWNT-FETs have recently been demonstrated. The first DNA sensor based on a SWNT-FET was reported by Star and colleagues [33], who used nonspecifically adsorbed probe ssDNA-immobilized network SWNT-FET transistors. In this case, changes in conductance were observed upon hybridization of the target ssDNA, with the change in conductance being smaller for mismatched targets. DNA sensors that employ a Schottky contact barrier have also been reported [34]. As modulation of the Schottky barrier affects the electronic properties of SWNT-FETs more significantly than channel doping, Tang *et al.* immobilized probe ssDNA on a Au electrode to monitor the hybridization of a target ssDNA [34]. In this case, the probe ssDNA was immobilized on the Au surface via thiol (SH) linkers, so that any delayed hybridization due to the helical wrapping of SWNTs with probe ssDNA could be prevented.

Virus-specific RNA has been detected with a SWNT-FET. Dastagir and colleagues immobilized peptide nucleic acid (PNA) on a SWNT through nonspecific adsorption, and measured the evolution of its gate transfer characteristics in the presence of a target hepatitis C virus–RNA interaction [35]. The conductance increased only with the complementary RNA target, whereas decreases in the conductance or no change in the conductance were observed with mismatched RNA or DNA.

It is also possible to measure surface proteins or carbohydrates on pathogens. For example, Zhang et al. used a ligand–receptor interaction to monitor adenovirus [36]. The adenovirus protein, Ad12 Knob, and its complementary receptor, the "Coxsackie virus and adenovirus receptor" (CAR), was immobilized on SWNTs via covalent binding. Air-oxidized SWNTs were used in these experiments, since carboxylic acid that can be used to covalently immobilize receptor proteins becomes abundant as a result of this process. It was confirmed that the proteins retained their original functionality after immobilization through carbodiimide (EDAC)-mediated activation, by using labeled antibodies. The electronic sensor response was measured by using a CAR-bound SWNT-FET. Exposing the CAR-SWNT device to the Knob protein suppressed the on-current of the device greatly, thereby demonstrating that adenovirus can be detected with a SWNT-FET via CAR-Knob specificity.

Takeda and colleagues used an anti-hemagglutinin antibody to fabricate sensors for influenza virus [37]. Influenza hemagglutinin is found on the surface of influenza viruses, and is responsible for the binding of viruses to infected cells. A CNT biosensor for the detection of hepatitis B virus (HBV) has been demonstrated by Oh et al. [38]. Here again, the SWNT-FET was used as an active transducer, with the binding of HBV antigen with HBV antibodies immobilized on the SWNT surface being monitored in real time. The conductance increased with increases in the concentration of the HBV antigen, and the gate threshold voltage became more positive after the interaction. The sensitivity of the SWNT sensor was found to be comparable to that of ELISA, although the detection time was much shorter.

Further, there are approaches that directly measure virions (virus particles) or bacterial cells. For example, Patolsky and colleagues demonstrated the binding of single virus particles and the subsequent changes in the electronic properties of Si nanowire transistors [4]. Attempts have also been made to monitor the binding of virus particles using SWNT-FETs. Figure 9.9 shows an atomic force microscopy (AFM) image of a SWNT-FET sensor after reacting with H3N1 influenza virus, and the electrical transport characteristics before and after the binding of the virus. For selective detection of the virus, polyclonal antibodies specific to the H3N1 influenza virus were immobilized on the sidewalls of the SWNT by using non-covalent binding linkers. As shown in the AFM image, two virus particles were bound on the SWNT-FET after interaction, although the change in electrical characteristics as a result of virus binding was rather small. As noted by Grüner [39], the virus particle is a sphere with a diameter of 80–120 nm, so the area of surface-bound proteins that make direct contact with the SWNT is relatively small.

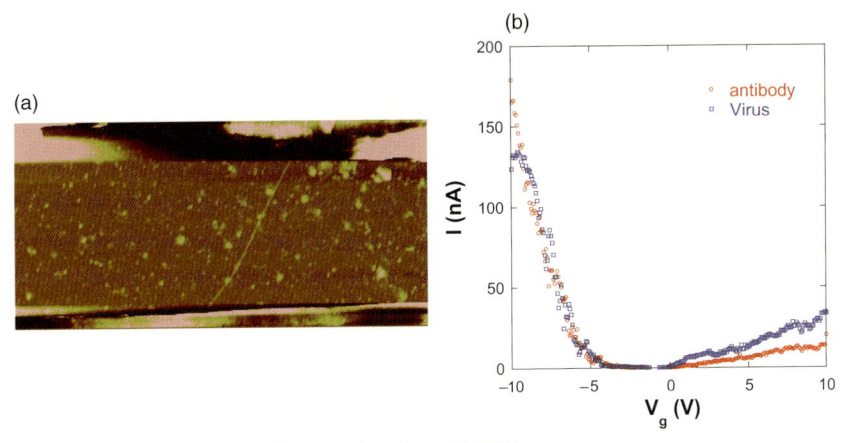

Figure 9.9 SWNT-FET sensor for virus detection. (a) AFM image of the device after reaction with H3N1 influenza virus. The dotted circles represent influenza virus particles; (b) Evolution of the electronic transfer characteristics upon virus interaction. Only a small increase in the conductance was observed in the n-channel, whereas there was no change in the conductance of the p-channel.

So and colleagues demonstrated the detection of *E. coli* by using a SWNT-FET [5]. RNA aptamers specific to *E. coli* were first chosen as the receptors, and then immobilized on the sidewalls of the SWNTs by using noncovalent binding linkers. Unlike the case of virus detection, the adsorption of a single *E. coli* cell was found to result in large changes in the electrical characteristics. In fact, the adsorption of a single *E. coli* cell was found to decrease the on-current by more than 50%, whereas no change in the conductance arose for *Salmonella* or for SWNT-FETs without aptamers. The sensing mechanism was explained in terms of scattering due to bound *E. coli*, since a single *E. coli* cell can cover the entire SWNT-FET; typically, the *E. coli* cell is 2~4 µm long and 0.5 µm wide, and the channel length of the SWNT-FET is 5 µm. However, these results suffered from a lack of consistency for diluted *E. coli* solutions, since the number of *E. coli* in each aliquot could vary. In microbiological studies, statistical methods have long been used to minimize such errors; for example, the most probable number (MPN) method was developed in 1960, and is still commonly in use. In this method, serial dilutions of the sample solutions are prepared, and each dilution is incubated in at least three Petri dishes. Dilution continues until no more bacterial colonies are found, at which point it is possible to estimate the titer of the bacteria in the original solution from the MPN table and the dilution factors. The procedure of a typical MPN method, with three Petri dishes, is shown in Figure 9.10.

Subsequently, So and colleagues adapted the MPN method for use with SWNT-FET sensors to improve reliability and sensitivity. In this case, instead of three Petri dishes, three SWNT-FET sensors were used for each diluted sample solution,

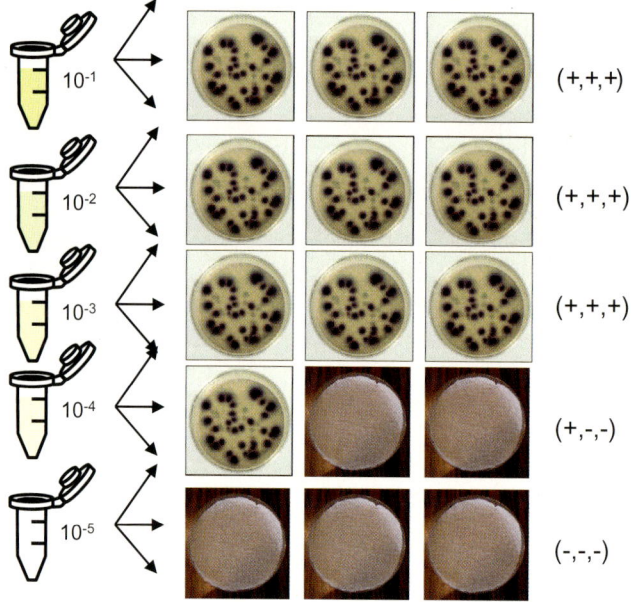

Figure 9.10 Conventional MPN method. To estimate the titer of the original solution, three sets of experiments are chosen before extinction (10^{-3}, 10^{-4}, 10^{-5} in this particular example), and their counts compared with the MPN table. In the 10^{-3} diluted solution, all three Petri dishes showed positive results, only one dish in the 10^{-4} diluted sample, and no more positives in the 10^{-5} diluted solution. This (3, 1, 0) counts corresponded to 0.43 in the MPN table. The estimated titer of E. coli in the solution is then:
$0.43 \times 10^4 = 4.3 \times 10^3 \, \text{ml}^{-1}$.

and the results compared with the MPN table. In this example, there was a decrease in conductance in all three transistors with the original solution [~10^5 colony-forming units (cfu) ml^{-1}], there was a decrease in conductance of the two sensors with a 1/10-diluted solution, and no change in conductance was observed for the 1/100-diluted solution. From the MPN table, the estimated titer of E. coli obtained using the aptamer-functionalized SWNT-FET was approximately ~10^3 cfu ml^{-1}, which was two orders of magnitude smaller than that of the original solution. This discrepancy was attributed to statistical errors that arose during sampling because the sample sizes were very small (~3 µl), and to the small sensor surface area. In order to increase the sensitivity, the sensor surface area can be increased by using larger arrays of SWNT-FETs, or by guiding the bacteria to the sensor surface by force.

9.6
The Future of Nanotube Biosensors

Following discussions of the various types of SWNT–FET sensor, it is clearly possible to develop highly sensitive biosensors for diverse purposes by using SWNT-

FETs and suitable receptors. However, several issues must first be overcome before SWNT-FET biosensors can be applied on a commercial basis. First, as control over the chiralities of SWNTs is not yet possible, a lack of uniformity and a consequent reduction in credibility means that SWNT-FET biosensors are not yet ready for commercial application. Second, since the sensing principle relies on the extra charges of the target molecules, measurement can only be performed in a "clean" environment. Although much current research is addressing this problem, sample purification steps or signal amplification are currently needed in order to process biological samples. Finally, because of the small sensing area, detection in very dilute samples and single-molecule detection are limited by the diffusion time of the target molecules. However, as these issues are currently being actively studied worldwide, it is likely that multipurpose, point-of-care SWNT biosensors might be achieved by the next decade.

References

1 Tans, S.J., Verschueren, A.R.M. and Dekker, C. (1998) Room-temperature transistor based on a single carbon nanotube. *Nature*, **393**, 49–52.

2 Kong, J., Franklin, N.R., Zhou, C., Chapline, M.G., Peng, S., Cho, K. and Dai, H. (2000) Nanotube molecular wires as chemical sensors. *Science*, **287**, 622–5.

3 Cui, Y., Wei, Q., Park, H. and Lieber, C.M. (2001) Nanowire nanosensors for highly sensitive and selective detection of biological and chemical species. *Science*, **293**, 1289–92.

4 Patolsky, F., Zheng, G., Hayden, O., Lakadamyali, M., Zhuang, C. and Lieber, C.M. (2004) Electrical detection of single viruses. *Proceedings of the National Academy of Sciences of the United States America*, **101**, 14017–22.

5 So, H.-M., Park, D.-W., Jeon, E.-K., Kim, Y.-H., Kim, B.S., Lee, C.-K., Choi, S.Y., Kim, S.C., Chang, H. and Lee, J.-O. (2008) Detection and titer estimation of Escherichia coli using aptamer-functionalized single-walled carbon-nanotube field-effect transistors. *Small*, **4**, 197–201.

6 Britto, P.J., Santhanam, K.S.V. and Ajayan, P.M. (1996) Carbon nanotube electrode for oxidation of dopamine. *Bioelectrochemistry and Bioenergetics*, **41**, 121–5.

7 Heinze, S., Tersoff, J., Martel, R., Derycke, V., Appenzeller, J. and Avouris, P. (2002) Carbon nanotubes as Schottky barrier transistors. *Physical Review Letters*, **89**, 106801–4.

8 Chen, R.J., Choi, H.C., Bangsaruntip, S., Yenilmez, E., Tang, X., Wang, Q., Chang, Y.L. and Dai, H. (2004) An investigation of the mechanisms of electronic sensing of protein adsorption on carbon nanotube devices. *Journal of the American Chemical Society*, **126**, 1563–8.

9 Heller, I., Janssens, A.M., Männik, J., Minot, E.D., Lemay, S.G. and Dekker, C. (2008) Identifying the mechanism of biosensing with carbon nanotube transistors. *Nano Letters*, **8**, 591–5.

10 Bradley, K., Briman, M., Star, A. and Grüner, G. (2004) Charge transfer from adsorbed proteins. *Nano Letters*, **4**, 253–6.

11 Bestman, K., Lee, J.-O., Wiertz, F.G.M., Heering, H.A. and Dekker, C. (2003) Enzyme-coated carbon nanotubes as single-molecule biosensors. *Nano Letters*, **3**, 727–30.

12 Star, A., Gabriel, J.-C.P., Bradley, K. and Grüner, G. (2003) Electronic detection of specific protein binding using nanotube FET devices. *Nano Letters*, **3**, 459–63.

13 Wang, S.G., Wang, R., Sellin, P.J. and Zhang, Q. (2004) DNA biosensors based on self-assembled carbon nanotubes. *Biochemical and Biophysical Research Communications*, **325**, 1433–7.

14 Zelada-Guillén, G.A., Riu, J., Duzgun, A. and Rius, F.X. (2009) Immediate

detection of living bacteria at ultralow concentrations using a carbon nanotube based potentiometric aptasensor. *Angewandte Chemie International Edition*, **48**, 7334–7.

15 Chen, R.J., Zhang, Y., Wang, D. and Dai, H. (2001) Noncovalent sidewall functionalization of single-walled carbon nanotubes for protein immobilization. *Journal of the American Chemical Society*, **123**, 3838–9.

16 Chen, R.J., Bangsaruntip, S., Drouvalakis, K.A., Kam, N.W.S., Shim, M., Li, Y., Kim, W., Utz, P.J. and Dai, H. (2003) Carbon nanotubes as biocompatible materials and highly specific electronic biosensors. *Proceedings of the National Academy of Sciences of the United States America*, **100**, 4984–9.

17 Lo, Y.-S., Nam, D.H., So, H.-M., Chang, H., Kim, J.-J., Kim, Y.H. and Lee, J.-O. (2009) Oriented immobilization of antibody fragments on Ni-decorated single-walled carbon nanotube devices. *ACS Nano*, **3**, 3649–55.

18 Quinn, B.M., Dekker, C. and Lemay, S.G. (2005) Electrodeposition of noble metal nanoparticles on carbon nanotubes. *Journal of the American Chemical Society*, **127**, 6146–7.

19 Loh, Y.-S., Lee, K.-J., So, H.-M., Chang, H., Kim, J.-J. and Lee, J.-O. (2009) Sensing mechanism of metal-decorated single-walled carbon nanotube field effect transistor sensors. *Physica Status Solidi B*, **246**, 2824–7.

20 So, H.-M., Won, K., Kim, Y.H., Kim, B.-K., Ryu, B.H., Na, P.S., Kim, H. and Lee, J.-O. (2005) Single-walled carbon nanotube biosensors using aptamers as molecular recognition elements. *Journal of the American Chemical Society*, **127**, 11906–7.

21 Lee, J.-O., So, H.-M., Jeon, E.-K., Chang, H., Won, K. and Kim, Y.H. (2008) Aptamers as molecular recognition elements for electrical nanobiosensors. *Analytical and Bioanalytical Chemistry*, **390**, 1023–32.

22 Maehashi, K., Matsumoto, K., Takamura, Y. and Tamiya, E. (2009) Aptamer-based label-free immunosensors using carbon nanotube field-effect transistors. *Electroanalysis*, **21**, 1285–90.

23 Cella, L.N., Sanchez, P., Zhong, W., Myung, N.V., Chen, W. and Mulchandani, A. (2010) Nano aptasensor for protective antigen toxin of anthrax. *Analytical Chemistry*, **82**, 2042–7.

24 Kim, J.P., Lee, B.Y., Hong, S. and Sim, S.J. (2008) Ultrasensitive carbon nanotube-based biosensors using antibody-binding fragments. *Analytical Biochemistry*, **381**, 193–8.

25 Park, J.-S., Cho, M.K., Lee, E.J., Ahn, K.-Y., Lee, K.E., Jung, J.H., Cho, Y., Han, S.-S., Kim, Y.K. and Lee, J.A. (2009) Highly sensitive and selective diagnostic assay based on virus nanoparticles. *Nature Nanotechnology*, **4**, 259–64.

26 Kim, J.P., Lee, B.Y., Lee, J., Hong, S. and Sim, S.J. (2009) Enhancement of sensitivity and specificity by surface modification of carbon nanotubes in diagnosis of prostate cancer based on carbon nanotube field effect transistors. *Biosensors and Bioelectronics*, **24**, 3372–8.

27 Star, A., Joshi, V., Han, T.-R., Altoé, M.V.P., Grüner, G. and Stoddart, J. (2004) Electronic detection of the enzymatic degradation of starch. *Organic Letters*, **6**, 2089–92.

28 Park, D.-W., Kim, Y.-H., Kim, B.S., So, H.-M., Won, K., Lee, J.-O., Kong, K. and Chang, H. (2006) Detection of tumor markers using single-walled carbon nanotube field effect transistors. *Journal of Nanoscience and Nanotechnology*, **6**, 3499–502.

29 Li, C., Currelli, M., Lin, H., Lei, B., Ishikawa, F.N., Datar, R., Cote, R., Thomson, M. and Zhou, C. (2005) Complementary detection of prostate-specific antigen using In_2O_3 nanowires and carbon nanotubes. *Journal of the American Chemical Society*, **127**, 12484–5.

30 Martinez, M.T., Tseng, Y.-C., Salvador, J.P., Marco, M.P., Ormategui, N., Loinaz, I. and Bokor, J. (2010) Electronic anabolic steroid recognition with carbon nanotube field-effect transistors. *ACS Nano*, **4**, 1473–80.

31 Zhou, X., Moran-Mirabal, J.M., Craighead, H.G. and McEuen, P.L. (2007) Supported lipid bilayer/carbon nanotube hybrids. *Nature Nanotechnology*, **2**, 185–90.

32 Kim, T.H., Lee, S.H., Lee, J., Song, H.S., Oh, E.H., Park, T.H. and Hong, S. (2009)

Single-carbon-atomic-resolution detection of odorant molecules using a human olfactory receptor-based bioelectronic nose. *Advanced Materials*, **21**, 91–4.

33 Star, A., Tu, E., Niemann, J., Gabriel, J.-C.P., Joiner, C.S. and Valcke, C. (2006) Label-free detection of DNA hybridization using carbon nanotube network field-effect transistors. *Proceedings of the National Academy of Sciences of the United States America*, **103**, 921–6.

34 Tang, X., Bangsaruntip, S., Nakayama, N., Yenilmez, E., Chang, Y.I. and Wang, Q. (2006) Carbon nanotube DNA sensor and sensing mechanism. *Nano Letters*, **6**, 1632–636.

35 Dastagir, T., Forzani, E.S., Zhang, R., Amlani, I., Nagahara, L.A., Tsui, R. and Tao, N. (2007) Electrical detection of hepatitis C virus RNA on single wall carbon nanotube-field effect transistors. *Analyst*, **132**, 738–40.

36 Zhang, Y.-B., Kanungo, M., Ho, A.J., Freimuth, P., van der Lelie, D., Chen, M., Khamis, S.M., Datta, S.S., Johnson, A.T., Misewich, J.A. and Wong, S.S. (2007) Functionalized carbon nanotubes for detecting viral proteins. *Nano Letters*, **7**, 3086–91.

37 Takeda, S., Ozaki, H., Hattori, S., Ishii, A., Kida, H. and Mukasa, K. (2007) Detection of influenza virus hemagglutinin with randomly immobilized anti-hemagglutinin antibody on a carbon nanotube sensor. *Journal of Nanoscience and Nanotechnology*, **7**, 752–6.

38 Oh, J., Yoo, S., Chang, Y.W., Lim, K. and Yoo, K.H. (2009) Carbon nanotube-based biosensor for detection hepatitis B. *Current Applied Physics*, **9**, e229–31.

39 Grüner, G. (2006) Carbon nanotube transistors for biosensing applications. *Analytical and Bioanalytical Chemistry*, **384**, 322–35.

# 10
Environmental Impact of Fullerenes

Naohide Shinohara

10.1
Introduction

Fullerenes are carbon allotropes composed entirely of carbon, in the form of hollow spheres or ellipsoids, first discovered by Kroto *et al.* in 1985 [1]. Although fullerenes have been considered typical examples of nanomaterials, in reality they may exist as crystals much larger than 100 nm, and have many properties that differ somewhat from those of other nanomaterials. Fullerene C_{60}, a representative fullerene, has a form that resembles a soccer ball with a diameter of approximately 1 nm. It has been used in small amounts for sporting goods and cosmetics, while in larger amounts it serves as a candidate substance in many nanotechnology applications in industrial and medical fields, such as energy conversion and drug delivery. The potential of C_{60} to induce various adverse effects must be considered, however, because current information regarding environmental and health risk is limited.

In this chapter, a comprehensive summary is provided of the environmental hazard assessment of fullerenes, together with the basic patterns of possible routes of exhaust emissions, and predictions of their environmental fate. In addition, details are provided of the toxicity of particles, as this is known to depend on the solvent type and particle size.

10.2
Methods Used to Prepare Fullerene Suspensions

10.2.1
Solubility of Fullerene

Unlike graphite and diamond, fullerene is slightly soluble in organic solvents such as toluene and carbon disulfide (solubility of 2.2–2.8 mg ml^{-1} and 5.2–7.9 mg ml^{-1}, respectively) [2, 3]. Although C_{60} is relatively insoluble in polar solvents such as water, acetone, tetrahydrofuran (THF), and ethanol, its solubility in aromatic hydrocarbons is relatively high [2]. Details of the solubility of C_{60} and C_{70} are

Table 10.1 Solubility of C_{60} in solvents.

C_{60}

Solvent	Solubility (mg ml^{-1})	Temperature (°C)	Reference	Solvent	Solubility (mg ml^{-1})	Temperature (°C)	Reference
Nitromethane	0.000	20–25	Ruoff et al. [2]	Hexane	0.03650	24.5	Heymann [4]
Acetonitrile	0.000	20–25	Ruoff et al. [2]		0.040	30	Sivaraman et al. [3]
Tetrahydrofuran	0.000	20–25	Ruoff et al. [2]		0.043	20–25	Ruoff et al. [2], Ruoff et al. [6]
1,2-Ethanediol	0.00000	24.5	Heymann [4]	1-Hexanol	0.04200	24.5	Heymann [4]
1,2,3-Propanetriol	0.00000	24.5	Heymann [4]	1-Octanol	0.04700	24.5	Heymann [4]
Water	0.000013[a] (Calculation)	25	Heymann et al. [5]	Hexane	0.04850	24.5	Heymann [4]
				Cyclohexane	0.051	30	Sivaraman et al. [3]
Methanol	0.000	20–25	Ruoff et al. [2]	Pyridine	0.3	24.5–25.5	Scrivens and Tour [7]
	0.000035	24.5	Heymann [4]		0.89	20–25	Ruoff et al. [2]
1,3-Propanediol	0.000900	24.5	Heymann [4]	Nonane	0.06230	24.5	Heymann [4]
Ethanol	0.001	20–25	Ruoff et al. [2]	Decane	0.070	30	Sivaraman et al. [3]
	0.00080	24.5	Heymann [4]		0.071	20–25	Ruoff et al. [2]
Acetone	0.001	20–25	Ruoff et al. [2]	Dodecane	0.091	30	Sivaraman et al. [3]
Nitroethane	0.002	20–25	Ruoff et al. [2]	Tetradecane	0.126	30	Sivaraman et al. [3]
Cyclopentane	0.002	20–25	Ruoff et al. [2]	Chloroform	0.16	20–25	Ruoff et al. [2]

				C_{70}			
Solvent	Solubility (mg ml^{-1})	Temperature (°C)	Reference	Solvent	Solubility (mg ml^{-1})	Temperature (°C)	Reference
n-Butylbenzene	1.9	24.5–25.5	Scrivens and Tour [7]	Acetone	0.0019	30	Sivaraman et al.[8]
Iodobenzene	2.1	24.5–25.5	Scrivens and Tour [7]	Pentane	0.0020	30	Sivaraman et al.[8]
cis-Decalin	2.2	20–25	Ruoff et al. [2]	Isopropyl alcohol	0.0021	30	Sivaraman et al.[8]
1,3-Dichlorobenzene	2.4	24.5–25.5	Scrivens and Tour [7]	Hexane	0.013	30	Sivaraman et al.[8]
Ethylbenzene	2.6	24.5–25.5	Scrivens and Tour [7]	Octane	0.042	30	Sivaraman et al.[8]
Toluene	2.150	30	Sivaraman et al. [3]	Heptane	0.047	30	Sivaraman et al.[8]
	2.8	20–25	Ruoff et al. [2], Ruoff et al. [6]	Decane	0.053	30	Sivaraman et al.[8]
	2.9	24.5–25.5	Scrivens and Tour [7]	Cyclohexane	0.080	30	Sivaraman et al.[8]
Bromobenzene	2.8	24.5–25.5	Scrivens and Tour [7]	Dichloromethane	0.080	30	Sivaraman et al.[8]
	3.3	20–25	Ruoff et al. [2]	Dodecane	0.098	30	Sivaraman et al.[8]
Decalin (cis:trans = 3:7)	4.6	20–25	Ruoff et al. [2]	Carbon tetrachloride	0.121	30	Sivaraman et al.[8]
1,2,3-Trimethylbenzene	4.7	24.5–25.5	Scrivens and Tour [7]	Benzene	1.300	30	Sivaraman et al.[8]
Xylene	5.2	20–25	Ruoff et al. [2]	Toluene	1.406	30	Sivaraman et al.[8]
1,1,2,2-Tetrachloroethane	5.3	20–25	Ruoff et al. [2]	Mesitylene	1.472	30	Sivaraman et al.[8]
Anisole	5.6	20–25	Ruoff et al. [2]	p-Xylene	3.985	30	Sivaraman et al.[8]

Table 10.1 Continued

C_{60}

Solvent	Solubility (mg ml^{-1})	Temperature (°C)	Reference	Solvent	Solubility (mg ml^{-1})	Temperature (°C)	Reference
2-Propanol	0.00210	24.5	Heymann [4]	Dichloromethane	0.254	30	Sivaraman et al. [3]
1,4-Butanediol	0.00217	24.5	Heymann [4]		0.26	20–25	Ruoff et al. [2]
2-Butanol	0.00360	24.5	Heymann [4]	Benzonitrile	0.41	20–25	Ruoff et al. [2]
Pentane	0.004	30	Sivaraman et al. [3]	Carbon tetrachloride	0.32	20–25	Ruoff et al. [2]
	0.005	20–25	Ruoff et al. [2]		0.447	30	Sivaraman et al. [3]
	0.00320	24.5	Heymann [4]	1,2-Dichloromethane	0.50	20–25	Ruoff et al. [2]
1-Propanol	0.00410	24.5	Heymann [4]	Fluorobenzene	0.59	20–25	Ruoff et al. [2]
1,5-Pentanediol	0.00444	24.5	Heymann [4]	Nitrobenzene	0.80	20–25	Ruoff et al. [2]
1-Butanol	0.00940	24.5	Heymann [4]	N-Methyl-2-pyrolidone	0.89	20–25	Ruoff et al. [2]
1,1,2-Trichlorotrifluoroethane	0.014	20–25	Ruoff et al. [2]	tert-Butylbenzene	0.9	24.5–25.5	Scrivens and Tour [7]
o-Cresol	0.014	20–25	Ruoff et al. [2]	sec-Butylbenzene	1.1	24.5–25.5	Scrivens and Tour [7]
2-Pentanol	0.01800	24.5	Heymann [4]	Tetrachloroethylene	1.2	20–25	Ruoff et al. [2]
2-Methyl-pentane	0.01840	24.5	Heymann [4]	iso-Propyl benzene	1.2	24.5–25.5	Scrivens and Tour [7]

				C₇₀			
Solvent	Solubility (mg ml⁻¹)	Temperature (°C)	Reference	Solvent	Solubility (mg ml⁻¹)	Temperature (°C)	Reference
1,2,3,4-Tetramethylbenzene	5.8	24.5–25.5	Scrivens and Tour [7]	Carbon disulfide	9.875	30	Sivaraman et al.[8]
p-Xylene	5.9	24.5–25.5	Scrivens and Tour [7]	1,2-Dichloromethane	36.210	30	Sivaraman et al.[8]
Chlorobenzene	5.7	24.5–25.5	Scrivens and Tour [7]				
	7.0	20–25	Ruoff et al. [2]				
2-Methylthiophene	6.8	20–25	Ruoff et al. [2]				
Carbon disulfide	5.160	30	Sivaraman et al. [3]				
	7.9	20–25	Ruoff et al. [2], Ruoff et al. [6]				
Quinoline	7.2	24.5–25.5	Scrivens and Tour [7]				
o-Xylene	8.7	24.5–25.5	Scrivens and Tour [7]				
1,2,4-Trichlorobenzene							
	8.5	20–25	Ruoff et al. [2]				
	10.4	24.5–25.5	Scrivens and Tour [7]				
1,2-Dibromobenzene	13.8	24.5–25.5	Scrivens and Tour [7]				
1,3-Dibromobenzene	13.8	24.5–25.5	Scrivens and Tour [7]				

Table 10.1 Continued

C_{60}

Solvent	Solubility (mg ml^{-1})	Temperature (°C)	Reference	Solvent	Solubility (mg ml^{-1})	Temperature (°C)	Reference
Dichlorodifluoroethane	0.020	20–25	Ruoff et al. [2]	Fluorobenzene	1.2	24.5–25.5	Scrivens and Tour [7]
Octane	0.025	30	Sivaraman et al. [3]	trans-Dekalin	1.3	20–25	Ruoff et al. [2]
	0.01990	24.5	Heymann [4]	m-Xylene	1.4	24.5–25.5	Scrivens and Tour [7]
3-Methyl-pentane	0.02510	24.5	Heymann [4]	1,3,5-Trimethylbenzene	0.997	30	Sivaraman et al. [3]
Isooctane	0.026	30	Sivaraman et al. [3]		1.5	20–25	Ruoff et al. [2]
3-Pentanol	0.02900	24.5	Heymann [4]		1.7	24.5–25.5	Scrivens and Tour [7]
Tetrahydrothiophene	0.030	20–25	Ruoff et al. [2]	Trichloroethylene	1.4	20–25	Ruoff et al. [2]
1-Pentanol	0.03000	24.5	Heymann [4]	Benzene	1.440	30	Sivaraman et al. [3]
Cyclohexane	0.036	20–25	Ruoff et al. [2]		1.5	24.5–25.5	Scrivens and Tour [7]
Thiophene	0.4	24.5–25.5	Scrivens and Tour [7]		1.7	20–25	Ruoff et al. [2]
Dioxane	0.041	30	Sivaraman et al. [3]	n-Propyl benzene	1.5	24.5–25.5	Scrivens and Tour [7]

a) Considering the F-T size effect, the estimated solubility becomes smaller.

				C_{70}			
Solvent	Solubility (mg ml^{-1})	Temperature (°C)	Reference	Solvent	Solubility (mg ml^{-1})	Temperature (°C)	Reference
Tetralin	16	20–25	Ruoff et al. [2]				
1,2,4-Trimethylbenzene	17.9	24.5–25.5	Scrivens and Tour [7]				
1,2,3,5-Tetramethylbe	20.8	24.5–25.5	Scrivens and Tour [7]				
1,2-Dichlorobenzene	27	20–25	Ruoff et al. [2]				
	24.6	24.5–25.5	Scrivens and Tour [7]				
1-Methylnaphthalene	33	20–25	Ruoff et al. [2]				
	33.2	24.5–25.5	Scrivens and Tour [7]				
1-Bromo-2-methylnaphthalene	34.8	24.5–25.5	Scrivens and Tour [7]				
Dimethylnaphthalene	36	20–25	Ruoff et al. [2]				
1-Phenylnaphthalene	50	20–25	Ruoff et al. [2]				
1-Chloronaphthalene	51	20–25	Ruoff et al. [2]				

summarized in Table 10.1 [2–8]. For their toxicological testing, both functionalization [9, 10] and inclusion with γ-cyclodextrin [11, 12] have been used to dissolve fullerenes in water.

10.2.2
Aqueous Suspensions of Fullerenes

C_{60} spontaneously forms stable nanoscale (25–500 nm) aggregates/agglomerates in solutions, including water, acetonitrile, ethanol, and acetone [13, 14]. The color of the suspension is yellow, in contrast to the solution of fullerene in toluene, which is purple. Various methods to prepare aqueous dispersions of fullerenes have been proposed to evaluate the toxicity of C_{60} particles towards aquatic organisms and rodents, including:

- The preparation of an aqueous dispersion of C_{60} by mixing water into a solution of C_{60} in an organic solvent such as THF [15], ethanol [16], toluene, or chloroform [17] and subsequently removing the solvent.
- The dispersal of C_{60} mixed with polyvinylpyrrolidone (PVP) [18].
- The dispersal of C_{60} mixed with a surfactant such as Tween 80, and grinding with a ball mill or bead mill [19–21].
- The dispersal of C_{60} by grinding with sugar candy and polyoxyethylene hydrogenated caster oil [22].
- The dispersal of C_{60} by stirring an aqueous solution for a period in excess of two weeks [23, 24].
- The dispersal of C_{60} by adding a dimethylsulfoxide (DMSO)-C_{60} solution to fresh water [25].

10.2.3
Toxicity of Aqueous Fullerene Suspensions as a Factor of the Dispersion Method

The effects of fullerene suspensions depend in part on the dispersion methods used to generate them. Oberdörster [26] reported that exposure for 48 h to a $0.5\,\text{mg}\,\text{l}^{-1}$ aqueous suspension of 30- to 100-nm C_{60} nanoparticles (nano-C_{60} suspension), which was prepared using THF (THF-nC_{60}), induced significant lipid peroxidation in the brains of juvenile largemouth bass (*Micropterus salmoides*). Sayes et al. [27] reported that THF-nC_{60} is cytotoxic due to its ability to induce lipid peroxidation. Although, previously, lipid peroxidation was shown to be induced by nano-C_{60} suspensions, this effect may not be attributable to C_{60} itself. THF-nC_{60} has been reported to be inappropriate for some toxicity tests of C_{60}, and is known to be more toxic toward daphnia [28, 29] and larval zebrafish [30] than a nano-C_{60} suspension prepared by stirring and sonication. The toxic effects of THF-nC_{60} on aquatic organisms were considered to be mediated by a THF-degradation product (γ-butyrolactone) rather than by fullerene C_{60} [30, 31]. Although the data were not reported, a suspension of C_{60} nanoparticles prepared by sonication (sonicated-nC_{60}) was shown to be slightly more toxic than a suspension of C_{60} nanoparticles prepared by long-term water stirring (water-stirred nC_{60}) [32].

10.3
Toxicological Data Relating to Fullerenes

10.3.1
Toxicological Effects of C_{60} on Fish

As THF-nC_{60} has obvious toxic properties, the results of tests using THF-nC_{60} do not accurately reflect the toxicity of the C_{60} nanoparticle itself. Thus, despite all previous studies of toxicity of C_{60} on fish being included in Table 10.2, only those that did not use THF-nC_{60} are summarized at this point.

For adult and larval fish, no obvious physical effects or deaths were reported in the following studies. The fathead minnow (*Pimephales promelas*) and Japanese medaka (*Oryzias latipes*) showed no obvious physical effects after 48 or 96 h of exposure to a 0.5 mg l^{-1} suspension of water-stirred-nC_{60} (LC_{50}: >0.5 mg l^{-1}) [29, 32]. Similarly, no deaths or physical effects were observed among Japanese medaka exposed for 96 h to a 2.15 mg l^{-1} suspension of C_{60} prepared using sugar candy and polyoxyethylene hydrogenated caster oil (sugar-candy-nC_{60}) [22]. The common carp (*Cyprinus carpio*), when exposed to a 4.5 mg l^{-1} suspension of C_{60} nanoparticles (50th percentile: 35 nm) prepared in Tween 80 solution by grinding with a bead mill (Tween80-nC_{60}), also demonstrated no deaths or adverse effects [21]. The survival of larval zebrafish (*Danio rerio*) was not reduced by exposure to water-stirred-nC_{60} [30].

In fathead minnows (*Pimephales promelas*) exposed to 0.5 mg l^{-1} water-stirred-nC_{60}, lipid peroxidation was elevated in the brain tissue and statistically significantly increased in the gills [29]. In that study, an upregulation of the CYP2 (cytochrome P450) family of isozymes, which may function to metabolize fullerene or to repair lipid peroxidation-induced damage, was also observed in the liver. Although, Oberdörster *et al.* [32] reported no changes in *CYP2* gene expression in fathead minnows exposed to 0.5 mg l^{-1} of water-stirred-nC_{60}, expression of the peroxisomal lipid transport protein PMP70 was significantly reduced, suggesting possible alterations in the acyl-CoA pathway. In contrast, no lipid peroxidation was observed in the brain tissue of common carp exposed to 4.5 mg l^{-1} of Tween 80-nC_{60} [21]. Since, in Japanese medaka, the PMP70 levels were not affected by fullerene exposure [32], species-specific differences in lipid peroxidation may be a factor in fullerene toxicity. In a report by Henry *et al.* [30], no changes in gene expression related to lipid peroxidation or peroxisomal lipid transport were observed in larval zebrafish exposed to water-stirred-nC_{60}. Indeed, the study results indicated that water-stirred-nC_{60} had only minimal effects on the gene expression of larval zebrafish, although a longer exposure to C_{60} may have resulted in different effects. In embryonic zebrafish, a 0.2 mg l^{-1} suspension of C_{60} particles prepared with

Table 10.2 Toxicity tests of fullerene C_{60} particles on fish.

Reference	Species	Preparation of suspensions	Particle size	Concentration
Oberdörster [26]	Juvenile largemouth bass (*Micropterus salmoides*)	THF method	30–100 nm [Reported value in previous study]	0.5, 1.0 mg l^{-1}
Zhu et al. [29]	Fathead minnow (*Pimephales promelas*)	THF method Water-stirred method (at least 2 months)	10–200 nm [Reported value in previous study]	0.5 mg l^{-1}
Oberdörster et al. [32]	Fathead minnow (*Pimephales promelas*) Japanese medaka (*Oryzias latipes*)	Water-stirred method (at least 2 months)	10–200 nm [Reported value in previous study]	0.5 mg l^{-1}
Henry et al. [30]	Larval zebrafish (*Danio rerio*)	THF method Water-stirred method (7 days)	50–300 nm [Dark-field microscopy observation]	THF: 0.625, 1.25 mg l^{-1}* Stirring: unknown
Zhu et al. [33]	Zebrafish Embryos (*Danio rerio*)	THF method	100 nm [TEM observation]	1.5 mg l^{-1}
Usenko et al. [25]	Embryonic zebrafish (*Danio rerio*)	DMSO-sonication method	Mode size: 0.1 mg l^{-1}: approx. 150 nm 0.2 mg l^{-1}: approx. 850 nm [DLS measurement]	0.1–0.5 mg l^{-1}
Seki et al. [32]	Japanese medaka (*Oryzias latipes*)	Sugar candy method	Average: 174 nm Range: 20–2000 nm [DLS measurement]	2.15 mg l^{-1} [HPLC]
Shinohara et al. [21]	Common carp (*Cyprinus carpio*)	Tween 80-grinding method	Median: 37 nm 95%ile: 95 nm [DLS measurement]	4.5 ± 0.1 mg l^{-1} [HPLC]

CYP, cytochrome P; GSH, Glutathione; GSSG, 5,5′-dithiobis(2-nitrobenzoic acid)-oxidized GSH; LPO, Lipid hydroperoxides; TBA, thiobarbituric acid.

Exposure duration	Endpoint	Results
48 h	Oxidative stress in brain (TBA assay) (GSSG recycling assay)	Significant lipid peroxidation was found in brains at 0.5 mg l^{-1} of THF-nC_{60}. GSH (glutathione) was also marginally depleted in the gills
48 h	Mortality Gene expression	Fish exposed to THF-nC_{60} died within 18 h. Exposure to water-stirred-nC_{60} can induce CYP2 isozymes
96 h	Gene expression (CYP in the mRNA and protein and PMP70)	Neither the mRNA nor protein-expression levels of cytochrome P450 isozymes were changed. The peroxisomal lipid transport protein PMP70 was significantly reduced in fathead minnow, but not in medaka
72 h	Mortality Gene expression	Survival was reduced in THF-nC_{60} but not in water-stirred-nC_{60}. Gene expression changes were also observed in fish exposed to THF-nC_{60}
96 h	Development Mortality Hatching rates Pericardial edema	THF-nC_{60} at 1.5 mg l^{-1} delayed zebrafish embryo and larval development, decreased survival and hatching rates, and caused pericardial edema
72 h (24 h to 96 h post-fertilization)	Morphological response Cellular response	Significant increases in malformations, pericardial edema, and mortality were observed at 0.2 mg l^{-1}. Both necrotic and apoptotic cellular death were induced throughout the embryo at 0.1 and 0.2 mg l^{-1}, respectively
96 h	Mortality	No adverse effects were observed at 2.15 mg l^{-1}
48 h	Oxidative stress in brain (LPO assay) (TBA assay) Translocation to brain	LPO assay confirmed the absence of lipid peroxidation after exposure to 4.5 mg l^{-1} of Tween-nC_{60}. Tween-nC_{60} did not translocate to the brain

DMSO (DMSO-C_{60}) significantly increased malformations, pericardial edema, and mortality [25]. Both, necrotic and apoptotic cellular deaths were induced throughout the embryo at 0.1 and 0.2 mg l^{-1} concentrations.

These results are listed in Table 10.2, in addition to the tests using THF-nC_{60} [21, 22, 25, 26, 29, 30, 32, 33]. In previous studies of the toxicity of C_{60} on fish, none of the fish died nor showed any physical adverse effects after exposure to 0.5 to 4.5 mg l^{-1} C_{60} particles for 48 to 96 h. As these studies have involved acute and/or subacute toxicity assessments of C_{60}, studies on the effects of chronic exposure to C_{60} have still to be carried out, and further research is required before any firm conclusions can be drawn regarding the effects of C_{60} nanoparticles on lipid peroxidation in all types of fish. Unfortunately, only one study has been conducted to date on embryonic fish, which may be differentially affected by exposure to C_{60} particles and necessitate low levels of C_{60} at the egg-laying sites. A further assessment of the effects of exposure on embryonic fish is necessary to determine their risk.

10.3.2
Toxicological Effects of C_{60} on Invertebrates

The results of previous studies of the toxicity of C_{60} on invertebrates are summarized below, except for those in which THF-nC_{60} was used. Data acquired from previous studies of C_{60} toxicity on invertebrates, including those with THF-nC_{60}, are listed in Table 10.3 [22, 28, 29, 32, 34–38].

In several studies, the mortality of *Daphnia magna* exposed to nC_{60} did not reach 50%, even at the maximum concentrations tested (2.25–35 mg l^{-1}) [22, 29, 32]. In contrast, in *Daphnia magna* the LC_{50} of sonicated-nC_{60} was 7.9 mg l^{-1} [28], and that of shaken-C_{60} particles was 10.5 mg l^{-1} [38]. In the former study, however, the toxicity of nC_{60}, may have been overestimated, as sonicated-nC_{60} was reported to have a higher toxicity than water-stirred nC_{60} [32]. The adhesion of C_{60} particles in the gut tract, as observed by Zhu *et al.* (Figure 10.1) may also occur in the natural environment, in which the C_{60} particles can also aggregate into large particles in water. The other effects on *D. magna* reported previously included a delay in molting, a reduction in offspring production (at 2.5 mg l^{-1}) [32], and immobilization (at 2.25 mg l^{-1} (EC_{10}) [22] and 9.34 mg l^{-1} (EC_{50}) [38]). Additionally, the existence of C_{60} aggregates could either increase or decrease the toxicity of other chemical compounds, due to their adsorption onto C_{60} particles. For example, the addition of aggregated C_{60} increased the toxicity of 6 mg l^{-1} phenanthrene on *D. magna* more than 10-fold, and reduced the toxicity of pentachlorophenol (PCP) by 25% [39].

For other invertebrates (*Hyalella azteca, Harpacticoid copepod, Thamnoplatyurus platyurus, Ceriodaphnia dubia, Daphnia pulex*), 50% mortality was not observed at the maximum concentrations (3.09–463 mg l^{-1}) of C_{60} particles tested [32, 34, 35]. The exception was a report by Klaper *et al.* [36], who observed a significantly increased mortality in *Daphnia pulex* exposed to 100 mg l^{-1} or 500 mg l^{-1} water-stirred-nC_{60}.

10.3.3
Toxicological Effects of C_{60} on Algae

Details of studies of C_{60} toxicity on algae, as described below, are listed in Table 10.4 [22, 34].

A dose-dependent growth inhibition (3.52%, 10.0%, 21.9%, and 35.6%) was observed in *Pseudokirchneriella subcapitata* exposed to nC_{60} of 0.0551, 0.203, 0.694, and 2.27 mg l^{-1}, respectively [22]. However, the 25% inhibition concentration for the same algae was reported by others to be >100 mg l^{-1} [34], the difference being considered due to particle size. The C_{60} particle sizes used by Seki *et al.* [22] were nanoscale (mean 174 nm; range 20–2000 nm), while those used by Blaise *et al.* [34] were micron-scale. As algae are capable of taking in smaller particles (nano-sized) more easily than larger particles (micron-sized), the burden of C_{60} may be higher when it is in the form of smaller particles.

10.3.4
Toxicological Effects of C_{60} on Bacteria and Soil Microbes

Details of studies of C_{60} toxicity on bacteria and soil microbes, as described below, are listed in Table 10.5 [14, 34, 35, 40–50].

Although many studies have been conducted to examine the exposure of Gram-negative (*Escherichia coli*) and Gram-positive bacteria (*Bacillus subtilis* or *Shewanella oneidensis*) to C_{60} [14, 35, 40–42, 46, 47, 50], THF-nC_{60} was most often used. As with THF-nC_{60} (0.9 mg l^{-1}), water-stirred-nC_{60}, sonication-nC_{60}, and PVP-nC_{60} each induced growth inhibition in *B. subtilis* at concentrations of 0.5 to 0.95 mg l^{-1} [41]. No effects on the growth of *E. coli* and *S. oneidensis* were observed following exposure to 20 mg l^{-1} of C_{60} particles [42]; the size of these particles is most likely on the micron scale, based on the preparation method used for that study. When comparing the toxicity of C_{60} particles of different sizes, but prepared by the same method, smaller C_{60} aggregates (~2 nm) showed a higher antibacterial activity than larger C_{60} aggregates (142 nm) [41]. Based on these results, it would appear that C_{60} is less toxic in the general aquatic environment than in the laboratory. This hypothesis is consistent with results obtained by Kang *et al.* [50], who noted that water-stirred-nC_{60} induced toxicity in monoculture experiments at a significantly higher rate than was observed in environmental samples.

Although the toxicity of C_{60} particles did not differ greatly between Gram-positive and Gram-negative bacteria, a reduced toxicity for *B. subtilis* has been reported by some groups [40, 46, 50]. Such tolerance of *B. subtilis* is presumed to be the result of physiological differences in the cell membrane of this species, which is known to possess increased levels of iso- and anteiso-branched fatty acids [44].

In most studies involving microbes in soil or sludge, no significant effects were observed on the structure and function of the microbial community, respiration, or biomass after exposure to 1–50 mg kg^{-1} soil of nC_{60} or 1000–50 000 mg kg^{-1} soil of C_{60} for 14 to 180 days [45, 46, 48].

Table 10.3 Toxicity tests of fullerene C_{60} particles on invertebrates.

Reference	Species	Preparation of suspension	Particle size	Concentration
Zhu et al. [29]	Daphnia magna	THF method Water-stirred method	10–200 nm [Reported value in previous study]	THF: 0.005–2.0 mg l^{-1} Stirred: 0.5–35 mg l^{-1}
Oberdörster et al. [32]	Daphnia magna	Water-stirred method	10–200 nm [Reported value in previous study]	48, 96 h test: 0.5–35 mg l^{-1} 21-day test: 0.5–5.0 mg l^{-1}
	Hyalella azteca			0.5–7.0 mg l^{-1}
	Harpacticoid copepod			3.75–22.5 mg l^{-1}
Lovern and Klaper [28]	Daphnia magna	THF method Sonication method	THF: 10–20 nm Sonication: 20–100 nm [TEM observation]	THF: 0.04–0.88 mg l^{-1} Sonication: 0.2–9 mg l^{-1} [TEM observation]
Lovern et al. [37]	Daphnia magna	THF method	10–20 nm [TEM observation]	0.26 mg l^{-1} [TEM observation]
Blaise et al. [34]	Hydra attenuate	Mixed in rotator (24 h)	–	<500 mg l^{-1} (Starting concn.)
	Thamnoplatyurus platyurus			
Seki et al. [32]	Daphnia magna	Sugar candy method	Average: 174 nm Range: 20–2000 nm [DLS measurement]	2.25 mg l^{-1} [HPLC]
Gao et al. [35]	Ceriodaphnia dubia	THF method	Purchased particle size: Average: 35.8 nm Particle size in suspension: Average: >100 nm [DLS measurement]	DI water: 0.83 mg l^{-1} River water: 0.038–3.09 mg l^{-1}
Zhu et al. [38]	Daphnia magna	Vigorously shaken	Purchased particle size: <200 nm [Reported value from the manufacturer]	0.5–100 mg l^{-1}
Klaper et al. [36]	Daphnia pulex	THF method Water-stirred method (24 h)	Average: 100 nm Range: 20–200 nm [TEM observation]	0.5–500 mg l^{-1}

CAT, catalase; GST, glutathione-S-transferase; LOEC, lowest observed effect concentration; NOEC, no observed effect concentration.

Exposure duration	Endpoint	Results
48 h	Mortality	LC_{50} were 0.8 and >35 mg l^{-1} for THF-nC_{60} and water-stirred-nC_{60}, respectively
48, 96 h 21 days	Mortality Offspring	LC_{50} was not observed at 35 mg l^{-1} in 48 h or 96 h test. At 2.5 and 5.0 mg l^{-1} significant delay in molting and significantly reduced offspring production were observed in the 21-day test
96 h	Mortality	LC_{50} and LOEC were >7.0 mg l^{-1}
96 h	Mortality	LC_{50} and LOEC were >22.5 mg l^{-1}
48 h	Mortality	LC_{50}, LOEC, and NOEC were 0.46, 0.26, and 0.18 mg l^{-1} of THF-nC_{60}. LC_{50}, LOEC, and NOEC were 7.9, 0.5, and 0.2 mg l^{-1} of sonicated-nC_{60}, respectively
60 min	Behavioral abnormality	THF-nC_{60} cause a significant increase in hopping frequency, heart rate, and appendage movement. There was no change in postabdominal claw curling
96 h	Acute sublethality indicated by morphological changes	EC_{50} was between 10 to 100 mg l^{-1}
24 h	Mortality	LC_{50} was >463 mg l^{-1}
48 h	Mortality Immobilization	Only 10% of D. magna showed the adverse effects (immobilization and lethargic symptom) at 2.25 mg l^{-1}. Thus, LC_{50} was >2.25 mg l^{-1}
48 h	Mortality	LC_{50} was >3.09 mg l^{-1}
48 h	Mortality Immobilization	LC_{50} and EC_{50} were 10.5 and 9.34 mg l^{-1}, respectively
24 h	Mortality GST and CAT analysis	Mortality and GST were significantly increased by water-stirred-nC_{60} at 100 and 500 mg l^{-1} (mortality: <50%), and by THF-nC_{60} at >5.0 mg l^{-1} (mortality: >50%), respectively

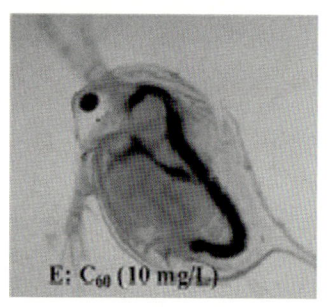

Figure 10.1 Dead *Daphnia magna* following 48 h exposure to 10 mg l^{-1} C60. Reproduced with permission from Ref. [38].; © 2009.

10.3.5
Toxicological Effects of C$_{60}$ on Other Organisms

Liu *et al.* [51] tested the effects of C$_{60}$, and also of carbon black (CB), single-walled carbon nanotubes (SWNTs) and multi-walled carbon nanotubes (MWNTs), on the larval and adult stages of *Drosophila melanogaster*. The dietary uptake of C$_{60}$, delivered via the food, at the larval stage had no detectable effect on egg-to-adult survival. The weakly adherent C$_{60}$ particles could be removed by grooming, and did not reduce either locomotor function or survival, unlike CB and SWNTs.

10.4
Possible Emission Sources of C$_{60}$

Fullerenes are currently used as additives in metals, resins, and cosmetics, with small amounts also being used as additives in engine oils and ski waxes. In the future, large amounts of fullerene may be used as electrode materials for solar cells, fuel cells, and capacitors, and also in medicine (Figure 10.2).

The first consideration is the possible emission of C$_{60}$ from a fullerene-manufacturing facility. As C$_{60}$ is synthesized in a vacuum and purified under wet conditions or within sealed equipment, possible emissions may occur during soot collection, product weighing and packaging, and in the maintenance and cleaning of equipment [53–55]. Under current manufacturing conditions, Fujitani *et al.* [55] detected the highest reported fullerene concentrations (peak concentrations) in an occupational environment: $<1 \times 10^8$ nm^3 cm^{-3} (1×10^4 particles cm^{-3}) for nano-sized particles (<50 nm) and $<5 \times 10^{10}$ nm^3 cm^{-3} (number concentration too low to be identified) (see Figure 10.3). In a pilot plant for the synthesis of Li@C$_{60}$, the peak concentration (30-min average) of C$_{60}$ in the working environment was 0.66 µg m^{-3}, with the smaller particles (<250 nm) comprising 4.3–30 wt% of C$_{60}$, and larger particles (>2.5 µm) comprising 26–54 wt% of C$_{60}$ [53]. As the exhaust

Table 10.4 Toxicity tests of fullerene C_{60} particles on algae.

Reference	Species	Preparation of suspension	Particle size	Concentration	Exposure duration	Endpoint	Results
Blaise et al. [34]	Pseudokirchneriella subcapitata	Mixed in rotator (24 h)	–	<500 mg l^{-1} (starting concn.)	72 h	Growth inhibition	IC$_{25}$ was >100 mg l^{-1}
Seki et al. [32]	Pseudokirchneriella subcapitata	Sugar candy method	Average: 174 nm Range: 20–2000 nm [DLS measurement]	0.0187–2.27 mg l^{-1} [HPLC]	72 h	Growth inhibition	EC$_{50}$ was >2.27 mg l^{-1}. The growth rate was significantly inhibited at 0.0551 mg l^{-1}. Thus, NOEC was estimated to be 0.0178 mg l^{-1}

Table 10.5 Toxicity tests of fullerene C_{60} particles on bacteria and microbes.

Reference	Species	Preparation of suspension	Particle size	Concentration
Lyon et al. [40]	Escherichia coli Bacillus subtilis	THF method	Aqueous suspension: 3–78 nm Phosphate-buffered saline: 35–11 384 nm Luria broth: 9–1052 nm	0.04–4.8 mg l^{-1}
Fortner et al. [14]	Escherichia coli DH5R Bacillus subtilis CB315	THF method (concentrated by ultrafiltration)	Average: 25–500 nm (depends on the pH, mixing ratio, and ionic strength)	0.04–4 mg l^{-1}
Lyon et al. [41]	Bacillus subtilis CB310	THF method Water-stirred method Sonication method PVP method	Mean diameter: THF: 39.1, 75.6, 97.4 nm Sonication: <2 nm Stirred: <2, 142.3 nm PVP: <2 nm [DLS measurement]	[Spectrophotometer]
Tang et al. [42]	Escherichia coli W3110 Shewanella oneidensis MR-1	Dried C_{60} was added directly to the medium	–	0–20 mg l^{-1}
Tong et al. [43]	Microbials in surface soil	THF method Granular form	–	THF-nC_{60}: 1 mg kg^{-1} soil Granular: 1000 mg kg^{-1} soil
Fang et al. [44]	Pseudomonas putida F1 Bacillus subtilis CB310	THF method	Average: 95 nm Range: 50–200 nm [DLS measurement]	P. putida: 0.01, 0.5 mg l^{-1} B. subtilis: 0.01, 0.75 mg l^{-1}
Blaise et al. [34]	Vibrio fischeri (Microtox®) 11 microbial species in MARA assay	Mixed in rotator (24 h)-	–	<500 mg l^{-1} (starting concn.)
Johansen et al. [45]	Microbials in clay loam	Water-stirred method (2 months)	50 nm to µm-size [DLS measurement]	0, 5, 25, 50 mg kg^{-1} soil
Lyon et al. [46]	Escherichia coli Bacillus subtilis	THF method	–	10 mg l^{-1}

Exposure duration	Endpoint	Results
3 h	Association with bacteria	THF-nC_{60} associated with both E. coli and B. subtilis, albeit more strongly with the former. Antimicrobial properties were displayed for E. coli and B. subtilis at 0.5 to 1.0 mg l^{-1} and 1.5 to 3.0 mg l^{-1}, respectively.
Approx. 40 h (administered early in the exponential growth phase)	Growth Aerobic respiration	THF-nC_{60} inhibit microbes at relatively low concentrations, indicated by lack of growth (0.4 mg l^{-1}) and decreased aerobic respiration rates (4 mg l^{-1})
	Growth inhibition	THF-nC_{60} had a MIC of 0.09 ± 0.01 mg l^{-1}, sonication-nC_{60} had a MIC of 0.7 ± 0.3 mg l^{-1}, water-stirred-nC_{60} had a MIC of 0.5 ± 0.13 mg l^{-1}, and PVP–C_{60} had a MIC of 0.95 ± 0.35 mg l^{-1}
40 h	Growth Substrate uptake	No effects were observed after the exposure to C_{60} at 20 mg l^{-1}
180 days	Respiration (basal and glucose-induced) Community size Community structure	Both C_{60} and nC_{60} had little impact on the structure and function of the soil microbial community and microbial processes
14 h	Membrane lipid composition Phase behavior	The MIC of THF-nC_{60} for B. subtilis was between 0.5 and 0.75 mg l^{-1}, whereas that for P. putida was between 0.25 and 0.5 mg l^{-1}. B. subtilis significantly increased the levels of iso- and anteiso-branched fatty acid at 0.01 mg l^{-1}
20 min 18 h	Acute sublethal light inhibition Growth inhibition.	20-min IC25 was >100 mg l^{-1}. 18-h MTC was >100 mg l^{-1}
14 days	Total respiration, microbial biomass, number, diversity of bacteria	Respiration and microbial biomass were unaffected by the C_{60}. The number of fast-growing bacteria was decreased by three- to four-fold. Protozoans seemed not to be very sensitive to C_{60}
15 min	Antibacterial activity (ROS production) (Lipid peroxidation) (Protein oxidation)	Neither ROS production nor ROS-mediated damage is found in THF-nC_{60}-exposed bacteria

Table 10.5 Continued

Reference	Species	Preparation of suspension	Particle size	Concentration
Lyon and Alvarez [47]	Escherichia coli K12 Bacillus subtilis 168	THF method	–	10 mg l^{-1}
Nyberg et al. [48]	Anaerobic digester sludge	–	–	MeOH/EtOH solution: 0.321 mg kg^{-1} biomass Suspension: 8.6 mg kg^{-1} biomass Toluene solution: 30 000 mg kg^{-1} biomass o-Xylene solution: 50 000 mg kg^{-1} biomass
Velzeboer et al. [49]	Vibrio fischeri (Microtox®)	THF method	–	1 mg l^{-1}
Gao et al. [35]	Escherichia coli bacteria (MetPLATE bioassay)	THF method	Purchased particle size: Average: 35.8 nm Particle size in suspension: Average: >100 nm [DLS measurement]	DI water: 0.83 mg l^{-1} River water: 0.038–3.09 mg l^{-1}
Kang et al. [50]	Escherichia coli Pseudomonas aeruginosa Bacillus subtilis Staphylococcus epidermis	Water-stirred method (5 weeks)	Average: 360 ± 210 nm Range: tens of nm to several μm [SEM and TEM observations]	–

MIC, minimum inhibitory concentration.

gas from the manufacturing facility was properly controlled, with measures such as high-efficiency particulate air (HEPA) filtration, emissions from similarly regulated manufacturing facilities can be considered to be low. Since water is not used in any process (organic solvents are used for extraction), no C_{60} should be present in the discharged water, except in the event of an accident. Similarly, emissions during transportation can be considered nonexistent, except in the case of accidents, because transportation is conducted in sealed cases.

Exposure duration	Endpoint	Results
1 h	Cell health and function. Oxidative damage in cellular component	THF-nC_{60} does not puncture the cell nor release ROS or toxic products, but instead exerts toxicity as a particle via a chemical interaction upon direct contact
150 days	Activity. Community structure	C_{60} fullerenes have no significant effect on the anaerobic community over an exposure period of a few months. Substantial community shifts were not observed due to treatment with C_{60}
15-min	Median effective concentration	Measurable effects were not induced by C_{60} at $1\,mg\,l^{-1}$. Thus, EC_{50} was $>1\,mg\,l^{-1}$
1.5 h	Bacterial-enzyme toxicity	All C_{60}-based aqueous suspensions gave no positive response to this test. Thus, LC_{50} was $>3.09\,mg\,l^{-1}$
1 h	Cytotoxicity	Compared to the other monocultures, *B. subtilis* experienced little cytotoxicity. Aq-nC_{60} induced toxicity in monoculture experiments at a significantly higher rate than observed in environmental samples

Metal and resin products with fullerene additives are used for tennis and badminton racquets, in bowling balls, golf clubs, and eyeglass frames. Although HEPA filters are currently rarely installed in these factories, the concentration in the working environment is quite low, even during the mixing of fullerene into the metal or resin. Thus, emission into the atmospheric and aquatic environments is considered to be low. During the processing (cutting, polishing) stages, as a result of attrition or damage, or during disposal by shredding, resin and metal particles containing fullerene may possibly be emitted. Emission into water under

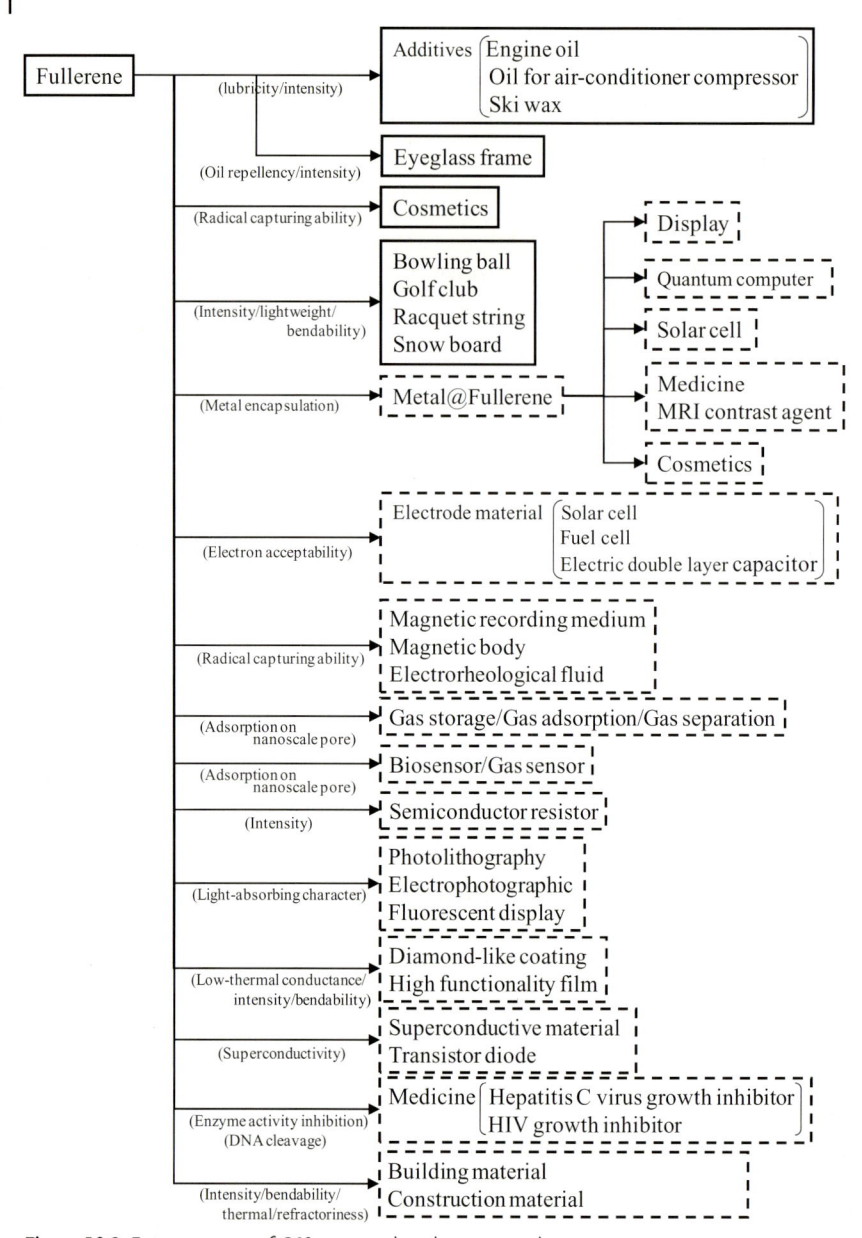

Figure 10.2 Future usages of C60 assumed at the present day.

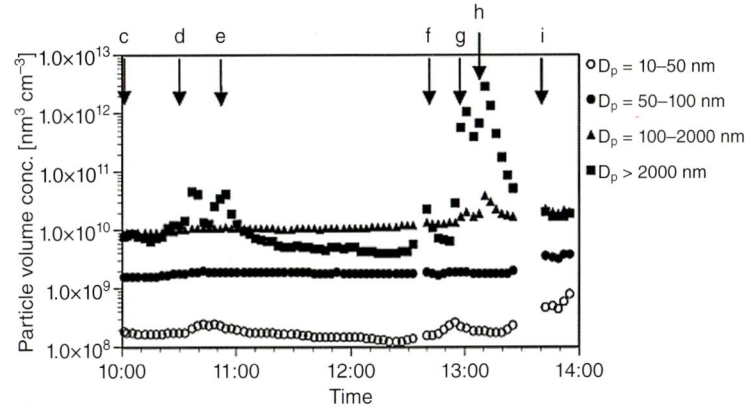

Figure 10.3 Time series of calculated particle volume concentrations in the C60 manufacturing factory. The letters at the top of the figure indicate the following events: c = start measuring indoor air during non-work period; d = start bagging operation; e = end bagging operation and all workers leave the room; f = start using vacuum cleaner to remove dust from floor; g = start moderate agitation; h = start extreme agitation; i = start measuring outdoor air. D_p, particle diameter. Reproduced with permission from Ref. [55].; © 2008.

these circumstances is not a risk, because fullerene is insoluble when fixed in resin or metal.

Fullerene used in solar cells and fuel cells is rarely emitted, even into the working environment, because the processes are sealed and/or wet. Fullerene may be emitted into the working environment during recovery and cleaning, although only at concentrations considered to be low. Thus, environmental emissions from this type of application are likely to be minimal. During transportation and use, no fullerene can be emitted because it is fixed in the electrode material. However, at the time of disposal by crushing, a small amount of fullerene might be emitted to the atmosphere with resin or metal.

Similar to fullerene additives used with resins and metals, fullerenes used for engine oils and waxes are rarely emitted to the atmosphere or aquatic environments during their processing and transportation. During use, small amounts of fullerenes present in engine oils might be emitted to the atmosphere through gaps in the engine, while small amounts of fullerenes included in ski waxes may be retained on the snow and transported into the soil and/or water.

In the case of cosmetics, very little fullerene is emitted to the atmosphere during manufacturing and transportation, although small amounts of soluble and dispersed fullerenes may be emitted into the water during manufacture, depending on the treatment facility. During use as cosmetics, small amounts of fullerenes may be emitted into the atmosphere from the surface of skin after drying, while following their use, most fullerenes in cosmetics will be washed away and transported to sewage treatment facilities. During sewage treatment, although most fullerene particles are likely to form deposits, soluble fullerenes may be emitted into aquatic environments. When used in medicines and drugs, the emission behavior of fullerenes will depend on whether they are in dry or liquid forms, but ultimately will be similar to that observed in the manufacture of cosmetics. Following their use, most fullerenes included in medicines would be eliminated by the body, transported to a sewage treatment facility, and ultimately – if soluble – would reach an aquatic environment.

The possibilities for fullerene emission from any type of product over the course of its life cycle into the atmosphere and environmental water sources, with the exception of possible accidents, are listed in Table 10.6.

In addition, C_{60} and C_{70} have been identified, using time-of-flight mass spectrometry (TOF-MS), in the particulates emitted from coal-burning power plants [56]. Both, fullerenes and CNTs were observed in nanoparticles emitted from gas–air and propane–air kitchen stove-top flame exhausts, in natural gas–air water heater roof-top exhausts, and other common fuel–gas combustion sources, using transmission electron microscopy (TEM) [57].

10.5
The Environmental Fate of C_{60}

The environmental fate of C_{60} is shown schematically in Figure 10.4. Fullerenes emitted into aquatic environments may either be transported (and thus diluted) some distance downstream, or deposited at the bottom of the body of water. C_{60} particles can easily be deposited in environmental water sources (especially seawater), because the C_{60} particles readily form aggregates in ion-enriched water where the electrolyte concentration is at or above 0.01 M NaCl [58] or 0.05 M NaCl [14]. According to Fortner et al. [14], these aggregates would remain stable for 15 weeks in water containing ≤0.01 M NaCl. Moreover, the particle size of C_{60} was shown to increase depending on the electrolyte concentration [14, 58, 59]. Although the particle size of C_{60} nanoparticles was increased at lower pH values, it did not differ in the range of pH 5 to 9 [14], which indicated that the pH of environmental water should not affect the aggregation and deposition of C_{60} nanoparticles. However, a re-dispersion might be promoted in areas of rushing water, because the C_{60} particles can be stably dispersed by long-term stirring [23, 24]. In natural water bodies such as rivers and lakes, natural organic matter (NOM) may also promote the disaggregation of nC_{60}. For example, Xie et al. [60] reported that the particle size of toluene-nC_{60} was decreased in the presence of NOM in a concentration-dependent manner, whereas that of THF-nC_{60} was not affected by NOM. At typical concentrations of NOM found in natural water sources, a 10 mg l^{-1} stable C_{60} suspension can be prepared by stirring within 10 days [59]. Based on these findings, the retention ratio of deposited nC_{60} on the bottom sediment may vary depending on conditions. Typically, the deposition of nC_{60} in porous media was increased as the ionic strength increased [58, 61], and was also shown to vary with the type of media [61].

10.5 The Environmental Fate of C_{60}

Table 10.6 Emission possibilities during the life cycle of C_{60}.

Product(s)	Pathway	Manufacture	Transportation	Consumption	Disposal
C_{60}	Exhaust air	Almost nothing	Nothing	–	–
	Discharged water	Nothing	Nothing	–	–
Sporting goods/ Resin product	Exhaust air	Possible (little) during processing[a]	Nothing	Possible (little) by friction[a]	Possible (little) during crushing[a]
	Discharged water	Possible (little) during washing[a]	Nothing	Nothing	Nothing
Solar battery/ Fuel battery	Exhaust air	Nothing	Nothing	Nothing	Possible (little) during crushing[a]
	Discharged water	Nothing	Nothing	Nothing	Nothing
Wax/Engine oil	Exhaust air	Nothing	Nothing	Little	Nothing
	Discharged water	Nothing	Nothing	Little	Nothing
Cosmetics	Exhaust air	Nothing	Nothing	Little	–
	Discharged water	Little[b]	Nothing	Almost during facial cleansing	–
Drugs/Medicines	Exhaust air	Nothing	Nothing	Nothing	–
	Discharged water	Little[b]	Nothing	Almost via excrement	–

a) Emission could occur mixed with resin or metal particles.
b) Although no C_{60} is emitted in waste solution at present, a little C_{60} could be emitted in waste solution in future.

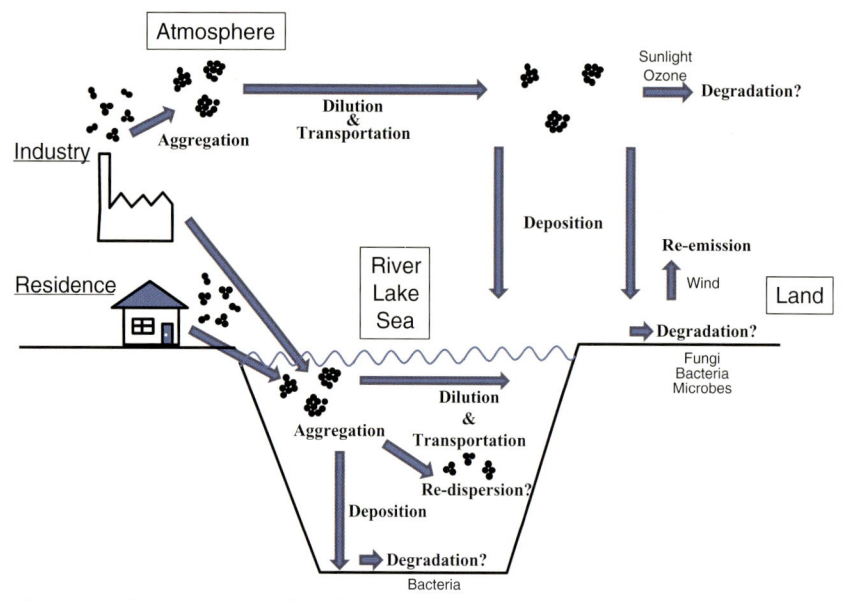

Figure 10.4 The environmental fate of C60.

Small aquatic organisms such as *Daphnia magna* intake C_{60} particles [38, 39] and deposit them at the bottom of water bodies. Thus, the concentrations of C_{60} particles in higher-level predators that feed on these small aquatic organisms and benthic organisms may be increased. Jafvert and Kulkarni [62] estimated the bioconcentration factors for C_{60} as 4.51, 5.37, and 5.71 in cod, earthworms, and salmon, respectively, on the basis of the experimentally measured log K_{OW}, 6.67. However, as these values do not apply to aggregated C_{60} particles but only to C_{60} solution, they cannot be used to estimate the environmental fate of C_{60} particles [63]. Although the emission of fullerenes from manufacturing facilities into atmospheric environments has been estimated as being extremely low, the emitted particles (which may be mainly micron-sized) can be transported some distance, with according dilution, by the wind. Moreover, during such transport the particles can also be deposited in water and on the land. Fullerenes emitted or deposited onto the soil can, in turn, be partially re-rolled to the atmosphere or transported into the ground or surface water by rainwater.

In any case, the rate of degradation of C_{60} particles is considered to be much slower than that of their deposition. The transformation of aqueous nC_{60} with ozone or sunlight has been characterized in several studies. The C_{60} concentrations and particle size of THF-nC_{60} and sonicated-nC_{60} were decreased upon exposure to sunlight, lamp light (300–400 nm), or ultraviolet light (254 nm) irradiation, but remained unchanged under conditions of dark, deoxygenation, or N_2 saturation [64, 65]. After irradiation, the new products had different absorption characteristics from nC_{60}, and also demonstrated reduced antibacterial effects on *E. coli* compared to nC_{60} [65]. These results indicated that nC_{60} is transformed by O_2 under sunlight, and that such transformation was not affected by the presence of fulvic acid, by the pH, or by the preparation method used [64]. THF-nC_{60}, when reacted with ozone, resulted in an aggregate dissolution concurrent with the formation of oxidized fullerene with hydroxyl and hemiketal functionalities ($C_{60}O_x(OH)_y$, where $x + y \leq 29$) [66]. One study which reported that $C_{60}(OH)_{19-27}$ was degraded to CO_2 in 32 weeks by two white-rot basidiomycetes (*Phlebia tremellosa* and *Trametes versicolor*) [67] also suggested that C_{60} could be degraded by microorganisms in the soil.

Clearly, future research into the environmental fate of C_{60} is required to supplement the limited data that are currently available.

10.6
Fullerenes in the Environment

Fullerenes are generated by the combustion of carbon or high-energy-state carbon, achieved by means such as lightning, the collision of a meteoroid, or a massive fire. Details of fullerenes identified in the environment are provided as follows.

C_{60} has been identified at levels of 10–20 ppt in rock from the Permo-Triassic boundary (250 million years ago) in Japan [68, 69], and at 0.058–5.4 ppb in clay from the Cretaceous–Tertiary boundary (65 million years ago) in New Zealand [70]. Both, C_{60} and C_{70} have been detected at the Cretaceous–Tertiary boundary worldwide, including the United States, Turkmenistan, Denmark, and Spain [5, 70–72]. These fullerenes are thought to have been generated by a massive wildfire, under anoxic conditions [68, 69].

Both, C_{60} and C_{70} were detected in shock-produced breccias (1–10 ppm) [72–74], shungite [75–77], fulgurite [78], and bitumens from pillow lavas (0.2–0.3 ppm) [79, 80]. These fullerenes have also been identified in coal [81, 82], dinosaur egg fossil [83], and Chinese ink sticks (C_{60}: 0.009–1.4 ppm, C_{70}: 0.030–1.29 ppm) [84, 85]. C_{60} and C_{70} were also detected in the Allende meteorite (C_{60}: 100 ppm) [73, 86], Tagish Lake meteorite [87], and Murchison meteorite [87]. The content rates of C_{60}, C_{70}, and higher-order fullerene differed among meteorites [87].

10.7
Conclusion

The toxicity of fullerenes prepared without organic solvents was undetectable or quite low in many experiments. However, an insufficient number of investigations has been conducted into the chronic effects of fullerene on the environment. Although little is known regarding the environmental fate of fullerenes, the emission of large amounts of fullerenes into the atmosphere or aquatic environments is unlikely, except in the event of a large accident or fire. Therefore, as far as we are aware, the risk of C_{60} fullerene on the environment can be assumed to be low at present and for the near future.

References

1 Kroto, H.W., Heath, J.R., O'Brien, S.C., Curl, R.F. and Smalley, R.E. (1985) C_{60}–buckminsterfullerene. *Nature*, **318** (6042), 162–3.

2 Ruoff, R.S., Malhotra, R., Huestis, D.L., Tse, D.S. and Lorents, D.C. (1993) Anomalous solubility behavior of C_{60}. *Nature*, **362**, 140–1.

3 Sivaraman, N., Dhamodaran, R., Kaliappan, I., Srinivasan, T.G., Vasudeva Rao, P.R. and Mathews, C.K. (1992) Solubility of C_{60} in organic solvents. *Journal of Organic Chemistry*, **57**, 6077–9.

4 Heymann, D. (1996) Solubility of C_{60} in alcohols and alkanes. *Carbon*, **34**, 627–31.

5 Heymann, D., Korochantsev, A., Nazarov, M.A. and Smit, J. (1996) Search for fullerenes C_{60} and C_{70} in Cretaceous-Tertiary boundary sediments from Turkmenistan, Kazakhstan, Georgia, Austria, and Denmark. *Cretaceous Research*, **17** (3), 367–80.

6 Ruoff, R.S., Tse, D.S., Malhotra, R. and Lorents, D.C. (1993) Solubility of C_{60} in a variety of solvents. *Journal of Physical Chemistry*, **97**, 3379–83.

7 Scrivens, W.A. and Tour, J.M. (1993) Potent solvents for C_{60} and their utility for the rapid acquisition of C^{13} NMR data for fullerenes. *Journal of the Chemical Society, Chemical Communications*, **15**, 1207–9.

8 Sivaraman, N., Dhamodaran, R., Kaliappan, I., Srinivasan, T.G., Vasudeva Rao, P.R. and Mathews, C.K. (1994) Solubility of C_{60} and C_{70} in organic solvents, in *Recent Advances in the Chemistry and Physics of Fullerenes and Related Materials* (eds K.M. Kadish and R.S. Ruoff), The Electrochemical Society of USA, Pennington, NJ, pp. 156–65.

9 Manolova, N., Rashkov, I., Beguin, F. and Vandamme, H. (1993) Amphiphilic derivatives of fullerenes formed by polymer modification. *Journal of the Chemical Society, Chemical Communications*, **23**, 1725–7.

10 Tokuyama, H., Yamago, S., Nakamura, E., Shiraki, T. and Sugiura, Y. (1993) Photoinduced biochemical-activity of fullerene carboxylic-acid. *Journal of the American Chemical Society*, **115**, 7918–19.

11 Andersson, T., Nilsson, K., Sundahl, M., Westman, G. and Wennerstrom, O. (1992) C_{60} embedded in gamma-cyclodextrin–a water-soluble fullerene. *Journal of the Chemical Society, Chemical Communications*, **8**, 604–6.

12 Priyadarsini, K.I., Mohan, H., Tyagi, A.K. and Mittal, J.P. (1994) Inclusion complex of gamma-cyclodextrin-C_{60} formation, characterization, and photophysical properties in aqueous solutions. *Journal of Physical Chemistry*, **98** (17), 4756–9.

13 Alargova, R.G., Deguchi, S. and Tsujii, K. (2001) Stable colloidal dispersions of fullerenes in polar organic solvents. *Journal of the American Chemical Society*, **123**, 10460–7.

14 Fortner, J.D., Lyon, D.Y., Sayes, C.M., Boyd, A.M., Falkner, J.C., Hotze, E.M., Alemany, L.B., Tao, Y.J., Guo, W., Ausman, K.D., Colvin, V.L. and Hughes, J.B. (2005) C_{60} in water: nanocrystal formation and microbial response. *Environmental Science and Technology*, **39**, 4307–16.

15 Deguchi, S., Alargova, R.G. and Tsujii, K. (2001) Stable dispersions of fullerenes, C_{60} and C_{70}, in water: preparation and characterization. *Langmuir*, **17**, 6013–17.

16 Dhawan, A., Taurozzi, J.S., Pandey, A.K., Shan, W.Q., Miller, S.M., Hashsham, S.A. and Tarabara, V.V. (2006) Stable colloidal dispersions of C_{60} fullerenes in water: evidence for genotoxicity. *Environmental Science and Technology*, **40**, 7394–401.

17 Sera, N., Tokiwa, H. and Miyata, N. (1996) Mutagenicity of the fullerene C_{60}-generated singlet oxygen dependent formation of lipid peroxides. *Carcinogenesis*, **17**, 2163–9.

18 Yamakoshi, Y., Yagami, T., Fukuhara, K., Sueyoshi, S. and Miyata, N. (1994) Solubilization of fullerenes into water with polyvinylpyrrolidone applicable to biological tests. *Journal of the Chemical Society, Chemical Communications*, **4**, 517–18.

19 Gharbi, N., Pressac, M., Hadchouel, M., Szwarc, H., Wilson, S.R. and Moussa, F. (2005) [60]Fullerene is a powerful antioxidant *in vivo* with no acute or

subacute toxicity. *Nano Letters*, **5**, 2578–85.

20 Endoh, S., Maru, J., Uchida, K., Yamamoto, K. and Nakanishi, J. (2009) Preparing samples for fullerene C_{60} hazard tests: stable dispersion of fullerene crystals in water using a bead mill. *Advanced Powder Technology*, **20**, 567–75.

21 Shinohara, N., Matsumoto, T., Gamo, M., Miyauchi, A., Endo, S., Yonezawa, Y. and Nakanishi, J. (2009) Is lipid peroxidation induced by the aqueous suspension of fullerene C_{60} nanoparticles in the brains of *Cyprinus carpio*? *Environmental Science and Technology*, **43**, 948–53.

22 Seki, M., Fujishima, S., Gondo, Y., Inoue, Y., Nozaka, T., Suemura, K. and Takatsuki, M. (2008) Acute toxicity of fullerene C_{60} in aquatic organisms. *Kankyo Kagakkaishi*, **21** (1), 53–62.

23 Mchedlov-Petrossyan, N.O., Klochkov, V.K. and Andrievsky, G.V. (1997) Colloidal dispersions of fullerene C_{60} in water: some properties and regularities of coagulation by electrolytes. *Journal of the Chemical Society, Faraday Transactions*, **93**, 4343–6.

24 Brant, J.A., Labille, J., Bottero, J.Y. and Wiesner, M.R. (2006) Characterizing the impact of preparation method on fullerene cluster structure and chemistry. *Langmuir*, **22**, 3878–85.

25 Usenko, C.Y., Harper, S.L. and Tanguay, R.L. (2007) *In vivo* evaluation of carbon fullerene toxicity using embryonic zebrafish. *Carbon*, **45** (9), 1891–8.

26 Oberdörster, E. (2004) Manufactured nanomaterials (fullerenes, C_{60}) induce oxidative stress in the brain of juvenile largemouth bass. *Environmental Health Perspectives*, **112**, 1058–62.

27 Sayes, C.M., Gobin, A.M., Ausman, K.D., Mendez, J., West, J.L. and Colvin, V.L. (2005) Nano-C_{60} cytotoxicity is due to lipid peroxidation. *Biomaterials*, **26**, 7587–95.

28 Lovern, S.B. and Klaper, R. (2006) Daphnia magna mortality when exposed to titanium dioxide and fullerene (C_{60}) nanoparticles. *Environmental Toxicology and Chemistry*, **25**, 1132–7.

29 Zhu, S., Oberdörster, E. and Haasch, M.L. (2006) Toxicity of an engineered nanoparticle (fullerene, C_{60}) in two aquatic species, *Daphnia* and fathead minnow. *Marine Environmental Research*, **62**, S5–S9.

30 Henry, T.B., Menn, F.M., Fleming, J.T., Wilgus, J., Compton, R.N. and Sayler, G.S. (2007) Attributing effects of aqueous C_{60} nano-aggregates to tetrahydrofuran decomposition products in larval zebrafish by assessment of gene expression. *Environmental Health Perspectives*, **115**, 1059–65.

31 Kovochich, M., Espinasse, B., Auffan, M., Hotze, E.M., Wessel, L., Xia, T., Nel, A.E. and Wiesner, M.R. (2009) Comparative toxicity of C_{60} aggregates toward mammalian cells: role of tetrahydrofuran (THF) decomposition. *Environmental Science and Technology*, **43** (16), 6378–84.

32 Oberdörster, E., Zhu, S.Q., Blickley, T.M., McClellan-Green, P. and Haasch, M.L. (2006) Ecotoxicology of carbon-based engineered nanoparticles: effects of fullerene (C_{60}) on aquatic organisms. *Carbon*, **44** (6), 1112–20.

33 Zhu, X., Zhu, L., Li, Y., Duan, Z., Chen, W. and Alvarez, P.I.J. (2007) Development toxicity in zebrafish embryos after exposure to manufactured nanomaterials: buckminsterfullerene aggregates (nC_{60}) and fullerol. *Environmental Toxicology and Chemicals*, **25** (5), 976–9.

34 Blaise, C., Gagne, F., Ferard, J.F. and Eullaffroy, P. (2008) Ecotoxicity of selected nano-materials to aquatic organisms. *Environmental Toxicology*, **23** (5), 591–8.

35 Gao, J., Youn, S., Hovsepyan, A., Llaneza, V.L., Wang, Y., Bitton, G. and Bonzongo, J.C.J. (2009) Dispersion and toxicity of selected manufactured nanomaterials in natural river water samples: effects of water chemical composition. *Environmental Science and Technology*, **43** (9), 3322–8.

36 Klaper, R., Crago, J., Barr, J., Arndt, D., Setyowati, K. and Chen, J. (2009) Toxicity biomarker expression in daphnids exposed to manufactured nanoparticles: changes in toxicity with functionalization. *Environmental Pollution*, **157** (4), 1152–6.

37 Lovern, S.B., Strickler, J.R. and Klaper, R. (2007) Behavioral and physiological change in *Daphnia magna* when exposed to nanoparticle suspensions (Titanium dioxide, Nano-C_{60} and $C_{60}H_xC_{70}H_x$).

Environmental Science and Technology, **41**, 4465–70.

38 Zhu, X.S., Zhu, L., Chen, Y.S. and Tian, S.Y. (2009) Acute toxicities of six manufactured nanomaterial suspensions to *Daphnia magna*. Springer, *Journal of Nanoparticle Research*, **11** (1), 67–75.

39 Baun, A., Sorensen, S.N., Rasmussen, R.F., Hartmann, N.B. and Koch, C.B. (2008) Toxicity and bioaccumulation of xenobiotic organic compounds in the presence of aqueous suspensions of aggregates of nano-C_{60}. *Aquatic Toxicology*, **86** (3), 379–87.

40 Lyon, D.Y., Fortner, J.D., Sayes, C.M., Colvin, V.L. and Hughes, J.B. (2005) Bacterial cell association and antimicrobial activity of a C60 water suspension. *Environmental Toxicology and Chemistry*, **24** (11), 2757–62.

41 Lyon, D.Y., Adams, L.K., Falkner, J.C. and Alvarez, P.J.J. (2006) Antibacterial activity of fullerene water suspensions: effects of preparation method and particle size. *Environmental Science and Technology*, **40** (14), 4360–6.

42 Tang, Y.J.J., Ashcroft, J.M., Chen, D., Min, G.W., Kim, C.H., Murkhejee, B., Larabell, C., Keasling, J.D. and Chen, F.Q.F. (2007) Charge-associated effects of fullerene derivatives on microbial structural integrity and central metabolism. *Nano Letters*, **7** (3), 754–60.

43 Tong, Z.H., Bischoff, M., Nies, L., Applegate, B. and Turco, R.F. (2007) Impact of fullerene (C_{60}) on a soil microbial community. *Environmental Science and Technology.*, **41** (8), 2985–91.

44 Fang, J., Lyon, D.Y., Wiesner, M.R., Dong, J. and Alvarez, P.J.J. (2007) Effect of a fullerene water suspension on bacterial phospholipids and membrane phase behavior. *Environmental Science & Technology.*, **41** (7), 2636–42.

45 Johansen, A., Pedersen, A.L., Jensen, K.A., Karlson, U., Hansen, B.M., Scott-Fordsmand, J.J. and Winding, A. (2008) Effects of C60 fullerene nanoparticles on soil bacteria and protozoans. *Environmental Toxicology and Chemistry*, **27** (9), 1895–903.

46 Lyon, D.Y., Brunet, L., Hinkal, G.W., Wiesner, M.R. and Alvarez, P.J.J. (2008) Antibacterial activity of fullerene water suspensions (nC_{60}) is not due to ROS-mediated damage. *Nano Letters*, **8**, 1539–43.

47 Lyon, D.Y. and Alvarez, P.J.J. (2008) Fullerene water suspension (nC60) exerts antibacterial effects via ROS-independent protein oxidation. *Environmental Science and Technology.*, **42** (21), 8127–32.

48 Nyberg, L., Turco, R.F. and Nies, L. (2008) Assessing the impact of nanomaterials on anaerobic microbial communities. *Environmental Science and Technology.*, **42** (6), 1938–43.

49 Velzeboer, I., Hendriks, A.J., Ragas, A.M.J. and Van de Meent, D. (2008) Aquatic ecotoxicity tests of some nanomaterials. *Environmental Toxicology and Chemistry*, **27** (9), 1942–7.

50 Kang, S., Mauter, M.S. and Elimelech, M. (2009) Microbial cytotoxicity of carbon-based nanomaterials: Implications for river water and waste water effluent. *Environmental Science and Technology*, **43** (7), 2648–53.

51 Liu, X.Y., Vinson, D., Abt, D., Hurt, R.H. and Rand, D.M. (2009) Differential toxicity of carbon nanomaterials in *Drosophila*: larval dietary uptake is benign, but adult exposure causes locomotor impairment and mortality. *Environmental Science and Technology.*, **43** (16), 6357–63.

52 Shinohara, N., Gamo, M. and Nakanishi, J. (2009) Risk assessment of manufactured nanomaterials -fullerene (C_{60})- interim report issued on October 16, 2009. Executive Summary. http://www.aist-riss.jp/main/modules/product/nano_rad.html?ml_lang=en (accessed 1st February 2010).

53 Shinohara, N., Ogura, I. and Gamo, M. (2009) Exposure to fullerene C_{60} particles in C_{60} handling pilot plant. *4th International Conference on Nanotechnology & Occupational and Environmental Health*, Helsinki, Finland.

54 Yeganeh, B., Kull, C.M., Hull, M.S. and Marr, L.C. (2008) Characterization of airborne particles during production of carbonaceous nanomaterials. *Environmental Science and Technology*, **42** (12), 4600–6.

55 Fujitani, Y., Kobayashi, T., Arashidani, K., Kunugita, N. and Suemura, K. (2008) Measurement of the physical properties

of aerosols in a fullerene factory for inhalation exposure assessment. Taylor & Francis, *Journal of Occupational and Environmental Hygiene*, **5** (6), 380–9.

56. Utsunomiya, S., Jensen, K.A., Keeler, G.J. and Ewing, R.C. (2002) Uraninite and fullerene in atmospheric particulates. *Environmental Science and Technology*, **36** (23), 4943–7.

57. Murr, L.E. and Soto, K.F. (2005) A TEM study of soot, carbon nanotubes, and related fullerene nanopolyhedra in common fuel-gas combustion sources. *Materials Characterization*, **55** (1), 50–65.

58. Brant, J., Lecoanet, H. and Wiesner, M.R. (2005) Aggregation and deposition characteristics of fullerene nanoparticles in aqueous systems. *Journal of Nanoparticle Research*, **7**, 545–53.

59. Li, Q.L., Xie, B., Hwang, Y.S. and Xu, Y.J. (2009) Kinetics of C_{60} fullerene dispersion in water enhanced by natural organic matter and sunlight. *Environmental Science and Technology.*, **43** (10), 3574–9.

60. Xie, B., Xu, Z., Guo, W. and Li, Q. (2008) Impact of natural organic matter on the physicochemical properties of aqueous C_{60} nanoparticles. *Environmental Science and Technology*, **42**, 2853–9.

61. Wang, Y.G., Li, Y.S., Fortner, J.D., Hughes, J.B., Abriola, L.M. and Pennell, K.D. (2008) Transport and retention of nanoscale C_{60} aggregates in water-saturated porous media. *Environmental Science and Technology*, **42** (10), 3588–94.

62. Jafvert, C.T. and Kulkarni, P.P. (2008) Buckminsterfullerene's (C_{60}) octanol-water partition coefficient (K_{ow}) and aqueous solubility. *Environmental Science and Technology*, **42** (16), 5945–50.

63. Isaacson, C.W., Kleber, M. and Field, J.A. (2009) Quantitative analysis of fullerene nanomaterials in environmental systems: a critical review. *Environmental Science and Technology.*, **43** (17), 6463–74.

64. Hou, W.C. and Jafvert, C.T. (2009) Photochemical transformation of aqueous C_{60} clusters in sunlight. *Environmental Science and Technology*, **43**, 362–7.

65. Lee, J., Cho, M., Fortner, J.D., Hughes, J.B. and Kim, J.H. (2009) Transformation of aggregated C_{60} in the aqueous phase by UV irradiation. *Environmental Science and Technology*, **43**, 4878–83.

66. Fortner, J.D., Kim, D.I., Boyd, A.M., Falkner, J.C., Moran, S., Colvin, V.L., Hughes, J.B. and Kim, J.H. (2007) Reaction of water-stable C_{60} aggregates with ozone. *Environmental Science and Technology*, **41**, 7497–502.

67. Schreiner, K.M., Filley, T.R., Blanchette, R.A., Bowen, B.B., Bolskar, R.D., Hockaday, W.C., Masiello, C.A. and Raebiger, J.W. (2009) White-rot Basidiomycete-mediated decomposition of C_{60} fullerol. *Environmental Science and Technology*, **43**, 3162–8.

68. Chijiwa, T., Arai, T., Sugai, T., Shinohara, H., Kumazawa, M., Takano, M. and Kawakami, S. (1999) Fullerenes found in the Permo-Triassic mass extinction period. *Geophysical Research Letters*, **26** (6), 767–70.

69. Becker, L., Poreda, R.J., Hunt, A.G., Bunch, T.E. and Rampino, M. (2001) Impact event at the Permian-Triassic boundary: evidence from extraterrestrial noble gases in fullerenes. *Science*, **291**, 1530–3.

70. Heymann, D., Chibante, L.P.F., Brooks, R.R., Wolbach, W.S. and Smalley, E. (1994) Fullerenes in the cretaceous-tertiary boundary layer. *Science*, **265**, 645–7.

71. Heymann, D., Wolbach, W.S., Chibante, L.P.F., Brooks, R.R. and Smalley, E. (1994) Search for extractable fullerenes in clays from the Cretaceous/Tertiary boundary of the Woodside Creek and Flaxbourne River site, New Zealand. *Geochimica et Cosmochimica Acta*, **58** (16), 3531–4.

72. Becker, L., Poreda, R.J. and Bada, J.L. (1996) Extraterrestrial helium trapped in fullerenes in the Sudbury Impact Structure. *Science*, **272**, 249–52.

73. Becker, L., Bada, J.L., Winans, R.E., Hunt, J.E., Bunch, T.E. and French, B.M. (1994) Fullerenes in the 1.85-billion-year-old Sudbury Impact Structure. *Science*, **265**, 642–4.

74. Becker, L., Poreda, R.J. and Bunch, T.E. (2000) Fullerenes: an extraterrestrial carbon carrier phase for noble gases. *Proceedings of the National Academy of Sciences of the United States of America*, **97**, 2979–83.

75. Buseck, P.R., Tsipursky, S.J. and Hettich, R. (1992) Fullerenes from the geological

environment. *Science*, **257**, 215–17.
76 Buseck, P.R., Galdobina, L.P., Kovalevski, V.V., Rozhkova, N.N., Valley, J.W. and Zaidenberg, A.Z. (1997) Shungites: the C-rich rocks of Karelia, Russia. *Canadian Mineralogist*, **35**, 1363–78.
77 Parthasarathy, G., Srinivasan, R., Vairamani, M., Ravikumar, K. and Kunwar, A.C. (1998) Occurrence of natural fullerenes in low grade metamorphosed Proterozoic shungite from Karelia, Russia. *Geochimica et Cosmochimica Acta*, **62** (21–22), 3541–4.
78 Daly, T.K., Buseck, P.R., Williams, P. and Lewis, C.F. (1993) Fullerenes from a Fulgurite. *Science*, **259** (5101), 1599–601.
79 Jehilička, J., Ozawa, M., Slanina, Z. and Osawa, E. (2000) Fullerenes in solid bitumens from pillow lavas of Precambrian age (Mitov, Bohemian massif). *Fullerene Science and Technology*, **8** (4-5), 449–52.
80 Jehilička, J., Svatoš, A., Frank, O. and Uhlik, F. (2003) Evidence for Fullerenes in Solid bitumen from pillow of proterozoic age from Mítov (Bohemian Massif, Czech Republic). *Geochimica et Cosmochimica Acta*, **67** (8), 1495–506.
81 Fang, P.H. and Wong, R. (1997) Evidence for fullerene in a coal of Yunnan, Southwestern China. *Materials Research Innovations*, **1**, 130–2.
82 Osawa, E. (1999) Natural fullerenes – will they offer a hint to the selective synthesis of fullerenes? *Fullerene Science and Technology*, **7**, 637–52.
83 Wang, Z.X., Li, X.P., Wang, W.M., Xu, X.J., Zi, C.T., Huang, R.B. and Zheng, L.S. (1998) Fullerenes in the fossil of dinosaur egg. *Fullerene Science and Technology*, **6** (4), 715–20.
84 Yamasaki, A., Iizuka, T. and Osawa, E. (1995) Fullerenes in Chinese Ink Sticks (Sumi). *Fullerene Science and Technology*, **3**, 529–43.
85 Osawa, E., Hirose, Y., Kimura, A., Shibuya, M., Kato, M. and Takezawa, H. (1997) Seminatural occurrence of fullerenes. *Fullerene Science and Technology*, **5** (5), 1045–55.
86 Becker, L., Bunch, T.E. and Allamandola, L.J. (1999) Higher fullerenes in the Allende meteorite. *Nature*, **400**, 227–8.
87 Pizzarello, S., Huang, Y., Becker, L., Poreda, R., Nieman, R.A., Cooper, G. and Williams, M. (2001) The organic content of the Tagish Lake Meteorite. *Science*, **293**, 2236–9.

11
Computational Tools for the Biomedical Application of Carbon Nanomaterials
Leela Rakesh

11.1
Introduction

Historically, cancer has been one of the deadliest diseases of humankind with, according to the American Cancer Society, over 500 000 patients dying in 2008 alone. Clearly, while cancer is one of the leading causes of death worldwide [1–9], the condition itself has many causes that include environmental hazards (e.g., exposure to the ultraviolet (UV) rays from the sun, excess pollution), as well as hereditary traits or health issues (such as smoking, alcohol abuse, and microbial infection). Other factors include the site of origin, the cause of the malignancy, and the degree of tumor spread. Although a variety of drugs and treatment techniques is at hand, the complexity of cancer means that no universal remedy exists to effect a complete cure with no, or minimal, side effects. In recent years, among the wide variety of methods and techniques that has been reported among the anticancer research literature, numerous successful examples have been identified in the area of Pt-based anticancer research.

Although, in the past, platinum has been used successfully in cancer treatment techniques (mainly because of its nuclear-specific biological activity), certain debilitating toxic effects demonstrated by Pt-based drugs have triggered the need for an improved Pt-based drug formulation and, indeed, extensive research in this area is currently under way in several laboratories [2–6]. Despite the several forms of toxicity brought about by these drugs, the overall biological activity of Pt-based drugs cannot be denied. Yet today, many other techniques are available, an example being that of nanoparticle-based treatment which, although becoming increasingly popular, has raised concerns relating to both its toxicity and efficacy. Of course, the question of whether one will "strike gold" will depend purely on how successfully these techniques can be applied to completely eradicate the root cause and remove the cancerous cells, without adverse side effects. Nonetheless, investigations should – and undoubtedly will – be continued in these areas until a highly effective, target-specific and nontoxic formulation has been developed.

In the case of nanoparticle cancer research, one of the best methods of devising an efficient drug-delivery system would be to design a nanodrug capsule of such specificity that it would bind only to the target cell. Bearing such an idea in mind, it should then be possible to identify different means of incorporating potent, water-insoluble drugs to alleviate any disease. Clearly, this would be of great interest to humanity, and also would lead to major reductions in healthcare costs. Currently, many research groups are targeting such an idea by incorporating nanoparticles with a variety of drugs, whether simply by physical loading, by covalent functionalization, or by growing nanocapsules via various mechanical and chemical syntheses [6–9]. Some of these applications and methodologies are described in the following subsections.

Interest in the use of nanotechnology to detect diseases such as cancer, HIV, and diabetes mellitus has undergone a rapid expansion during recent years. Indeed, today a wide variety of molecular machines, devices and modern computer modeling is being used in both industrial and medical applications. Yet, questions often arise as to the extent of information that can be obtained from biology, and which machines and devices derived from industrial applications might be useful for mimicking biological functions. To achieve this goal, it is first necessary to recognize the effects of downsizing or miniaturizing these industrial machines to fit biological needs. Consequently, research groups are seeking to imitate Nature and biological (under normal/abnormal conditions) functioning, so as to deliver drugs directly to targeted cells and to create rapidly functioning and inexpensive diagnostic tools such as biosensors/biomarkers for diabetes, cancer, influenza and AIDs, by using nanocarriers or nanocontainers.

> "Nanotechnology has been in the forefront, especially in the development of nanoparticles and carbon nanotubes (CNTs). The structural optical, magnetic and electronic properties of these materials leads them to many industrial and biomedical applications. One fast-growing nanotechnology-based application is the incorporation of nanoparticles with the respective drugs, and drug delivery. This has shown promise in minimizing harmful cells to the surrounding tissues and reducing the side effects, especially in invasive cancer treatments such as chemotherapy and surgery [9–14]. An approach to targeting such cells is to coat the surface of nanoparticles with biological or biocompatible molecules that bind to receptor molecules found in tumor tissues. This 'smart-bomber'-type technology can be used effectively to detect tumors or foreign bodies in the human circulatory system [14]. To understand the diverse and dynamic properties in applications such as this, computational studies can be conducted without the arduous experimental voyage. There are many computational software, tool kits, and modeling software packages available for designing drugs and materials. Such innovations will bring together material

and life scientists into one umbrella, bridging the gap between these fields, thereby saving time, efforts and cost by identifying designs that will work in the biological environment. Such computer-aided designs will help researchers sort through combinations of drugs and devices. Simulations from the predictions can then be used to determine the nature of those nanoparticles which are most biologically safe, active, and compatible for the release of drugs to the targeted tissues, tissue repair, biosensing, and biomarking. The protein-detecting capability of the software could help in designing biodegradable, diagnostically smart nanoparticles that can latch onto molecular biomarkers to reveal their locations in the body" [15], according to Leroy Hood [16].

The need for understanding the properties of nanoparticles are: (i) size, (ii) shape, and (iii) surface functionality [17–19]. One of the fascinating characteristics of nanoparticles comes from the surface of the material governing the properties of the bulk. Such phenomena can be viewed by the bending of bulk copper or such similar materials transpire with movement of copper atoms and clusters of size around the 50 nm scale. Copper nanoparticles smaller than 50 nm are considered superhard materials that do not exhibit the same ductility, bendability and malleability as bulk copper. But, in the case of ferroelectric materials at room temperature, a size of less than 10 nm usually leads to a switch in their direction of magnetization, rendering them unusable for memory storage or any other similar application [20, 21]. It is very important to understand why gold nanoparticles often seem to appear deep red to black in solution, but this highlights the peculiar characteristics of nanoparticles. There is, therefore, a need to study the classifications [15, 19, 22–24] such as size, shape, orientation, and packing. Many semisolid to soft particles have been manufactured, and the packing ability criteria for biomedical applications has also been verified by mathematical and computer simulation models [10, 17, 19]. Nanomaterials, such as carbon nanotubes, are actively being perceived for many such applications. Their mechanical, electronic and optical properties are also currently being explored for applications in molecular electronics, biomedical, biosensing, optoelectronics, and energy conversion" [15, 20–27] .

Adapted from Ref. [15], with permission
from Future Medicine Ltd.

The most important applications for carbon nanotubes (CNTs) lie in the area of biomedical science and biotechnology [6, 20, 21, 25], as CNTs and nanomaterials of different sizes and shapes, when loaded with peptides, proteins and drugs, are able to translocate across the cell membrane. This major advantage has attracted increasing attention for the design of novel delivery systems using nanomaterials. For instance, CNTs have been proposed as a nanoscale vehicle for drug delivery, and have formed the basis of the development of a new *in situ* human organ pressure monitor, which can pass through the blood barrier [15, 24, 28–35]. Consequently, CNTs can be used effectively as vehicles to deliver drugs to targeted cells (e.g., cancer/infectious cells) with unprecedented accuracy and efficiency.

In such a drug-delivery process, the drug molecules may either be attached covalently or functionalized noncovalently to a CNT, and then are released towards the targeted areas. For this, the drug molecules may also be placed inside the CNT, or cleverly wrapped around it by using a biologically active surfactant for delivery. Molecular dynamics (MD) simulations have demonstrated that a DNA molecule can be spontaneously encapsulated, and the molecules inside a CNT have been shown to use a variety of techniques [15, 19, 28, 29, 34–40]. The inter- and intramolecular interactions between a guest molecule and a CNT, in the presence of biosurfactant, would be of great interest when investigating molecular mechanisms during the drug- delivery process. However, by using a dissipative particle dynamics simulation, it was also shown that the molecule would remain preferentially inside the CNT, in the ground state [31]. In fact, an energy barrier would prevent the molecule from being encapsulated or wrapped, both inside and outside of the CNT [15, 40]. Hence, there is a need to search for an effective actuation mechanism that will expel the molecule from the SWNTs and/or MWNTs for drug delivery, if indeed it does prove preferable to encapsulate the drug inside the CNT. It appears logical that the drugs would be carried either inside or outside the nanotubes, depending on which technique is available, and which is the more effective to eject the drug or deliver it to the target cell, without causing damage to any normal cells. Although, such a study could be conducted *in vivo* once the effectiveness had been proven *in vitro*, this is beyond the scope of the simulation systems. Nonetheless, such investigations will surely be undertaken in the very near future.

Nanostructures such as gold nanoparticles (AuNPs), cadmium, tellurium and gold nanorods (AuNRs) represent fascinating nanoscale materials that are currently being used for a wide variety of biomedical applications. Example of these include *in vivo* drug delivery, hormone replacement therapy, the biological detections of viruses, medical imaging, and the biological detection of proteins [17–22, 25–27, 40–49]. Typically, these applications require the nanoparticles to be well suspended in aqueous solution, as an effective detection will rely on surface passivation by the molecules of choice, as well as by amphiphilic polymer coatings, bilayer lipids, and surface silanization [40]. Currently, many investigations are ongoing with regards to the synthesis of stabilized AuNPs in thiol ligands and pyrene-containing and phospholipid-coating moieties, many of which have been extensively investigated in relation to their efficacy of binding with CNTs using stabilized AuNP–thiol ligands in phospholipid-coating moieties [15, 22–24, 26, 27, 42, 46–48, 50–52].

Important factors such as van der Waals forces, charge transfer, π–π ordering and stacking, and/or hydrophobic forces, represent the main interactions involved in the case of AuNPs [9]. Today, the noncovalent functionalization of nanomaterials using polymeric amphiphiles is becoming popular in biomedical applications such as Raman infrared scattering and near-infrared fluorescence, both of which are employed in biomedical imaging, mainly because the inherent properties of the original material are protected, such that no physical property damage is caused [48]. A micellization process can be achieved by the appropriate addition of an aqueous surfactant such as a lipid, polysaccharide, protein, DNA, γ-polyglutamic acid or poly(ethylene glycol) (PEG) in SWNTs. Reports have also been made suggesting that an increased *in vivo* circulation in body fluids can be achieved if the nanomaterial surfaces are densely coated with PEG [50–53]. Currently, several techniques are being investigated to overcome the dispersion of CNTs in structural and functional polymers, through covalent, electrostatic, or noncovalent modification [6, 9, 15, 23, 24, 28, 29, 35–40, 47, 48, 50]. In the past, a side-wall functionalization of CNTs has been employed, using CNTs with hydroxo and amino groups that include low-generation dendrimers and polyester amines, in addition to a physical loading of the hydrophobic end of the surfactant or biosurfactant species attached to a hydrophobic drug. Such covalent functionalization and/or physical loading of the surfactant to CNTs is often used for biomedical and industrial applications, due to efficient solubility characteristics.

Many case studies have been conducted in an attempt to understand the role of vitamins during tumor formation and in treatment options [49, 54]. Based on the potent interactions that occur under biological conditions, the decision was taken to monitor the effects on the nanoparticle carrying drugs in the presence of calcitriol (vitamin D2). Whilst the discovery of vitamin D and the subsequent elimination of rickets represented a major achievement in medicinal science [55], many other studies have been conducted concerning the medical benefits of fat-soluble vitamin A and associated exposure to UV light [56]. It was concluded, correctly, by McCollum that vitamin D and its derivatives represent the most valuable hormonal activity in the human body [57], as they had been used to treat not only many illnesses but also single nuclear type 2 receptors to assist in the suppression or activation of targeted genes [58]. Finally, investigations were also conducted into the interaction of nanoparticle–drug complexes with single-stranded DNA (ssDNA) having 20 adenine–thymine and guanine–thymine base pairs.

Unfortunately, these materials may demonstrate many disadvantages in *in vivo* systems if they are not incorporated under the appropriate conditions [50–53, 59]. It is known, fundamentally, that most inorganic nanomaterials are insoluble in physiological buffers, and that it is often necessary to carry out a functionalization, using either surfactants or thiols, in order to attain the biocompatibility that in turn improves bioavailability under various operating conditions. The results of a number of experimental studies have suggested that several surfactant coatings might prove to be unstable in serum not only during the processing but also when extracting any excess of the coating molecule [50–53].

Some results have indicated that, following intravenous injection, the type of interaction, and the strength of the adsorptive binding to the nanoparticles, will vary with different surfactants and body distribution patterns [59]. PEGylation is one of the most common strategies used to impart functionality, and to improve the water-solubility and biocompatibility of the effective drugs developed today, most of which are hydrophobic in nature [48, 50–53]. Currently, a lack of slow release uptake by the reticuloendothelial system (RES) and short blood circulation times represent the major problems associated with the direct functionalization of nanomaterials [60]. A sufficient blood circulation time is critical for both imaging and *in vivo* release of the excipient coating; moreover, it also improves the circulatory behavior of the SWNTs for *in vivo* applications such as tumor targeting, imaging, and chemotherapy. Consequently, a long half-life of the drug carrier in the blood is an important characteristic for functionalized nanotubes if the bioavailability of the system under investigation is to be improved [53, 59–66]. Typically, nanomaterial inclusions are slowly but consistently cleared from the blood circulation by uptake into the macrophages of the RES, and in turn are accumulated in the spleen; notably, recent studies have indicated that an extended circulation is achieved only if a rapid RES uptake is circumvented [48, 50, 53]. Nonetheless, other results have indicated that a physical loading/coating rather than a chemical functionalization of drugs onto SWNTs can delay RES uptake, and thus increase the blood circulation time [9]. Consequently, in the present studies, various combinatorial aspects of the loading of drugs were investigated, within a PEGylated-lipophilic environment and using functionalized and unfunctionalized SWNTs, via MD simulations. Whilst previously, the application of CNTs in biomedical, molecular electronics or computers has been predicted most often on the basis of theory and simulations, experiments are currently under way to complement these computational efforts.

Although the therapeutic use of the antitumor agent paclitaxel (PTX) is limited initially by its poor solubility and rapid destruction, the adverse side effects that it causes lead to major problems in the drug's effective application in patients. Amphiphilic copolymer micelle carriers possess significant potential for improving drug solubility and stability and, indeed, various reports have determined that SWNTs, when functionalized by amphiphilic polymers, have a high stability in aqueous solution but lose their functionality at different serum pH-values and temperatures [9, 15, 22–30]. Moreover, the polymer-coated SWNTs exhibit remarkably long blood circulation times after intravenous injection in mice, surpassing all existing values [9, 48, 50–53, 59–70]. Such a very long circulation time in the blood will greatly delay clearance of the nanomaterials via the RES, and is a desirable property for *in vivo* nanomaterials applications, for example in drug delivery and imaging [50, 51, 68, 69]. Moreover, PEG is nontoxic and has been approved by the US Food and Drug Administration (FDA) for use as excipient/carrier fluid in different pharmaceutical formulations, foods, and cosmetics [52]. Most PEGs with molecular weights >1000 Da are removed rapidly from the body without being altered, with the clearance rate being inversely proportional to the polymer's molecular weight [53]. This property, combined with the hydrophilic availability of PEGs with a wide range of end-functions, contributes to drug delivery, tissue engineering skin scaffolds, surface functionalization, and many other applications [59, 60]. The branched PEGylation of PTX via an ester linkage, as in SWNT–PTX conjugates, affords the water-solubility of PTX, while PEGylation favors a prolonged blood circulation. However, the high hydrophobicity of PTX reduces the hydrophilicity and biological inertness of branched PEG-functionalized SWNTs, causing significantly shortened blood circulation half-lives of the SWNT–PTX formulation compared to PEGylated SWNTs without PTX attachment. Such higher hydrophobicity led to an increased nonspecific protein absorption on the nanotube conjugates, which caused an accelerated uptake by macrophages in the RES [38, 43, 59–64].

During the past 10 years, major advances have been made in the use of liposomes, conjugates, nanoparticles, and microspheres/microcylinders as vehicles to deliver pharmacological agents and enzymes to the sites of disease. The main advantages of liposome-encapsulated and carrier fluid–solid-mediated drugs are: (i) an increased solubility; (ii) a prolonged duration of exposure; (iii) a selective delivery of the entrapped drug to the site; (iv) an improved therapeutic index; and (v) the potential to overcome any resistance associated with a regular anticancer agent [68, 69, 71, 72]. From a theoretical standpoint, it is well known that the chemical activity of these agents may accumulate in tumor and tissues, due to the low pH values and chemical potentials which enhance the permeation and retention effects. However, experimental evidence of their interactions and toxicity must be closely examined by incorporating various combinatorial aspects of nanoparticle size and shapes, liposomes, and conjugates with the drugs. Indeed, this entire process may be "fast-tracked" by initially studying the various interaction effects using computer simulations, after which only the promising candidates would be verified experimentally.

Consequently, the present studies involved the computer simulation of several new PEG-grafted unbranched and branched polymers coupled to lipophilic species such as cholesterol or phospholipids (PEGylated phospholipids/cholesterol), with covalent and noncovalent functionalization of SWNTs by the physical loading of PTX, the aim being to understand the hydrophobic and van der Waals interaction forces that clearly provide an enhanced aqueous solubility and improved biocompatibility. Thus, MD simulations were conducted in order to understand the interaction between physically loaded PTX and covalent and/or noncovalent functionalized SWNT carrying branched and linear PEG–phospholipids (e.g., dipalmitoylphosphatidylcholine; DPCC) and cholesterol.

Biological nanoparticles such as liposomes are currently used as drug-delivery systems for anticancer agents and vaccines, and have been characterized using a variety of analytical techniques [36, 48, 50–53, 59–68]. Although the incorporation of nanoparticles into biomedical applications may be closely related to toxic effects, the potential for toxicity can be minimized (or even eliminated) by their careful incorporation via clever manipulation or functionalization with chemical entities.

Many of the health-risk issues associated with nanoparticles can also be examined using computer models [7, 19–21, 25, 48, 70, 73–79]. Simulations not only complement conventional experiments but also enable the acquisition of new data that cannot be obtained in other ways.

Whilst today the two main simulation techniques available are MD and Monte Carlo (MC) [19–23, 25–27, 32, 33, 47, 52, 74, 80], in this chapter attention will be focused MD simulations. Simulations serve as a "bridge" between the microscopic length and time scales and the macroscopic world of the laboratory. The aim is to provide a "guess" at the interactions between molecules, but to obtain exact predictions of the bulk properties to a high degree of accuracy, subject to the limitations imposed by computer budgets and computational power. MD simulation consists of the numerical, step-by-step, solution of the classical equations of motion known as Newton's second law. It provides information concerning the structural determination of ultra-small particles directly on their support material; such data are essential for a fundamental understanding of the physical and chemical properties of these particles before they can be considered for any application. Molecular modeling represents one of the fastest ways to model much of this information, and will surely revolutionize computational biology.

In these research investigations, SWNTs, with their versatile physico-chemical features, have been investigated as an efficient platform for both the detection and targeted therapy of various infectious diseases and cancer. The objective of these studies was how MD simulations could be used to explain the various interactions and motions of nanoparticles with and without drug molecules, and how efficiently they can be used to transport drugs to the target sites. By using modeling, the aim is to answer several questions:

- How does a single nanoparticle molecule interact or adsorb on different surface?
- What happens when more molecules or different types of molecule are added?
- Or, given several different possible materials, which will show the strongest binding for a drug under biological environment?

Such a layer-by-layer or molecule-by-molecule build up approach is essential to understand the power of the implementation of nanoparticles, and their application in various demanding environments.

11.2
Simulation Methods

11.2.1
Background

The initial question might be, "Why should modeling be carried out?" In fact, scientists have for many years been simulating responses from chemical reactions, and investigating the properties and interactions of molecules and materials by using computational tools and interfaces. This is made possible by the availability of supercomputers, hardware power, and advanced software–tools that can be used to complement, devise, design and direct some (or all) of the experimental information, and in turn help to alleviate (or even replace) many arduous experimental procedures. Likewise, impossible or unimaginable reactions can also be recognized before valuable laboratory time and resources are wasted. For example, reactions difficult to study experimentally, as well as the time taken to complete and/or understand the toxicity of the chemicals, can be studied by using computer simulations, and insights obtained into their mechanical and chemical properties. Both, theoretical modeling and computer simulations have long provided insights into the verification and explanation of what has been completed and observed experimentally. In the case of nanoscale systems, simulations and theoretical predictions have provided details of the original properties that have led to the design of new materials and systems, to the conceptualization of new devices, and to nanotechnology applications. For the present simulation, Material Studio (MS) modeling was used, in conjunction with Accelrys software [74], and many of these concepts are employed at this point. The use of this software provides many benefits for the polymer, and for nano- and biomaterials modeling. Typically, MS modeling offers the widest range of simulation tools, as well as the forcefield such as universal, PCFF, compass and dreiding force fields for polymer and biopolymer applications. Since these approaches all involve large numbers of atoms, the simulation tools of choice are atomistic modeling and mesoscale methods. For example, the Discover technique is a classical simulation tool for small- to medium-sized molecules and periodic systems, and its combination with the compass forcefield allows the prediction of highly accurate properties of these molecules. It also allows the study of condensed phase systems, as reported by several research groups [6, 9, 15, 28–30, 74–80].

The compass forcefield parameters are used by organic as well as inorganic materials, such as metals and transition metal oxides, and alkali and alkaline earth halides. Three different bonding mechanisms of these materials include: (i) metallic; (ii) semi-ionic; and (iii) ionic bonds. This is of particular importance for nanomaterials and biomaterials research, which typically covers a wide range of materials with organic and inorganic interfaces and surfaces. For polymeric systems, Amorphous Cell (AC) is the best and is a condensed-phase simulation tool in the MS studio. Because of the specified density criteria, pure fluids, solutions and liquid crystalline phase and mixtures are constructed using AC. However, if the system requires longer timescales and length than conventional methods, it becomes necessary to utilize the concepts of mesoscale methods using dissipative particles dynamics (DPD) and MesoDyn. These offer the ability to study phases in terms of their kinetics and structures that are critical to a materials' performance, which is in-between the molecular and the bulk stage. Subsequently, once the resulting structures have been obtained they can be used to feed into finite-element methods.

11.2.2
Molecular Modeling

Nanoparticles or nanotubes are known not to reach their target cells if they are >100 nm in size, as they become trapped in the bloodstream; however, if the particles are <10 nm in size they are retained in the organs, such as the kidneys and liver. Consequently, whilst the ideal nanomaterial must be sizes in-between the above dimensions, as would be appropriate for any simulation experiment, a careful choice must be made in order to obtain the best benefit for biomedical applications. The present MD simulations were conducted at room temperature in order to investigate the interaction of several new PEG-grafted unbranched and branched polymers coupled to lipophilic species such as cholesterol or phospholipids (PEGylated phospholipids/cholesterol), by covalent and noncovalent functionalization of SWNTs with the physical loading of PTX. In order to obtain the lowest energy conformation, structures of each of the above-mentioned species were built (Figure 11.1a–e), with (13, 0) SWNT being used as the starting structure. Following geometric optimization, the resultant most stable structures were subjected to a systematic variation of two side-chain torsional angles in order to study their conformational flexibility. The axial side-chain orientation was slightly more stable than the equatorial form. Typically, two weak intramolecular hydrogen bonds contributed to the stabilization of the axial conformer, whereas in the equatorial conformer only one hydrogen bond was detected. An 8 ps MD simulation at 300 K suggested that, at realistic temperatures, the molecule would be flexible enough to undergo internal motions (rotations, vibrations), rendering questionable the biological significance of mere conformational properties.

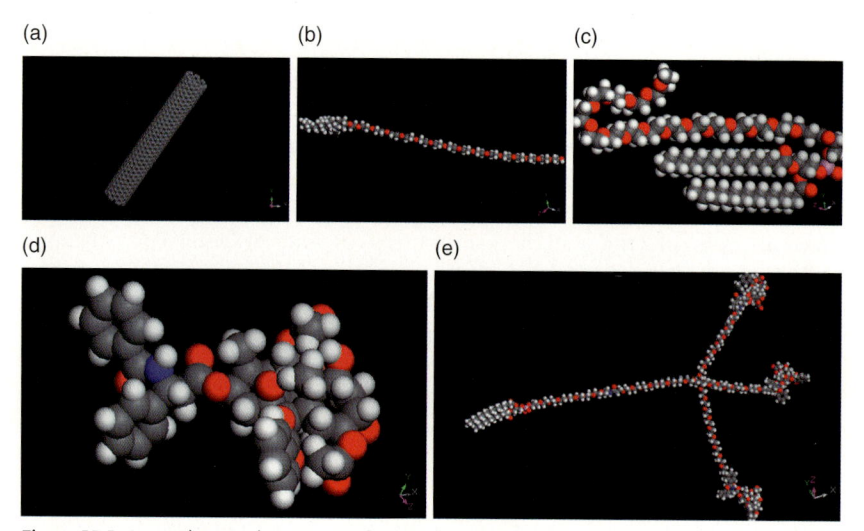

Figure 11.1 (a–e) The initial structure of SWNT, branched PEGylated cholesterol and phospholipids (DPCC), PTX, and PEGylated phospholipids-*co*-PTX.

In order to obtain a stable structure, the system under consideration must be minimized under dynamics conditions, using annealing and quenching procedures, successively. Thus, for the molecular model building process an annealing and quenching scheme was used which employed successive steps of energy minimization of MD runs for each of the molecules under consideration. When the system had reached its equilibrium condition, the production run would proceed by again running MD simulations at 300 K, using MS studio Forcite code and compass force field [9, 45–48, 59, 67, 70, 73–76]. All of the structures were minimized for 800 steps to remove any close contacts, followed by a 200 ps period of high-temperature MD (500 K) to create a range of structures that fully explored the conformational space. Due to limits of computational time, 10 frames from the trajectory were used for further investigations. Each of the 10 structures was minimized for 500 steps to remove any close contacts, followed by 150 ps of equilibrium, followed by a 100 ps production run at 0.2 fs time step at 300 K.

The final minimized structures of each of the molecules are shown in Figure 11.1a–e. The variation of density was also determined as a function of the distance from the center of mass of the SWNT [6, 15, 28, 29, 32]. In general, two sets of simulation experiments were conducted. Each minimized structure of lipophilic species was then: (i) covalently functionalized with SWNTs at the terminal ends, while PTX was physically loaded; and (ii) PTX-functionalized branched lipophilic species were un-covalently linked or physically loaded with SWNT. The energy minimization was then re-commenced until it reached the minimum energy according to the above procedure. Often, these system consists of more than 100–200 atoms, excluding the hydrogen per unit cell of PM3N structure. The computational data analysis of the lowest minimum energy structures is provided in Section 11.3 (Results and Discussion). The minimized structure was then used with the energy-minimized structure of PTX for physical loading (also provided in Section 11.3).

11.3
Results and Discussions

The antitumor drug, PTX, with and without PEGylated cholesterol and physically loaded onto SWNTs, is shown in Figure 11.2a–c. It shows a very strong interaction with the CNTs surfaces (noncovalent SWNT functionalized), as can be seen from the figures (initial and final energy-minimized structures). In order to understand the role of PEGylation in the SWNT drug delivery, a series was made of PEGylated lipids un-covalent and covalent functionalization with SWNT and the PTX (mostly) and Irinotecan drugs, which are mainly used for cancer therapy. Various analyses are described below in an attempt to understand the critical number of drug molecules that can be carried via SWNTs. Some results are also provided as to how these complexes behave in the presence of vitamin D, with and without 20-base pair (guanine–thymine) ssDNA [71].

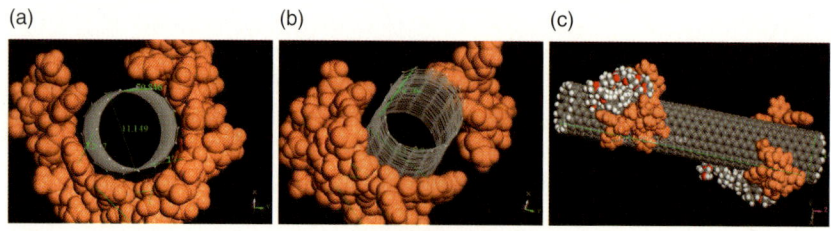

282 | *11 Computational Tools for the Biomedical Application of Carbon Nanomaterials*

Figure 11.2 Energy-minimized structure of SWNT without PEGylated cholesterol and PTX (physically loaded). (a) Initial starting structure; (b) After 20 ps, with a time step of 0.2 fs; (c) With PEGylated cholesterol.

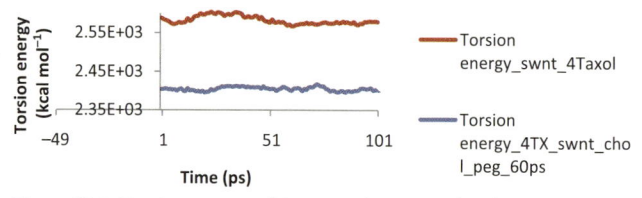

Figure 11.3 Torsion energy of SWNT without PEGylated cholesterol and PTX (physically loaded)

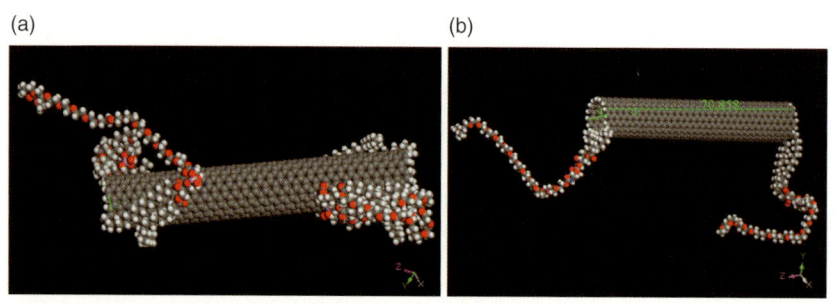

Figure 11.4 Energy-minimized structures of PEGylated DPCC-functionalized SWNT. (a) 1 CH_2 of DPCC functionalized SWNT; (b) 2 CH_2 of DPCC-functionalized SWNT.

Figure 11.2 shows the minimized structures of SWNT w/o PEGylated cholesterol and PTX (physically loaded), while Figure 11.3 shows its torsion energy barrier. Clearly, the torsion energy is much more lower in the presence of PEGylated cholesterol, which indicates the flexible nature of the complex system.

The typical energy-minimized structure of PEGylated phospholipids (1 versus 2 ch_2) (co)functionalized SWNT is shown in Figure 11.4a, b, where the minimum energies are 7.071444 and 7.008×10^3 kcal mol^{-1}, respectively.

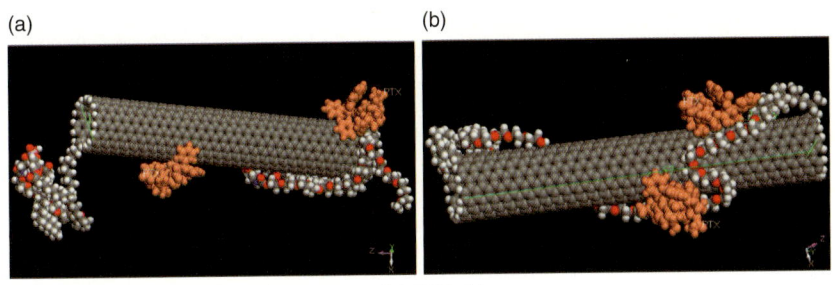

Figure 11.5 Energy-minimized structures of SWNT. (a) -co-PEGylated phospholipids; (b) PEGylated cholesterol with PTX (physically loaded) after 20 ps, with a time step of 0.2 ps.

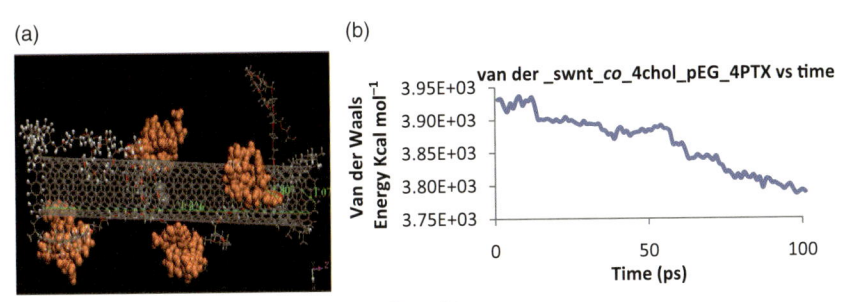

Figure 11.6 Energy-minimized structures of SWNTs. (a) PEGylated cholesterol with four PTX (physically loaded) after 100 ps with a time step of 1 fsp; (b) Plot of van der Waals energy versus time.

Figure 11.5a, b shows the energy-minimized (7.485261/7.377103 e + 003 kcal mol^{-1}) structures of PTX-loaded (physically) SWNT-co-PEGylated phospholipids and PEGylated cholesterol. The minimized-energy forms demonstrate some remarkable differences (the interaction with PTX, and its viability to deliver drugs to targeted sites will be further investigated).

Figure 11.6a, b shows the energy minimized (8.316935 × 10^3 kcal mol^{-1}) structure of PTX loaded SWNT-co-4PEGylated cholesterol, along with its van der Waals energy.

Figure 11.7a, b shows the energy-minimized (8.419014 e + 003 kcal mol^{-1}) structure of 4PTX loaded SWNT-co-6PEGylated phospholipids, along with its van der Waal's energy.

Figure 11.8 shows the nonbonded energy difference with and without PTX. Figure 11.9 shows that the average probability of finding PTX in the nearby vicinity of SWNT is much closer in the presence of PEGylated phospholipids. This also suffices the role of interaction of physically loaded PTX in the presence of PEGylated phospholipids-co-SWNT, and can be seen as a favorable environment for drug delivery. A similar paradigm is seen for PEGylated cholesterol-co-SWNT (not shown in the figure).

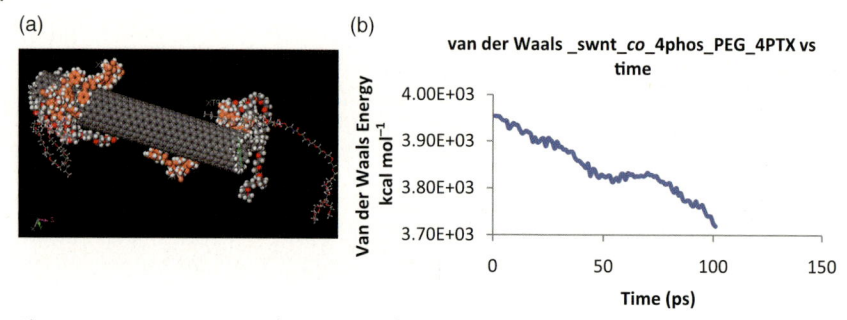

284 | *11 Computational Tools for the Biomedical Application of Carbon Nanomaterials*

Figure 11.7 Energy-minimized structures of SWNT. (a) SWNT-*co*-PEGylated phospholipids (with PTX loaded physically) after 100 ps, with a time step of 1 fs; (b) Plot of van der Waals energy versus time.

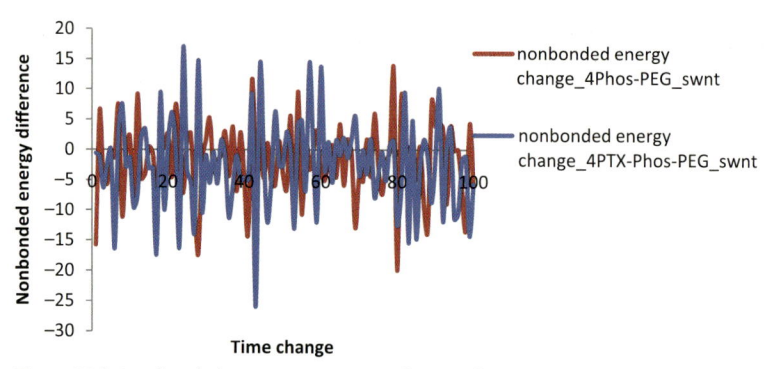

Figure 11.8 Nonbonded energy versus time change of SWNT-*co*-PEGylated phospholipids with physically loaded PTX.

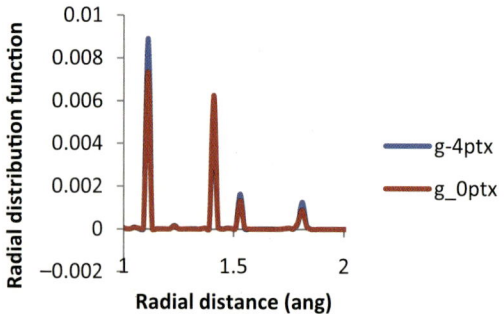

Figure 11.9 Radial distribution function versus radial distance of SWNT-*co*-PEGylated phospholipids with PTX (loaded physically).

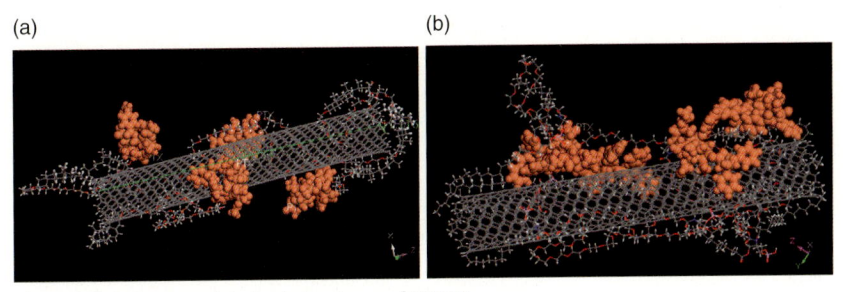

Figure 11.10 Energy-minimized structures of SWNT-*co*- (a) 6PEGylated phospholipids and (b) 6PEGylated cholesterol with four PTX (physically loaded) after 20 ps, with a time step of 0.2 ps.

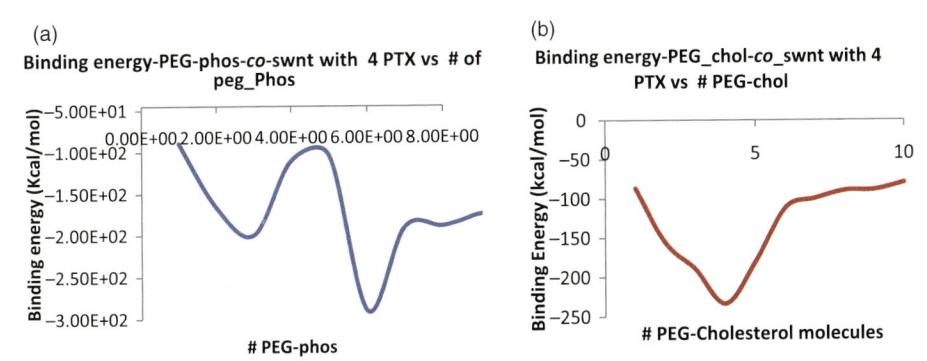

Figure 11.11 Binding energies of SWNT -*co*- (a) PEGylated phospholipids and (b) PEGylated cholesterol with four PTX (physically loaded) after 100 ps, with a time step of 1 fs.

Figure 11.10a,b shows the energy-minimized ($8.482\,885 \times 10^3$ kcal mol^{-1}) structure of four molecules of PTX loaded onto SWNT-*co*-6PEGylated cholesterol and 6PEGylated phospholipids.

Figure 11.11a,b shows that each of the SWNTs could carry at least four to six PTX molecules in the presence of lipophilic lipid carriers, for a better binding characteristics.

In the following simulation, various combinatorial complexations of PEGylated lipophilic lipid carried with PTX and physically loaded onto SWNT were carried out (Figure 11.12a,b). Figure 11.12a shows the energy-minimized (9.897095 e + 003) structure of 12 molecules of PTX and PEGylated phospholipids physically loaded on to SWNTs. Figure 11.12b shows the energy-minimized (1.208649 e + 004) structure of the complex with cholesterol molecules instead of PTX. The minimized total energy showed large differences in the case of the drug-carrying cholesterol molecules, although this may have been due to their interaction with potential differences, which in turn suggests that phospholipids may be better

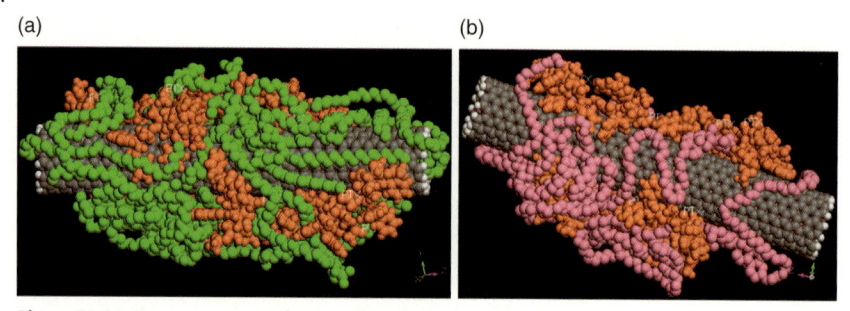

Figure 11.12 Energy-minimized structures of SWNT -unco- (a) 12 PEGylated phospholipids and (b) 12PEGylated cholesterol with 12 PTX (physically loaded) after 100 ps, with a time step of 1 fs.

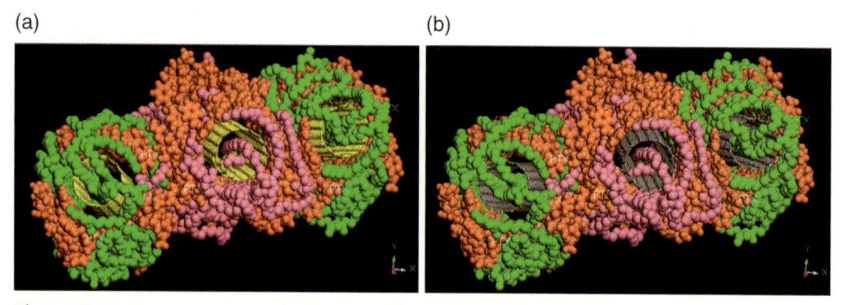

Figure 11.13 Energy-minimized structures of 12 PEGylated cholesterol and two sets of 12 PEGylated phospholipids -unco-SWNT with 36 PTX (min energy: 4.422382 e + 004 kcal mol^{-1}).

candidates for drug delivery. However, experimental evidence is required to confirm this final proposition.

In order to assess the maxim number of drug molecules loaded physically onto SWNTs, various permutation combinations of PEGylated cholesterol and phospholipids were investigated. As can be seen from the energy-minimized structures (Figure 11.13a,b), 36 drug molecules can be carried by each SWNT if one set of 12 PEGylated cholesterol and two sets of 12 PEGylated phospholipids are used.

Although these results require further investigation, it is worth reinvestigating such interactions experimentally, if possible. Such an investigation would provide an insight as to the maximum number of SWNTs and PEGylated lipophilic lipid required to carry certain amount of drug to the targeted site. However, this would require not only experimental study but also further computational investigations, using supercomputer facilities, and will be pursued in the future.

11.3.1
Branched PEGylated DPCC Functionalized PTX Physical Loading on SWNTs

In this section, the investigations were extended to determine how branched PEGylated lipophilic lipid would influence PTX loading onto SWNTs. Two different conformations of the branched PEGylated-DPCC (phospholipids) are shown in Figure 11.14a,b.

Figure 11.15a–c shows the initial and final minimized structure of two molecules of branched PEGylated DPCC functionalized PTX on to SWNT 20 ps versus 100 ps with a 0.2 fs time step size ($8.013821/7.939024$ e + 003 kcal mol^{-1}). This clearly shows that the final minimized structure interacts in different ways with the nanotube, first wrapping around the nanotube and then fitting tightly to its end.

Figure 11.16a shows that the drug molecules approach within less than 5 distance to the vicinity of the SWNT during a 100 ps dynamic simulation time. It also clearly shows the interaction of, and affinity towards, the SWNT. Figure 11.16b also confirms this interaction via the mean-square displacement of the drug to the SWNT. During the first few picoseconds, the drug moves slowly towards the SWNT, but during the intermediate period it moves faster and finally slowly stabilizes.

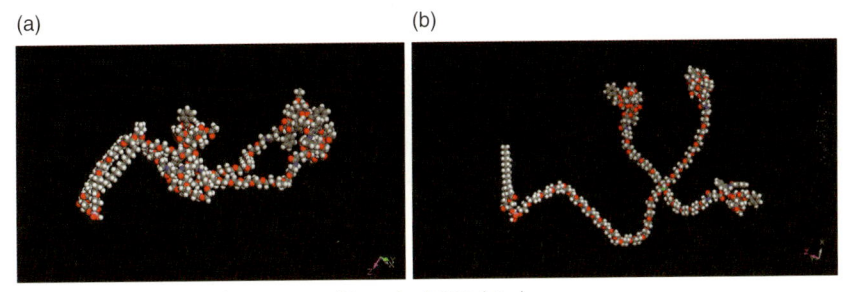

Figure 11.14 Minimized structures of branched PEGylated DPCC-functionalized PTX (1.110947 e + 003 kcal mol^{-1} and 1.184309 e + 003 kcal mol^{-1}).

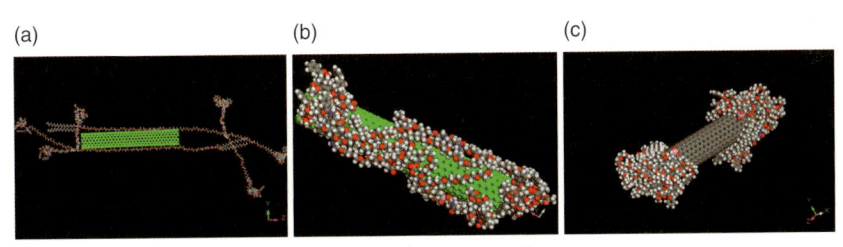

Figure 11.15 (a–c) Initial and final minimized structures of branched PEGylated DPCC functionalized PTX onto SWNT.

(a)

g(r)_RPT3_2ptx-branch-peg-dpcc_SWNT

[Plot: Radial distribution function-g(r) vs Radial distance (ang), y-axis from -0.2 to 0.8, x-axis from 0 to 10]

(b)

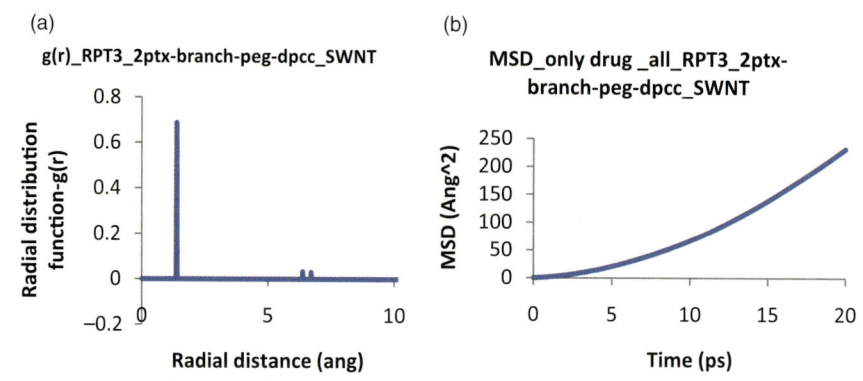

MSD_only drug _all_RPT3_2ptx-branch-peg-dpcc_SWNT

[Plot: MSD (Ang^2) vs Time (ps), y-axis from 0 to 250, x-axis from 0 to 20]

Figure 11.16 (a,b) Radial distribution function and mean square displacement (MSD): two branched PEGylated DPCC functionalized PTX onto SWNT after 100 ps, with 0.2 fs time step.

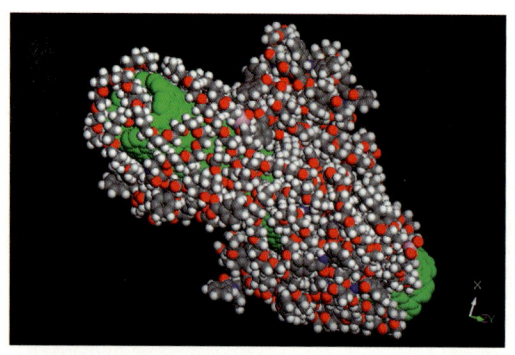

Figure 11.17 Minimized structure of branched 4 PEGylated DPCC functionalized PTX (physically loaded) with SWNT (9.631652 e + 003 kcal mol^{-1}).

Figure 11.17 shows the minimized structure of the branched four-PEGylated DPCC-functionalized PTX physically loaded with SWNT; the minimum energy here is approximately 9.631652 e + 003 kcal mol^{-1}. Figure 11.18 shows the mean square displacement (<MSD) of these molecules in the presence of the nanotube, which is an indirect measure of diffusivity or its motion towards the nanotube. Figure 11.19 shows the minimized structure of branched three-PEGylated DPCC-functionalized PTX with SWNT, and has a minimum energy of 9.200065 e + 003 kcal mol^{-1}. The unbranched three- and four-PEGylated DPCC-functionalized PTX with nanotube has a much higher minimum energy (data not shown) than the branched counterpart. Figure 11.20 shows MSD versus time and radial distribution versus radial distance for each of these. All of these results confirm that the branched uncovalent functionalized DPCCs carrying drugs

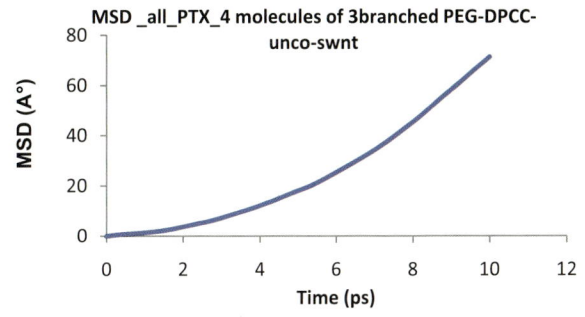

Figure 11.18 Minimized structure of branched PEGylated DPCC functionalized PTX (physically loaded) with SWNT (9.631652 e + 003 kcal mol^{-1}).

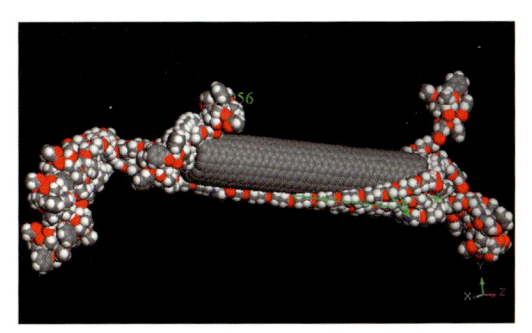

Figure 11.19 Minimized structure of branched 3 PEGylated DPCC functionalized PTX with SWNT (9.200065 e + 003 kcal mol^{-1}).

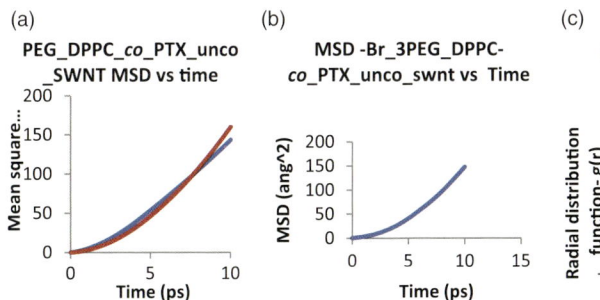

Figure 11.20 MSD versus time and radial distribution function versus radial distance (Å) (branched 3 PEGylated DPCC functionalized PTX with SWNT).

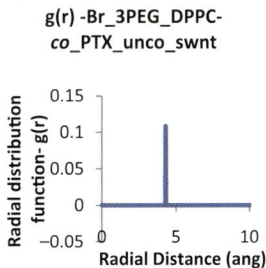

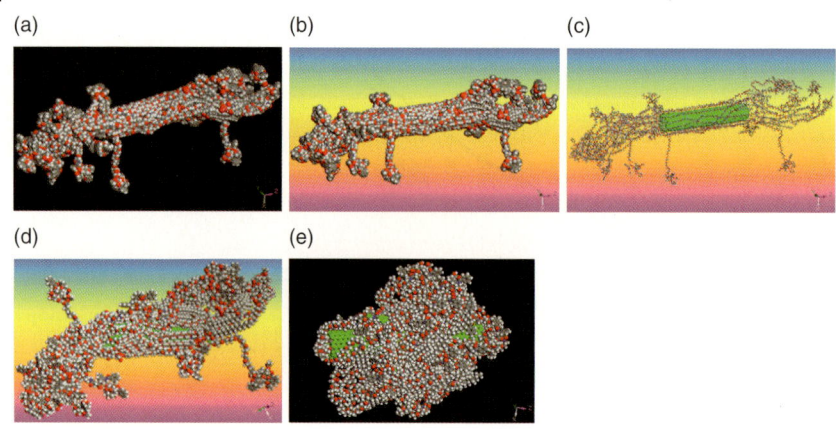

Figure 11.21 (a–e) Initial to intermediate to final energy-minimized structures of branched 9 PEGylated DPCC functionalized PTX (physically loaded) on to SWNT.

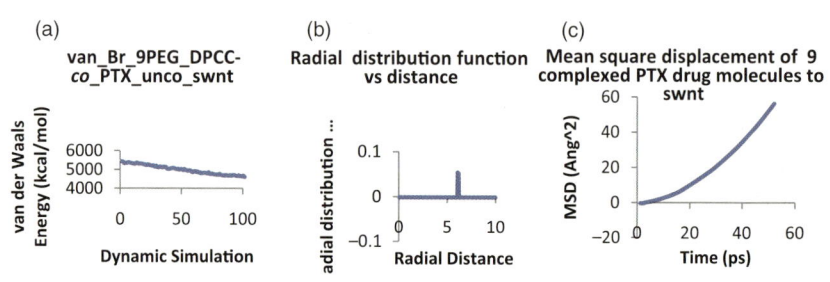

Figure 11.22 (a) Van der Waals energy; (b) Radial distribution function versus radial distance; (c) Mean square displacement (9 PTX complexed with SWNT).

are effective in the presence of CNTs, and carry the drugs both steadily and effectively.

Figures 11.21 and 11.22 show the energy-minimized structure of various numbers of PTX-complexed molecules with SWNTs. The van der Waals energy and mean square displacement only for nine PTX complex molecules with SWNT is shown in Figure 11.22a. The dynamics behavior of the mean square displacements of these complexes shows that, initially, the movement of PTX species towards SWNT decreased slowly, then decreased steadily for a shorter period, but then increased rapidly until the binding energy of the system reached an optimum; subsequently, the movement was stabilized. These findings indicate that the complexed PTX initially approached the SWNT slowly during the simulation time (this may be the time lag required to establish the π–π interaction of the lipophilic group with the SWNT). Electrostatic group repulsions of the coated SWNTs provided the

11.3 Results and Discussions | 291

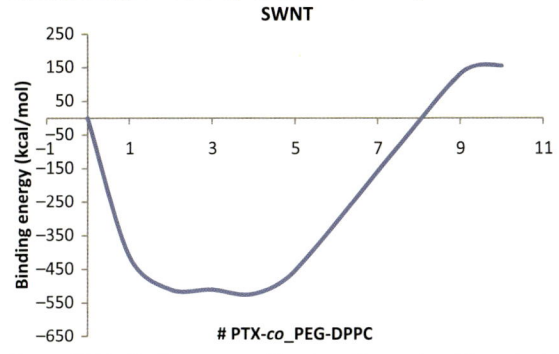

Figure 11.23 Binding energy. Number of complexed branched PEGylated DPCC functionalized PTX with SWNT.

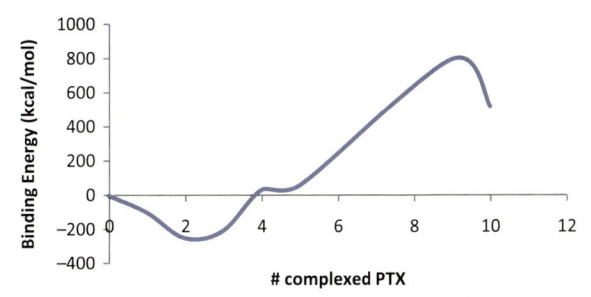

Figure 11.24 Binding energy. Number of complexed functionalized PTX branched PEGylated DPCC covalent functionalized SWNT.

dispersibility and adhesion, after which the electrostatic forces became more prominent and eventually pulled the PTX towards the center of the SWNT. It is possibly for this reason that the binding energy became most negative (as can be seen in Figure 11.23).

As can be seen from the binding energy diagram (Figure 11.23), approximately 12 drug molecules (four molecules of PTX carried by each branch of PEG) can be inserted with each nanotube to the targeted site if the lipophilic end is branched. Insertion also depends on the size of the nanotubes, but this will be further investigated in the future. A similar result is also observed above (see Figure 11.13) if the PEGylated lipophilic molecules are physically loaded with PTX with certain concentration of SWNTs. Such observations appear to concur with experimental data reported elsewhere [41].

Figure 11.24 shows how the binding energy changes as the number of complexed functionalized PTX-branched PEGylated DPCC covalent functionalized

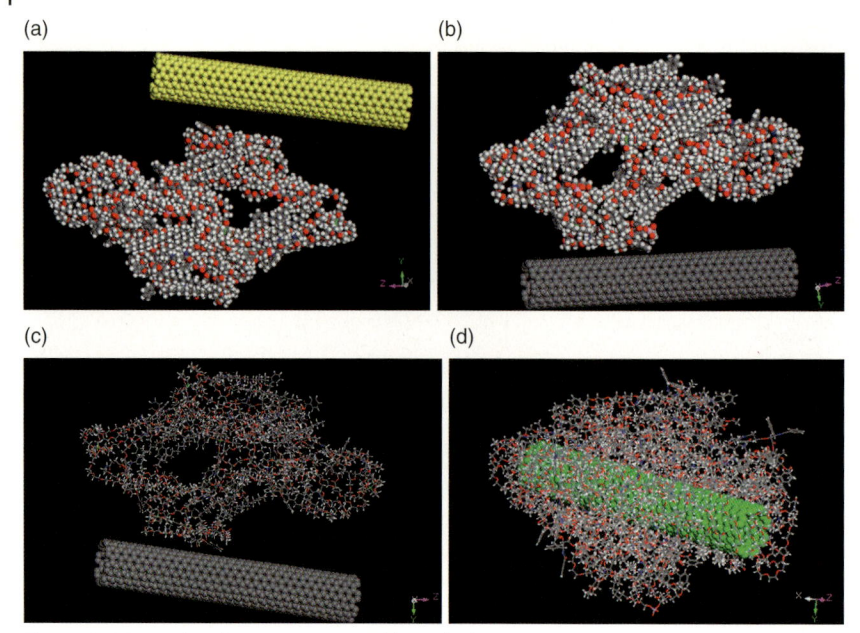

Figure 11.25 Initial and energy-minimized branched 9 PEGylated DPCC functionalized PTX (physically loaded) with SWNT.

SWNT is increased. As can be seen, the covalent-bound PTX complex can carry fewer drug molecules than un-covalent PTX complex (as observed above). However, this may be due to a restriction of forming a free lipophilic groups π–π interaction with the SWNT.

In order to determine how PTX-complexed branched PEGylated DPCC wrap around the SWNT, the latter were removed from the system (see Figure 11.25). Subsequently, the PTX-complexed branched PEGylated DPCC was seen to be coiled around the SWNT, with the PTX side group facing the SWNT and the PEG exposed to the outer region. This computer simulation clearly indicated that the drug molecules effectively lay around the SWNT.

11.3.2
Interaction of Irinotecan-co-br-PEG-DPCC-unco-SWNT and with ssDNA (Adenine-Thymine (AT))

Figure 11.26 (the green and gray colors represent the same types of SWNT, but are color-coded for clarity) shows that, as the dynamic time during the simulation was increased, the branched PEGylated DPCC-functionalized Irinotecan was physically loaded onto SWNTs wrapped around the CNT. However, in the presence of ssDNA, the complexed PTX–CNT system was shown to have moved towards the ssDNA. The data in Figure 11.27 also confirm similar findings, that is, as the

11.3 Results and Discussions | 293

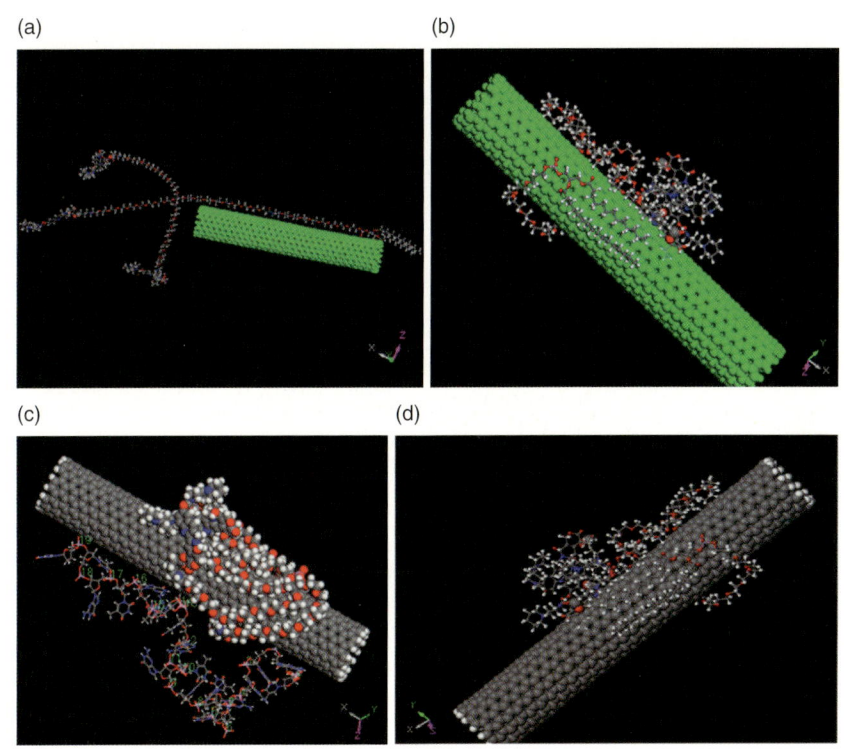

(a) (b)

(c) (d)

Figure 11.26 Initial (a,b) and final energy-minimized structure of br-PEG-DPCC functionalized Irinotecan (physically loaded) with SWNT with (c) or without (d) 20-mer ssDNA (guanine–thymine).

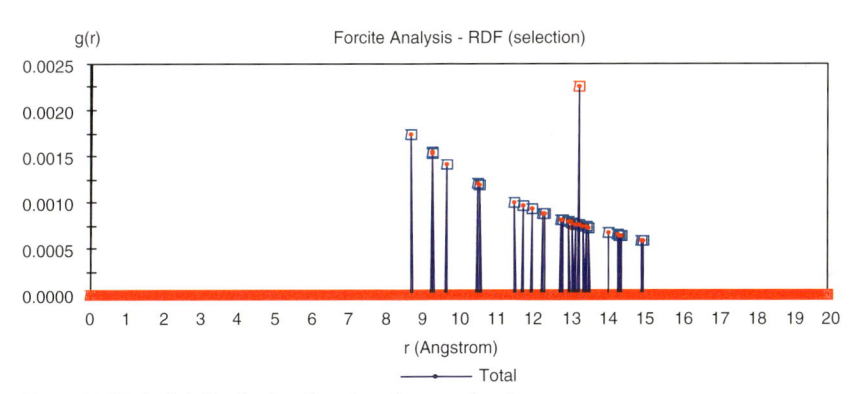

Figure 11.27 Radial distribution function of one molecule br-PEG-DPCC functionalized Irinotecan (physically loaded) with SWNT.

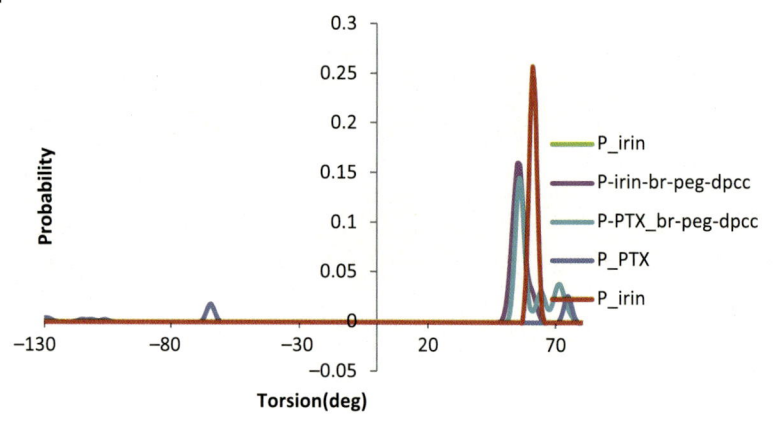

Figure 11.28 Torsion angle versus probability distribution of one molecule of branched (br)-PEG-DPCC functionalized Irinotecan and PTX without each other.

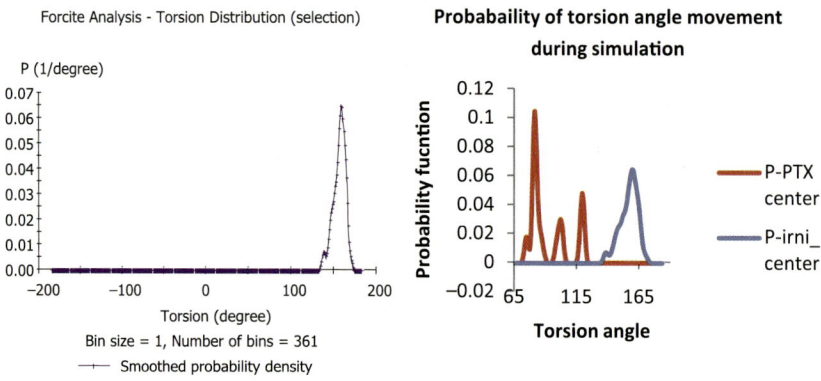

Figure 11.29 Torsion angle versus probability distribution of one molecule branched (br)-PEG-DPCC functionalized Irinotecan and PTX, with and without each other, from the center of the mass.

dynamic simulation time progressed, the drug molecules moved towards the SWNT, approaching as close as 3.5 . This demonstrates the strong interaction of nanotubes with the complexed PEGylated Irinotecan drug.

Figures 11.28 and 29 demonstrate the flexibility of the molecules during their interaction and binding processes. Often, the torsion angles between certain functional groups are very important in identifying certain peculiar characteristics of a molecule under investigation. The information acquired also indicated that the torsion potentials play an important role in dictating ligand or molecule conformations in both the free and bound states. When observing the flexibility of the molecule(s) via the torsion angle, the measurements showed that the Irinotecan

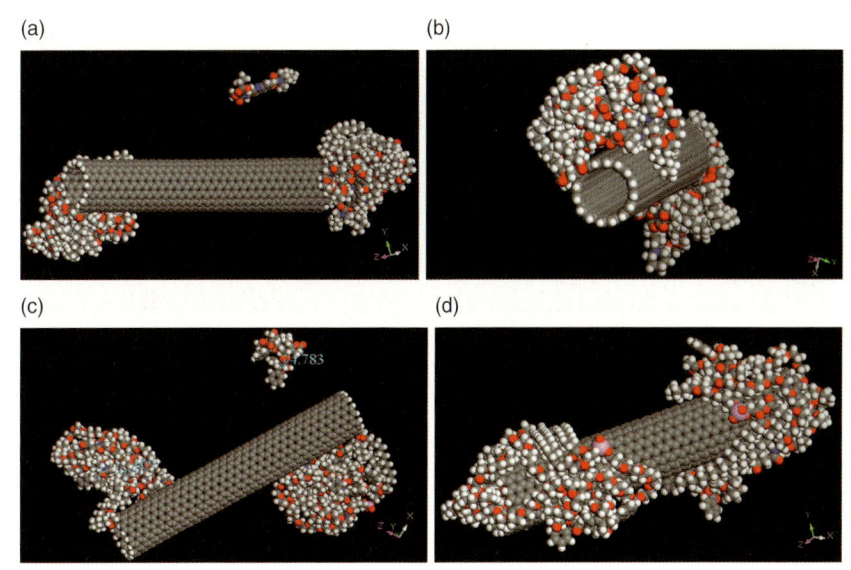

Figure 11.30 Energy-minimized two molecules of branched PEGylated DPCC functionalized Irinotecan and PTX without SWNT functionalization (co/unco-SWNT) (7.484603/7.380 e + 003, 8.443547e + 003/7.95 e + 03 kcal mol^{-1}) along with the respective Irinotecan and PTX molecules (top of the complex; 119/377 kcal mol^{-1}).

molecule torsion was within the limits of most biological systems, though the PTX torsion was much greater than that of Irinotecan. This may be the reason why Irinotecan is much more preferred for nanoparticle interaction with biological systems.

As can be seen from Figure 11.30a–d, two complexed molecules of branched PEGylated DPCC functionalized Irinotecan and PTX physically loaded and covalent functionalized SWNT. The minimum energy of each of these complexes, along with their interactions, showed that physically loaded complexed drugs would be much more interactive or effective than would functionalized SWNTs. Similar results have been confirmed by Lay et al. [66], using an experimental approach.

Figure 11.31a–d shows how the SWNT carrying the physically loaded branched PEGylated DPCC drug molecules interacted with 20mer adenine–thymine and guanine–thymine (ssDNA). The interactions are clearly shown, irrespective of PTX or Irinotecan, although the Irinotecan-carried SWNT had a faster interaction than PTX (as evidenced by the radial distribution function). From the torsion angle distribution function (Figures 11.28 and 11.29), however, it is clear that the angle is much smaller for Irinotecan than for PTX. This torsion angle influences the winding and unwinding process of the DNA close to the drug molecules, which in turn may reduce the DNA replication process during cancerous or malignant stages. Due to torsion distribution disparity, it might be conjecture

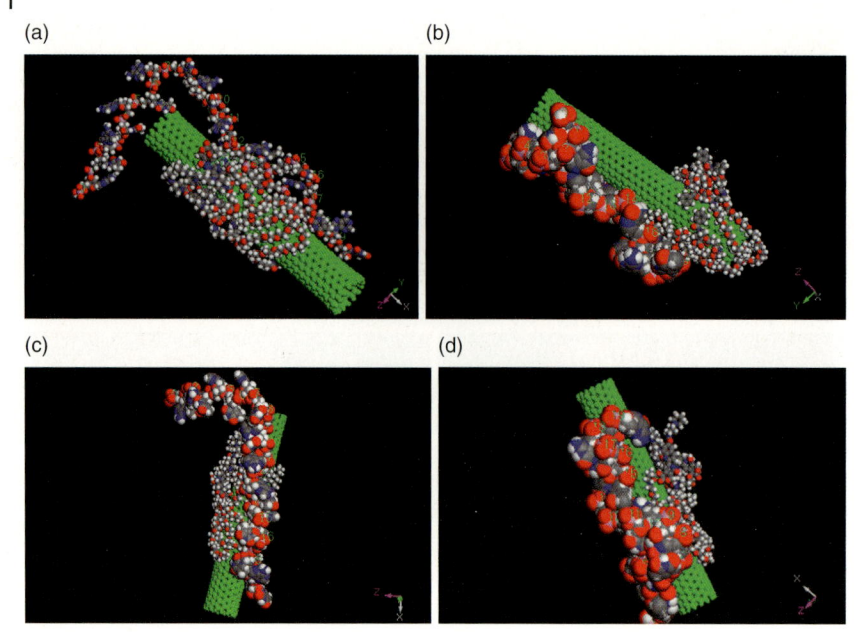

Figure 11.31 (a–d) Energy-minimized one molecule of branched PEGylated DPCC functionalized PTX (physically loaded) onto SWNT in the presence of ssDNA [adenine–thymine (AT) and guanine–thymine (GT)] and PEGylated DPCC functionalized Irinotecan (physically loaded) with SWNT in the presence of ssDNA (AT and GT).

that an SWNT carrying Irinotecan would be more effective than PTX, due to its smaller rotational angle distribution, which is sufficient to prevent fast DNA replication and thus suppress the cancer growth process. Figure 11.32a,b shows a covalent functionalized SWNT complexed with PTX and Irinotecan, and also highlights the great interaction in the presence of ssDNA. These investigations will be continued in order to further clarify the situation in the presence of double-stranded DNA.

11.3.3
Interaction of Nanotube with 20-Base Pair Guanine–Thymine -ssDNA in the Presence of Calcitriol

An interesting interaction of nanotubes with ssDNA in the presence of calcitriol (vitamin D2) (Figure 11.33a–c), a hormonally active form of vitamin D, is reported in this section. As can be seen from Figures 11.34 and 11.35, the interaction of SWNTs with ssDNA occurs much more rapidly in the presence of calcitriol than in its absence.

Figure 11.36a–c shows the Irinotecan- and PTX-complexed covalent SWNT with ssDNA in the presence of calcitriol. Although both have the tendency to wrap

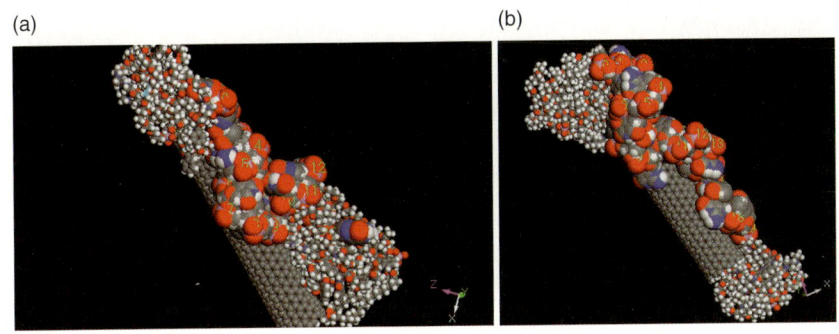

Figure 11.32 (a,b) Energy-minimized two molecules of branched PEGylated DPCC functionalized PTX and Irinotecan with covalent functionalized SWNT in the presence of ssDNA (GT).

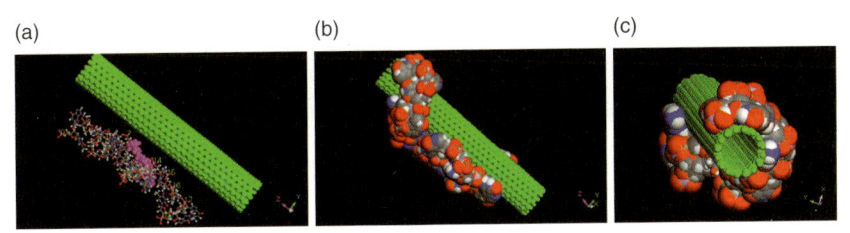

Figure 11.33 (a) Initial and (b) energy-minimized structures of SWNT, ssDNA, calcitriol; (c) Cross-sectional view (parallel and un-covalent).

around the nanotube, the radial distribution of the respective complexed PTX and Irinotecan shows clearly how well each is bound to the SWNT in the presence of calcitriol.

Figures 11.37 and 38 show that interaction of Irinotecan-complexed SWNT in the presence of calcitriol is much more favored with ssDNA than that of PTX. As the dynamic time progresses, the radial distribution function as a function of the distance from the drug to the SWNT is decreased, and can be seen to be in the close vicinity of the SWNT; in other words, Irinotecan appears to be closer than PTX. This clearly shows that the Irinotecan interaction within the same environment is much more prominent than that of PTX. Nanoparticle interaction with Irinotecan in the presence of a PEG complex has been studied experimentally by several groups, and similar results obtained [68, 69, 71, 72].

Figure 11.39 shows the final energy-minimized structure of (using the same starting geometry) covalent functionalized SWNT with Irni complexes (branched PEG-DPCC, ssDNA and calcitriol). This shows clearly that interaction in the presence of calcitriol is more prominent than without, and is worthy of further investigation.

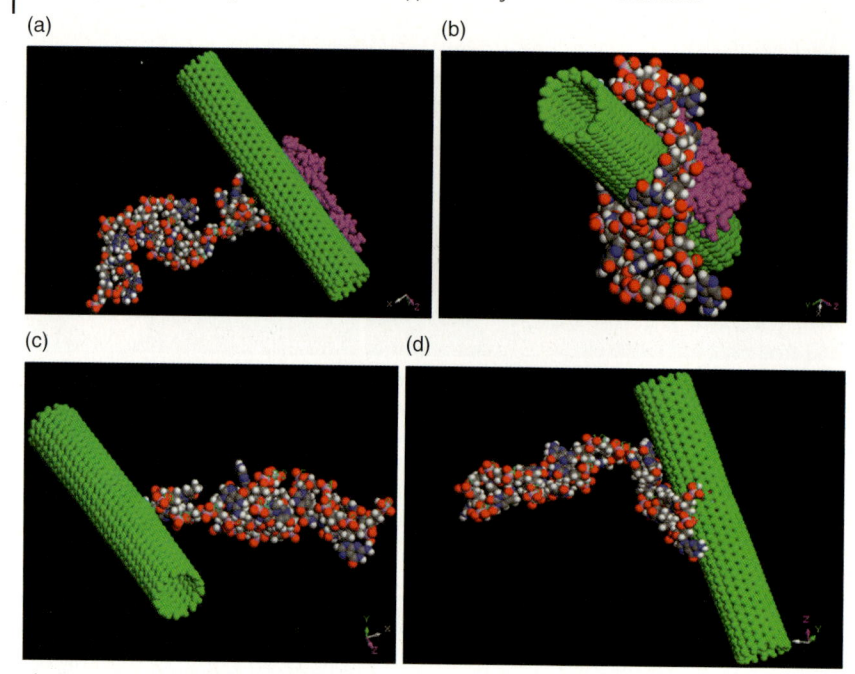

Figure 11.34 (a–d) Initial and final energy-minimized structures of SWNT-uncovalent-ssDNA (perpendicular), and without uncovalent calcitriol.

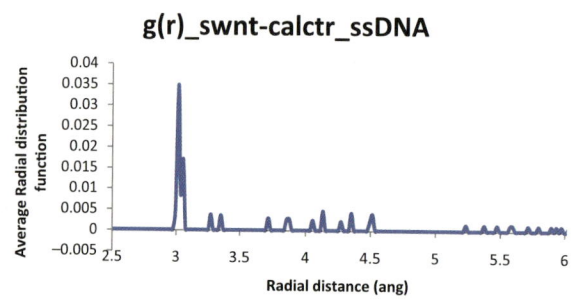

Figure 11.35 Average radial distribution function of ssDNA and SWNT in the presence of calcitriol.

11.4
Future Perspectives

The future perspectives for computational identification and analysis with the help of many nanodevices are limitless and "explosive." Based on the advancements of computer hardware and software, computer modeling will undoubtedly help medical practitioners and research groups to identify biocompatible and safe nanomaterials and nanotools for the early detection of diseases.

11.4 Future Perspectives | 299

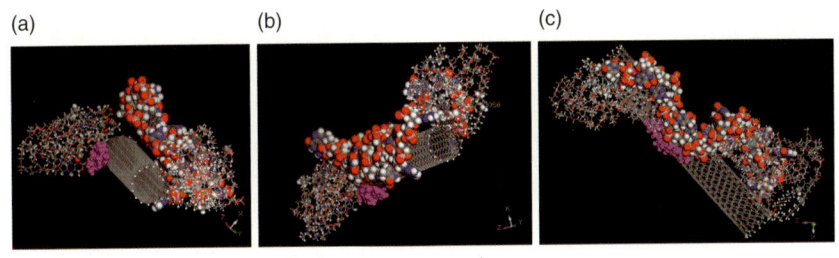

Figure 11.36 (a) Initial and (b) final energy-minimized Irin-*co*-br3-PEG-*co*-DPCC-*co*-SWNT; (c) Final energy-minimized two molecules of PTX-*co*-br3-PEG-*co*-DPCC-*co*-SWNT (1.332130 e + 004 kcal mol^{-1}) interaction of 20-mer ssDNA (GT) in the presence of calcitriol.

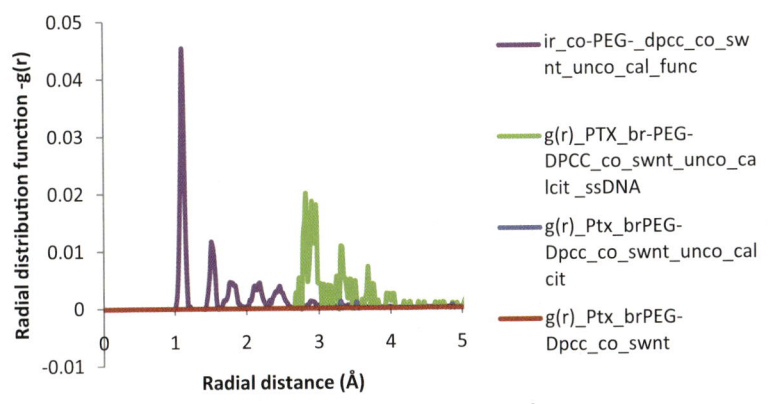

Figure 11.37 Radial distribution function. Two molecules of branched (3) PEGylated DPCC functionalized PTX and Irinotecan with covalent functionalized SWNT interaction of ssDNA (GT) without calcitriol.

Most illness can be eradicated through a "clean water" technology using nanotechnology, which in turn reduces healthcare costs. In short, such a transformation would revolutionize everything from manufacturing to medicine, to energy production. Such an initiative would help in the design and simulation of various stable biocompatible molecules with drugs, using MD techniques for faster optimization, and this in turn would help in the synthesis and testing only of promising biocompatible structures in the laboratory. At the same time, computer modeling studies can be used to improve the molecular interactions under various conditions.

Electrostatic Energy vs dynamics simulation time

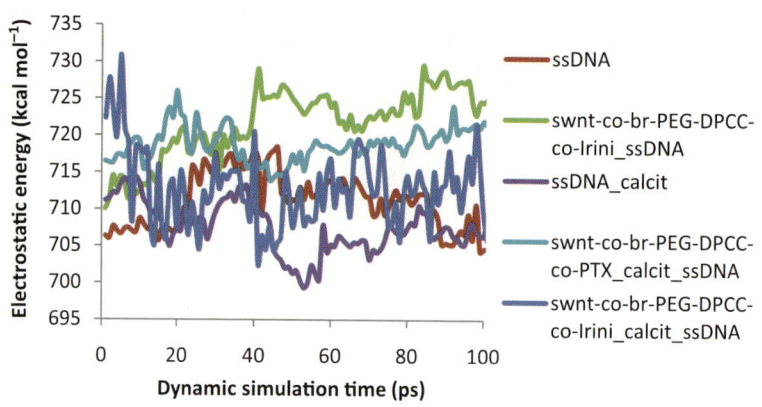

Figure 11.38 Electrostatic energy. Two molecules of branched (3) PEGylated DPCC functionalized PTX and Irinotecan with covalent functionalized SWNT interaction of ssDNA (GT) without calcitriol.

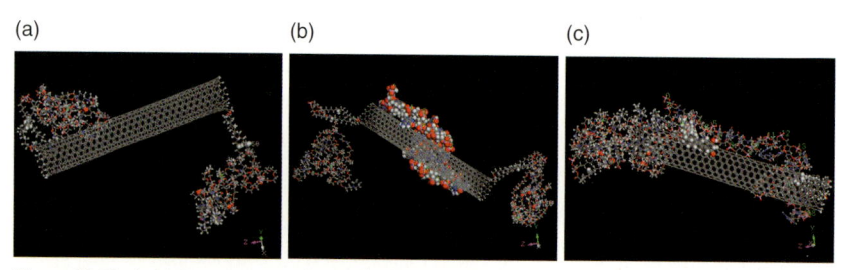

Figure 11.39 (a–c) Dynamic simulation final structure of (a) SWNT-*co*-br-PEG-Irni; (b) SWNT-*co*-br-PEG-Irni with ssDNA; (c) SWNT-*co*-br-PEG-Irni with ssDNA and calcitriol.

11.5
Executive Summary

- Molecular dynamics and mesoscale simulations can be used to predict the characteristics of the interaction of a variety of nanoscale smart materials with biological entities.

- The synthesis and characterization of a novel nanoscale antitumor drug can be efficiently developed through a series of MD and mesoscale simulations.

- In this chapter, attention is focused mainly on the development of PEGylated-lipophilic polymers with or without functionalized SWNT, with physical loading of PTX and Irinotecan as drug-delivery carriers for anticancer therapeutics. The various simulation production and characterization of these materials, complexed with SWNTs and controlled-release strategies, are discussed. A brief overview is also presented of various targeted drug-delivery approaches in cancer, using SWNTs with and without covalent functionalization chemistry.

- The structure and function of PEGylated lipophilic biological molecules covalent and noncovalent functionalization of SWNT with physically loaded or functionalized PTX with PEG and its adsorption characteristics have been investigated.

- PEGylation with lipophilic molecules may help to achieve the processability and properties of CNTs, to benefit their biomedical application.

- Branched PEGylated DPCC-functionalized drug molecules physically loaded to SWNTs seem better than their functionalized counterparts.

- If the drug molecules are not functionalized, then they may be physically loaded and localized near SWNTs. Branched hydrophilic polymers would have the added advantage of carrying more drug molecules, which can bound by a cleavable linker.

- The complexation of these compounds in the presence of ssDNA seems to be more interactive. However, although this type of combination should be tested experimentally, it is beyond the scope of this simulation.

- It has been indicated that, in the presence of calcitriol (vitamin D2), hormonally active compounds will help to interact with these complexes much more rapidly, but this must also be tested experimentally.

Acknowledgments

The author acknowledges the partial financial support provided by the departments of Chemistry and Mathematics, Science of Advanced Materials, the College of Science and Technology, the School of Graduate studies at CMU and the Dow Chemical Company for the completion of these studies. The author wishes to thank Professors S. Hirschi, B. Howell, A. Dumitrascu, J. Falender, S. Juris, C. Lee, A. Sharma, M. Chai, A. Mellinger, A. Mueller, A. Jensen, and D. Mohanty for their valuable discussions. Thanks are also expressed to undergraduate students E. Shaw, P. Chhetri, M. Lalko, O. Fadiran, and graduate students N. Almedia, T. Arthanayaka, C. Ampadu, E. Goold, M. Kujawski (U. Maryland) and A. Abu-Shams for their valuable support.

References

1 Edwards, B.K., Ward, E., Kohler, B.A., Eheman, C., Zauber, A.G., Anderson, R.N., Jemal, A., Schymura, M.J., Lansdorp-Vogelaar, I., Seeff, L.C., van

1. Ballegooijen, M., Luuk Goede, S. and Ries, L.A.G. (2009) Annual report to the nation on the status of cancer, 1975-2006, featuring colorectal cancer trends and impact of interventions risk factors, screening, and treatment to reduce future rates. *Cancer*, **116** (3), 544–73.
2. Kelland, L.R. and Farrell, N.P. (2000) *Platinum-Based Drugs in Cancer Therapy*, Humana Press, New Haven, and references therein.
3. Weiss, R.B. and Christian, M.C. (1993) New cisplatin analogues in development: a review. *Drugs*, **46**, 360–77.
4. Gordon, M. and Hollander, S. (1993) Review of platinum anticancer compounds. *Journal of Medicine*, 209–65.
5. Carraher, C., Scott, W. and Giron, D. (1985) Progress in biomedical polymers, in *Bioactive Polymeric Systems* (eds C. Gebelein and C. Carraher), Plenum, New York, Chapter 20, 3–15.
6. Howell, B.A., Fan, D. and Rakesh, L. (2008) Nanoscale dendrimer-platinum conjugates as multivalent antitumor drugs, in *Inorganic and Organometallic Macromolecules, Design & Applications* (eds A.S. Abd-El-Aziz, C.E. Carraher Jr, C.U. Pittmon Jr and M. Zeldin), Springer Science, Ch. 11, p. 269 and references therein.
7. Kam, N.W.S. and Dai, H.J. (2005) Carbon nanotubes as intracellular protein transporters: generality and biological functionality. *Journal of the American Chemistry Society*, **127**, 6021–6.
8. Raffa, V., Ciofani, G., Vittorio, O., Riggio, C. and Cuschieri, A. (2010) Physicochemical properties affecting cellular uptake of carbon nanotubes. *Nanomedicine*, **5** (1), 89–97, and references therein.
9. Chambers, G., Carroll, C., Farrell, G.F., Dalton, A.B., McNamara, M., Panhuis, M. and Byrne, H.J. (2003) Characterization of the interaction of gamma cyclodextrin with single-walled carbon nanotubes. *Nano Letters*, **3** (6), 843–6, and references therein.
10. Couvreur, P., Kante, B., Grislain, L., Roland, M. and Speiser, P. (1982) Toxicity of poly-alkyl-cyanoacrylate nanoparticles II: doxorubicin-loaded nanoparticles. *Journal of Pharmaceutical Sciences*, **71** (7), 790–2 and references therein.
11. Daxiang, C., Tian, F., Cengiz, S.O., Wang, M. and Gao, H. (2005) Effect of single wall carbon nanotubes on human HEK293 cells. *Toxicology Letters*, **155**, 73–85.
12. Böttcher, C., Endisch, C., Fuhrhop, J.H., Catterall, C. and Eaton, M. (1998) High yield preparation of oligomeric C-type DNA toroids and their characterization by cryoelectron microscopy. *Journal of the American Chemical Society*, **120**, 12–17.
13. Jawalkar, S.S., Raju, K.V., Halligudi, S.B., Sairam, M. and Aminabhavi, T.M. (2007) Molecular modeling simulations to predict compatibility of poly(vinyl alcohol) and chitosan blends: a comparison with experiments. *Journal of Physical Chemistry B*, **111** (10), 2431–9.
14. Chawla, J.S. and Amiji, M.M. (2002) Biodegradable poly(caprolactone) nanoparticles for tumor-targeted delivery of tamoxifen. *International Journal of Pharmaceutics*, **249** (1–2), 127–38.
15. Kujawski, M., Howell, B., Chai, M., Mueller, A., Fan, D., Ravi, S., Slominski, C. and Rakesh, L. (2008) Computer-aided design of nanoscale smart materials for biomedical application. *Journal of Nanomedicine*, **3** (5), 719–39, and references therein.
16. Hood, L. (1998) Biomarker implications. *Human Genome News*, **9** (3), 401–21 and references therein.
17. Keppler, B. (ed.) (1993) *Metal Complexes in Cancer Chemotherapy*, VCH, Weinheim.
18. Daigle, C.C., Chalupa, D.C., Gibb, F.R., Morrow, P.E., Oberdörster, G. and Utell, M.J. (2003) Ultrafine particle deposition in humans during rest and exercise. *Inhalation Toxicology*, **15**, 539–52.
19. Mansfield, M., Rakesh, L. and Tomalia, D. (1996) Packing algorithms. *Chemical Physics*, **105**, 3245–9.
20. Martin, C.R. and Kohli, P. (2003) The emerging field of nanotubes biotechnology. *Nature Reviews Drug Discovery*, **2**, 29, 37 and references therein.
21. Chen, J., Perebeinos, V., Freitag, M., Tsang, J., Fu, Q. and Liu, J. (2005) Bright infrared emission from electrically induced excitons in carbon nanotubes.

Science, **310** (5751), 1171–4 and references therein.

22 Ikeda, N. and Miyasaka, T. (2007) Plastic and solid-state dye-sensitized solar cells incorporating single-wall carbon nanotubes. *Chemistry Letters*, **36** (3), 466–7.

23 Svensson, K., Olin, H. and Olsson, E. (2004) Nanopipettes for metal transport. *Physical Review Letters*, **93**, 145901.

24 Liu, F. (2008) In situ pressure monitor and associated methods, Patent, Pub. No. WO/2008/ 127797, October 23, 2008, International Application No. PCT/ US2008/055525.

25 Borghetti, J., Derycke, V., Lenfant, S., Chenevier, P., Filoramo, A. and Goffman, M. (2006) Optoelectronic switch and memory devices based on polymer-functionalized carbon nanotubes transistors. *Advanced Materials*, **18** (19), 2535–40.

26 Heller, I., Janssens, A.M., Mannik, J., Minot, E.D., Lemay, S.G. and Dekker, C. (2008) Identifying the mechanism of biosensing with carbon nanotube transistors. *Nano Letters.*, **8** (2), 591–5.

27 Wei, J.Q., Jia, Y., Shu, Q.K., Gu, Z.Y., Wang, K.L. and Zhuang, D.M. (2007) Double-walled carbon nanotube solar cells. *Nano Letters*, **7** (8), 2317–21.

28 Rakesh, L. and Lee, C. (2009) Molecular modeling of interaction between diabetic drug and antioxidant in controlling sucrose. *Conference Proceedings of the American Institute of Physics*, Vol. **1** (1168), pp. 827–31.

29 Chai, M., Ravi, S. and Rakesh, L. (2009) *Investigation of Colloidal Suspension of SWNT & Cyclodextrin Using AFM & MD Simulation*, ACS, Ch. 28, ISBN # 978-0-24.

30 Kujawski, M., Rakesh, L., Gala, K., Jensen, A., Fahlman, B., Feng, Z. and Mohanty, D. (2007) Molecular dynamics simulation of PAMAM dendrimer-fullerene conjugates: generation one through four. *Journal of Nanoscience and Nanotechnology*, **7**, 4/5.

31 Longhurst, M.J. and Quirke, N. (2007) Temperature-driven pumping of fluid through single-walled carbon nanotubes. *Nano Letters*, **7**, 3324–8.

32 Dai, Y., Tang, C. and Guo, W. (2008) Simulation studies of a "nanogun" based on carbon nanotubes. *Nano Research*, **1** (2) 176–83, and references therein.

33 Král, P. and Tománek, D. (1999) Laser-driven atomic pump. *Physical Review Letters*, **82**, 5373–6.

34 Insepov, Z., Wolf, D. and Hassanein, A. (2006) Nanopumping using carbon nanotubes. *Nano Letters*, **6**, 1893.

35 Wang, Q. (2009) Atomic transportation via carbon nanotubes. *Nano Letters*, **9**, 245–9.

36 Ulbricht, H., Moos, G. and Hertel, T. (2003) Interaction of C60 with carbon nanotubes and graphite. *Physical Review Letters*, **90**, 095501.

37 Lee, J.U., Huh, J., Kim, K.H., Park, C. and Jo, W.H. (2007) Aqueous suspension of carbon nanotubes via non-covalent functionalization with oligothiophene-terminated poly(ethylene glycol). *Carbon*, **45**, 1051–7.

38 Moghimi, S.M. and Szebeni, J. (2003) Stealth liposomes and long circulating nanoparticles: critical issues in pharmacokinetics, opsonization and protein-binding properties. *Progress in Lipid Research*, **42**, 463–78.

39 Molineux, G. (2002) PEGylation: engineering improved pharmaceuticals for enhanced therapy. *Cancer Treatment Reviews*, **28** (Suppl. A), 13–16.

40 Roberts, M.J., Bentley, M.D. and Harris, J.M. (2002) Chemistry for peptide and protein PEGylation. *Advanced Drug Delivery Reviews*, **54**, 459–76.

41 Liu, Z., Cai, W.B., He, L.N., Nakayama, N., Chen, K., Sun, X.M., Chen, X.Y. and Dai, H.J. (2007) In vivo biodistribution and highly efficient tumor targeting of carbon nanotubes in mice. *Nature Nanotechnology*, **2** (1), 47–52.

42 Kryachko, E.S. and Remacle, F. (2005) Complexes of DNA bases and gold clusters Au-3 and Au-4 involving nonconventional N-H center dot center dot center dot Au hydrogen bonding. *Nano Letters*, **5** (4), 735–9.

43 Nath, N. and Chilkoti, A. (2004) Label-free colorimetric biosensing using nanoparticles. *Journal of Fluorescence*, **14** (4), 377–89.

44 Fan, H.Y., Leve, E.W., Scullin, C., Gabaldon, J., Tallant, D., Bunge, S., Boyle, T., Wilson, M.C. and Brinker, C.J.

(2005) Surfactant-assisted synthesis of water-soluble and biocompatible semiconductor quantum dot micelles. *Nano Letters*, **5** (4), 645–8 and references therein.
45 Markowitz, M.A., Dunn, D.N., Chow, G.M. and Zhang, J. (1999) The effect of membrane charge on gold nanoparticle synthesis via surfactant membranes. *Journal of Colloid and Interface Science*, **210** (1), 73–85. 53.
46 Chen, R.J., Zhang, Y.G., Wang, D.W. and Dai, H.J. (2001) Noncovalent sidewall functionalization of single-walled carbon nanotubes for protein immobilization. *Journal of the American Chemical Society*, **123** (16), 3838–9.
47 Xiaoming Sun, Z.L., Welsher, K., Robinson, J.T., Goodwin, A., Zaric, S. and Dai, A.H. (2008) Nano-graphene oxide for cellular imaging and drug delivery. *Nano Research*, **1**, 203–12, and references therein.
48 Lupulescu, A. (1990) *Hormones and Vitamins in Cancer Treatment*, CRC Press.
49 Whyte, M.P., Haddad, J.G., Jr, Walters, D.D., and Stamp, T.C.B. (1979) Vitamin D bioavailability: serum 25-hydroxyvitamin D levels in man after oral, subcutaneous, intramuscular, and intravenous vitamin D administration. *Journal of Clinical Endocrinology and Metabolism*, **48**, 906–11.
50 Sinani, V.A., Gheith, M.K., Yaroslavov, A.A., Rakhnyanskaya, A.A., Sun, K., Mamedov, A.A., Wicksted, J.P. and Kotov, N.A. (2005) Aqueous dispersions of single-wall and multiwall carbon nanotubes with designed amphiphilic polycations. *Journal of the American Chemical Society*, **127** (10), 3463–72 & references therein.
51 Fuertges, F. and Abuchowski, A. (1990) The clinical efficacy of poly (ethylene glycol)-modified proteins. *Journal of Controlled Release*, **11**, 139.
52 Working, P.K., Newman, M.S., Johnson, J. and Cornacoff, J.B. (1997) Safety of poly(ethylene glycol) and poly(ethylene glycol) derivatives, in *Polyethylene Glycol Chemistry and Biological Applications* (eds J.M. Harris and S. Zalipsky), American Chemical Society, Washington DC, p. 45 and references therein.

53 Mahato, R.I. (2005) *Biomaterials for Delivery and Targeting of Proteins and Nucleic Acids*, CRC Press.
54 Steenbock, H. (1924) The induction of growth promoting and calcifying properties in a ration by exposure to light. *Science*, **60**, 224–5.
55 Steenbock, H. and Black, A. (1924) Fat-soluble vitamins. XVII. The induction of growth-promoting and calcifying properties in a ration by exposure to ultraviolet light. *Journal of Biology Chemistry*, **61**, 405–22.
56 McCollum, E.V., Simmonds, N., Becker, J.E. and Shipley, P.G. (1922) An experimental demonstration of the existence of a vitamin which promotes calcium deposition. *Journal of Biological Chemistry*, **53**, 293–8.
57 De Luca, H.F. (2004) Overview of general physiologic features and functions of vitamin D. *American Journal of Clinical Nutrition*, **80** (Suppl.), 1689S–96S.
58 Daniel, L., Manthey, B., Kreuter, J., Speiser, P. and Deluca, P.P. (1984) Distribution and elimination of coated polymethyl [2-14C]methacrylate nanoparticles after intravenous injection in rats. *Journal of Pharmaceutical Science*, **73** (10), 1433–7.
59 Araujo, L., Löbenberg, R. and Kreuter, J. (1999) Influence of the Surfactant concentration on the body distribution of nanoparticles. *Journal of Drug Targeting*, **6** (5), 373–85.
60 Singh, R., Pantarotto, D., Lacerda, L., Pastorin, G. and Klumpp, C. (2006) Tissue biodistribution and blood clearance rates of intravenously administered carbon nanotube radiotracers. *Proceedings of the National Academy of Sciences of the United States of America*, **103**, 3357–62.
61 Shim, M., Kam, N., Chen, R., Li, Y. and Dai, H. (2002) Functionalization of carbon nanotubes for biocompatibility and biomolecular recognition. *Nano Letters*, **2**, 285–8.
62 Kam, N.W.S., Liu, Z. and Dai, H.J. (2005) Functionalization of carbon nanotubes via cleavable disulfide bonds for efficient intracellular delivery of siRNA and potent gene silencing. *Journal of the American Chemical Society*, **127**, 12492–3.

63 Zhuang, L., Davis, C., Weibo, C., Lina, H., Chen, X. and Hongjie, D. (2008) Circulation and long-term fate of functionalized, biocompatible single-walled carbon nanotubes in mice probed by Raman spectroscopy. *Proceedings of the National Academy of Sciences of the United States of America*, **105** (5), 1410–15.

64 Liu, Z., Chen, K., Davis, C., Sherlock, S., Cao, Q., Chen, X. and Dai, H. (2008) Drug delivery with carbon nanotubes for in vivo cancer treatment. *Cancer Research*, **68** (16), 6652–60.

65 Chatterjee, T., Yurekli, K., Hadjiev, V.G. and Krishnamoorti, R. (2005) Single-walled carbon nanotube dispersions in poly(ethylene oxide). *Advanced Functional Materials*, **15** (11), 1832–8 and references therein.

66 Yang, S.T., Fernando, K.A.S., Liu, J.H., Wang, J., Sun, H.F., Liu, Y.F., Chen, M., Huang, Y.P., Wang, X., Wang, H.F. and Sun, Y.P. (2008) Covalently PEGylated carbon nanotubes with stealth character in vivo. *Small*, **4** (7), 940–4.

67 Liu, Z., Fan, A.C., Rakhr, K., Sherlock, S., Goodwin, A., Chen, X., Yang, Q., Felsher, D.W. and Dai, H. (2009) Supramolecular stacking of doxorubicin on carbon nanotubes for in vivo cancer therapy. *Angewandte Chemie International Edition*, **48** (41), 7668–72 and references therein.

68 Williams, J., Lansdown, R., Sweitzer, R., Romanowski, M., LaBell, R., Ramaswami, R. and Unger, E. (2003) Nanoparticle drug delivery system for intravenous delivery of topoisomerase inhibitors. *Journal of Controlled Release*, **91** (1–2), 28, 67–172 and references therein.

69 Pisani, J. and Bonduelle, Y. Opportunities and barriers in the biosimilar market: Evolution or revolution for generic companies? Available at: http://www.ableindia.org/biosimilars.pdf; ww.pwc.com/uk/strategy (accessed February 2010).

70 Ashiuchi, M., Nawa, C., Kamei, T., Song, J.S., Hong, S.P., Sung, M.H., Soda, K., Yagi, T. and Misono, H. (2001) Physiological and biochemical characteristics of poly-γ-glutamate synthetase complex of *Bacillus subtilis*. *European Journal of Biochemistry*, **268** (22), 6003–3.

71 McKeage, M. and Kelland, L. (1992) New platinum drugs, in *Molecular Drug-DNA Interactions* (eds S. Neidle and M. Waring), Macmillan, New York, pp. 169–212.

72 Machiday, Y., Onishi, H. and Kato, Y. (2003) Efficacy of nanoparticles containing irinotecan prepared using poly(DL-lactic acid) and poly(ethylene glycol)-poly(propylene glycol)-poly(ethylene glycol) against M5076 tumor in the early liver metastatic stage. *STP Pharma Sciences*, **13** (4), 225–30.

73 Parveen, S. and Sahoo, S.K. (2006) Nanomedicine – Clinical applications of polyethylene glycol conjugated proteins and drugs. *Clinical Pharmacokinetics*, **45** (10), 965–88 & references therein.

74 Material Studio (software) (2005) Accelrys Inc., San Diego.

75 Li, Z., Chen, Y., Li, X., Kamins, T.I., Nauka, K. and Williams, R.S. (2004) Sequence-specific label-free DNA sensors based on silicon nanowires. *Nano Letters*, **4**, 245.

76 Langer, R. (2007) Cancer nanotechnology: small, but heading for the big time. *Nature Reviews Drug Discovery*, **6**, 174–5. http://www.nature.com/nrd/journal/v6/n3/full /nrd2285.html.

77 Ostiguy, C., Lapointe, G., Trottier, M., Ménard, L., Cloutier, Y., Boutin, M., Antoun, M. and Normand, C. (2008) Studies and Research Projects, Health Effects of Nanoparticles, Second edition. IRSST, Communications Division, Montréal, Report R-589.

78 Srivastava, D., Menon, M. and Cho, K. (2001) Computational nanotechnology with carbon nanotubes and fullerenes. *Computing in Science and Engineering*, **3** (4), 42 and references therein.

79 Zheng, M., Jagota, A., Strano, M.S., Santos, A.P., Barone, P., Chou, S.G., Diner, B.A., Dresselhaus, M.S., McLean, P.S. and Onoa, G.B. (2003) Structure-based carbon nanotube sorting by sequence-dependent DNA assembly. *Science*, **302**, 1545–8.

80 Lay, C.L., Liu, H.Q., Tan, H.R. and Liu, Y. (2001) Delivery of paclitaxel by physically loading onto poly(ethylene glycol) (PEG)-graft-carbon nanotubes for potent cancer therapeutics. *Nanotechnology*, **21**, 065101.

Part Two
Overview of Applications in Cancer

12
Carbon Nanotubes for Cancer Therapy

William H. Gmeiner

12.1
Introduction

The past decade has resulted in only a moderate demonstration of tangible progress in the collective efforts of the medical and academic communities, together with the pharmaceutical industry, to significantly reduce the incidence, prevalence, and mortality of cancer. Today, the public remains fearful of a diagnosis of cancer, and skeptical that preclinical results presented as "breakthroughs" will ever have a significant effect on cancer treatment in humans. The magnitude of the problem remains staggering: in 2009, a total of 1 479 350 men and women was diagnosed with cancer, and 562 340 deaths resulted from the condition (Figure 12.1) [1]. Moreover, several types of cancer, including pancreatic cancer and glioblastoma multiforme, remain essentially incurable (ACS Facts & Figures 2009) [2].

The overall age-adjusted incidence and mortality rates due to all malignancies have declined only slightly during the past decade, despite the introduction of targeted therapeutics into the clinic that take advantage of the significant new insights into the molecular basis for cancer etiology and progression (Figure 12.1). Clearly, there is a need for the continued development of new technologies that can impact cancer detection and treatment.

12.1.1
Limitations of Current Therapy Options

Cancer is currently treated by a combination of surgery, radiation, and chemotherapy. Surgery can be curative if the primary tumor is resectable, and if metastatic disease is either not detectable or the metastases do not become life-threatening. Either radiation or chemotherapy can be used in a neoadjuvant setting to reduce tumor mass prior to surgery. Radiation and/or chemotherapy may also be used postoperatively in an adjuvant setting to eradicate any occult disease. For some malignancies (e.g., prostate cancer), radiation therapy provides an alternative to surgery. Instances of chemotherapy alone being curative are

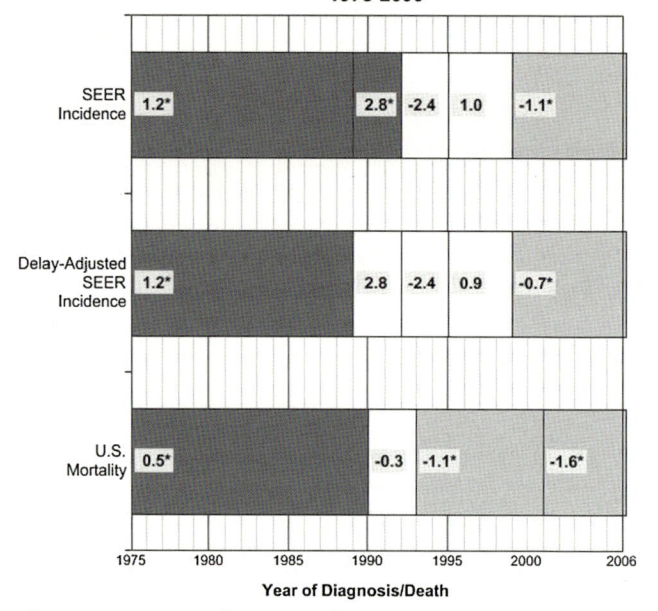

Figure 12.1 Depiction of recent trends in cancer incidence and mortality. Despite the introduction of new targeted therapies for cancer treatment, incidence and mortality have declined only slightly. Data generated from the NCI SEER web site.

relatively rare, however. For example, testicular cancer can be successfully treated with combination chemotherapy regimens that include platinated drugs [3]. Chemotherapy is frequently utilized in an adjuvant setting to reduce the risk of recurrent disease, and often also for the palliation of advanced disease.

12.1.2
Developing Nanomaterials for Cancer Treatment

Nanotechnology is the development, investigation, and manipulation of materials of the nanometer-sized dimension (1 nanometer is one-billionth of a meter). As the diameters of typical human cells range between 10 and 100 µm, nanometer-scale materials are much smaller than human cells, or even smaller than organelles within human cells, such as mitochondria. In this chapter, attention is focused on the recent developments that have employed carbon nanotubes (CNTs) for cancer treatment. Physically, CNTs range from 2 to 40 nm in diameter, and may be up to a few microns in length. Thus, the lengths of CNTs will generally greatly exceed their widths, so that the CNTs are deemed to be "quasi-one-dimensional (1-D)" objects [4]. The nanometer-sized diameter of CNTs, in conjunction with their high length:width ratio (i.e., high aspect ratio) and tubular shape, confers physical properties that differ dramatically differ from those of other forms of carbon, such as graphite and diamond. The most notable benefits of CNTs include their high electrical and thermal conductivities, and high tensile strength [5]. Currently, CNTs are undergoing extensive investigation in preclinical models to determine their potential for improving cancer treatment. In particular, three properties of CNTs are of particular value for cancer treatment:

- The efficient conversion of light into thermal energy (the photothermal effect) [6–8].
- The accumulation of CNTs into tumor tissues, via an enhanced permeability and retention (EPR) effect [9].
- The transport of CNTs across the plasma membrane, which results in the cellular internalization of CNTs and any associated "cargo" (e.g., drugs) [10–12].

These three important properties of CNTs are currently being used to develop the CNT-based nanomaterials that are expected to impart a marked improvement in cancer treatment.

12.1.3
CNTs: Physical Properties, Manufacture, and Chemical Modifications

Carbon is able to exist in several forms or structures (allotropes), including diamond and graphite – that occur naturally – and buckyballs and CNTs – that are man-made. CNTs are composed of sp^2-hybridized carbon, as occurs in graphite. A single-walled carbon nanotube (SWNT) can be visualized as being composed of a single graphene layer rolled into a cylinder (Figure 12.2). In general, SWNTs have diameters ranging from 1 to 10 nm, and are capped at one end. They can also be either metallic or semi-conducting, depending on how the two sides of the graphene sheet are "joined" (i.e., the CNT chirality; Figure 12.3). Although SWNTs can be visualized as being formed from a graphene layer, they are produced via growth processes from low-molecular-weight carbon stock, using methods such as chemical vapor deposition (CVD), arch discharge, and plasma, pulsed laser vaporization; alternatively, a Co-Mo catalyst (CoMoCat) can be used [5, 13]. Those CNTs produced via a CVD process, such as high-pressure CO conversion (HiPCO) have a higher purity than those produced by alternative methods. Unfortunately, the CVD-produced CNTs also have more wall defects [14] which reduce the very high electrical conductivity that is characteristic of CNTs [15]. The SWNTs can also be nested together in a "Russian doll" fashion, to produce multi-walled carbon nanotubes (MWNTs) [16], where the multiple layers are held together by van der Waals forces. An MWNT composed of two layers is referred to as a double-walled nanotube (DWNT) [17]. Both, SWNTs and MWNTs have been used for photothermal and drug-delivery applications in the treatment of cancer [7, 9, 18].

The van der Waals forces that promote the nesting together of SWNTs to form MWNTs can also cause the aggregation of individual SWNTs or MWNTs.

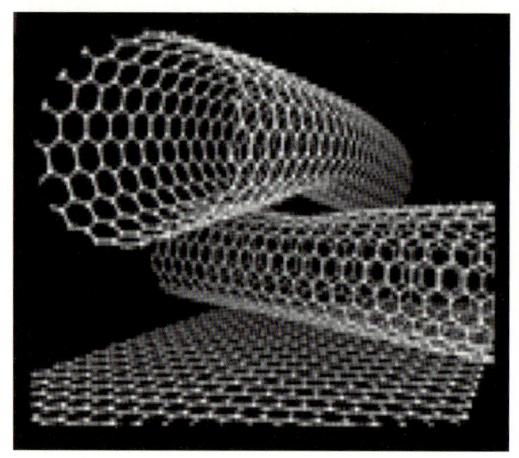

Figure 12.2 Visualization of SWNTs from an "end-on" perspective, and also from one side. The CNTs are shown above a single graphene sheet. The SWNTs are composed of hexagonal arrays of sp^2-hybridized carbon. Graphic courtesy of Aloysius Hepp (NASA).

Typically, CNT aggregates do not have the same favorable properties as individualized CNTs for cancer treatment applications. Furthermore, CNT aggregates may cause serious problems of toxicity that the individualized CNTs do not cause [19, 20]. CNTs are often coated with an amphiphilic molecule such as a phospholipid-poly(ethylene glycol) (PL-PEG) [21] or single-stranded DNA [7, 22, 23] to promote their aqueous solubility and inhibit CNT aggregation. CNT suspensions prepared with DNA, sodium deoxycholate, and carboxymethylcellulose were found to be of the highest quality [24]. Alternatively, CNTs can be oxidized using a strong acid to generate the carboxylate groups that confer aqueous solubility and also provide functionality for the chemical conjugation of CNTs [25].

12.2
Hyperthermia for Cancer Treatment

Hyperthermia is a type of cancer treatment in which the body tissues are exposed to supraphysiological temperatures, either to kill the cancer cells or to render them more sensitive to the effects of radiation and chemotherapy. Hyperthermia generally increases the blood flow, and thus may enhance the effectiveness of chemotherapy and radiation therapy. The localization of a hyperthermic treatment to regions of tissues composed predominantly of malignant cells is essential for therapeutic success, as hyperthermia may either damage or kill both nonmalignant and malignant cells, or perhaps sensitize nonmalignant cells to chemotherapy or radiation therapy. One fundamental challenge to the use of nanotubes

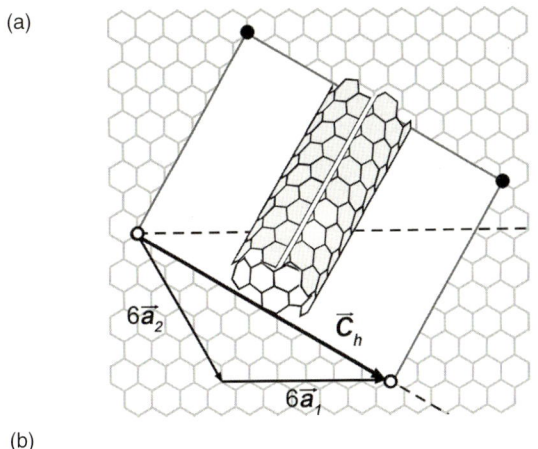

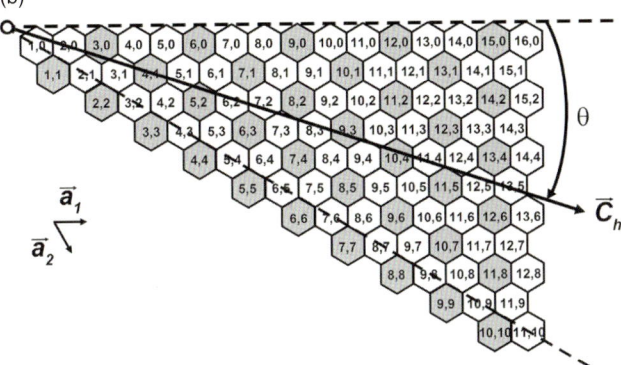

Figure 12.3 (a) Construction of a nanotube from a graphene sheet, where the white region represents the area used to form the nanotube. Two corresponding lattice points overlap (open or filled circles) to form a (6,6) nanotube with a chiral angle of 30°; (b) Depiction of possible SWNT (n, m) structures represented on a single graphene sheet. Semiconducting SWNTs are shown in white, and metallic SWNTs are shaded. Reprinted with permission from Ref. [4]; © 2008, American Chemical Society.

and other nanoparticles to facilitate hyperthermic cancer treatment is the development and implementation of strategies that will lead to the selective accumulation of nanotubes in malignant tissues, either through the EPR effect [9, 26, 27] or by the active targeting of proteins expressed at elevated levels by malignant cells [28, 29].

The early models of cell killing by heat suggested that heat-induced cellular inactivation was a two-step process. In the first step, heating produced nonlethal lesions, while in the second step heating converted the nonlethal lesions into lethal events [30]. The assumptions of this model were that:

- Nonlethal lesions are produced at random, and at a rate that is constant during heat exposure at a constant temperature.

- The number of nonlethal lesions per cell remains constant following heat treatment (i.e., there is no reversion or repair).
- The conversion of a nonlethal lesion into a lethal event occurs at random, and at a rate that is constant during heat exposure.
- The conversion of one nonlethal lesion to a lethal lesion leads to cell death.

This model of hyperthermia provided a reasonable fitting for the experimental clonogenic survival data for cells treated initially at an elevated temperature, followed by further treatment at a lower – but still supraphysiological – temperature. Alternatively, the "single-hit, multi-target" and the "linear-quadratic" models that were developed to rationalize the effects of radiation-induced cellular damage have been adapted to study the cellular effects of hyperthermia treatment [31].

Regardless of the model used for hyperthermia analysis, the results of *in vitro* studies have shown that the rate of cell killing is exponential, and is dependent on the cell-type, the temperature, and the duration of exposure. For example, very high levels of cell killing (survival fractions <0.001) are observed at a temperature of 45 °C for 1 h, whereas temperatures <42 °C for up to 8 h have little effect on clonogenic survival [32]. In the ablative approaches in current clinical use (e.g., radiofrequency ablation; RFA), the malignant tissues are often heated to >55 °C for brief time periods (>1 s), leading to coagulative necrosis and immediate cell death [33]. While higher temperatures are achieved during the implementation of currently used ablative treatments, the subtle temperature increases that determine cell death or survival are important for designing and implementing ablative strategies using CNTs or other nanoparticles [34]. Notably, the dissipation of heat from individual CNTs is very important when considering to what extent the very high temperatures achieved at the CNT surface are capable of inducing cell death [35].

At a molecular level, the principal targets for hyperthermia treatment are those proteins that undergo denaturation at temperatures as low as 42 °C [36]. Hyperthermia may also cause damage to the cytoskeleton [37] and to the plasma membrane [38]. Several types of biological macromolecule also undergo thermal transitions, and are potential targets for hyperthermia treatment. For example, lipids undergo temperature-dependent transitions below 37 °C, while DNA/RNA typically undergo unfolding and denaturation processes at temperatures above 70 °C. Many proteins undergo thermally induced denaturation processes at temperatures above 40 °C, matching the observed thermal effects of reduced clonogenic survival in this temperature range. Previous studies using differential scanning calorimetry (DSC) have revealed seven groups of proteins that underwent thermal transitions between 45 °C and 87 °C [36]. There is a good correspondence between the cell-killing effects of hyperthermia and the denaturation of the most easily denatured group of proteins (i.e., those that denature at <45 °C). An Arrhenius analysis of temperature-dependent survival revealed that the activation energy associated with lethal hyperthermia events is approximately 150 kcal mol^{-1}, a value that is also consistent with the heat of inactivation for proteins [39]. Cellular defense against stress processes, including thermally induced protein denaturation, may also counter the effects of thermal insult. In this regard, the elevated expression of heat-shock proteins (HSPs) in tumor cells relative to adjacent normal tissues decreases the effectiveness of hyperthermia for tumor treatment, while rendering the adjacent normal tissue fully vulnerable to the effects of hyperthermia [40, 41].

Tissue injury due to hyperthermia has been described as a two-phase process [42, 43]:

- In the first phase, proteins are denatured, leading to a loss of functionality for key cellular organelles, such as the mitochondria. RNA synthesis and transport is also disrupted in this phase.

- In the second phase, prolonged elevated temperatures cause alterations in apoptotic signaling and vascular injury, as well as cytokine release and immunomodulatory effects that have long-term damaging effects at a tissue and organism level.

In some respects, this two-phase description of thermal lethality is reminiscent of the very early models that predicted a two-step process of thermally induced cell death [30]. The more recent view of hyperthermia as a two-phase process, however, takes into account the effects of hyperthermia at a tissue and organism level, as well as a cellular level, as being important for the biological response to hyperthermia treatment.

12.2.1
Current Ablative Technologies

Ablative technologies, such as the percutaneous administration of ethanol [44], have been used in cancer treatment for several decades [45]. In particular, RFA is currently in widespread use for treating primary and metastatic cancers. Both, RFA and other ablative approaches (such as laser ablation [46]) require a knowledge of the distribution of malignant tissue (metastatic lesions or primary tumor), based on imaging studies. For RFA, the metastatic lesion is first penetrated by needles, through which electrodes are then deployed. Subsequent activation of the electrodes results in heating of the malignant tissue through frictional heat caused by the movement of ions [47]. As a consequence, tissue temperatures of 50–100 °C are achieved, which results in coagulative necrosis [48]. Unfortunately, RFA has serious adverse side effects, with a mortality rate of 0.5% and complication rates of between 8% and 35% [45].

Since the development of RFA, several additional focal ablative technologies have been developed, and are today in widespread use for cancer treatment:

- *Microwave ablation* is conceptually similar to RFA, except that a microwave antenna rather than an electrode is placed into the tumor. During treatment, water molecules interact with microwaves and heat the tissue in the vicinity of the microwave antenna [49, 50]. Both, RFA and microwave ablation are limited by "heat-sink" effects, whereby ablation close to large blood vessels will result in heat dissipation and a reduced effectiveness of the ablative therapy.

- **Cryoablation** is a controlled freezing rather than a controlled heating of tissue [48]. Cooling to approximately −40 °C is achieved by using liquid nitrogen or argon; the subsequent tissue freezing causes the cellular membranes to fracture and the proteins to denature.

- **High-intensity focused ultrasound** also elevates tissue temperatures locally, and can be used for ablative therapy [33].

A recent trend has been to use focal ablative therapies for localized malignancies for indications such as prostate cancer, where although total resection is feasible it may reduce the patient's quality of life. The localization of CNTs or other nanoparticles to malignant cells through targeting approaches has the potential to markedly enhance the precision of ablative technologies, and also to greatly reduce the side effects.

12.2.2
Use of CNTs for Hyperthermia Treatment

Several recently reported studies have demonstrated the value of using CNTs to ablate tumor tissue in animal models of human cancer, following irradiation with a near-infrared (NIR) laser [7, 51, 52]. The general principle employed in all of these studies has been that human tissues are relatively transparent to NIR radiation (700–1100 nm) [53], and that CNTs are strong absorbers of NIR and have a very high thermal stability. CNTs are quasi-1-D objects with well-defined energy states (van Hove singularities). NIR irradiation (700–1400 nm) excites the E_{11} transition of the CNT with the frequency for maximal absorption which is dependent on the dimensions of the CNT [4]. The magnitude of NIR absorption for SWNTs is length-dependent with E_{11}, as well as E_{22} and other van Hove singularities displaying more intense absorption for samples of longer CNTs relative to samples of shorter CNTs. The energy of these transitions is, however, not length-dependent [54]. Upon excitation of E_{11} or other transitions, the CNT may emit a photon and fluoresce. As the fluorescence of SWNTs occurs in the NIR region, SWNT fluorescence is currently undergoing development for use in tumor imaging (*vide infra*) [55, 56]. Although the quantum yields for the fluorescence of SWNTs are very low (<1%) [57], a majority of the absorbed energy is dissipated as heat (photothermal effect) [7].

The thermal stability of CNTs exceeds that of alternative forms of carbon, including graphite and diamond. Zewail and coworkers investigated the timescale of heating and cooling for CNTs in various solvents [6]. For this, the CNTs were heated using a pulsed NIR laser at 1400 nm (the conditions for pulsed excitation were 400–700 nJ per pulse, with a 120 fs pulse duration). Heating occurred as rapidly as 16 ps in water, with nanotube temperatures of up to 4000 K being achieved by repetitive pulsing. The CNTs proved to be extremely photostable and amenable to repetitive excitation, without photobleaching [4]. Nanotube cooling also occurred rapidly, and was seen to depend on the nature of the solvent, and its temperature. For nanotubes in water (at 22 °C), the recovery time for cooling following a single pulse was 526 ps.

The NIR irradiation of CNTs results in a substantial heating of the bulk solution. For example, Dai and coworkers reported a >20 °C temperature increase in bulk solution temperature upon 40 s irradiation of a 25 μg ml^{-1} solution of SWNTs at 808 nm (1.4 W cm^{-2}) [18]. The excitation of E_{11} for SWNTs appears to be highly sensitive to wavelength, since excitation with a 1064 nm laser (3 W cm^{-2}) resulted in a much more modest temperature increase of just a few degrees [51]. The Gmeiner laboratory recently reported a systematic study evaluating the effects of CNT concentration, irradiation time, and power on the extent of bulk solution temperature increase (Figure 12.4) [7]. This study was conducted with MWNTs that had been solubilized by encasement with ssDNA. The data revealed that increases in bulk solution temperature were linear with respect to both the time and the power of irradiation over the range evaluated (20–80 s; 2–4 W cm^{-2}). Thus, no long-term degradation in photothermal effects was observed following an extended irradiation of the MWNTs. These linear relationships make for straightforward trade-offs between the time and power of irradiation in order to obtain the targeted extent of heating. Unfortunately, the concentration-dependence of heating is nonlinear with regards to saturation in the rate of increase for bulk solution temperature at concentrations of DNA-encased MWNTs above ~60 μg ml^{-1}. The observed saturation effects most likely result from the physical occlusion of NIR irradiation, as MWNTs of relatively large diameter (~40 nm) will either absorb or scatter most of the irradiated NIR radiation at or above saturating concentrations. Although continuous laser irradiation has been used for the thermal ablation of malignant tissue, and is well-suited for annihilating larger tumor masses, a pulsed laser irradiation that generates much higher localized temperatures may be more effective for eradicating micrometastases or circulating tumor cells [58].

The heat emitted by CNTs following NIR irradiation – or by other means, such as radiofrequency (RF) excitation (*vide infra*) – is sufficient to achieve biologically significant endpoints for drug delivery and thermal ablation applications. For example, Dai and coworkers demonstrated not only the photothermal delivery of a small interfering RNA (siRNA) that was attached noncovalently to CNTs, but also the induction of cell death resulting from hyperthermia following the NIR irradiation of CNTs internalized by cancer cells [8, 18]. It was shown recently at the present author's laboratory, that the NIR irradiation of DNA-encased MWNTs injected intratumorally into prostate tumor xenografts in nude mice would result in a complete eradication of the tumor mass within a few days (Figure 12.5) [7]. Importantly, tumor eradication was achieved with no damage to any adjacent tissues, and all mice receiving the MWNT + NIR treatment were completely cured. Likewise, Torti and coworkers showed that MWNT + NIR treatment resulted in a survival benefit in subcutaneously implanted RENCA kidney tumors [51], with the survival benefit correlating with the quantity of MWNT injected into the tumor.

The use of CNTs to absorb NIR irradiation and to selectively heat malignant tissues has great potential for cancer treatment. This approach, however, is limited by the poor penetration of NIR irradiation through human tissues [46, 53], the inability to localize CNTs specifically to malignant tissues, and problems in controlling the extent of heating so as to limit damage to nonmalignant tissues. While human tissues do not substantially absorb NIR irradiation, the energy is effectively

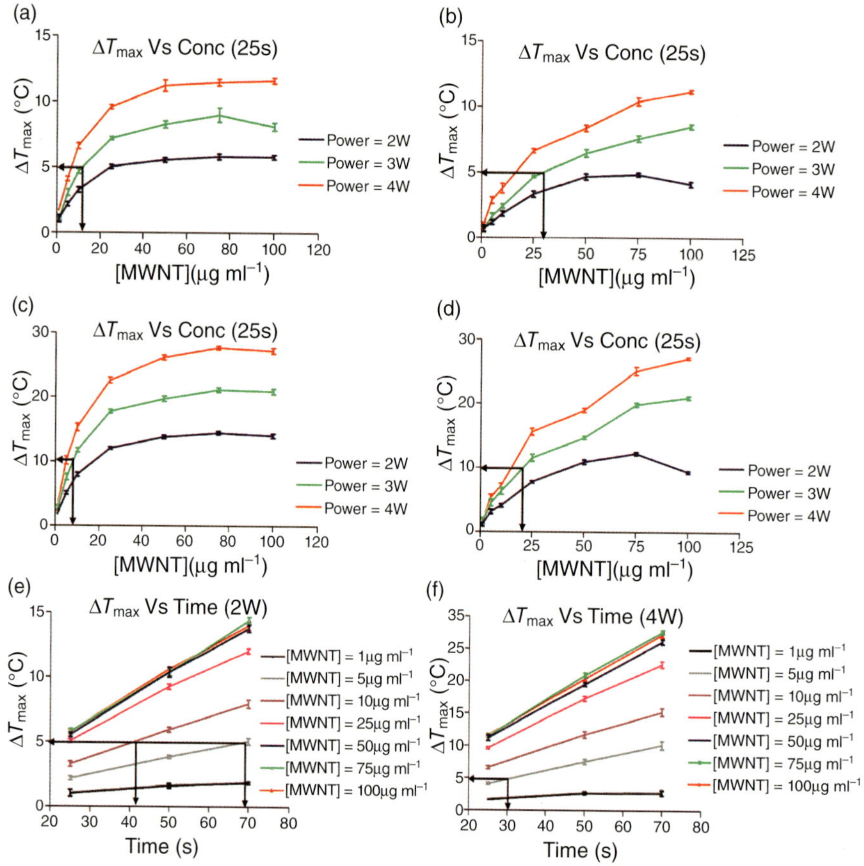

Figure 12.4 Representative plots for *in vitro* heating experiments with DNA-encased MWNTs. (a–d) Concentration-dependent heating of DNA-encased (a,c) and non-DNA-encased (b,d) MWNTs. The MWNT concentration required for a 5 °C increase in solution temperature upon 3 W irradiation for 25 s is indicated by horizontal and vertical arrows for DNA-encased MWNTs in panel (a), and for non-DNA-encased MWNTs in panel (b). Similarly, arrows denote the concentrations required for a 10 °C temperature increase upon 3 W irradiation for 70 s in (c) DNA-encased MWNTs and (d) non-DNA-encased MWNTs; (e,f) Range of conditions suitable for a 5 °C temperature increase upon irradiation of DNA-encased MWNTs. Reprinted with permission from Ref. [7]; © 2009, American Chemical Society.

scattered while passing through the tissues, such that the average depth of optical penetration has been reported as <1 cm [59] or 1–2 cm [46, 53]. Thus, until technologies for transmitting NIR radiation more deeply into targeted malignant tissues have been developed, its use must be limited to the treatment of superficial malignancies.

These limitations of NIR radiation have resulted in studies to investigate the use of alternative frequency irradiation to stimulate heat production from CNTs. For

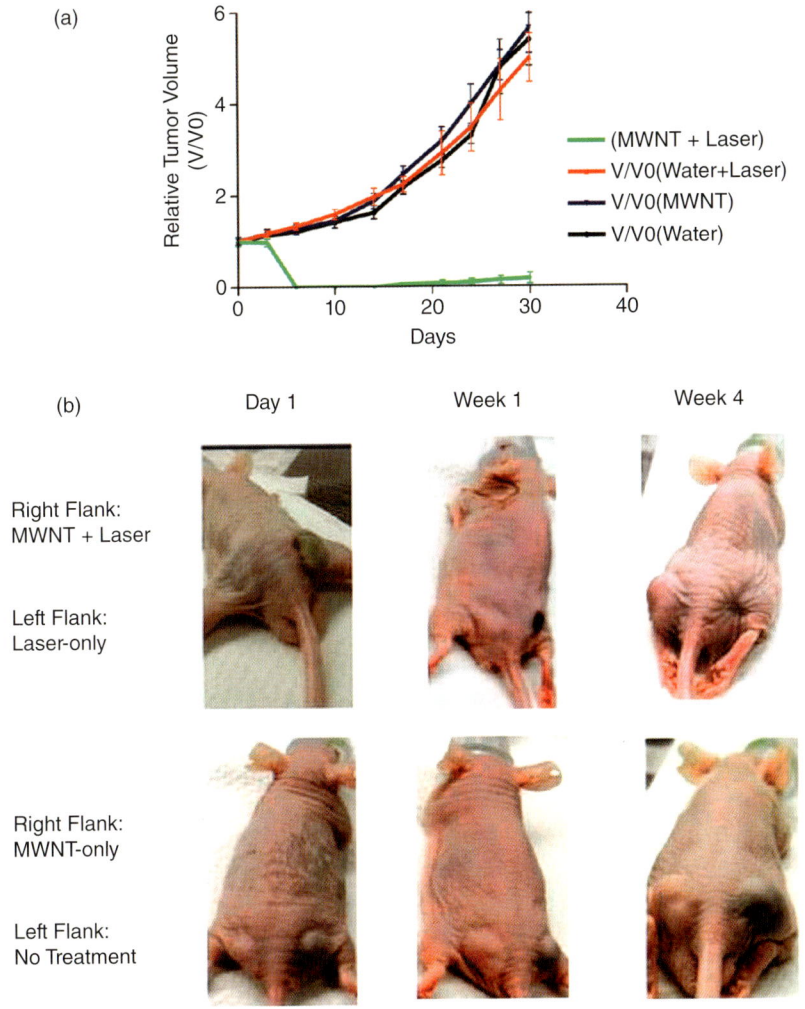

Figure 12.5 (a) Plot of the relative volume for the four tumor groups evaluated in the *in vivo* study. The initial tumor volume was ca. 225 mm^3; there was no significant difference in tumor volume among the four groups at baseline. Tumors injected with DNA-encased MWNTs and irradiated with a NIR laser at 1064 nm were completely eradicated within six days following treatment for all eight animals. The region where the tumor had been was completely healed over by day 24. The tumor groups that received MWNTs-only, laser irradiation only, or no treatment, grew at similar rates to one another throughout the study; (b) Upper row: Photographs from one animal for which the right flank tumor was treated with both MWNTs and irradiated with a NIR laser at day 1, week 2, and week 4 following treatment. Lower row: Photographs from one animal for which the right flank tumor was treated with MWNTs-only at the same time points. Reprinted with permission from Ref. [7]; © 2009, American Chemical Society.

example, Gannon and coworkers have shown that high-powered RF fields (100–800 W) at 13.56 MHz resulted in substantial increases in bulk solution temperature when applied to solutions of CNTs (50–500 μg ml^{-1}) [60]. In this case, the mechanism of heating was resistive conductivity, while the Kentera polymer used to solubilize the CNTs contributed significantly to the solution heating. Microwave frequencies also induce electrical currents in CNTs, which results in resistive heating [61]. In principle, the selective localization of CNTs to malignant tissues could enhance the heating of malignant tissues following microwave irradiation relative to nonmalignant tissues not containing CNTs, resulting in a therapeutic benefit.

12.3
CNTs for Drug Delivery

Numerous compounds have been identified that are highly cytotoxic towards cancer cells in tissue culture, and several of these (e.g., 5-fluorouracil, cisplatin, paclitaxel, PTX) have displayed appropriate therapeutic indices and have been approved for cancer chemotherapy. Nonetheless, the role of chemotherapy in cancer treatment remains limited, with only moderate increases in lifespan associated with the adjuvant [62] and neo-adjuvant [63] use of chemotherapy to treat many malignancies. It is likely, however, that the performance of many drugs currently used to treat cancer could be improved through the use of drug-delivery systems that would increase the levels of active drugs in tumor cells.

One reason why many of the current drugs are nonoptimal is that, in general, these small-molecular-weight anticancer drugs are rapidly cleared from the body. For example, the half-life of 5-fluorouracil, which is used widely to treat colorectal cancer and other malignancies, is approximately 15 min [64]. However, other factors often also limit the effectiveness of chemotherapy, including the toxicity of the drug and/or its metabolites to nonmalignant cells [65], the development of resistance in malignant cells [66], ineffective subcellular routing following cellular internalization [67], and inefficient cellular uptake [68].

To date, a variety of drug-delivery systems have been developed to increase the efficacy of anticancer drugs, including liposomes [69–71], microspheres [72, 73], and polymeric drug-delivery systems [74, 75]. Recently, considerable interest has been shown in the use of various nanoparticles, including CNTs [76, 77], for drug-delivery applications. Each of these approaches has potential advantages relative to systemically administered drugs in terms of an increased half-life. Likewise, many such approaches will result in an enhanced tumor localization through the EPR effect [9, 27]. As with other nanodelivery systems, CNTs must overcome a series of barriers prior to delivery and cellular uptake into cancer cells, including the avoidance of uptake by the cells of the reticuloendothelial system (RES) [78]. Once taken up into the tumor cells, the CNTs must release their drug cargo into the appropriate subcellular compartment for drug activity [18, 29]. At present, a variety of approaches is being evaluated to develop CNT-based drug-delivery

12.3.1
Localization of CNTs to Malignant Tissues

The localization of CNTs in malignant tissues can be accomplished by direct injection if the malignancy can be imaged, and if the extent of heating can be controlled to avoid damage to adjacent, nonmalignant tissues. Unfortunately, as many malignancies are not amenable to direct injection, it is important to develop strategies for localizing CNTs to malignant tissues following intravenous injection, so that the full therapeutic potential of CNTs for hyperthermia and drug-delivery applications can be realized. Frequently, CNTs may localize to malignant tissue as a consequence of their size and the inherent "leakiness" of tumor tissue, a phenomenon known as the EPR effect [27]. CNTs may also be localized to malignant tissues by conjugation with a ligand (e.g., folic acid), that has a high affinity for a particular cell-surface protein that is expressed at elevated levels in malignant cells compared to nonmalignant cells [29, 79]. In addition, CNTs may be chemically modified and localized to targeted regions thought to contain malignant tissue, by physical means [80].

The dimensions of the CNTs used for therapeutic purposes are typically several hundreds of nanometers in length, with diameters ranging from a few nanometers to 40–50 nm, depending on the type of CNT (e.g., SWNT, MWNT) [7, 18]. These dimensions are suitable for passage through the fenestrations which frequently are present in tumor vasculature, but absent from the vasculature of nonmalignant tissues. Tumors typically have low levels of lymphatic clearance and a slow venous return that leads to the accumulation of nanoparticles in the interstitium of the lesion [27]. In a recent study, the treatment of tumor-bearing mice with CNTs conjugated with PTX resulted in increased tumor drug levels relative to the delivery of PTX with a standard clinical formulation [9]. The conjugation of PTX to CNTs increased the drug's half-life from 18.8 min to 81.4 min. Unconjugated CNTs typically have half-lives in the serum of a few hours [81]. The conjugation of PTX to CNTs also led to changes in the drug's biodistribution, with an increased tumor uptake evident, but also an increased localization in the liver and spleen, which harbor cells of the RES. While these results validated nanotube delivery to the tumor, based on the EPR, they also demonstrated the propensity for CNTs to accumulate in the liver, spleen and at other sites. Unfortunately, the latter property may raise concern for toxic effects being associated with CNT treatments (*vide infra*).

Malignant cells frequently express elevated levels of proteins on their plasma membranes that are expressed only at lower levels in nonmalignant cells. Frequently, such proteins are expressed due to the higher nutrient requirements of tumor cells (e.g., folate receptor; FR-α) [29]. Alternatively, proteins that regulate interaction with the extracellular matrix (e.g., $α_vβ_3$ integrins) [82, 83] or receptor tyrosine kinases (e.g., epidermal growth factor receptor; EGFR) [28] that drive cellular proliferation, may display elevated expressions in malignant cells. The conjugation of CNTs with ligands that bind to proteins that show an elevated expression on malignant cells has been shown to localize CNTs to tumor cells both *in vitro* and *in vivo*, and in some cases also to increase the cellular internalization of the conjugated CNT. When Lippard and coworkers conjugated SWNTs with folic acid and Pt(IV) complexes, they were able to demonstrate an increased cellular internalization into FR-α-expressing cancer cells, and a more than eightfold enhancement in cytotoxicity relative to cisplatin [29]. Subsequently, Bhirde *et al.* conjugated the EGF peptide to SWNTs, and demonstrated a specific uptake into cancer cells expressing the EGFR [28]. The knockdown of EGFR with siRNA inhibited the cellular uptake of the CNT:EGF conjugate. The internalized CNTs had a perinuclear distribution in EGFR-expressing cancer cells, with EGF-conjugated CNTs displaying a 2.5-fold greater tumor localization relative to nonconjugated CNTs, while the conjugates displayed a substantial antitumor activity.

12.3.2
Drug Delivery Using CNTs

The potential of CNTs for drug-delivery applications, as well as for the delivery of radionuclides and other cytotoxic substances, was recognized shortly after their initial characterization [84]. Conjugation or complex formation of drugs with CNTs is expected to increase the plasma half-life relative to that of the free drug, and also to alter the drug's biodistribution. Previously, CNTs have been described as "longboat delivery systems" [85], in reference to their high aspect ratios. While CNTs may offer potential solutions to the problem of short half-lives, undesirable biodistribution profiles, and inefficient cellular uptake for small-molecule pharmaceuticals, the use of CNTs for drug-delivery applications continues to present new challenges in the development of appropriate new chemistries for drug–CNT complex formation, the accumulation of these complexes in RES cells and other sites, and other new challenges associated with the preclinical and eventual clinical development of these materials.

As CNTs present a hydrophobic surface, three basic strategies have been used to modify the CNT surface for use in drug-delivery applications:

i. Chemical modification to permit the covalent attachment of drugs through intervening linkers.
ii. A layered assembly or co-assembly of PEG or other surfactants and drugs on the CNT surface.
iii. The noncovalent attachment of hydrophobic drugs to the CNTs.

One approach that has been used to prepare CNT–drug complexes, based on second of the strategies above, is the noncovalent association of an amine-terminated PEGylated phospholipid with CNTs, to both solubilize and disperse the CNTs and also to permit the covalent attachment of a chemotherapeutic drug and targeting moieties [9, 18, 29]. The chemotherapeutic agent is derivatized to include a terminal carboxylate moiety that can then be coupled to the amine-terminated PEG to produce a drug-loaded CNT. The PEG used to solubilize the

CNTs and permit drug conjugation to the solubilized CNTs can, however, cause adverse immune responses, including the elicitation of anti-PEG IgM responses, and may also decrease cellular internalization [86–90]. Serum protein may also rapidly displace surfactants from the CNTs [91] although, depending on the conditions utilized, the surfactant can remain bound to the CNT surface for several months [92]. While PEG decreases the opsonization of coated CNTs by cells of the RES [93], PEG-derivatized drugs have been shown to activate the complement system [94]. An alternative drug conjugation approach based on strategy (i) is to activate the CNT surface with strong acid; this will produce carboxylate groups that can be linked covalently to chemically modified chemotherapeutics through short spacer moieties. This strategy has been used to deliver camptothecin analogs to cancer cells [86]. A related approach has been to covalently attach methotrexate to CNTs, following derivatization of the CNT surface, through a cycloaddition reaction [95]. Perhaps the simplest strategy for modifying CNTs for drug delivery is to directly bind hydrophobic drugs, such as doxorubicin (DOX), to the hydrophobic surfaces of CNTs via π–π stacking [as in strategy (iii)]. Dai and coworkers have shown that DOX remained bound via π-stacking to CNTs for periods sufficient for drug delivery [96], although the branched PEG used may have activated the complement system and solubilized the CNT, thereby enhancing the plasma half-life of the modified CNTs. Kostarelos and coworkers showed DOX to be associated with pluronic solubilized MWNTs, and that the resulting noncovalent CNT–DOX complexes were cytotoxic towards cancer cells [97].

The cellular uptake of CNT–drug complexes is important for the antitumor activity of these materials. Typically, cellular uptake into the RES occurs by phagocytosis, and the efficiency of this process is decreased by complexation with PEG [9, 21]. As the RES uptake of CNT–drug complexes is generally undesirable, many groups have used PEG to solubilize the CNTs. CNT length also is a factor in RES uptake [98, 99]. The mechanism of cellular uptake of CNTs into non-RES cells requires additional clarification, and may be cell-type-dependent. Dai and coworkers have reported that CNTs are internalized by HeLa cells, by endocytosis through clathrin-coated pits [11], while Kostarelos and coworkers determined that CNTs, under certain conditions, may act as "nanoneedles" and enter the cells without entry into any vesicular compartment [12]. A recent study concluded that, whilst MWNT bundles enter cells via an endocytotic mechanism, single MWNTs were found to enter cells by direct penetration [10]. Subsequently, once internalized, a single CNT could enter the nucleus or be recruited into lysosomes, and eventually be excreted. Additional studies evaluating the mechanisms of cellular internalization and the subcellular routing of CNT–drug complexes will be important for maximizing the antitumor activities of these materials.

12.4
Imaging Using CNTs

The propensity of CNTs to accumulate in tumor tissues via the EPR, or based upon conjugation strategies with monoclonal antibodies (mAbs) [100], aptamers

[101], or small molecules having a high affinity for proteins expressed at elevated levels on tumor cells [85], has resulted in investigations into the use of CNTs for the imaging of malignancies, in addition to their use for cancer treatment. The imaging of CNTs is also important for determining their biodistribution, and thus also for investigations of CNT toxicity that must be completed prior to their widespread use for cancer treatment. CNTs have intrinsic properties that can be utilized for detection based upon Raman scattering [102], photoluminescence [92, 103, 104], and photoacoustic imaging [105, 106]. CNTs can also be conjugated with radionuclides or other agents for imaging purposes [107]. Semiconducting SWNTs display NIR fluorescence at frequencies that depend on the dimensions of the CNT [55]. The excitation of either E_{11} in the NIR or E_{22} in the visible region can cause NIR fluorescence which occurs with a large Stokes shift at 1100–1400 nm, a region of low autofluorescence [104]. Although the efficiency of NIR fluorescence for SWNTs is nonoptimal [57], no photobleaching is observed with CNTs [108]. The surfactant and the extent of defects in the CNT structure may have a marked influence on CNT fluorescence [92]. Previously, NIR fluorescence has been used to study the biodistribution of CNTs, both in cells [108] and *in vivo* [56, 91, 92]. Alternatively, the Raman scattering of CNTs can be used for high-resolution imaging of CNTs in tissues and *in vivo* [103, 109]. The photoacoustic imaging of CNTs, following excitation with NIR radiation, has recently been demonstrated and used to detect tumors *in vivo* [105, 106]. Photoacoustic imaging can also be combined with high-frequency ultrasound imaging [110].

12.5
CNT-Related Toxicity

The preceding sections have highlighted the enormous potential of CNTs for photothermal therapy, drug delivery, and imaging applications. However, the realization of this potential depends heavily on any toxic effects that might result from CNT treatments being minimal. Currently, the toxicity of CNTs is being actively investigated, and the bulk of evidence acquired to date is that CNTs prepared and used in animal studies for therapeutic and imaging applications have caused no serious toxic effects [109, 111]. Although evidence indicates that CNTs may remain unchanged in tissues for many months following *in vivo* administration [111], more recent studies have suggested that CNTs may be susceptible to long-term enzymatic degradation [112, 113]. CNTs are excreted via the biliary and renal pathways [9, 19, 109], and imaging methods such as Raman scattering [102], CNT fluorescence [92, 103, 104], and photoacoustic imaging [105, 106] have proved useful for quantifying the long-term biodistribution of intact CNTs *in vivo* [109]. While the administration of CNTs under certain conditions can result in serious toxicities, including the induction of mesothelioma [114, 115], the conditions used to induce these effects (including the intraperitoneal injection of large quantities of nondispersed CNTs, or the intra-scrotal injection of MWNTs) were not considered relevant to any of the therapeutic applications described to date [116]. Long CNTs (5–15 μm in length), when administered by inhalation, were localized in the macrophages and caused long-term immunosuppression [98, 117]. The mechanism of such immuosuppression involved an activation of the cyclooxygenase enzymes in the spleen, in response to signals in the lungs [118]. Although a potential concern for occupational exposure to CNTs, the conditions used for these studies were not relevant to any described therapeutic uses of CNTs for cancer treatment. CNTs have also been reported to cause oxidative damage in treated cells [119], and to result in increased levels of 8-oxo-dG, a biomarker of oxidative damage, in the liver and lungs following the oral administration of CNTs [120]. Recent *in vivo* studies evaluating the antitumor response as a primary endpoint have, however, not identified any organ toxicities associated with CNT treatment over several months after CNT administration, whether by intravenous or intratumoral injection [51, 52]. The results of these studies indicate that the use of CNTs in humans for cancer treatment is likely to be well tolerated.

12.6
Summary and Future Perspective

The past few years have led to several important advances in the use of CNTs for the treatment of cancer in preclinical models. Specifically, advances in the photothermal treatment of cancer, drug delivery and tumor imaging have brought CNTs to the threshold of clinical utility. Photothermal treatment has been demonstrated following the intratumoral injection of CNTs into xenografted and subcutaneously implanted tumors. The advantages of photothermal ablation, relative to alternative ablative technologies, include the propensity of CNTs to accumulate in tumor tissues either through the EPR effect, or as a result of molecular targeting following intravenous administration. It is likely that future preclinical studies will strive to demonstrate photothermal ablation following the intravenous administration of CNTs, and will maximize the photothermal effects obtained from minimal power NIR irradiation. Such studies may facilitate the translation of CNTs for photothermal applications to treat cancer in humans. Drug-delivery applications using CNTs have demonstrated significant therapeutic advantages, including an increased bioavailability, relative to conventional formulations. Tumor localization via the EPR and/or molecular targeting approaches has been demonstrated in animal models. It is likely that drug delivery using CNTs will result in enhanced tumor drug levels in humans treated with drug-conjugated CNTs. A variety of chemical modifications have been developed that aim to minimize uptake into the RES, and maximize the release of active drug into the targeted tumor cells. Identifying those formulations of CNTs which have the greatest potential for cancer treatment will represent an important objective in the near future. To date, there is no clear evidence that any of the approaches used for CNTs in photothermal treatment or drug-delivery applications will result in any toxic effects that would limit clinical development. However, as the large majority of these studies have been conducted in rodents, their successful translation into clinical trials will require any toxicity issues first to be addressed in larger animal species. CNTs may also provide a unique handle to follow biodistribution using NIR fluorescence, Raman scattering, or photoacoustic imaging. The amenability of CNTs to *in vivo* analysis, using spectral and imaging modalities, will provide opportunities for improved detection, as well as advantages in refining protocols for improved cancer treatments. In summary, with preclinical studies using CNTs for photothermal and drug-delivery applications and imaging having proved to be successful, there is little doubt that CNTs will also be used successfully to treat cancer in the near future.

Acknowledgments

The author is grateful for financial support from NIH-NCI 1U01 CA102532 and the North Carolina Biotechnology Center. Thanks are also given to Supratim Ghosh for help in obtaining some of the references cited in this work.

Abbreviations

ACS	American Cancer Society
CNT	carbon nanotube
CoMoCat	Co-Mo catalyst
CVD	carbon vapor deposition
DOX	doxorubicin
DSC	differential scanning calorimetry
DWNT	double-walled carbon nanotube
EGFR	epidermal growth factor receptor
EPR	enhanced permeability and retention effect
FR	folate receptor
HiPCO	high-pressure carbon monoxide conversion
HSP	heat shock protein
i.v.	intravenous
MWNT	multi-walled carbon nanotube
NIR	near infrared
PEG	polyethylene glycol
PL	phospholipid
PTX	paclitaxel
RES	reticuloendothelial system
RF	radiofrequency
RFA	radiofrequency ablation
ssDNA	single-stranded DNA
SWNT	single-walled carbon nanotube

References

1. National Cancer Institute. Surveillance, Epidemiology and End Results, seer.cancer.gov/ (accessed 10 January 2010).
2. American Cancer Society Facts & Figures (2009) www.cancer.org/ (accessed 10 January 2010).
3. Troost, M.M., Sternberg, C.N. and de Wit, R. (2009) Management of good risk germ-cell tumours. *British Journal of Urology International*, **104**, 1387–91.
4. Carlson, L.J. and Krauss, T.D. (2008) Photophysics of individual single-walled carbon nanotubes. *Accounts of Chemical Research*, **41**, 235–43.
5. Saito, N. et al. (2009) Carbon nanotubes: biomaterial applications. *Chemical Society Reviews*, **38**, 1897–903.
6. Mohammed, O.F., Samartzis, P.C. and Zewail, A.H. (2009) Heating and cooling dynamics of carbon nanotubes observed by temperature-jump spectroscopy and electron microscopy. *Journal of the American Chemical Society*, **131**, 16010–11.
7. Ghosh, S. et al. (2009) Increased heating efficiency and selective thermal ablation of malignant tissue with DNA-encased multiwalled carbon nanotubes. *ACS Nano*, **3**, 2667–73.
8. Kam, N.W., Liu, Z. and Dai, H. (2005) Functionalization of carbon nanotubes via cleavable disulfide bonds for efficient intracellular delivery of siRNA and potent gene silencing. *Journal of the American Chemical Society*, **127**, 12492–3.
9. Liu, Z. et al. (2008) Drug delivery with carbon nanotubes for *in vivo* cancer treatment. *Cancer Research*, **68**, 6652–60.
10. Mu, Q., Broughton, D.L. and Yan, B. (2009) Endosomal leakage and nuclear translocation of multiwalled carbon nanotubes: developing a model for cell uptake. *Nano Letters*, **9**, 4370–5.
11. Kam, N.W., Liu, Z. and Dai, H. (2006) Carbon nanotubes as intracellular transporters for proteins and DNA: an investigation of the uptake mechanism and pathway. *Angewandte Chemie International Edition*, **45**, 577–81.
12. Lara Lacerdaa, S.R., Pratoc, M., Biancod, A. and Kostarelosa, K. (2007) Cell-penetrating CNTs for delivery of therapeutics. *Nanotoday*, **2**, 38.
13. Li, Z., Zheng, L., Yan, W., Pan, Z. and Wei, S. (2009) Spectroscopic characteristics of differently produced single-walled carbon nanotubes. *ChemPhysChem*, **10**, 2296–304.
14. Eklund, P.C. (2006) Carbon Nanotube Manufacturing 2006: Introduction and Overview. Proceedings, WTEC Workshop, International R&D of Carbon Nanotube Manufacturing and Applications, November 2006. National Science Foundation, Arlington, VA.
15. Tien, L.-G., Tsai, C.-H., Li, F.-Y. and Lee, M.-H. (2008) Influence of vacancy defect density on electrical properties of armchair single wall carbon nanotube. *Diamond and Related Materials*, **17**, 563–6.
16. Koziol, K., Shaffer, M. and Windle, A. (2005) Three-dimensional internal order in multiwalled carbon nanotubes grown by chemical vapor deposition. *Advanced Materials*, **17**, 760–3.
17. Qi, H., Qian, C. and Liu, J. (2007) Synthesis of uniform double-walled carbon nanotubes using iron disilicide as catalyst. *Nano Letters*, **7**, 2417–21.
18. Kam, N.W., O'Connell, M., Wisdom, J.A. and Dai, H. (2005) Carbon nanotubes as multifunctional biological transporters and near-infrared agents for selective cancer cell destruction. *Proceedings of the National Academy of Sciences of the United States of America*, **102**, 11600–5.
19. Lacerda, L. et al. (2008) Carbon-nanotube shape and individualization critical for renal excretion. *Small*, **4**, 1130–2.
20. Ichihara, G., Castranova, V., Tanioka, A. and Miyazawa, K. (2008) Re: induction of mesothelioma in p53+/− mouse by intraperitoneal application of multi-wall carbon nanotube. *Journal of Toxicological Sciences*, **33**, 381–2; author reply 382–4.
21. Liu, Z., Tabakman, S.M., Chen, Z. and Dai, H. (2009) Preparation of carbon nanotube bioconjugates for biomedical applications. *Nature Protocols*, **4**, 1372–82.

22 Zheng, M. et al. (2003) DNA-assisted dispersion and separation of carbon nanotubes. *Nature Materials*, **2**, 338–42.

23 Cathcart, H., Nicolosi, V., Hughes, J.M., Blau, W.J., Kelly, J.M., Quinn, S.J. and Coleman, J.N. (2008) Ordered DNA wrapping switches on luminescence in single-walled nanotube dispersions. *Journal of the American Chemical Society*, **130**, 12734–44.

24 Haggenmueller, R. et al. (2008) Comparison of the quality of aqueous dispersions of single wall carbon nanotubes using surfactants and biomolecules. *Langmuir*, **24**, 5070–8.

25 Li, Y., Zhang, X., Luo, J., Huang, W., Cheng, J., Luo, Z., Li, T., Liu, F., Xu, G., Ke, X., Li, L. and Geise, H.J. (2004) Purification of CVD synthesized single-wall carbon nanotubes by different oxidation treatments. *Nanotechnology*, **15**, 1645–9.

26 Cho, K., Wang, X., Nie, S., Chen, Z.G. and Shin, D.M. (2008) Therapeutic nanoparticles for drug delivery in cancer. *Clinical Cancer Research*, **14**, 1310–16.

27 Maeda, H., Bharate, G.Y. and Daruwalla, J. (2009) Polymeric drugs for efficient tumor-targeted drug delivery based on EPR-effect. *European Journal of Pharmaceutics and Biopharmaceutics*, **71**, 409–19.

28 Bhirde, A.A. et al. (2009) Targeted killing of cancer cells *in vivo* and *in vitro* with EGF-directed carbon nanotube-based drug delivery. *ACS Nano*, **3**, 307–16.

29 Dhar, S., Liu, Z., Thomale, J., Dai, H. and Lippard, S.J. (2008) Targeted single-wall carbon nanotube-mediated Pt(IV) prodrug delivery using folate as a homing device. *Journal of the American Chemical Society*, **130**, 11467–76.

30 Jung, H. (1986) A generalized concept for cell killing by heat. *Radiation Research*, **106**, 56–72.

31 Jordan, A., Schmidt, W. and Scholz, R. (2000) A new model of thermal inactivation and its application to clonogenic survival data for WiDr human colonic adenocarcinoma cells. *Radiation Research*, **154**, 600–7.

32 Dewhirst, M.W., Viglianti, B.L., Lora-Michiels, M., Hanson, M. and Hoopes, P.J. (2003) Basic principles of thermal dosimetry and thermal thresholds for tissue damage from hyperthermia. *International Journal of Hyperthermia*, **19**, 267–94.

33 Haar, G.T. and Coussios, C. (2007) High intensity focused ultrasound: physical principles and devices. *International Journal of Hyperthermia*, **23**, 89–104.

34 Salloum, M., Ma, R. and Zhu, L. (2009) Enhancement in treatment planning for magnetic nanoparticle hyperthermia: optimization of the heat absorption pattern. *International Journal of Hyperthermia*, **25**, 309–21.

35 Avedisian, C.T., Cavicchi, R.E., McEuen, P.L. and Zhou, X. (2009) Nanoparticles for cancer treatment: role of heat transfer. *Annals of the New York Academy of Sciences*, **1161**, 62–73.

36 He, X., Wolkers, W.F., Crowe, J.H., Swanlund, D.J. and Bischof, J.C. (2004) In situ thermal denaturation of proteins in dunning AT-1 prostate cancer cells: implication for hyperthermic cell injury. *Annals of Biomedical Engineering*, **32**, 1384–98.

37 Tucker, N.R. and Shelden, E.A. (2009) Hsp27 associates with the titin filament system in heat-shocked zebrafish cardiomyocytes. *Experimental Cell Research*, **315**, 3176–86.

38 Bischof, J.C. et al. (1995) Dynamics of cell membrane permeability changes at supraphysiological temperatures. *Biophysical Journal*, **68**, 2608–14.

39 Simanovskii, D.M., Mackanos, M.A., Irani, A.R., O'Connell-Rodwell, C.E., Contag, C.H., Schwettman, H.A. and Palanker, D.V. (2006) Cellular tolerance to pulsed hyperthermia. *Physical Review E*, **74**, 011915 1–7.

40 Rylander, M.N. et al. (2006) Optimizing heat shock protein expression induced by prostate cancer laser therapy through predictive computational models. *Journal of Biomedical Optics*, **11**, 041113.

41 Rylander, M.N., Feng, Y., Bass, J. and Diller, K.R. (2007) Heat shock protein expression and injury optimization for laser therapy design. *Lasers in Surgery and Medicine*, **39**, 731–46.

42 Everts, M. (2007) Thermal scalpel to target cancer. *Expert Review of Medical Devices*, **4**, 131–6.

43 Nikfarjam, M., Muralidharan, V. and Christophi, C. (2005) Mechanisms of focal heat destruction of liver tumors. *Journal of Surgical Research*, **127**, 208–23.

44 Monchik, J.M., Donatini, G., Iannuccilli, J. and Dupuy, D.E. (2006) Radiofrequency ablation and percutaneous ethanol injection treatment for recurrent local and distant well-differentiated thyroid carcinoma. *Annals of Surgery*, **244**, 296–304.

45 Liapi, E. and Geschwind, J.F. (2007) Transcatheter and ablative therapeutic approaches for solid malignancies. *Journal of Clinical Oncology*, **25**, 978–86.

46 Gough-Palmer, A.L. and Gedroyc, W.M. (2008) Laser ablation of hepatocellular carcinoma – a review. *World Journal of Gastroenterology*, **14**, 7170–4.

47 Higgins, H. and Berger, D.L. (2006) RFA for liver tumors: does it really work? *Oncologist*, **11**, 801–8.

48 Kutikov, A., Kunkle, D.A. and Uzzo, R.G. (2009) Focal therapy for kidney cancer: a systematic review. *Current Opinion in Urology*, **19**, 148–53.

49 Liang, P. and Wang, Y. (2007) Microwave ablation of hepatocellular carcinoma. *Oncology*, **72** (Suppl. 1), 124–31.

50 Simon, C.J., Dupuy, D.E. and Mayo-Smith, W.W. (2005) Microwave ablation: principles and applications. *Radiographics*, **25** (Suppl. 1), S69–83.

51 Burke, A. *et al.* (2009) Long-term survival following a single treatment of kidney tumors with multiwalled carbon nanotubes and near-infrared radiation. *Proceedings of the National Academy of Sciences of the United States of America*, **106**, 12897–902.

52 Moon, H.K., Lee, S.H. and Choi, H.C. (2009) *In vivo* near-infrared mediated tumor destruction by photothermal effect of carbon nanotubes. *ACS Nano*, **3**, 3707–13.

53 Konig, K. (2000) Multiphoton microscopy in life sciences. *Journal of Microscopy*, **200**, 83–104.

54 Fagan, J.A. *et al.* (2007) Length-dependent optical effects in single-wall carbon nanotubes. *Journal of the American Chemical Society*, **129**, 10607–12.

55 Bachilo, S.M. *et al.* (2002) Structure-assigned optical spectra of single-walled carbon nanotubes. *Science*, **298**, 2361–6.

56 Leeuw, T.K. *et al.* (2007) Single-walled carbon nanotubes in the intact organism: near-IR imaging and biocompatibility studies in *Drosophila*. *Nano Letters*, **7**, 2650–4.

57 Aprile, C., Martin, R., Alvaro, M., Scaiano, J.C. and Garcia, H. (2009) Near-infrared emission quantum yield of soluble short single-walled carbon nanotubes. *ChemPhysChem*, **10**, 1305–10.

58 Biris, A.S. *et al.* (2009) Nanophotothermolysis of multiple scattered cancer cells with carbon nanotubes guided by time-resolved infrared thermal imaging. *Journal of Biomedical Optics*, **14**, 021007.

59 Faris, F. *et al.* (1991) Non-invasive *in vivo* near-infrared optical measurement of the penetration depth in the neonatal head. *Clinical Physics and Physiological Measurement*, **12**, 353–8.

60 Gannon, C.J. *et al.* (2007) Carbon nanotube-enhanced thermal destruction of cancer cells in a noninvasive radiofrequency field. *Cancer*, **110**, 2654–65.

61 Vazquez, E. and Prato, M. (2009) Carbon nanotubes and microwaves: interactions, responses, and applications. *ACS Nano*, **3**, 3819–24.

62 Benson, A.B., 3rd (2007) New approaches to assessing and treating early-stage colon and rectal cancers: cooperative group strategies for assessing optimal approaches in early-stage disease. *Clinical Cancer Research*, **13**, 6913s–20s.

63 Bathe, O.F. *et al.* (2009) A phase II experience with neoadjuvant irinotecan (CPT-11), 5-fluorouracil (5-FU) and leucovorin (LV) for colorectal liver metastases. *BMC Cancer*, **9**, 156.

64 Finch, R.E., Bending, M.R. and Lant, A.F. (1979) Plasma levels of 5-fluorouracil after oral and intravenous administration in cancer patients. *British Journal of Clinical Pharmacology*, **7**, 613–17.

65 Wang, J. *et al.* (2009) Systems toxicology study of doxorubicin on rats using ultra performance liquid chromatography

coupled with mass spectrometry based metabolomics. *Metabolomics*, **5**, 407–18.

66 Liu, F.S. (2009) Mechanisms of chemotherapeutic drug resistance in cancer therapy – a quick review. *Taiwanese Journal of Obstetrics and Gynecology*, **48**, 239–44.

67 Bareford, L.M. and Swaan, P.W. (2007) Endocytic mechanisms for targeted drug delivery. *Advanced Drug Delivery Reviews*, **59**, 748–58.

68 Hembruff, S.L. et al. (2008) Role of drug transporters and drug accumulation in the temporal acquisition of drug resistance. *BMC Cancer*, **8**, 318.

69 Ho, E.A. et al. (2010) Characterization of cationic liposome formulations designed to exhibit extended plasma residence times and tumor vasculature targeting properties. *Journal of Pharmaceutical Sciences*, **99** (6), 2839–53.

70 Obata, Y., Tajima, S. and Takeoka, S. (2010) Evaluation of pH-responsive liposomes containing amino acid-based zwitterionic lipids for improving intracellular drug delivery in vitro and in vivo. *Journal of Controlled Release*, **142** (2), 267–76.

71 Demirgoz, D., Garg, A. and Kokkoli, E. (2008) PR_b-targeted PEGylated liposomes for prostate cancer therapy. *Langmuir*, **24**, 13518–24.

72 Lee, K.H. et al. (2010) Doxorubicin-loaded quadrasphere microspheres: plasma pharmacokinetics and intratumoral drug concentration in an animal model of liver cancer. *Cardiovascular and Interventional Radiology*, **33** (3), 576–82.

73 Wang, X., Yucel, T., Lu, Q., Hu, X. and Kaplan, D.L. (2010) Silk nanospheres and microspheres from silk/PVA blend films for drug delivery. *Biomaterials*, **31** (6), 1025–35.

74 Gaucher, G., Marchessault, R.H. and Leroux, J.C. (2010) Polyester-based micelles and nanoparticles for the parenteral delivery of taxanes. *Journal of Controlled Release*, **143** (1), 2–12.

75 Park, J.H., Saravanakumar, G., Kim, K. and Kwon, I.C. (2010) Targeted delivery of low molecular drugs using chitosan and its derivatives. *Advanced Drug Delivery Reviews*, **62** (1), 28–41.

76 Ahmed, M., Jiang, X., Deng, Z. and Narain, R. (2009) Cationic glyco-functionalized single-walled carbon nanotubes as efficient gene delivery vehicles. *Bioconjugate Chemistry*, **20**, 2017–22.

77 Lay, C.L., Liu, H.Q., Tan, H.R. and Liu, Y. (2010) Delivery of paclitaxel by physically loading onto poly(ethylene glycol) (PEG)-graft-carbon nanotubes for potent cancer therapeutics. *Nanotechnology*, **21**, 065101.

78 Li, S.D. and Huang, L. (2009) Nanoparticles evading the reticuloendothelial system: role of the supported bilayer. *Biochimica et Biophysica Acta*, **1788**, 2259–66.

79 Ko, S., Liu, H., Chen, Y. and Mao, C. (2008) DNA nanotubes as combinatorial vehicles for cellular delivery. *Biomacromolecules*, **9**, 3039–43.

80 Yang, F., de Fu, L., Long, J. and Ni, Q.X. (2008) Magnetic lymphatic targeting drug delivery system using carbon nanotubes. *Medical Hypotheses*, **70**, 765–7.

81 Singh, R. et al. (2006) Tissue biodistribution and blood clearance rates of intravenously administered carbon nanotube radiotracers. *Proceedings of the National Academy of Sciences of the United States of America*, **103**, 3357–62.

82 Schottelius, M., Laufer, B., Kessler, H. and Wester, H.J. (2009) Ligands for mapping alphavbeta3-integrin expression in vivo. *Accounts of Chemical Research*, **42**, 969–80.

83 Villa, C.H. et al. (2008) Synthesis and biodistribution of oligonucleotide-functionalized, tumor-targetable carbon nanotubes. *Nano Letters*, **8**, 4221–8.

84 Iijima, S. (1991) Helical microtubules of graphitic carbon. *Nature*, **354**, 56–8.

85 Feazell, R.P., Nakayama-Ratchford, N., Dai, H. and Lippard, S.J. (2007) Soluble single-walled carbon nanotubes as longboat delivery systems for platinum(IV) anticancer drug design. *Journal of the American Chemical Society*, **129**, 8438–9.

86 Wu, W. et al. (2009) Covalently combining carbon nanotubes with anticancer agent: preparation and antitumor activity. *ACS Nano*, **3**, 2740–50.

87 Zeineldin, R., Al-Haik, M. and Hudson, L.G. (2009) Role of polyethylene glycol integrity in specific receptor targeting of carbon nanotubes to cancer cells. *Nano Letters*, **9**, 751–7.

88 Zolnik, B.S., Gonzalez-Fernandez, A., Sadrieh, N. and Dobrovolskaia, M.A. (2010) Minireview: nanoparticles and the immune system. *Endocrinology*, **151** (2), 458–65.

89 Wang, X., Ishida, T. and Kiwada, H. (2007) Anti-PEG IgM elicited by injection of liposomes is involved in the enhanced blood clearance of a subsequent dose of PEGylated liposomes. *Journal of Controlled Release*, **119**, 236–44.

90 Ishida, T., Wang, X., Shimizu, T., Nawata, K. and Kiwada, H. (2007) PEGylated liposomes elicit an anti-PEG IgM response in a T cell-independent manner. *Journal of Controlled Release*, **122**, 349–55.

91 Cherukuri, P. et al. (2006) Mammalian pharmacokinetics of carbon nanotubes using intrinsic near-infrared fluorescence. *Proceedings of the National Academy of Sciences of the United States of America*, **103**, 18882–6.

92 Welsher, K. et al. (2009) A route to brightly fluorescent carbon nanotubes for near-infrared imaging in mice. *Nature Nanotechnology*, **4**, 773–80.

93 Liu, Z. et al. (2008) Multiplexed multicolor Raman imaging of live cells with isotopically modified single walled carbon nanotubes. *Journal of the American Chemical Society*, **130**, 13540–1.

94 Hamad, I., Hunter, A.C., Szebeni, J. and Moghimi, S.M. (2008) Poly(ethylene glycol)s generate complement activation products in human serum through increased alternative pathway turnover and a MASP-2-dependent process. *Molecular Immunology*, **46**, 225–32.

95 Pastorin, G. et al. (2006) Double functionalization of carbon nanotubes for multimodal drug delivery. *Chemical Communications*, 1182–4.

96 Liu, Z. et al. (2009) Supramolecular stacking of doxorubicin on carbon nanotubes for *in vivo* cancer therapy. *Angewandte Chemie International Edition*, **48**, 7668–72.

97 Ali-Boucetta, H. et al. (2008) Multiwalled carbon nanotube-doxorubicin supramolecular complexes for cancer therapeutics. *Chemical Communications*, 459–61.

98 Kostarelos, K. (2008) The long and short of carbon nanotube toxicity. *Nature Biotechnology*, **26**, 774–6.

99 Sato, Y. et al. (2005) Influence of length on cytotoxicity of multi-walled carbon nanotubes against human acute monocytic leukemia cell line THP-1 *in vitro* and subcutaneous tissue of rats *in vivo*. *Molecular Biosystems*, **1**, 176–82.

100 Xiao, Y. et al. (2009) Anti-HER2 IgY antibody-functionalized single-walled carbon nanotubes for detection and selective destruction of breast cancer cells. *BMC Cancer*, **9**, 351.

101 Huang, Y.F., Sefah, K., Bamrungsap, S., Chang, H.T. and Tan, W. (2008) Selective photothermal therapy for mixed cancer cells using aptamer-conjugated nanorods. *Langmuir*, **24**, 11860–5.

102 Zavaleta, C. et al. (2008) Noninvasive Raman spectroscopy in living mice for evaluation of tumor targeting with carbon nanotubes. *Nano Letters*, **8**, 2800–5.

103 Hartschuh, A., Qian, H., Meixner, A.J., Anderson, N. and Novotny, L. (2005) Nanoscale optical imaging of excitons in single-walled carbon nanotubes. *Nano Letters*, **5**, 2310–13.

104 Smith, A.M., Mancini, M.C. and Nie, S. (2009) Bioimaging: second window for *in vivo* imaging. *Nature Nanotechnology*, **4**, 710–11.

105 Xiang, L. et al. (2009) Photoacoustic molecular imaging with antibody-functionalized single-walled carbon nanotubes for early diagnosis of tumor. *Journal of Biomedical Optics*, **14**, 021008.

106 De la Zerda, A. et al. (2008) Carbon nanotubes as photoacoustic molecular imaging agents in living mice. *Nature Nanotechnology*, **3**, 557–62.

107 McDevitt, M.R. et al. (2007) PET imaging of soluble yttrium-86-labeled carbon nanotubes in mice. *PLoS ONE*, **2**, e907.

108 Cherukuri, P., Bachilo, S.M., Litovsky, S.H. and Weisman, R.B. (2004) Near-infrared fluorescence microscopy

of single-walled carbon nanotubes in phagocytic cells. *Journal of the American Chemical Society*, **126**, 15638–9.

109 Liu, Z. *et al.* (2008) Circulation and long-term fate of functionalized, biocompatible single-walled carbon nanotubes in mice probed by Raman spectroscopy. *Proceedings of the National Academy of Sciences of the United States of America*, **105**, 1410–15.

110 Harrison, T. *et al.* (2009) Combined photoacoustic and ultrasound biomicroscopy. *Optics Express*, **17**, 22041–6.

111 Schipper, M.L. *et al.* (2008) A pilot toxicology study of single-walled carbon nanotubes in a small sample of mice. *Nature Nanotechnology*, **3**, 216–21.

112 Allen, B.L. *et al.* (2008) Biodegradation of single-walled carbon nanotubes through enzymatic catalysis. *Nano Letters*, **8**, 3899–903.

113 Allen, B.L. *et al.* (2009) Mechanistic investigations of horseradish peroxidase-catalyzed degradation of single-walled carbon nanotubes. *Journal of the American Chemical Society*, **131**, 17194–205.

114 Takagi, A. *et al.* (2008) Induction of mesothelioma in p53+/− mouse by intraperitoneal application of multi-wall carbon nanotube. *Journal of Toxicological Sciences*, **33**, 105–16.

115 Sakamoto, Y. *et al.* (2009) Induction of mesothelioma by a single intrascrotal administration of multi-wall carbon nanotube in intact male Fischer 344 rats. *Journal of Toxicological Sciences*, **34**, 65–76.

116 Oberdorster, G., Elder, A. and Rinderknecht, A. (2009) Nanoparticles and the brain: cause for concern? *Journal of Nanoscience and Nanotechnology*, **9**, 4996–5007.

117 Mitchell, L.A. *et al.* (2007) Pulmonary and systemic immune response to inhaled multiwalled carbon nanotubes. *Toxicological Sciences*, **100**, 203–14.

118 Mitchell, L.A., Lauer, F.T., Burchiel, S.W. and McDonald, J.D. (2009) Mechanisms for how inhaled multiwalled carbon nanotubes suppress systemic immune function in mice. *Nature Nanotechnology*, **4**, 451–6.

119 Pacurari, M. *et al.* (2008) Raw single-wall carbon nanotubes induce oxidative stress and activate MAPKs, AP-1, NF-kappaB, and Akt in normal and malignant human mesothelial cells. *Environmental Health Perspectives*, **116**, 1211–17.

120 Folkmann, J.K. *et al.* (2009) Oxidatively damaged DNA in rats exposed by oral gavage to C60 fullerenes and single-walled carbon nanotubes. *Environmental Health Perspectives*, **117**, 703–8.

13
Cancer Treatment with Carbon Nanotubes, Using Thermal Ablation or Association with Anticancer Agents

Roger G. Harrison, Luís F. F. Neves, Whitney M. Prickett and David Luu

13.1
Introduction

Since their discovery in 1991 by Iijima, carbon nanotubes (CNTs) have undergone intensive study in several different fields, including fairly recently for the treatment of cancer. The advanced physical properties of CNTs offer an innovative way to bypass significant challenges in delivering highly selective targeted anticancer therapy. In this chapter, a review is provided of up-to-date information regarding this important field of CNT research.

Carbon nanotubes have several properties that have been exploited in studies to treat cancer. They consist of graphene (a single layer of graphite), rolled into a cylindrical structure with lengths ranging from several hundred nanometers to several micrometers, and diameters ranging from 0.4 to 2 nm for single-walled carbon nanotubes (SWNTs), and from 2 to 100 nm for coaxial multiple-walled carbon nanotubes (MWNTs) [1]. One property of CNTs that has been used extensively in cancer treatment studies to date is their strong absorbance of light in the near-infrared (NIR) range (700–1400 nm). This strong absorbance of light leads to a heating of the nanotubes and, in turn, the destruction of cancer cells that have either taken up CNTs or are in their close vicinity. Limited investigations have also been conducted using a radiofrequency (RF) field to heat the nanotubes. In order to develop CNTs for use in cancer treatment, it has been important to be able to suspend the nanotubes in aqueous solutions [2]. Suspended nanotubes have been complexed in various ways with numerous anticancer agents, and these complexes have been used in studies to treat cancer.

In this chapter, progress on the use of CNTs to treat cancer will be organized into the two general types of treatment available, namely heating of the nanotubes by NIR or RF field radiation, and complexation of the nanotubes with anticancer agents.

13.2
Use of Nanotubes as Heated Particles

The initial investigations into the use of CNTs in cancer therapy were performed using NIR light to heat the nanotubes, with the objective of killing those cancer cells close to the nanotubes. Several studies have been carried out to date using NIR light, with both SWNTs and MWNTs. A study has also been carried out using a RF field to heat the nanotubes. Both, in vitro and in vivo studies have been carried out; the results of which studies are summarized in Table 13.1.

The use of CNTs to cause the thermal ablation of cancer cells was first investigated in 2005 by Kam et al. [3]. In this case, the SWNTs were noncovalently bound to single-stranded DNA (ssDNA) or to a moiety with a folic acid terminal group. The average length of the SWNTs was 150 nm. Following incubation of the functionalized nanotubes with cervical cancer cells at 37 °C, and washing to remove any SWNTs not associated with the cells, the majority of the CNTs were observed to have accumulated in the cytoplasm of the cells, with negligible translocation of the nanotubes into the nucleus. This internalization of the nanotubes was not observed when they were incubated with cells at 4 °C, which suggests the existence of an energy-dependent endocytosis process for the uptake found at 37 °C. Viable cells without CNTs were unaffected when irradiated with the NIR laser beam at a wavelength of 808 nm for up to $1050 \, J \, cm^{-2}$, while the cells with internalized CNTs showed cell death at an energy density of only $168 \, J \, cm^{-2}$. A control solution without SWNTs that was subjected to continuous irradiation at $168 \, J \, cm^{-2}$ showed a minimal temperature increase; however, application of the same irradiation profile to a solution containing the DNA-functionalized SWNTs caused the solution temperature to reach 70 °C. It was also shown that, after irradiating the cells with a pulsed beam (total energy density of $84 \, J \, cm^{-2}$), they were not damaged. However, it was observed (by using confocal microscopy) that DNA was released from the nanotubes and translocated into the cell nucleus. When functionalizing the nanotubes with folate, an anionic phospholipid conjugated to poly(ethylene glycol) (PEG) with folic acid at its terminus was used, where the hydrophobic tail of the phospholipid was adsorbed to the surface of the nanotubes. The complex was reported to bind to the cells with overexpressed folate receptors and to cause cell death after irradiation, but without killing cells that did not express abundant receptors on the cell surface.

A number of studies of the use of NIR light to kill cancer cells in vitro in the presence of SWNTs have subsequently been performed (see Table 13.1). Shao et al. [4] were able to obtain the complete destruction of breast cancer cells using SWNTs that were noncovalently functionalized with monoclonal antibodies to target either the IGF1R receptor or the HER2 receptor; in this case the SWNTs were observed, using both optical and confocal microscopy, to be internalized in the cells. A similar study was performed by Chakravarty et al. [5], where the monoclonal antibody was bound noncovalently to the SWNTs, and targeted either the CD22 receptor on human Burkitt's lymphoma cells, or the CD25 receptor on peripheral blood mononuclear cells. A significant cell killing was reported for the

Table 13.1 A summary of *in vitro* and *in vivo* studies using carbon nanotubes as heated particles for cancer therapy.

Type of nanotube	Type of cancer cell	Irradiation technique	Energy density (J cm^{-2})	Targeting moiety	Conjugation of targeting moiety to nanotubes	Testing in animals	Reference
SWNT	Cervical	NIR	168	Folic acid	Noncovalent	No	[3]
SWNT	Breast	NIR	144	Anti-HER2 Ab Anti-IGF1R Ab	Noncovalent	No	[4]
SWNT	Lymphoma	NIR	2100	Anti-CD-22 Ab	Noncovalent	No	[5]
SWNT	Lymphoma	NIR	2280	Anti-CD-22 Ab	Covalent	No	[6]
SWNT	Breast	NIR	600	Anti-HER2 Ab	Covalent	No	[7]
SWNT	Breast	NIR	60 and 120	Folic acid	Noncovalent	Mouse	[8]
SWNT	Liver	NIR	16	Folic acid	Noncovalent	No	[9]
MWNT	Kidney	NIR	720	None	–	No	[10]
MWNT	Cervical	NIR	NR	None	–	No	[11]
SWNT MWNT	Kidney	NIR	90	None	–	Mouse	[12]
MWNT	Colorectal	NIR	38	None	–	No	[13]
MWNT	Neuroblastoma	NIR	3780	Anti-GD2 Ab	Covalent	No	[14]
MWNT	Mouth	NIR	684	None	Noncovalent	Mouse	[15]
SWNT	Prostate	NIR	175	None	Noncovalent	Mouse	[16]
MWNT	Liver, pancreatic	RF	ND	None	–	Rabbit	[17]

Ab, antibody; ND, not determined; NIR, near-infrared light; NR, not reported; RF, radiofrequency field.

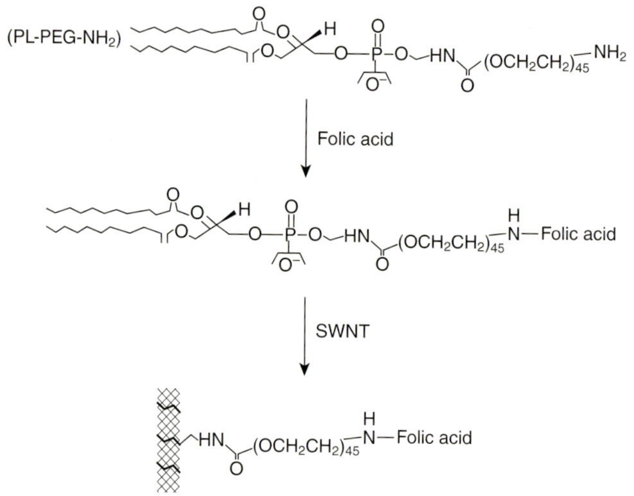

Figure 13.1 Steps in the preparation of a SWNT-folic acid conjugate. PL, phospholipid; PEG, polyethylene glycol. Modified and reproduced from Ref. [8], with permission from the Society of Photographic Instrumentation Engineers.

lymphoma cells when the SWNTs were targeted to the CD22 receptor, but these cells were not significantly killed using SWNTs targeted to the CD25 receptor. When a later study was performed by this same research group, but using covalent coupling of the antibody to the SWNTs [6], similar cell killing results were obtained, indicating that the chemical coupling of antibodies to the SWNTs did not destroy the optical properties necessary for converting NIR light into heat. Xiao et al. [7] used SWNTs covalently coupled to the anti-HER2 IgY antibody for targeting breast cancer cells; the SWNT–HER2 IgY antibody complex was found by confocal microscopy to be located almost entirely on the surface of the cells. Nearly complete killing of the breast cancer cells was obtained after NIR light treatment, but there was no significant killing of cells when the SWNTs were not targeted. Zhou et al. [8] conjugated SWNTs noncovalently with folic acid for targeting mouse mammary cancer cells. Steps in the preparation of this conjugate are shown in Figure 13.1. Significant cell killing was obtained at an energy density of 60 J cm^{-2}, but with no killing of cells not targeted with the SWNT–folate complex; however, at double this energy density there was significant cell killing in the absence of the complex. In a study conducted by Kang et al. [9], liver cancer cells were targeted by SWNTs with folate attached noncovalently, and the cells were subsequently destroyed almost completely by using a Q-switched pulsed NIR laser at a wavelength of 1064 nm, whereas more than 90% of the normal cells survived. Notably, this technique shows much promise, as a lower power is required than for continuous NIR treatment, such that the effects on normal cells are minimized.

Several studies have been performed using MWNTs instead of SWNTs to kill cancer cells *in vitro* using NIR light. Torti et al. [10] obtained almost complete killing of renal cancer cells incubated in MWNT suspensions. Similar results were obtained by Biris et al. [11] for cervical cancer cells, and by Burke et al. [12] for kidney cancer cells incubated in MWNT suspensions. By using optical microscopy, Biris et al. observed aggregations of MWNTs inside the cells a few hours after the nanotubes had been added to the culture medium. Levi-Polyachenko et al. [13] used NIR light to rapidly heat (in 10 s) colorectal cancer cells to 42 °C in the presence of MWNTs and oxaliplatin or mitomycin C; subsequently, the effectiveness of killing the cells was found to be the same as with a 2 h period of radiative heating to this same temperature in the presence of the same drugs. Wang et al. [14] covalently coupled MWNTs to an anti-GD2 monoclonal antibody for targeting a disialoganglioside (GD2) overexpressed on the surface of human neuroblastoma cells. After exposure of cells to the MWNT–antibody conjugate and washing, the cells were irradiated with NIR light; this resulted in near-complete cell killing, compared to no killing of cells exposed to MWNTs with no antibody attached. These results again indicate that sufficient optical absorption of the nanotubes remains after covalent conjugation for heat to be generated by their exposure to NIR light.

Both, SWNTs and MWNTs have been used in several studies of killing tumors *in vivo*. For example, when Zhou et al. [8] injected a SWNT–folate complex (previously described in the discussion of *in vitro* studies above) into the center of a breast tumor, and irradiated it with NIR light, 88% of the tumor cells were killed, compared to a 56% death rate without nanotubes injected. Moon et al. [15] injected nontargeted SWNTs into human epidermoid mouth tumor xenografts in nude mice, and obtained complete destruction of the tumor by 20 days after treatment with NIR light (Figure 13.2). The growth of tumors treated only with NIR light was similar to that of untreated tumors. Mice treated with SWNTs and NIR light were very healthy, and showed no toxic effects, abnormal behavior, or recurrence

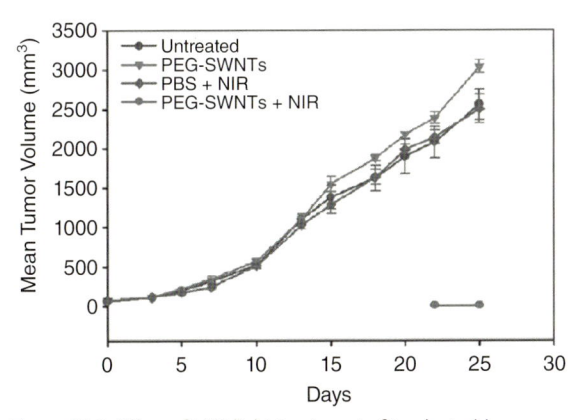

Figure 13.2 Effect of NIR light treatment of implanted human epidermoid mouth tumor xenografts in mice. PEG, polyethylene glycol. Reproduced with permission from Ref. [15]; © 2009, American Chemical Society.

of tumors over six months. A biodistribution study was also performed by Moon et al.; after 40 days from the initial treatment, the SWNT levels in the heart, lungs, liver, stomach, intestine, muscle, skin, and blood fell to near zero. Ghosh et al. [16] suspended MWNTs by adsorbing ssDNA, and then injected the suspension into human prostate tumor xenografts implanted in mice. Treatment of the tumor with NIR light resulted in a complete eradication of the tumors by six days after treatment, whereas tumors treated only with the laser or with MWNTs showed growth similar to that of the untreated tumors. Similar results were obtained by Burke et al. [12] when using suspended MWNTs that had been injected into murine kidney tumors implanted in nude mice, with the inhibition of tumor growth being dependent on the dose of the MWNTs.

A noninvasive RF field was used for the first time by Gannon et al. [17] to cause the thermal ablation of cancer cells that had been treated with suspended SWNTs. The thermal response of the nanotubes was evaluated by irradiating suspensions with different concentrations of nanotubes with radio waves at 13.56 MHz, emitted by a signal generator. The heating rate was determined to be $1.6\,\mathrm{K\,s^{-1}}$ for a SWNT concentration of $50\,\mathrm{mg\,l^{-1}}$, while the lowest concentration of nanotubes that caused enhanced bulk heating was $5\,\mathrm{mg\,l^{-1}}$. Complete destruction of human liver and pancreatic cancer cells was found when the cells were incubated in a suspension of SWNTs for 24 h, washed, and then exposed to a RF field for 2 min at 800 W; exposure for 1 min produced cytotoxicity rates of between 42% and 68%. With no SWNTs present, the cell lines exposed for 2 min to the RF field had death rates of 11–35%, although no cell death was found at a 1 min exposure time. Rabbits with liver tumors were treated with suspended SWNTs by intratumoral injection, and then with a continuous RF field for 2 min. At 48 h from the time of treatment, there was complete necrosis of the tumor, although some thermal injury to liver tissue was noted within a distance of 5 mm of the tumor. Control tumors that received RF treatment only did not have any evidence of tumor death.

13.3
Use of Anticancer Agents Associated with Nanotubes

A number of studies have been performed on treating cancer both *in vitro* and *in vivo* with CNTs associated with various anticancer agents. These agents include anticancer drugs, interfering RNA for gene therapy, proteins for immunotherapy, and cell toxins. Both SWNTs and MWNTs have been used in these studies, the details of which are summarized in Table 13.2.

In 2008, SWNTs with a taxoid anticancer drug covalently attached via a cleavable linker were used to treat leukemia cells *in vitro* by Chen et al. [18]. Biotin was also covalently attached to the SWNTs in order to bind to biotin receptors on the cancer cells. The linker was designed to contain a disulfide bond that would be stable in blood plasma, but readily cleavable by intracellular thiols such as glutathione and thioredoxin. By employing a fluorescein label, the taxoid was found (using confocal microscopy) to be internalized in the cells. The apparent cytotoxicity per taxoid of

Table 13.2 A summary of the significant *in vitro* and *in vivo* studies using carbon nanotubes associated with anticancer molecules for cancer therapy.

Type of nanotube	Type of cancer cell	Anticancer agent	Chemical linkage to anticancer agent	Targeting moiety	Testing in animals	Reference
SWNT	Leukemia	Taxoid	Covalent, cleavable	None	No	[18]
SWNT	Cervical	Doxorubicin	None	Folic acid	No	[19]
SWNT	Cervical, pancreatic	Etoposide	None	None	No	[20]
SWNT	Head and neck squamous	Cisplatin	Covalent	EGF	Mouse	[21]
SWNT, MWNT	Cervical, breast	Paclitaxel	None	None	No	[22]
SWNT	Skin	Doxorubicin	Covalent, cleavable	None	Mouse	[23]
MWNT	Bladder	Carboplatin	None	None	No	[24]
SWNT	Breast	Doxorubicin	None	None	No	[25]
MWNT	Stomach (*in vitro*), liver (*in vivo*)	HCPT	Covalent, cleavable	None	Mouse	[26]
SWNT	Breast	Paclitaxel	Covalent	None	Mouse	[27]
SWNT	Lymphoma	Doxorubicin	Noncovalent	None	Mouse	[28]
SWNT	Cervical, ovarian, lung	SiRNA	None	None	Mouse	[29]
MWNT	Lung	SiRNA	None	None	Mouse	[30]
SWNT	Leukemia	SiRNA	None	None	No	[31]
MWNT	Breast, liver	Antisense c-myc oligonucleotide	None	None	No	[32]
MWNT	Liver	Tumor lysate proteins	Covalent	None	Mouse	[33]
MWNT	Breast, cervical	Ricin A chain protein	None	Anti-HER2 Ab	No	[34]

EGF, epidermal growth factor; HCPT, 10-hydroxycamptothecin; SiRNA, small-interfering RNA.

the conjugate was substantially more than for the taxoid itself (52 nM for the conjugate versus 88 nM for the free taxoid).

Several other *in vitro* studies have been carried out with SWNTs associated with anticancer drugs. For example, Zhang *et al.* [19] suspended SWNTs by adsorbing a polysaccharide and then bound doxorubicin to the SWNTs to create a drug-delivery system with pH-triggered release; folic acid was also covalently attached to the polysaccharide for targeting folate receptors. This conjugate was substantially more cytotoxic towards cervical cancer cells incubated for 72 h compared to free doxorubicin. Mahmood *et al.* [20] studied the effect of the combination of SWNTs and the anticancer drug etoposide on the viability of cervical and pancreatic cancer cells. When the concentration of etoposide was held constant while increasing the SWNT concentration, the cell death rate was increased in both cancer cell lines, suggesting that the SWNTs had enhanced the apoptotic effect of etoposide. Bhirde *et al.* [21] covalently attached cisplatin and epidermal growth factor (EGF) to SWNTs that had been oxidized; the EGF was used for targeting the EGF receptors overexpressed on head and neck squamous cancer cells. The SWNT–cisplatin–EGF conjugate was substantially more cytotoxic towards the cancer cells than free cisplatin, at an eightfold lower concentration. Lay *et al.* [22] used SWNT–PEG and MWNT–PEG conjugates which had been physically loaded with paclitaxel (PTX) to treat human cervical cancer and breast cancer cells. Both PTX-loaded conjugates produced a high cytotoxicity towards both cancer cell lines, and also gave lower IC_{50} values than for free PTX. When Chaudhuri *et al.* [23] attached doxorubicin to SWNTs via a cleavable carbamate linker, the SWNT–doxorubicin conjugate was seen to be more cytotoxic towards murine melanoma cells than was free doxorubicin at the same concentrations. The conjugate was found to be internalized in the lysosomal compartment of the cells within 30 min of treatment.

Anticancer drugs have also been used in studies with MWNTs to treat cancer cells *in vitro*. Hampel *et al.* [24] used MWNTs carrying carboplatin to treat human bladder cancer cells. In order to incorporate the drug in the nanotubes, they were treated with nitric acid; this caused the nanotubes to open, such that capillarity became the driving force for drug uptake. The killing of cancer cells after incubation for 72 h with carboplatin-loaded nanotubes was found to depend on the dosage, with almost complete cell killing at the highest dose (nanotube concentration of 500 mg l^{-1}); unfortunately, however, it was not possible to compare the effects of the free drug with those of the drug-loaded nanotubes. Ali-Boucetta *et al.* [25] adsorbed the anticancer drug doxorubicin onto MWNTs that had been suspended using a block copolymer. The binding was attributed to π–π stacking of the aromatic chromophore of doxorubicin and the surface of the CNTs. The incubation of breast cancer cells for 24 h with the MWNT–doxorubicin complex resulted in a significantly greater number of cells being killed than was achieved with free doxorubicin at the same concentration. When Wu *et al.* [26] covalently coupled 10-hydroxycamptohecin to MWNTs using a cleavable ester linkage, the MWNT–hydroxycamptothecin conjugate caused 10–20% higher cell death compared to the free drug after 48 h, over the concentration range studied.

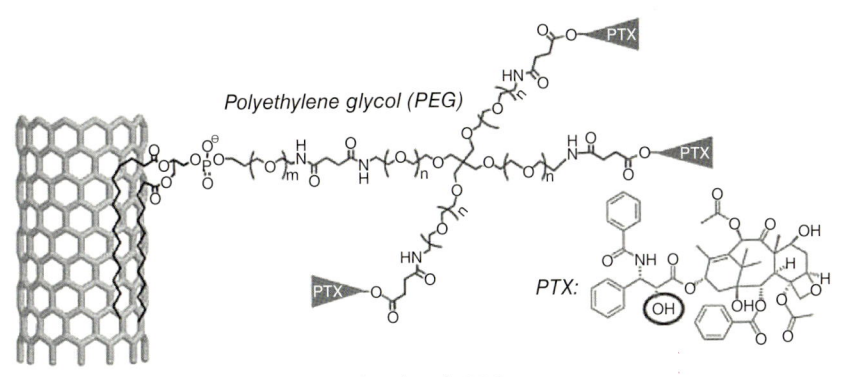

Figure 13.3 Schematic representation of paclitaxel (PTX) conjugated to SWNT using a branched PEG linked to an anionic phospholipid. Modified and reproduced from Ref. [27], with permission from the American Association for Cancer Research.

Both SWNTs and MWNTs complexed with anticancer drugs have been tested to treat tumors *in vivo*. The first report of such a treatment was in 2008 by Liu et al. [27], who suspended SWNTs using PEG and covalently linked PTX to the PEG (Figure 13.3). The administration of SWNT–PTX into the tail vein of mice with implanted mouse breast tumors was shown to inhibit tumor growth by 59% compared to untreated tumors. Moreover, such treatment was significantly more effective than that with PTX alone, or with PEG–PTX at the same dosage of PTX. In a biodistribution study, the SWNT–PTX complex was shown to be taken up at a much higher level in the tumor compared to PTX alone or PEG–PTX. This type of treatment allows for the desired damage to the cancer cells to be achieved, but has the advantage of using lower drug concentrations and therefore minimizing toxicity towards normal organs/tissues. The following year, the same group conducted a study in which lymphoma tumor xenografts in nude mice were treated with PEGylated SWNTs to which doxorubicin has been adsorbed via π–π stacking [28]. Those mice injected intravenously with the SWNT–doxorubicin complex had a significantly higher morbidity-free survival compared to mice receiving the same concentration of doxorubicin alone, or doxorubicin encapsulated in liposomes. Over a two-week treatment period, those mice treated with SWNT–doxorubicin had no mortalities, whereas treatment with doxorubicin and liposome-encapsulated doxorubicin resulted in 20% and 40% mortality, respectively.

A few other investigators have studied nanotubes complexed with anticancer drugs to treat cancer *in vivo*. For example, Bhirde et al. [21] treated nude mice with a SWNT–cisplatin–EGF conjugate (as discussed above for *in vitro* studies) via intravenous injection, and found a large reduction in tumor growth compared to a control group treated with SWNT–cisplatin over a period of 10 days after injection; this treatment resulted in the arrest of tumor growth (Figure 13.4). When Wu et al. [26] used a MWNT–hydroxycamptothecin conjugate (as discussed above

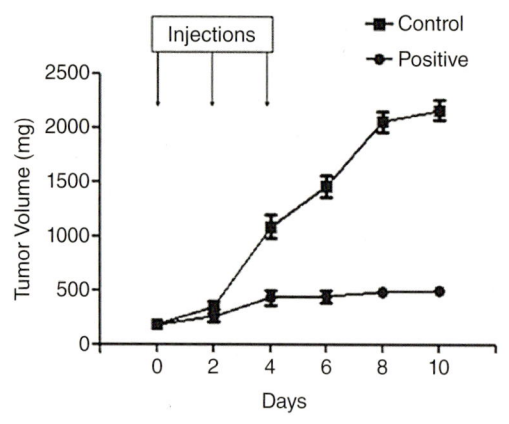

Figure 13.4 Effect of treatment with a SWNT–cisplatin–EGF conjugate, injected intravenously at the times indicated, on the growth of head and neck squamous tumors in nude mice. The control was the SWNT–cisplatin conjugate. Reproduced with permission from Ref. [21]; © 2009, American Chemical Society.

for *in vitro* studies) to treat implanted liver tumors in mice, tumor growth over 15 days in mice that had received the conjugate by intravenous injection was somewhat less than in mice that had received the free drug, but substantially less than in mice that received only MWNTs. Chaudhuri *et al.* [23] used a SWNT–doxorubicin conjugate administered intravenously (as discussed above for *in vitro* studies) to treat mice with implanted mouse melanoma tumors. Subsequently, whilst very little tumor growth occurred over 16 days after treatment with either the conjugate or with free doxorubicin, the untreated tumors showed a large amount of growth. However, the weight loss of mice treated with free doxorubicin was significantly greater than those treated with the conjugate, which indicated the presence of fewer toxic side effects following treatment with the conjugate.

The first studies using CNTs in gene therapy to treat cancer were conducted by Zhang *et al.* [29] in 2006. In this case, the SWNTs were functionalized with telomerase reverse transcriptase (TERP) small interfering RNA (siRNA). The SWNT–siRNA conjugate caused a major suppression of cell growth after six days of incubation in lung, cervical, and ovarian cancer cells compared to free siRNA or a mock SWNT–siRNA conjugate with an irrelevant RNA. Transcription of the *TERP* gene and protein expression of TERP were found to be greatly reduced in those cancer cell lines incubated for 24h with the SWNT–siRNA conjugate. The same conjugate was also injected into mouse lung tumors and human cervical cancer tumors implanted in mice, with the result that the conjugate gave a large suppression of tumor growth after seven days, compared to tumors injected with free siRNA or the mock SWNT–siRNA. In another study of gene therapy *in vivo*, Podesta *et al.* [30] complexed MWNTs with a proprietary toxic siRNA, and injected this complex into human epithelial lung tumors implanted into nude mice. At

27 days after tumor implantation, the complex had inhibited tumor growth significantly compared to tumors injected with free siRNA or unconjugated MWNTs. At day 50 after tumor implantation, 80% of the mice treated with the MWNT–siRNA complex had survived, but none of those that had not been treated survived.

Various other *in vitro* studies of using gene therapy to treat cancer have been carried out. For example, Wang *et al.* [31] used a complex of cyclin A_2 siRNA absorbed to SWNTs to treat human erythroleukemic cells. The growth of cells transfected with the SWNT–siRNA complex was much less over a period of three weeks compared to nontransfected cells, or cells transfected with free siRNA or mock SWNT–siRNA. Pan *et al.* [32] functionalized MWNTs with various dendrimers, and then complexed the MWNT–dendrimers with an antisense c-myc oligonucleotide (asODN). One of the MWNT–dendrimer–asODN complexes inhibited the growth *in vitro* of two human breast cancer cell lines and of a human liver cancer cell line to a greater extent than did a MWNT–asODN complex or dendrimer–asODN complex.

Tumor lysate proteins were covalently attached to MWNTs by Meng *et al.* [33] in a study to treat murine liver tumors implanted in mice. This conjugate was injected subcutaneously once, along with three subcutaneous injections of a tumor cell vaccine. This treatment resulted in a significantly higher 90-day survival of mice compared to treatment with the tumor cell vaccine alone. Some of the cured mice were injected with liver cancer cells again, and found to reject the tumor.

MWNTs have also been used by Weng *et al.* as a carrier for recombinant ricin A chain protein and anti-HER2 antibody to target cancer cells *in vitro* [34]. Although the ricin A chain catalytically inactivates ribosomes, it is generally nontoxic when outside the cells. After attaching the proteins to the MWNTs by adsorption, immunofluorescence confocal microscopy revealed that cells incubated with the MWNT–ricin A conjugate showed an intense fluorescence, while those incubated with only ricin A showed a much lower fluorescence. Incubation of the MWNT–ricin A–anti-HER2 antibody conjugate resulted in a death rate which was about twice as high for breast cancer cells that overexpressed the HER2 receptor, compared to normal Chinese hamster ovary cells. By comparison, incubation with the MWNT–ricin A conjugate gave equal death rates for these two cell types.

13.4
Summary

Within a remarkably short time, a significant body of research has been developed with regards to the use of CNTs in the treatment of cancer. In studies where nanotubes have been used as heated particles, both targeted and nontargeted nanotubes have been employed (see Table 13.1), with the targeting moieties having been attached either noncovalently or covalently to the nanotubes. It has been possible to achieve complete destruction of cancer cells by using NIR light exposure, with nanotubes either present in a suspension with the cells or attached to the cells, whether on the cell surface, or inside the cells. One particularly impressive *in vivo* result was obtained in mice by Moon *et al.* [15], where the complete destruction of implanted tumors was achieved by 20 days after NIR light treatment, and with no toxic effects or tumor recurrence after six months.

The energy density of the studies involving NIR light varied from 16 to 3780 J cm^{-2} (Table 13.1). The lowest value in this range was used with a Q-switched pulsed laser, although in only a single study. The next lowest energy value (38 J cm^{-2}) was used in a study where the cancer cells were heated to only 42 °C [13]. Yet, even excluding these two energy densities, the range is still large and is most likely a consequence of the various NIR light wavelengths and cancer cell types studied.

Another interesting finding was that cancer cells could be killed with NIR light, using nanotubes with antibodies covalently coupled to their surfaces [6, 7, 14]. However, in general there will still be more heat generated by nanotubes with moieties not covalently attached than with them covalently attached, which will lead to a lower energy density required and thus less harm to normal tissue. The one study in which a RF field is used to heat cancer cells in the presence of nanotubes is of major interest because the method is noninvasive [17]. Indeed, a major concern of using CNTs *in vivo* is their safety, although a biodistribution study to address this point [15] demonstrated near-zero levels of nontargeted SWNTs at 40 days after treatment with NIR light.

Both, *in vitro* and *in vivo* investigations have demonstrated that a greater cytotoxicity towards cancer cells can be achieved by using a nanotube–anticancer agent complex than by using the agent alone. In fact, in some *in vivo* studies the nanotube–anticancer agent complex was able to arrest tumor growth [21, 23]. One very important point derived from these *in vivo* studies was that the side effects of treatment with nanotube–anticancer agent complexes were significantly less than with agent alone, despite using the same dosage [23, 28].

Few studies of anticancer agents associated with nanotubes have involved targeting the nanotubes. Nonetheless, numerous approaches have been taken to complex the SWNTs with anticancer agents, including adsorption, capillary action, and even covalent linkage to the nanotube surface (with or without a cleavable linker). Invariably, however, the *in vivo* studies with anticancer drugs were performed using intravenous injection, while gene therapy studies employed intratumoral injection.

13.5
Future Perspective

Although the research conducted to date on the use of CNTs for cancer treatment has shown much promise, it has mostly focused on cancer cells *in vitro*. Hence, it will be of great interest to see how data acquired *in vitro* translate to tests in animals. The *in vivo* studies to date using heated nanotubes for treatment have all involved intratumoral injection for delivery of the SWNTs. However, this type of injection is practical only for primary tumors, and has seen only limited clinical use [35]. Consequently, in order to apply therapy that involves the heating of nanotubes for primary tumors as well as tumor metastases, there will be a need to develop a system by which nanotubes can be delivered systemically. Whilst most studies on the heating of nanotubes for cancer treatment have utilized NIR light, this has proved disadvantageous as NIR radiation must be delivered endoscopically for tumors located below the surface of the skin; this would require multiple insertions of the endoscope needle in the case of a large tumor. Moreover, the depth of penetration of NIR light in human tissues is relatively low; for example, its penetration into prostate tissue at 1064 nm was only 5 mm [36]. Thus, the use of a RF field would be advantageous for tumors beneath the skin, and further studies are needed using this approach.

Continued research should be carried out on the use of nanotubes associated with anticancer agents in order to translate the results of the most promising *in vitro* studies into animal testing. To date, the use of nanotubes for gene therapy has been limited by the need for administration via intratumoral injection. Hence, it would be advantageous to develop a method (or methods) capable of providing systemic delivery for gene therapy not only of the primary tumor but also of metastases.

Acknowledgments

Luís Neves received support from the U.S. Department of Energy-Basic Energy Sciences (DE-FG02-06ER64239) and the Foundation for Science and Technology in Portugal.

References

1 Lin, Y., Taylor, S., Li, H.P., Fernando, K.A.S., Qu, L.W., Wang, W., Gu, L.R., Zhou, B. and Sun, Y.P. (2004) Advances toward bioapplications of carbon nanotubes. *Journal of Materials Chemistry*, **14**, 527–41.

2 Harris, P.J.F. (2009) *Carbon Nanotube Science*, Cambridge University Press, Cambridge, UK.

3 Kam, N., O'Connell, M., Wisdom, J. and Dai, H. (2005) Carbon nanotubes as multifunctional biological transporters and near-infrared agents for selective cancer cell destruction. *Proceedings of the National Academy of Sciences of the United States of America*, **102**, 11600–5.

4 Shao, N., Lu, S., Wickstrom, E. and Panchapakesan, B. (2007) Integrated molecular targeting of IGF1R and HER2 surface receptors and destruction of breast cancer cells using single wall carbon nanotubes. *Nanotechnology*, **18**, 315101.

5 Chakravarty, P., Marches, R., Zimmerman, N., Swafford, A., Bajaj, P., Musselman, I., Pantano, P., Draper, R. and Vitetta, E. (2008) Thermal ablation of tumor cells with antibody-functionalized single-walled carbon nanotubes. *Proceedings of the National Academy of Sciences of the United States of America*, **105**, 8697–702.

6 Marches, R., Chakravarty, P., Musselman, I., Bajaj, P., Azad, R., Pantano, P., Draper, R. and Vitetta, E. (2009) Specific thermal ablation of tumor cells using single-walled carbon nanotubes targeted by covalently-coupled monoclonal antibodies.

International Journal of Cancer, **125**, 2970–7.

7 Xiao, Y., Xiugong, G., Oleh, T., Stephen, T., Aaron, U., David, H., Richard, C., Thomas, A., Somenath, M. and Ronak, S. (2009) Anti-HER2 IgY antibody-functionalized single-walled carbon nanotubes for detection and selective destruction of breast cancer cells. *BMC Cancer*, **9**, 351.

8 Zhou, F., Ou, Z., Wu, B., Resasco, D. and Chen, W. (2009) Cancer photothermal therapy in the near-infrared region by using single-walled carbon nanotubes. *Journal of Biomedical Optics*, **14**, 021009.

9 Kang, B., Yu, D., Dai, Y., Chang, S., Chen, D. and Ding, Y. (2009) Cancer-cell targeting and photoacoustic therapy using carbon nanotubes as bomb agents. *Small*, **5**, 1292–301.

10 Torti, S., Byrne, F., Whelan, O., Levi, N., Ucer, B., Schmid, M., Torti, F., Akman, S., Liu, J. and Ajayan, P. (2007) Thermal ablation therapeutics based on CNx multi-walled nanotubes. *International Journal of Nanomedicine*, **2**, 707–14.

11 Biris, A., Boldor, D., Palmer, J., Monroe, W., Mahmood, M., Dervishi, E., Xu, Y., Li, Z., Galanzha, E. and Zharov, V. (2009) Nanophotothermolysis of multiple scattered cancer cells with carbon nanotubes guided by time-resolved infrared thermal imaging. *Journal of Biomedical Optics*, **14**, 021007.

12 Burke, A., Ding, X., Singh, R., Kraft, R., Levi-Polyachenko, N., Rylander, M., Szot, C., Buchanan, C., Whitney, J. and Fisher, J. (2009) Long-term survival following a single treatment of kidney tumors with multiwalled carbon nanotubes and near-infrared radiation. *Proceedings of the National Academy of Sciences of the United States of America*, **106**, 12897–902.

13 Levi-Polyachenko, N., Merkel, E., Jones, B., Carroll, D. and Stewart, J., IV (2009) Rapid photothermal intracellular drug delivery using multiwalled carbon nanotubes. *Molecular Pharmacology*, **6**, 1092–9.

14 Wang, C., Huang, Y., Chang, C., Hsu, W. and Peng, C. (2009) *In vitro* photothermal destruction of neuroblastoma cells using carbon nanotubes conjugated with GD2 monoclonal antibody. *Nanotechnology*, **20**, 315101.

15 Moon, H., Lee, S. and Choi, H. (2009) *In vivo* near-infrared mediated tumor destruction by photothermal effect of carbon nanotubes. *ACS Nano*, **3**, 3707–13.

16 Ghosh, S., Dutta, S., Gomes, E., Carroll, D., D'Agostino, R., Jr, Olson, J., Guthold, M. and Gmeiner, W. (2009) Increased heating efficiency and selective thermal ablation of malignant tissue with DNA-encased multiwalled carbon nanotubes. *ACS Nano*, **3**, 2667–73.

17 Gannon, C., Cherukuri, P., Yakobson, B., Cognet, L., Kanzius, J., Kittrell, C., Weisman, R., Pasquali, M., Schmidt, H. and Smalley, R. (2007) Carbon nanotube-enhanced thermal destruction of cancer cells in a noninvasive radiofrequency field. *Cancer*, **110**, 2654–65.

18 Chen, J., Chen, S., Zhao, X., Kuznetsova, L., Wong, S. and Ojima, I. (2008) Functionalized single-walled carbon nanotubes as rationally designed vehicles for tumor-targeted drug delivery. *Journal of the American Chemical Society*, **130**, 16778–85.

19 Zhang, X., Meng, L., Lu, Q., Fei, Z. and Dyson, P. (2009) Targeted delivery and controlled release of doxorubicin to cancer cells using modified single wall carbon nanotubes. *Biomaterials*, **30**, 6041–7.

20 Mahmood, M., Karmakar, A., Fejleh, A., Mocan, T., Iancu, C., Mocan, L., Iancu, D., Xu, Y., Dervishi, E. and Li, Z. (2009) Synergistic enhancement of cancer therapy using a combination of carbon nanotubes and anti-tumor drug. *Nanomedicine*, **4**, 883–93.

21 Bhirde, A., Patel, V., Gavard, J., Zhang, G., Sousa, A., Masedunskas, A., Leapman, R., Weigert, R., Gutkind, J. and Rusling, J. (2009) Targeted killing of cancer cells *in vivo* and *in vitro* with EGF-directed carbon nanotube-based drug delivery. *ACS Nano*, **3**, 307–16.

22 Lay, C., Liu, H., Tan, H. and Liu, Y. (2010) Delivery of paclitaxel by physically loading onto poly(ethylene glycol) (PEG)-graft-carbon nanotubes for potent cancer therapeutics. *Nanotechnology*, **21**, 065101.

23 Chaudhuri, P., Soni, S. and Sengupta, S. (2010) Single-walled carbon nanotube-conjugated chemotherapy exhibits increased therapeutic index in melanoma. *Nanotechnology*, **21**, 025102.

24 Hampel, S., Kunze, D., Haase, D., Krämer, K., Rauschenbach, M., Ritschel, M., Leonhardt, A., Thomas, J., Oswald, S. and Hoffmann, V. (2008) Carbon nanotubes filled with a chemotherapeutic agent: a nanocarrier mediates inhibition of tumor cell growth. *Nanomedicine*, **3**, 175–82.

25 Ali-Boucetta, H., Al-Jamal, K., McCarthy, D., Prato, M., Bianco, A. and Kostarelos, K. (2008) Multiwalled carbon nanotube–doxorubicin supramolecular complexes for cancer therapeutics. *Chemical Communications*, 459–61.

26 Wu, W., Li, R., Bian, X., Zhu, Z., Ding, D., Li, X., Jia, Z., Jiang, X. and Hu, Y. (2009) Covalently combining carbon nanotubes with anticancer agent: preparation and antitumor activity. *ACS Nano*, **3**, 2740–50.

27 Liu, Z., Chen, K., Davis, C., Sherlock, S., Cao, Q., Chen, X. and Dai, H. (2008) Drug delivery with carbon nanotubes for *in vivo* cancer treatment. *Cancer Research*, **68**, 6652–60.

28 Liu, Z., Fan, A., Rakhra, K., Sherlock, S., Goodwin, A., Chen, X., Yang, Q., Felsher, D. and Dai, H. (2009) Supramolecular stacking of doxorubicin on carbon nanotubes for *in vivo* cancer therapy. *Angewandte Chemie International Edition*, **48**, 7668–72.

29 Zhang, Z., Yang, X., Zhang, Y., Zeng, B., Wang, S., Zhu, T., Roden, R., Chen, Y. and Yang, R. (2006) Delivery of telomerase reverse transcriptase small interfering RNA in complex with positively charged single-walled carbon nanotubes suppresses tumor growth. *Clinical Cancer Research*, **12**, 4933–9.

30 Podesta, J., Al-Jamal, K., Herrero, M., Tian, B., Ali-Boucetta, H., Hegde, V., Bianco, A., Prato, M. and Kostarelos, K. (2009) Antitumor activity and prolonged survival by carbon-nanotube-mediated therapeutic siRNA silencing in a human lung xenograft model. *Small*, **5**, 1176–85.

31 Wang, X., Ren, J. and Qu, X. (2008) Targeted RNA interference of cyclin A2 mediated by functionalized single-walled carbon nanotubes induces proliferation arrest and apoptosis in chronic myelogenous leukemia K562 cells. *ChemMedChem*, **3**, 940–5.

32 Pan, B., Cui, D., Xu, P., Ozkan, C., Feng, G., Ozkan, M., Huang, T., Chu, B., Li, Q. and He, R. (2009) Synthesis and characterization of polyamidoamine dendrimer-coated multi-walled carbon nanotubes and their application in gene delivery systems. *Nanotechnology*, **20**, 125101.

33 Meng, J., Duan, J., Kong, H., Li, L., Wang, C., Xie, S., Chen, S., Gu, N., Xu, H. and Yang, X. (2008) Carbon nanotubes conjugated to tumor lysate protein enhance the efficacy of an antitumor immunotherapy. *Small*, **4**, 1364–70.

34 Weng, X., Wang, M., Ge, J., Yu, S., Liu, B., Zhong, J. and Kong, J. (2009) Carbon nanotubes as a protein toxin transporter for selective HER2-positive breast cancer cell destruction. *Molecular Biosystems*, **5**, 1224–31.

35 Goldberg, E., Hadba, A., Almond, B. and Marotta, J. (2002) Intratumoral cancer chemotherapy and immunotherapy: opportunities for nonsystemic preoperative drug delivery. *Journal of Pharmacy and Pharmacology*, **54**, 159–80.

36 Newman, C. and Jacques, S.L. (1991) Laser penetration into prostate for various wavelengths. *Lasers in Surgery and Medicine*, **Suppl.** 3, 75–6.

14
Carbon Nanotubes for Targeted Cancer Therapy
Reema Zeineldin

14.1
Introduction

The concept of a "magic bullet" was coined by the nineteenth century scientist Paul Ehrlich, who was the first to use chemotherapeutics that specifically targeted microorganisms within an infection, without affecting the normal tissues of the body [1]. This magic bullet concept was later applied to cancer tissue, by employing targeting moieties such as monoclonal antibodies and ligands that could specifically recognize tumor-specific surface antigens or receptors. Such targeting moieties have recently been used in combination with a variety of nanocarriers as drug-delivery vehicles for the specific targeting and destruction of cancer cells, while sparing normal tissue the toxic effects of cancer chemotherapeutic agents. The development of efficient drug-delivery vehicles is very important for improving the efficacy of therapeutic agents, and also for enhancing the uptake of those drugs which suffer from a poor cellular uptake. It is more effective not only for drug-delivery vehicles specifically to target cancer cells, but also to deliver drugs intracellularly, where they are more potent.

Carbon nanotubes (CNTs) represent one of the new emerging drug-delivery nanocarriers that can efficiently transport a wide variety of adsorbed or chemically conjugated molecules inside cells. The specific advantages of using CNTs as drug-delivery vehicles include:

- Their high aspect ratio, which permits the addition of multiple copies of various moieties onto the surface of CNTs. These include targeting moieties, therapeutic agents, and imaging agents.

- Their ease of uptake by cells, and their ability to deliver a variety of cargoes.

- Their effective role in the photothermal or photoacoustic destruction of targeted cancer cells [2–4].

To date, several reviews have been produced regarding the use of CNTs for drug delivery, and of nanocarriers and their roles in targeted cancer therapy. No reviews have yet been produced, however, that specifically summarizes the findings with cancer-targeted CNTs and evaluates the targeting strategies or the targets. The aim of this chapter is to address this issue, first by introducing the subject of cancer and its conventional chemotherapy, followed by details of the advantages of nanocarriers in targeted drug delivery. Details of CNTs specifically as drug nanocarriers, and of their functionalization and cellular uptake, together with examples of their use in cancer targeting, are then provided. The chapter concludes with a summary and a discussion of future perspectives.

14.2
Cancer

Cancer is a malignant tumor or growth that results from an uncontrolled proliferation of abnormal cells that have accumulated several genetic changes. Some of these genetic changes may have affected tumor suppressor genes or oncogenes (e.g., growth factor receptors and signaling molecules), causing a loss of the regulation of cell proliferation and migration. Cancer may be localized in an organ, or it may spread (metastasize) to adjacent or nonadjacent organs through the bloodstream or lymphatic vessels. The invading cells of cancer dissociate from the primary tumor and degrade the extracellular matrix (ECM) by the action of specific proteases. They then break through the basement membrane to invade the surrounding tissue (Figure 14.1), and migrate into the bloodstream and the lymphatic vessels to metastasize to other parts of the body (Figure 14.1). Although, cancer is characterized by leaky blood vessels [5], the lymphatic system serves as the main pathway for the metastasis of tumors. In comparison to blood capillaries, the lymphatic capillaries have overlapping endothelial cells with reduced or no tight junctions, no pericytes (smooth muscle cells), no fibrous tissues, and a lower internal pressure. Each of these characteristics renders the lymphatic capillaries easily accessible by metastasizing cancer cells [6, 7].

The accumulated genetic changes in cancer cells lead to the suppression or aberrant expression of several proteins within the cell, or antigens on the surfaces of the cancer cells. Although the new antigens on cancer cells instigate an immune and inflammatory response within the patient, leading to accumulation of activated macrophages and immune cells at the tumor sites (Figure 14.1), this is insufficient to eliminate the tumors. In fact, the tumors are capable of immunosuppressing both local and systemic immunity [8]. The activated macrophages and inflammatory cells secrete growth factors and cytokines that further promote proliferation and tumorigenesis, especially with the increased expression of growth factor receptors on the surface of cancer cells. Cancers are also characterized by *hypoxia*, which causes the induction of growth factors that promote *angiogenesis*–that is, the formation of new blood vessels [5].

As the tumor microenvironment plays a major role in promoting tumorigenesis, to attack this microenvironment with cytotoxic agents – for example, by attacking

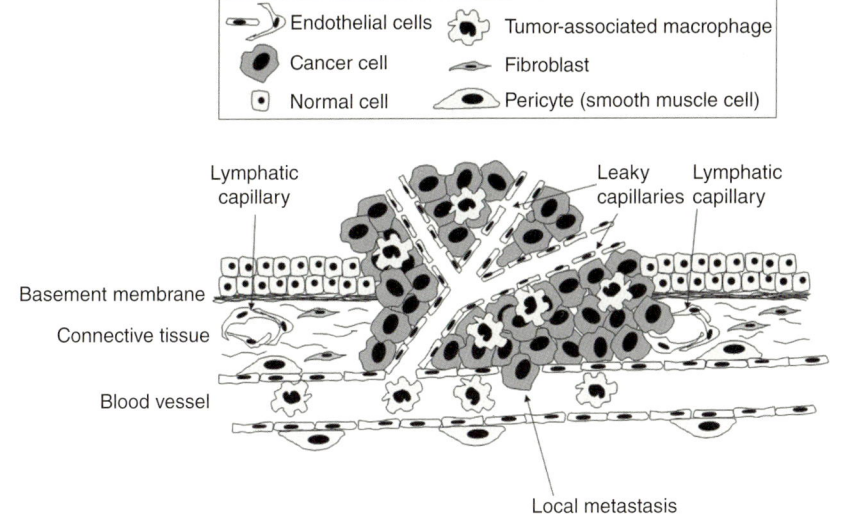

Figure 14.1 Schematic of cancer and its microenvironment. Cancer cells detach from the tumor and break through lymph and blood vessels to metastasize to other tissues. Cancers are characterized by angiogenesis and inflammation with the presence of inflammatory cells, including activated macrophages, in addition to leakiness of the blood vessels.

activated macrophages and/or the blood supply by targeting markers of angiogenesis – would lead to the destruction of cancer cells. The main aim when targeting cancer is to restrict any therapeutic activity to the tumor tissue and/or its microenvironment, thereby avoiding cytotoxic effects on normal tissues. In contrast, conventional cancer chemotherapy affects the four main cellular processes that occur in both cancer and non-cancer cells, namely DNA synthesis, RNA synthesis, protein synthesis, and cell division. Thus, to treat a patient with conventional chemotherapeutic agents may cause them to suffer several adverse side effects of the drug, including hair loss, a reduced formation of blood cells, and sterility [9]. The reason for these effects is that the chemotherapeutic agents are in fact toxic chemicals that kill cells nonspecifically; hence, they are cytotoxic to both cancer and non-cancer cells. For example, if a chemotherapeutic agent is designed to target dividing cells, then all normal dividing cells within the body, including hair cells, gastrointestinal cells, blood cells, and germ cells, will all be affected by the drug [9]. Likewise, if a chemotherapeutic agent is designed to target specific molecules in cells, whether they are dividing or not, then non-cancer cells will also be affected by the drug [9]. It is anticipated that, in medicine, nanotechnology has the potential to overcome not only the above-mentioned problems, but also those associated with conventional chemotherapy.

14.3
Conventional Cancer Chemotherapy versus Nanocarrier-Mediated Drug Delivery

14.3.1
Challenges with Chemical Compounds as Therapeutic Agents

In addition to their nonspecificity, which may leads to adverse side effects (as noted above), the use of conventional chemotherapeutic agents presents additional challenges:

- They often have a low aqueous solubility, which hinders their transport within the aqueous environment of the blood.
- They may undergo biochemical degradation by the liver, and are rapidly cleared from the body via the liver and the kidneys [9].
- They encounter a resistance to their therapeutic effect, because an efflux pump actively pumps the drugs out of the cells, thus reducing their efficacy [10].

Although such problems with chemotherapeutics may limit the dose, frequency, or duration of treatment, they can be overcome by using combination chemotherapy regimens that involve more than one drug, each with a different toxicity, mechanism of action, and target. Overall, however, it would be preferable to reduce the quantities of chemotherapeutic agents used by specifically targeting cancer cells, as this will reduce both the drug toxicities and biochemical degradation. This is exactly the situation where nanocarriers can play a role and, indeed, several forms of drug nanocarrier have now been developed (including CNTs) that are capable of delivering chemotherapeutic agents specifically to cancer cells, following their functionalization with a targeting moiety.

14.3.2
Advantages of Nanocarriers as Drug-Delivery Vehicles

The use of nanocarriers in a drug-delivery role has many advantages over the use of drugs directly as chemotherapeutic agents. These advantages of nanocarriers include:

- An ability to deliver poorly soluble drugs, either by enclosing them within the hydrophobic interfaces of the nanocarriers, or by acting as drug carriers in the bloodstream.
- Helping to stabilize their cargo by prolonging the circulation time as a result of drug encapsulation and protection from inactivation by metabolic enzymes. This is in addition to a reduced renal clearance [11] that enhances the pharmacokinetics and biodistribution of the therapeutic agents.
- Reducing the systemic toxicity of chemotherapeutic agents.
- Playing a multifunctional role, as several types of molecule can be loaded onto the nanocarrier, including drugs, imaging agents, targeting moieties (e.g., ligands or antibodies), and poly(ethylene glycol) (PEG).

- Helping to improve other technologies, such as the delivery of gene silencing inhibitory RNA (RNAi) by specifically targeting disease cells while facilitating the entry of RNAi into cells, as well as providing protection from cellular degradative nucleases [12].

- Helping to overcome drug resistance through more than one mechanism. One approach involves loading more than one drug onto the nanocarrier, while a second involves the nanocarrier avoiding recognition by the drug efflux pump [10]. The latter effect is a direct result of the increased size of the nanocarriers, and of active targeting when employed. The increased size of the nanocarrier promotes escape from the size-limited drug efflux pump, while active targeting promotes binding to the cell surface molecules that may or may not undergo endocytosis. If the target surface molecule does undergo endocytosis, the nanocarrier is directed to the endosomes and thus escapes the drug efflux pump. On the other hand, if the target surface molecule does not undergo endocytosis, the drug on the nanocarrier inhibits the surface molecule to which the nanocarriers are bound; thus, the therapeutic goal is achieved without internalization, which also escapes the drug efflux pump.

Each of the advantages of using nanocarriers in drug delivery will enhance the overall efficacy of a therapeutic agent. Yet, this applies also to CNTs, the main advantages of which are listed in the following section.

14.4
Carbon Nanotubes as Drug-Delivery Vehicles

The two types of CNT—single-walled carbon nanotubes (SWNTs) and multi-walled carbon nanotubes (MWNTs)—can be easily functionalized through acid oxidation to create carboxyl groups at defect sites or at the ends of the CNTs. These groups can then be further chemically modified to introduce new groups, or even to covalently attach a variety of molecules [13–19]. In addition, CNTs can be functionalized noncovalently through the adsorption of molecules onto their surfaces; alternatively, molecules may be encapsulated in the interior space of the CNTs [13, 16–19]. The functionalization of CNTs with various molecules, including surfactants, proteins, and PEG, helps to solubilize and disperse the CNTs [20–23]. The functionalization of CNTs with various cargo molecules also facilitates the transport of these molecules into cells. Today, CNTs are emerging as drug-delivery vehicles that are capable of efficiently transporting a variety of adsorbed or chemically conjugated molecules inside cells. To date, CNTs with lengths ≤1 µm have been used to transport into amino acids [24], nucleic acids [25–45], peptides and proteins [45–48], and therapeutic agents [49–53] into cells.

One advantage of using CNTs for drug delivery is that they are multivalent, high-aspect-ratio vehicles that can be tagged with a variety of molecules that includes targeting ligands or antibodies, and multiple drugs or imaging agents (Figure 14.2). The results of a recent study showed that, even when CNTs are coated with polymers such as PEG, they retain their ability to bind a large number

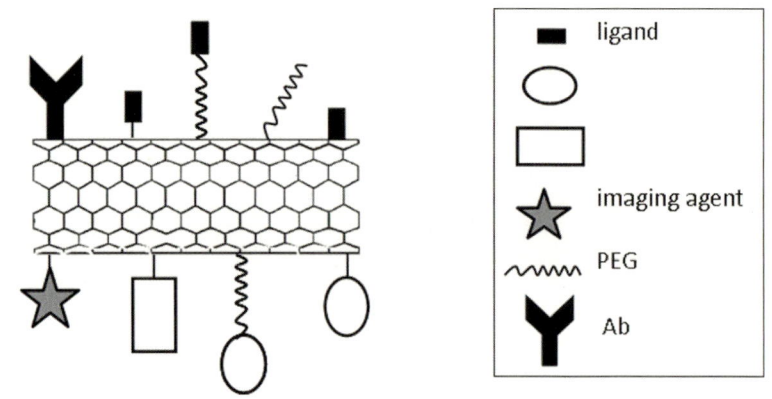

Figure 14.2 Multifunctional SWNT. CNTs can be easily functionalized with several molecules that may serve to specifically target the CNTs (such as ligands or antibodies), enhance the half-life of CNTs (such as PEG), or kill or image the cancer cells (such as therapeutic and imaging agents). These molecules may be either physically adsorbed or chemically conjugated to CNTs.

of drugs and imaging agents, because of their multi-functionality [54]. Loading a nanocarrier with multiple drugs represents a valuable strategy in cases of drug resistance, or when using multiple drugs to increase therapeutic effectiveness. Besides serving as a vehicle for multiple drugs, CNTs can also help to stabilize their cargo by extending their lifetime, as well as reducing the systemic toxicity. Functionalized CNTs also display a low cellular toxicity and are nonimmunogenic – both of which are ideal properties for a drug carrier. Furthermore, the intrinsic photothermal and photoacoustic transduction properties of CNTs makes them ideal for the destruction of targeted cancer cells [2–4].

14.5
Cellular Uptake of CNTs

Currently, limited information is available on the mechanism of internalization of CNTs (with length ≤1 µm), mainly because the debate has persisted as to whether the process is passive, or it involves clathrin-mediated endocytosis [25, 45, 46, 55–57]. There is some evidence that the mechanism is endocytosis-independent, where CNTs pass through the lipid bilayer of the cell membrane as nanoneedles, regardless of the functional groups on the CNTs and the cell type, even when endocytosis is inhibited [25, 46, 55]. In contrast, evidence that the uptake of CNTs involves clathrin-mediated endocytosis was supported by the blockade of endocytosis, which also blocked the uptake of transferrin, a control receptor which is known to be internalized via clathrin-mediated endocytosis [45, 56]. In addition, the microscopic tracking of trajectories of internalized DNA–

SWNTs in cells were consistent with receptor-mediated endocytosis rates [58]. There have been suggestions that the nature of the functional group (be it a simple carboxyl group or a complex one such as a polypeptide or nucleic acid) at the CNT surface may be the major determinant in the interaction with cells and the mechanism of cellular uptake. Conversely, a passive uptake of CNTs was reported to be independent of the functional group and cell type [55]. Another factor may be the length of the CNTs, as the various studies employed different CNTs lengths that could be roughly grouped into lengths of either 300–1000 nm or <300 nm. However, the evidence in general points towards the existence of both mechanisms – passive and endocytosis-mediated – depending on the targeting of CNTs to cells. When CNTs are functionalized with a ligand that targets a receptor on the cell, then the internalization of CNTs involves the mechanism specific to that receptor, whereas nontargeting CNTs seem to be taken up passively by the cells. Indeed, a recent study showed that fluorescein-crosslinked SWNTs were taken up passively by cells, while the same cancer cells took up biotin-crosslinked SWNTs by an endocytosis-mediated pathway [52]. It is possible that, in the latter case, the uptake was both passive and through endocytosis. Evidence also points to role for PEGylation in enhancing the specificity of receptor-mediated uptake. Fluorescein-functionalized SWNTs were taken up by cells, whereas the PEGylation of fluorescein-SWNTs prevented the uptake of these CNTs by the same cancer cells [59]. Furthermore, the introduction of a ligand to a receptor on the cancer cell surface, such as folic acid which binds to folate receptor α, or epidermal growth factor (EGF) which binds to the EGF receptor, caused a specific uptake by cancer cells expressing the corresponding receptor to the ligand [59].

In conclusion, CNTs efficiently transport various types of cargo into cells through two mechanisms, one being passive and the other endocytosis-dependent. However, functionalization with ligands promotes targeting to a specific receptor and uptake through endocytosis, while the PEGylation of CNTs enhances the specificity of this endocytosis-mediated uptake. In fact, for drug delivery the functionalization of CNTs with PEG is important so as to prolong the half-life of CNTs *in vivo*, in addition to reducing the nonspecific uptake of CNTs by nontargeted cells. In the case of CNTs, PEGylation represents an efficient method of dispersing CNTs in aqueous solutions. Because of the value of PEGylation of CNTs for dispersing CNTs, and for their use in drug delivery, further details of this process are provided in the following section.

14.6
Functionalization of CNTs with Polyethylene Glycol

CNTs are functionalized with PEG to enhance their dispersion in aqueous solutions, either by chemical conjugations or by physical adsorption [20, 27, 60, 61]. In the case of physical adsorption, phospholipid–PEG (PL-PEG) is employed, where the PL portion is adsorbed strongly and stably to the sidewalls of the CNTs, while the PEG moiety extends into the aqueous solution [27, 60]. In addition to dispersing CNTs, the PEGylation of CNTs plays another role when the CNTs are used as drug-delivery vehicles. The functionalization of drug-delivery vehicles with PEG of molecular weight (MW) between 1 and 40 kDa reduces not only their immunogenicity but also their phagocytosis by the reticuloendothelial system (RES), and this leads to a prolonged circulation time [62]. This effect is caused by PEG creating a hydrated hydrophilic cloud that repels interactions with proteins in the circulation or on cells [63]. An inhibition of the binding of opsonin proteins to drug-delivery vehicles in the circulation prevents these vehicles from being tagged for removal by the RES, and this in turn prolongs their circulation time so as to create "stealth" (protected) drug-delivery vehicles [63]. PEG can be chemically modified very easily to become covalently linked to a targeting moiety as well as therapeutic and imaging agents, thus providing a platform for the assembly of various molecules [63]. On the other hand, the heavy PEGylation of drug-delivery vehicles could mask ligands and interfere with their binding to the cancer cell surface molecules, leading in turn to a reduced targeting of the cancer cells [63]. The results of a recent study showed that coating CNTs with polymers such as PEG did not interfere with their ability to bind to a large number of drugs and imaging agents, which is probably due to the CNTs' multifunctionality and their high aspect ratio [54]. Previously, CNTs have been PEGylated with short, long, or branched PEGs, and their half lives have been evaluated *in vivo* in animal models. Covalently PEGylated SWNTs had a longer half-life *in vivo* than did nonPEGylated SWNTs (~10 h) [64]. Additionally, CNTs which had been PEGylated by the adsorption of PL-PEG to SWNTs had a half-life that was increased from 1.2 h for PL-PEG 2000 i.e., the MW of PEG is ~2000 Da) to 5 h for the higher-molecular-weight PEGs (5, 7, and 12 kDa [65]. In contrast, branched PEGs increased the half-life even further, to 24 h [65, 66].

In the case of CNTs, it has been found that to adsorb phospholipid-PEG (PL-PEG) would prevent the nonspecific binding of proteins to the surface of CNTs [61, 67, 68]. The usual method for functionalizing CNTs with PL-PEG involves their ultrasonication in a solution of PL-PEG for 1 h, or more [27, 60, 69, 70]. However, it was shown recently that such treatment caused the fragmentation of PL-PEG 2000, interfering with its ability to block the nonspecific uptake by cells, whereas unfragmented PL-PEG 2000 (ultrasonicated with SWNT for 10 min only) promoted the specific cellular uptake of targeted SWNTs to two distinct classes of receptors expressed by cancer cells [59]. This would explain the controversial findings of some reports, where PL-PEG 2000 adsorbed to SWNTs did not prevent the uptake of SWNTs either *in vitro* [27] or *in vivo* [60]. In addition, another study with PEG 6000 alone, without PL or CNTs, demonstrated the fragmentation of PEG 6000 in a time-dependent manner, with partial fragmentation occurring by 2 h and complete fragmentation by 8 h [71]. Thus, higher-molecular-weight PEG appears to remain intact if sonicated for 2 h or less. In conclusion, by combining the intact PEG functionalization of CNTs with a ligand to a tumor cell-surface marker leads to the specific recognition and targeting to tumor cells, while eliminating or minimizing nonspecific protein binding, and thus nonspecific uptake by non-cancerous cells which do not express the targeted cancer cell marker. This is important when specifically targeting cancer cells with drugs, to prevent the exposure of non-cancerous cells to cytotoxic drugs.

14.7 Targeting of Cancers

As stated in Section 14.1, the goal of targeting cancer is to restrict the therapeutic activity to tumor tissue or its microenvironment—that is, the elements that contribute to the tumor's maintenance and expansion through metastasis. Two types of targeting have been identified that are relevant to using CNTs as drug nanocarriers, namely "passive" and "active" targeting, both of which are described in the following subsections.

14.7.1 Passive Targeting

Passive targeting results from the fact that disease tissues including cancers exhibit large gaps in the vascular endothelium and, at the same time, lack effective lymphatic drainage (Figure 14.3) [72, 73]. This causes what is known as an enhanced permeability and retention (EPR) effect, where nanocarriers within the size range 20–400 nm leak through the endothelial gaps into the cancer tissue and

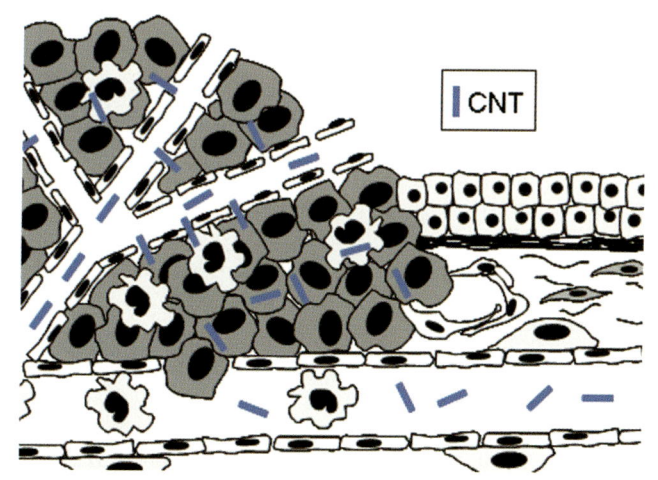

Figure 14.3 The enhanced permeability and retention (EPR) effect for CNTs. The leaky vessels in cancers and the defective lymphatic vessels cause retention of CNTs at the cancer tissue site, thus passively targeting the cancer. Note: The schematic representation is not drawn to scale. See Figure 14.1 for details of tissue components.

are retained there because of the defective lymphatic drainage; the net effect is a passive targeting of the cancer tissue [72]. When the drug is released from the nanocarriers at the cancer site, the net effect is a partitioning of the drug into the cancer, which leads to high local concentrations of the drug and more potent therapies. As short CNTs fit into the size range of nanocarriers that may display the EPR effect, they would – on that basis – be expected to be capable of passively targeting cancer tissue (Figure 14.3).

14.7.2
Active Targeting

The therapeutic active targeting of cancer cells involves discriminating between cancer cells and normal cells, by recognizing cell markers that are either unique to the cancer cells or are expressed differently between cancer cells and normal cells. Typically, the markers may be overexpressed by cancer cells, or have an altered structure, and in that case both the marker and the binding ligand (specificity and affinity, respectively) will determine the efficiency of the targeting. Tumors require an efficient blood supply to maintain nutrient and oxygen levels for tumor-cell survival, and inevitably new blood vessels are formed (angiogenesis) to provide these needs. Likewise, with tumors being classified as inflammatory conditions, the presence of cells of the immune system will contribute, via the secretion of growth factors, to promote tumor growth, proliferation, and even invasiveness (see Figure 14.1). As the tumor microenvironment plays a clear role in contributing to tumor progression, then to target the vasculature and stroma within the microenvironment will help to eliminate the cancer. Unfortunately, solid tumors have an irregular vasculature and a high interstitial pressure, and this serves as a major hurdle for drug accumulation in the tumor tissues, with the drugs tending to accumulate rather in adjacent normal tissues and causing a variety of toxic effects [74]. The targeting of cancer tissues with therapeutic agents reduces their toxicities towards other tissues as the drug is delivered only to the cancer tissue. Several targeting moieties have been used to either target markers on the cancer tissue or on the components of its microenvironment; these include antibodies, aptamers, ligands, and peptides. Such moieties have been used to target markers that are either specific to, or are overexpressed on, the cancer cells or blood vessels.

14.7.3
Trafficking of Targeted Drug-Delivery Vehicles

Drug-delivery vehicles that are targeted to receptors on cancer cells that undergo internalization will most likely undergo internalization with the targeted receptor. Receptors that are destined for degradation proceed to early endosomes, sorting endosomes, multivesicular bodies, then to late endosomes and lysosomes (Figure 14.4). In contrast, recycling receptors proceed through a distinct sorting compartment before returning to the plasma membrane for reuse (Figure 14.4). The localization of drug-delivery vehicle in these various endosomes can be conducive

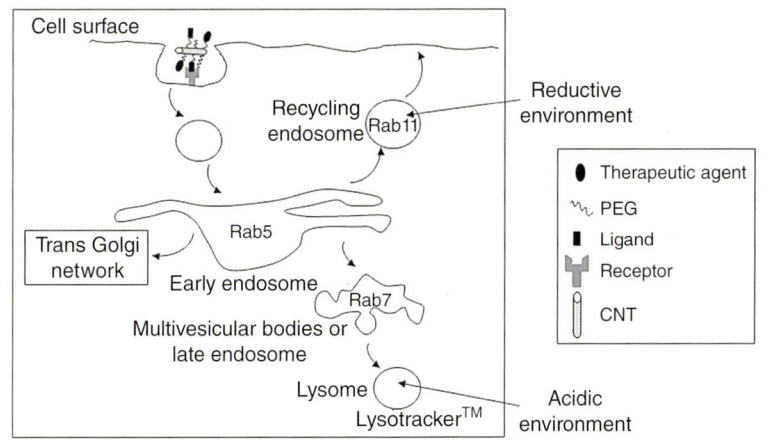

Figure 14.4 Intracellular trafficking of internalized CNTs. Targeted CNTs are taken up by cells through endocytosis, and may be directed to various endosomes. The destination of CNTs inside the cells can be determined by examining colocalization with endocytic markers, as shown in this figure. The internal environment of the endosome maybe be conducive to the release of the therapeutic load based on the use of crosslinkers that are acid labile or thiolytic, which can be cleaved in lysosomes or recycling endosomes, respectively.

to the release of the therapeutic load as a result of the acidic or reductive environment of the endosome. Another benefit for targeting specific receptors on the cells involves the fate of the internalized receptors; the compartment to which a receptor is directed may contain targets for drugs or, alternatively, its environment can be employed for releasing drugs.

14.8
Targeted Cancer Therapy Employing CNTs and a Critique of Current Studies

Studies using CNTs to target cancer cells or their microenvironment have either used a targeting moiety alone, or a targeting moiety in addition to PEG to enhance the specificity of targeting and increase the vehicle half-life in the circulation. PEGylation, as mentioned above, either blocks or reduces the nonspecific uptake of CNTs, which enhances targeting (Figure 14.5). In this section, details of the various cell markers targeted by CNTs so far, which include erbB family members, vitamins, angiogenic markers, and markers of lymphomas, leukemias or neuroblastomas, are summarized. It is important when evaluating studies of targeting cancers with CNTs to consider whether these CNTs are PEGylated by covalent conjugation or physical adsorption. If by physical adsorption, the question is was PL-PEG used or PEG alone? Likewise, if the adsorption involved sonication, then for how long was the sonication carried out? A 1 h sonication for PEG2000 [59], and more than 2 h sonication for higher-molecular-weight PEGs [71], will cause

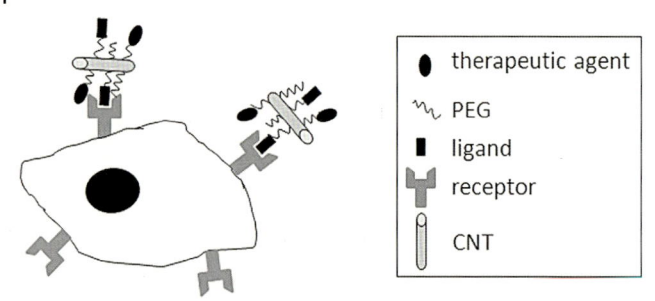

Figure 14.5 Specific receptor targeting on cancer cells by CNTs. The CNTs are functionalized with a ligand that recognizes and binds to a specific cancer cell receptor. The ligand may be a polypeptide, an antibody, or an aptamer. CNTs are also PEGylated to block their nonspecific uptake by cells that do not express the receptor, and to increase the half-life in the circulation.

fragmentation that will interfere with the specificity of the cellular uptake and clearance from circulation.

It is important that targeting studies include proper controls – that is, a cell control by which the selective uptake of targeted PEGylated CNTs can be examined by using cells that are positive for the targeted marker, and will also confirm any lack of, or significant reduction in, the uptake by cells that are marker-negative. It is also important to include a ligand control; that is, to functionalize CNTs with everything in addition to either excluding the targeting moiety or replacing it with a molecule to which it is closely related, but not close enough to bind to the receptor of the targeting ligand. Another important point is to evaluate *what* is being examined – the binding, the cellular uptake, or the therapeutic efficacy. Although binding is the first step in targeting, the intracellular localization must still be demonstrated to show that there will be an internal effect of the therapeutic agent. Nevertheless, if the targeting CNTs are used for photothermal or photoacoustic therapy, then binding to the cell surface is sufficient to induce the destruction of these cells. Existing studies with targeting cancer cells with CNTs are summarized in the following subsection.

14.8.1
erbB Family Members

Overexpression or mutations that affect the expression and/or activity of members of the erbB family of receptor tyrosine kinases occurs frequently in cancers [75–79]. The erbB family has four members: epidermal growth factor receptor (EGFR, also referred to as erbB1 or HER1); erbB2 (HER2); erbB3 (HER3); and erbB4 (HER4). Each receptor consists of a single polypeptide chain that has an extracellular domain, a transmembrane domain, and an intracellular catalytic domain. The extracellular domain binds to ligands, which activate the receptors by causing

14.8 Targeted Cancer Therapy Employing CNTs and a Critique of Current Studies

dimerization of two molecules and trans-autophosphorylation of specific tyrosine residues on the dimers. The only member of the family which has no known ligand is erbB2; this is activated through interaction and transphosphorylation with the other family members. Both, EGFR and erbB2 are good candidates for targeted therapy, because they are often aberrantly overexpressed in several cancers [75–79]. The drawback of targeting erbB family members is that they are not specifically expressed by cancer cells alone, but are also expressed by normal tissue. This means there is a likelihood of their being taken up by non-cancer tissues, although the targeting vehicles would accumulate more in cancer tissues because of the overexpression of these receptors.

CNTs functionalized with targeting moieties to EGFR or erbB2 were prepared by using antibodies to the receptors or a ligand to EGFR, and tested *in vitro*. Non-PEGylated MWNTs functionalized by adsorption to anti-erbB2 antibody and ricin A toxin caused increased cell death to erbB2-expressing cells in comparison to MWNTs functionalized with the antibody and denatured toxin that were tested with erbB2-expressing cells [80]. Another study employed non-PEGylated SWNTs that were covalently attached to chicken anti-erbB2 IgY, which shows a higher specificity and sensitivity than immunoglobulin Gs (IgGs) [81]. It was estimated that there were ten IgYs per nanotube (length 88 ± 44 nm). Subsequent imaging showed a localization of these functionalized SWNTs on the cell surface, and they selectively destroyed erbB2-positive cells only through near-infrared (NIR) radiation [81]. Targeting a cell marker that remains on the cell surface in the case of photothermal therapy, as in this study, represents a successful strategy. On the other hand, erbB2 is usually recycled to the cell surface through recycling endosomes [82]. This means that, although the majority of erbB2 remains on the cell surface, some should be detected intracellularly, though it is unclear why in this study the functionalized SWNTs remained on the cell surface without being internalized and recycled to the surface.

EGFR was also targeted by non-PEGylated SWNTs that were chemically conjugated to EGF (36 ± 10 EGF molecules per 100 nm length of SWNTs) and quantum dots (for imaging) or the chemotherapeutic agent cisplatin (for therapy) [83]. These CNTs showed an intracellular localization in EGFR-expressing, but not EGFR-suppressed, squamous cell carcinoma cells *in vitro* and in their xenografts *in vivo* [83]. The accumulation in tumor xenografts caused a regression of tumor growth, whereas control SWNTs without EGF were rapidly cleared from the circulation [83].

PEGylated CNTs were used to target EGFR or erbB2. SWNTs adsorbed to unfragmented PL-PEG2000 and to EGF promoted the intracellular uptake of SWNTs, while preincubation with the EGFR ligand, EGF inhibited the uptake [59]. This study did not evaluate the uptake of nanocarriers with EGFR-negative cells, which would confirm the specificity of targeting EGFR. In spite of this, the fact that preincubation with EGF (an EGFR-specific ligand) inhibited the uptake of nanocarriers is an indicator of the specificity of uptake through EGFR. An examination of the trafficking of this vehicle showed that it was trafficked to the lysosomal compartments (Figure 14.6), as would be expected for EGFR [82].

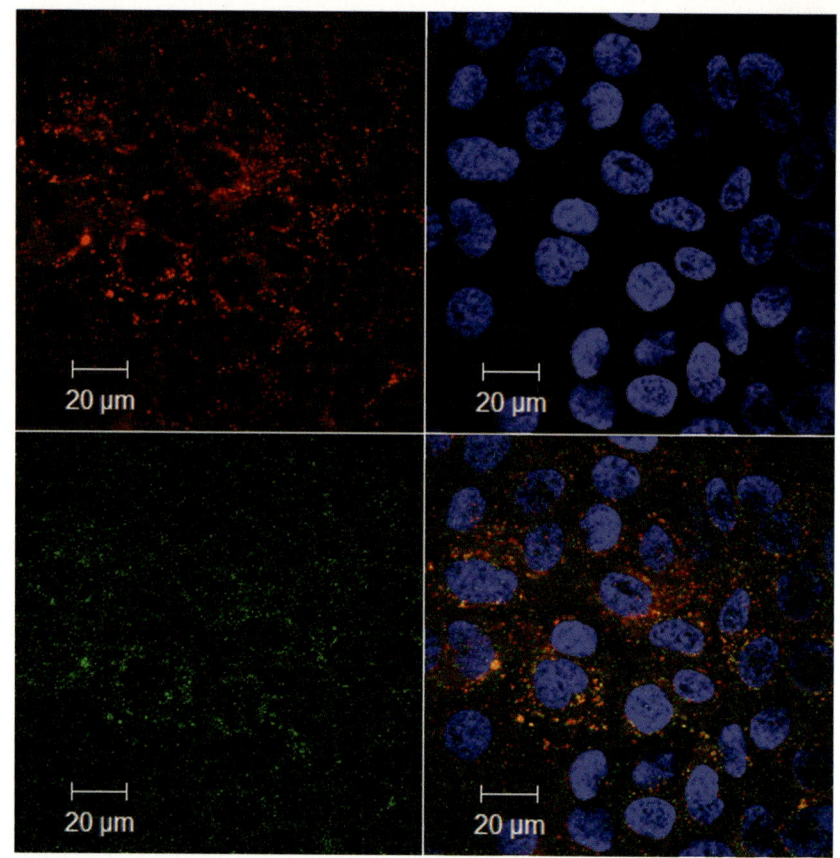

Figure 14.6 Trafficking of EGFR-targeted SWNT to lysosomes. Confocal microscopy images of OVCA 433 ovarian cancer cells that express EGFR after incubation for 15 h with SWNTs that were physically adsorbed to PL-PEG-fluorescein and EGF. The SWNTs final concentration was 20 μg ml^{-1}. The blue color represents DAPI staining for nuclei; the green staining detects the targeting SWNTs; the red is lysotracker™, which is a stain for lysosomes. The targeting SWNTs are colocalized with lysosomes, as detected by the yellow color in the merged image (lower right).

SWNTs functionalized by adsorption to PEG8000, and two antibodies to two different receptor tyrosine kinases that are expressed by breast cancer cells, were employed in another study [84]. One of the targeted receptors was the erbB member, erbB2, and the second was the insulin-like growth factor 1 receptor (IGF1R) [84]. The targeting of cells that expressed either receptor was included for the purpose of photothermal cell destruction, and the viability of cells was examined using Trypan blue exclusion (this dye is excluded *only* from living cells). The selective death only of receptor-positive cells was confirmed, but not of the positive cells targeted with SWNTs functionalized with PEG, nor of a ligand control, which was a mouse anti-human myeloma IgG [84].

SWNTs PEGylated by adsorption to high-molecular-weight PL-PEG (MW ≥5000 Da) were used to target either EGFR [85] or erbB2 [69, 85] after conjugating its antibody to PL-PEG. An imaging-based detection of the targeted CNTs showed the nanocarriers to be located selectively only inside EGFR- or erbB2-positive cells only. The imaging was conducted by using either the intrinsic NIR photoluminescence of CNTs [69], or by multi-color Raman spectroscopy that was isotope-dependent, using isotopically modified SWNTs [85].

14.8.2
Folate Receptor α

The expression of either folate receptor alpha (FRα) or folate receptor β (FRβ), both of which are 38 kDa GPI-anchored membrane glycoproteins, was detected at high frequency in cancers. FRα (also referred to as folate receptor 1, FOLR1) is overexpressed by epithelial tumors, while FRβ (also termed folate receptor 2, FOLR2) is overexpressed by non-epithelial tumors [86, 87]. (Note: Folic acid is also known as vitamin B9, but in this chapter the term "folate" refers to folic acid itself, and not to its derivatives tetrahydrofolate or dihydrofolate, which are usually referred to as folates.) These other derivatives are transported into cells via another receptor, which is the reduced folate carrier (RFC) that is ubiquitously present in all cells. Nevertheless, this carrier has a low affinity for folic acid (K_m ~100–400 µM) [86], whereas FRα has a high affinity for folic acid (K_d ~1 nM) [86]. It has been reported that conjugates of folic acid to drugs are not transported by the RFC, but rather by FRα or FRβ, which is useful for specific targeting when using drug-delivery vehicles [88]. In non-cancer tissues, FRα is expressed exclusively in the tissues of epithelial origin, with limited distribution to the kidneys, lungs, choroid plexus, and placenta [86]. The receptors in these tissues, except for the placenta, are localized in the apical membrane facing away from the blood (towards the urine, and airways) (Figure 14.7), and this causes FRα to be inaccessible to any folate conjugates administered either intravenously or intraperitoneally. However, as epithelial cancer cells that express FRα lose this polarity, FRα becomes expressed all over the cell surface, and also at high levels, which causes FRα to become accessible intravenously (Figure 14.7). In addition, in the case of ovarian cancers, where the cancer cells exist as a clump of free-floating cells in ascitic fluid in the peritoneal cavity, the FRα on the cancer cells become accessible after intraperitoneal therapy. Such findings have led to FRα becoming an ideal candidate for therapeutic targeting on cancer cells. It has been reported that activated macrophages associated with ovarian and other cancers express high levels of FRα or FRβ, and secrete cytokines and angiogenic factors that promote tumor growth [89–91]. These macrophages can also be targeted with folate conjugates in addition to cancer cells, and this is useful for regulating the microenvironment of the tumor. In fact, folate conjugates have been used for drug-delivery purposes to target folate receptors on ovarian cancer cells and their associated activated macrophages [89].

The potency of any FRα-targeted therapies depends on a knowledge of the binding kinetics of the ligand-nanocarriers to the cell-surface receptor. Various

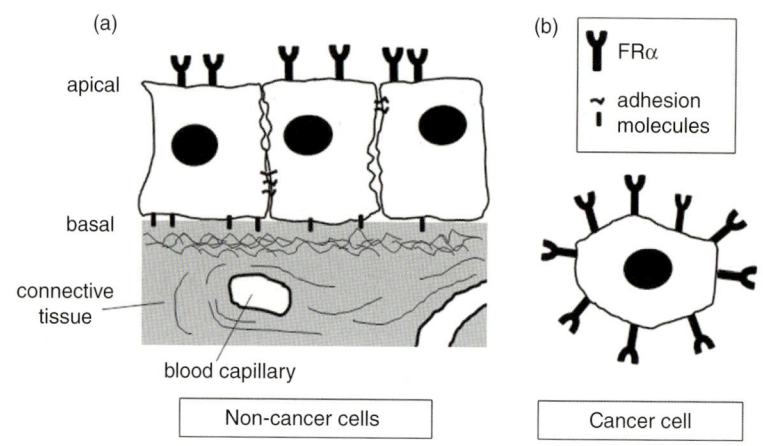

Figure 14.7 Localization of FRα on non-cancer epithelial cells and on cancer cells. (a) In normal epithelial cells, FRα is localized at the apical cell surface facing the urine and airways; (b) In cancer cells, the FRα is distributed all over the cell surface as the cell loses its polarity.

folate-conjugated nanocarriers, such as albumin, liposomes, and dendrimers, appear to display a higher affinity for binding FRα than free folate; this seems to result from the multivalent interactions with FRα on cancer cells [92–94]. CNTs, as high-aspect ratio nanocarriers, can also provide a similar effect because of their ability to bind several folate moieties.

The proof of the principal of targeting cancer cells expressing FRα was demonstrated with SWNTs coated with folate-crosslinked PL-PEG, where those cancer cells that expressed FR could internalize the SWNTs, but the other cells could not [3]. Since that study, several *in vitro* investigations have been reported in which FRα-expressing cancers were targeted with PEGylated or non-PEGylated CNTs to deliver various chemotherapeutic agents. For example, PEGylated CNTs were functionalized with a folate–PL-PEG-crosslinked to a platinum prodrug that was ultimately reduced intracellularly to generate active cisplatin [95]. Another study highlighted the importance of optimizing the amount of folate-crosslinked PL-PEG on SWNTs to prevent uptake through the ubiquitously expressed low-affinity receptor, RFC [59].

Among several studies that did not utilize PEG functionalization was an investigation in which PEG was replaced with a hydrophilic cloud created by the adsorption of polysaccharides onto SWNTs, followed by the covalent attachment of folate [96]. This vehicle was used to deliver the adsorbed chemotherapeutic agent doxorubicin to FRα-positive cells, the process being inhibited by preincubation with folate [96]. By using transmission electron microscopy (TEM), the SWNTs were shown to be localized in the lysosomes. Doxorubicin caused a greater degree of cytotoxicity in the nanocarrier-targeted cells than did the drug-free vehicle or free drug [96]. A separate study, in which polysaccharide-functionalized SWNTs were employed, with folate as a targeting moiety, demonstrated the cellular uptake of targeted cells, with the nanovehicle being used for *in vitro* photoacoustic therapy [2]. Another study employed neutralized dendrimers instead of PEG; in this case, the dendrimers were chemically conjugated to MWNTs to which folate or fluorescein had been covalently attached [97]. The amino termini of the dendrimers were modified to become neutral, which in turn led to a reduction in their nonspecific binding to the negatively charged cell membrane and allowed their specific uptake by FRα-expressing cells [97]. If the positive charges of the dendrimers were not neutralized, then the uptake of CNTs would have been enhanced, even in the absence of any targeting [29]. In a recent study, folate-functionalized CNTs were used to target FRα on circulating tumor cells for the purpose of photoacoustic imaging, after these cells had been magnetically captured through another cell marker [98]. In this case, the capturing step provided the initial scanning and selection steps, thus eliminating any need to use PEG or other molecules on CNTs to reduce nonspecific targeting (as noted above). The targeting of CNTs towards circulating tumor cells bears promise for the noninvasive destruction of these cells directly in the blood vessels.

Although FRα is known to transport folate through endocytosis, the precise mechanism employed has not yet been delineated [99]. Whilst FRα does not traffic to lysosomes, it tends to be recycled back to the cell surface through early endosomes, which have a reductive environment [100–102]. Recently, it was reported that the FRα trafficking pathway does not involve acidic compartments (pH < 6.5), which distinguishes them from other endocytic pathways [103]. FRα does not reach the Golgi apparatus during intracellular trafficking [102]; consequently, FRα undergoes endocytosis through an early endosome, and most likely is recycled through a mechanism that does not involve the Golgi apparatus. It would appear that FRα is useful for drugs that can be released by reduction, an example being the reduction of disulfide bridges. However, it has been reported that drugs conjugated to folate through an acid labile linkage can be delivered to cells, which would indicate that the receptor is rather trafficked to acidic endosomes [104, 105]. Whilst the destination of FRα may need to be determined for the particular vehicle with which it is being used, the trafficking of PEGylated CNTs functionalized with folate has not been evaluated to date.

14.8.3
Biotin Receptor

The accumulation of biotin (vitamin H or vitamin B7) in cancer cells has been reported, and is thought to result from an overexpression of the biotin receptor; however, the presence of biotin-specific receptors causing biotin accumulation in cancer cells has yet to be confirmed [88]. Non-PEGylated SWNTs crosslinked to biotin were used to target cancer cells *in vitro* that accumulate biotin, the aim being to deliver a paclitaxel prodrug that was attached covalently through a cleavable disulfide linker [52]. It was shown that only cells positive for biotin accumulation internalized the nanocarriers, and that the delivered taxoid was cytotoxic. The results of this confirmed that the uptake of biotin-functionalized SWNTs was endocytosis-mediated, and that the drug release was the result of disulfide reduction [52]. This suggested that the nanocarriers were either trafficked to endosomes with a reducing environment, or that the effect may have resulted from a reducing environment in the cytosol.

14.8.4
Integrins

Integrins $\alpha_v\beta_3$ and $\alpha_v\beta_5$ are among other markers that are expressed on the tumor vascular endothelium at higher levels than in other tissues, including resting endothelial cells [74]. These integrins are also overexpressed in epithelial cancers [106]. Integrins are heterodimers of α and β subunits that mediate the adhesion of cells to components of the ECM such as fibronectin, laminin, and collagen. The integrins play a role in initiating the angiogenesis associated with cancer, and are commonly targeted by arginine-glycine-aspartic acid (RGD) -containing peptides that serve as a recognition sequence for integrins [74]. As they are overexpressed on cancer cells and tumor vasculature, the integrins represent good candidates for targeting cancer with cytotoxic drugs, and its microenvironment with anti-angiogenic agents, in order to deprive the cancer of its blood and nutrient supply [106]. Although, RGD-functionalized drug-delivery vehicles accumulate in the tumor vasculature and cancer cells, these vehicles will be taken up to a lesser extent by other tissues that express $\alpha_v\beta_3$ and $\alpha_v\beta_5$ integrins or even $\alpha_v\beta_5$, $\alpha_v\beta_1$ integrins [107].

In an *in vitro* study, SWNTs were functionalized covalently with an RGD peptide and a single-stranded oligonucleotide that helped to create biocompatible SWNTs capable of self-assembly [108]. As a control, the RGD was replaced with RAD (where A is alanine), to demonstrate the specificity of binding to integrin $\alpha_v\beta_3$-positive cells by flow cytometry [108]. *In vitro* studies employing PEGylated CNTs to target integrin $\alpha_v\beta_3$ on cancer cells have included the use of SWNTs functionalized with RGD crosslinked to PL-PEG, either to deliver the cytotoxic agent doxorubicin specifically to integrin-positive cells [54], or for the purpose of imaging by multi-color Raman spectroscopy (that is isotope-dependent), using isotopically modified SWNTs [85]. In another *in vitro* study, an anti-integrin $\alpha_v\beta_3$ monoclonal antibody was employed instead of RGD to create an $\alpha_v\beta_3$-specific PEGylated-SWNT targeting vehicle by employing protein A that was crosslinked to PL-PEG to bind to the antibody [109].

In one *in vivo* study, an RGD peptide crosslinked to PL-PEG was employed to deliver radionuclides to xenografts of integrin $\alpha_v\beta_3$-positive cells in mice [60]. The radionuclides were attached by a chelating agent crosslinked to other PL-PEGs on the SWNTs, and imaging of mice was carried out using positron emission tomography (PET). The SWNTs functionalized with PL-PEG5400-RGD exhibited a high tumor uptake of 10–15% of the injected dose (ID) per gram of tissue, which was significantly higher than the uptake of control vehicles, which included SWNT–PL-PEG free of RGD, and PL-PEG-RGD [60]. Furthermore, injecting a high dose of RGD blocked the accumulation of functionalized nanocarriers in the tumors, and there was also a minimal accumulation of functionalized nanocarriers in xenografts of integrin $\alpha_v\beta_3$-negtive cancer cells [60]. These results were confirmed in another study with tumor mouse models targeted with SWNTs-PL-PEG-RGD, that used noninvasive Raman spectroscopy imaging. The advantage of Raman spectroscopy is that, by utilizing the inherent Raman signature of CNTs, it avoids the use of radiolabeling [110]. In another *in vivo* study, an anti-$\alpha_v\beta_3$ antibody linked to PL-PEG2000 through a protein was used for the photoacoustic imaging of xenografts of an $\alpha_v\beta_3$-positive cell line in mice [111].

Recently, a report was made in which a tetrameric RGD-synthetic peptide was used to show that a multimeric presentation caused an increased affinity, clustering and internalization of the integrin $\alpha_v\beta_3$ [112]. Due to their high aspect ratio, CNTs serve to present multi-RGDs; this most likely imitates multimeric presentation and may lead to a similar effect. However, this effect remains to be determined, along with details of the intracellular destination of cargo delivered by CNTs targeting integrins.

14.8.5
Markers for Lymphomas or Leukemias

Both, CD20 and CD22 are surface antigenic markers of human Burkitt's lymphoma that are expressed on the cell surface of B cells which have been targeted with CNTs for therapeutic purposes. In an *in vitro* study, PL-PEGylated SWNTs were used to target CD20 after conjugating its antibody to PL-PEG. The detection of targeted CNTs selectively inside CD20-positive cells only was achieved by imaging using the intrinsic NIR photoluminescence of CNTs [69]. In a further *in vitro* study, anti-CD22 antibodies were used to target lymphoma cancer cells with SWNTs for NIR thermal destruction [113]. In that case, the antibody was streptavidin bound, and also was bound to biotinylated PL-PEG2000 that was adsorbed onto SWNTs following a 10 min sonication. The specific binding of the functionalized nanocarriers to only CD22-positive cells was confirmed using flow cytometry, and the NIR thermal destruction was shown to be specific to targeted CD22-positive cells, with the use of an additional ligand control with an anti-CD25 antibody [113]. When the study was repeated with non-PEGylated SWNTs that had been covalently functionalized with antibodies, similar results were obtained [114].

The targeting of radiolabeled-anti-CD20 to a human Burkitt's lymphoma cell line *in vitro*, and to its xenograft *in vivo*, was carried out with SWNTs that had been covalently functionalized with a radionuclide chelator, and with the anti-CD20 agent, rituximab (Rituxan®; Genentech, Inc., San Francisco, CA, USA) [115]. The study included a ligand control by using functionalizing SWNTs with an anti-CD33 antibody. The CD20-targeting nanocarriers were bound selectively to CD20-positive cells only when the binding kinetics appeared to be diffusion-controlled, with increased binding efficiency associated with increases in both time and temperature [115]. Both, binding *in vitro* and accumulation in the tumor xenografts *in vivo* were higher for targeting SWNTs than for non-targeting SWNTs, though both were less than that of radiolabeled-Rituximab without SWNTs [115]. The accumulation of targeting nanocarriers in the liver and kidney was higher than for radiolabeled-Rituximab [115]. Further *in vivo* studies are required with PEGylated CNTs to examine the targeting and *in vivo* accumulation of CD20- or CD22-specific nanocarriers.

Likewise, CD3ε and CD28 are markers for T-cell leukemia that were targeted with PL-PEGylated SWNTs functionalized with antibodies to these markers, through biotin–streptavidin interactions [116]. The targeting nanocarriers were internalized and detected (using confocal microscopy) in the lysosomes in the marker-positive cells only [116]. In addition, these targeting nanocarriers were crosslinked to a fusegenic polymer that caused an escape of the nanocarriers from the lysosomes into the cytosol [116].

14.8.6
Disialoganglioside (GD2)

Disialoganglioside (GD2) is a glycosphingolipid that is overexpressed on the surface of neuroblastoma cells but is absent from normal tissue; consequently, it can serve as a candidate for targeted therapy [117]. An anti-GD2 antibody was crosslinked to MWNTs and to a fluorophor to target neuroblastoma cells *in vitro* [118]. Subsequent fluorescence microscopy studies demonstrated the uptake of targeting nanocarriers selectively by GD2-positive cells only, while NIR radiation caused destruction of the cells that specifically took up the targeting nanocarriers [117].

14.9
Summary and Future Perspective

The development of efficient drug-delivery vehicles is very important not only for improving the efficacy of therapeutic agents, but also for enhancing their pharmacokinetics, distribution, and bioavailability. Today, CNTs are emerging as new drug-delivery vehicles capable of efficiently transporting a variety of adsorbed or crosslinked molecules to the interior of cells (see Section 14.4). The targeting of cancer cells, or of the cancer microenvironment, permits a specific targeting of cancer cells while sparing the normal tissue from cytotoxicity of chemotherapeutic agents, thus reducing the adverse side effects of chemotherapy. In recent years, CNTs–whether SWNTs or MWNTs–have been used for the targeting of cancer cells (see Section 14.8), and such studies have highlighted the importance of employing PEGylated CNTs for targeted cancer therapy as PEG reduces nonspecific cellular uptake. Although, in the absence of PEGylation, CNTs will clearly transport any cargo nonspecifically into the cells, the presence of a targeting moiety further enhances such transport, even in the absence of PEG. Alternatives to PEGylation have also been evaluated, including the use of neutralized dendrimers or a hydrophilic polysaccharide cloud (see Section 14.8.2). Based on the results

14.9 Summary and Future Perspective

(a) Intact PEG — Blocks nonspecific uptake
(b) Receptor-specific uptake only
(c) Combined nonspecific and receptor-mediated uptake
(d) Nonspecific uptake

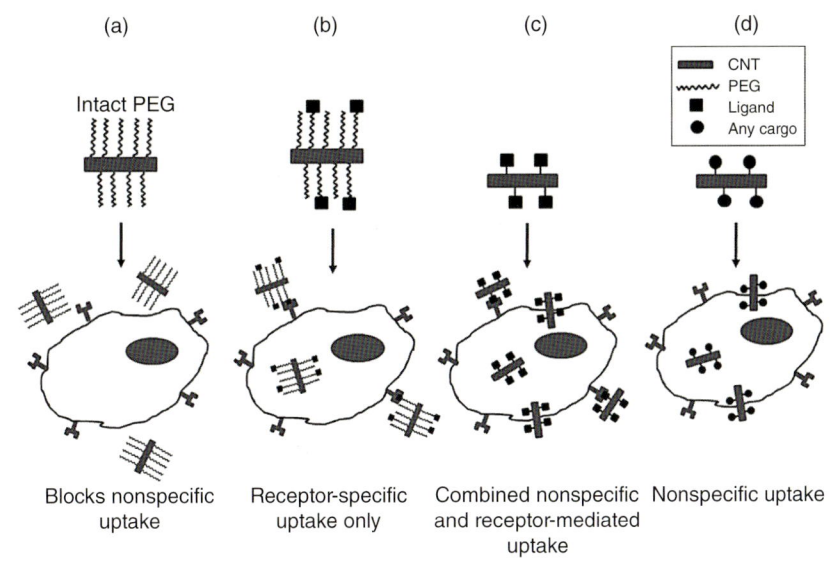

Legend: CNT, PEG, Ligand, Any cargo

Figure 14.8 The promotion of cellular uptake by CNTs. (a) Nonspecific uptake is blocked by PEGylating CNTs or using polysaccharides or neutralized dendrimers; (b) Receptor-specific uptake is promoted by functionalizing PEGylated CNTs with a receptor-specific ligand; (c) Receptor-mediated uptake and nonspecific uptake seem to be involved in the uptake of non-PEGylated but ligand-functionalized CNTs; (d) Nonspecific uptake is the mechanism involved in uptake of any nontargeted cargo.

of previous studies it is proposed that, for targeted therapy, CNTs should be functionalized by PEG or alternative molecules to block any nonspecific cellular uptake (Figure 14.8a). The crosslinking of a ligand to PEG will promote specific cellular uptake through receptors expressed on the cell surface (Figure 14.8b). In contrast, non-PEGylated CNTs that are functionalized with a ligand will be taken up by cells through receptor-mediated endocytosis and passive uptake (Figure 14.8c), whereas CNTs functionalized by any type of cargo that does not target a particular receptor on the cell will be taken up passively by the cells (Figure 14.8d).

Whilst the targeting of cancer cells with CNTs after functionalizing with an antibody or several proteins creates a large-sized nanocarrier, this may cause problems for the targeting of solid tumors, or it may even hinder cellular uptake. One alternative to using antibodies would be the use smaller *aptamers*; these are short nucleic acids that bind specifically to antigens, but without the bulky nature of antibodies [119]. A recent evaluation of the effect of the shape of carbon nanovectors on angiogenesis suggested that CNTs functionalized with doxorubicin would promote angiogenesis, affect the signaling pathways, and reduce the cytotoxicity of doxorubicin on endothelial cells [120]. It must be noted, however, that the CNTs were not PEGylated, and this may have been a factor in preventing the effect. Hence, further studies are required to evaluate such effects, in addition to investigating the effect of CNTs on the components of proteases and cell junctions.

Another type of targeting (not discussed in this chapter) involves the lymphatic system, the aim being to block the occurrence of cancer metastases via the lymphatic vessels. In a recent report it was demonstrated that, just like activated carbon, the MWNTs would accumulate selectively in draining lymph nodes in both healthy and lymph node-metastatic mice following subcutaneous or intravenous injection, and that this would render MWNTs plausible for targeting the lymphatic system in cancers [121]. In this case, the MWNTs were functionalized with magnetic nanoparticles, which permitted their monitoring by magnetic resonance imaging. Notably, a subcutaneous injection – unlike an intravenous injection – did not lead to any accumulation of magnetic MWNTs in the spleen and liver [121].

In general, nonbiodegradable nanocarriers and foreign particles would be expected to pass into the lymphatic system and ultimately be sequestered in the lymph nodes [122]. However, it must be noted that an overloading of the lymph nodes with foreign particles will lead to lymph node fibrosis and calcification and, eventually, to death, with each effect being dependent on the dose level and the duration of exposure. The spleen may also sequester foreign particles, and this may can lead to granulomas and splenomegaly [122].

One advantage of targeting of cancer cells is to enhance the rate of bioavailability (through passive and active targeting) and intracellular delivery (through active targeting involving endocytosis) of the drug. This means that a smaller drug dose could be loaded onto the CNT nanocarriers, although studies with animals must first be conducted to optimize the drug loading for the CNTs. To date, few investigations have been conducted to determine the fate of internalized CNTs or their therapeutic cargo within cells. The active targeting of cell-surface receptors that are internalized normally leads to receptor-mediated endocytosis via an endosomal pathway, during which a large proportion of the drug may be trapped in the endosome, or degraded. The release of drugs from CNTs and endosomes may employ linkages that are cleavable within the environment of the endosome to which they are directed, after which diffusion of the drug from the endosome would allow it to reach its target within the cell. The mechanisms used to release drugs from the CNTs include the cleavage of either thiolytic or acid-labile linkages [123]. Likewise, the escape of drugs from endosomes can involve membrane-penetrating peptides [124, 125] or proton sponges [126]:

- *Membrane-penetrating peptides*, such as HIV Tat-derived peptide, normally have basic amino acids that allow the peptides to penetrate through lipid bilayers [124, 125]. When fused to drugs, these peptides carry the drugs with them through the endosome's bilayer.

- *Proton sponges* are endosmolytically active polymers containing amine groups that become protonated in acidic endosomes. Replenishment of the protons through their influx into endosomes, along with chloride anions, leads to an increase in the osmotic pressure and rupture of the endosomes such that they release their content [126].

Currently, several other mechanisms are being evaluated for controlled release using stimuli-responsive "smart" polymers (for a review, see Ref. [127]). Following the release of a drug from the endosomes, it must still reach its target, either in the cytoplasm or in another organelle (e.g., the nucleus in the case of stable transfection of small interfering RNA). In order to achieve this goal, nuclear targeting moieties such as nuclear signaling peptide [128] or other organelle-targeting moieties can be fused to drugs or CNTs, as long as they are designed not to release their cargo before reaching their destination.

Clearly, future research is required to address the physical stability issues of CNTs as therapeutic nanocarriers, including settling, changing particle size, aggregation, and the stability of their biofunctional groups, in addition to evaluating the effect of conversion into dry forms and sterilization. Finally, there is a need for long-term animal studies to determine the clearance and fate of CNTs, as a step in the direction of clinical trials. Recently, it was shown that CNTs could be degraded enzymatically with horseradish peroxidase [129, 130], and studies are needed to evaluate the effects of various cellular peroxidases on CNTs, and whether these might be used to degrade the nanocarriers intracellularly. The challenge then would be to design CNTs that, following the delivery of their cargo, could be directed towards peroxidase-containing cellular compartments to be degraded and, ultimately, eliminated.

In conclusion, CNTs continue to show great promise as nanocarriers of therapeutic agents, or indeed as therapeutic agents themselves, when used for photothermal or photoacoustic therapy. An appropriate functionalization with moieties such as PEG, to reduce their nonspecific cellular uptake along with a cancer cell-specific ligand, would allow their use in cancer cell targeting while avoiding the nonspecific delivery of their cargo to non-cancer cells, thus sparing the latter cells from any cytotoxic effects. The aim of future research will be not only to improve the targeting of these nanocarriers, but also to enhance the intracellular delivery of their cargo to specific cellular compartments, with subsequent elimination of the CNTs from the body.

Acknowledgments

Funding for these studies was provided by the Massachusetts College of Pharmacy and Health Sciences Faculty Development Grant Program, an American Cancer Society Institutional Grant #IRG-92-024; by the University of New Mexico, Albuquerque, NM; and a Research Award from Health Sciences Center Research Allocation Committee-Nursing/Pharmacy/Allied Health, University of New Mexico, Albuquerque, NM.

References

1 Gensini, G.F., Conti, A.A. and Lippi, D. (2007) The contributions of Paul Ehrlich to infectious disease. *Journal of Infection*, 54, 221–4.

2 Kang, B., Yu, D., Dai, Y., Chang, S., Chen, D. and Ding, Y. (2009) Cancer-cell targeting and photoacoustic therapy using

carbon nanotubes as "bomb" agents. *Small*, **5**, 1292–301.

3 Kam, N.W., O'Connell, M., Wisdom, J.A. and Dai, H. (2005) Carbon nanotubes as multifunctional biological transporters and near-infrared agents for selective cancer cell destruction. *Proceedings of the National Academy of Sciences of the United States of America*, **102**, 11600–5.

4 Gannon, C.J., Cherukuri, P., Yakobson, B.I., Cognet, L., Kanzius, J.S., Kittrell, C., Weisman, R.B., Pasquali, M., Schmidt, H.K., Smalley, R.E. and Curley, S.A. (2007) Carbon nanotube-enhanced thermal destruction of cancer cells in a noninvasive radiofrequency field. *Cancer*, **110**, 2654–65.

5 Si, Z.C. and Liu, J. (2008) What "helps" tumors evade vascular targeting treatment? *Chinese Medical Journal*, **121**, 844–9.

6 Skobe, M. and Detmar, M. (2000) Structure, function, and molecular control of the skin lymphatic system. *Journal of Investigative Dermatology Symposium Proceedings*, **5**, 14–19.

7 Saintigny, P., Morère, J.F., Breau, J.L., Bernaudin, J.F. and Kraemer, M. (2007) Lymph node metastasis as a new target for cancer treatment. *Targeted Oncology*, **2**, 49–57.

8 Finn, O.J. (2008) Cancer immunology. *New England Journal of Medicine*, **358**, 2704–15.

9 Chu, E. and Sartorelli, A. (2009) Cancer chemotherapy, in *Basic and Clinical Pharmacology* (ed. B. Katzung, A.J. Trevor and S.B. Masters), Lange Medical Publications, Los Altos, CA, pp. 935–62.

10 Takara, K., Sakaeda, T. and Okumura, K. (2006) An update on overcoming MDR1-mediated multidrug resistance in cancer chemotherapy. *Current Pharmaceutical Design*, **12**, 273–86.

11 Longmire, M., Choyke, P.L. and Kobayashi, H. (2008) Clearance properties of nano-sized particles and molecules as imaging agents: considerations and caveats. *Nanomedicine*, **3**, 703–17.

12 Whitehead, K.A., Langer, R. and Anderson, D.G. (2009) Knocking down barriers: advances in siRNA delivery. *Nature Reviews. Drug Discovery*, **8**, 129–38.

13 Pastorin, G. (2009) Crucial functionalizations of carbon nanotubes for improved drug delivery: a valuable option? *Pharmaceutical Research*, **26**, 746–69.

14 Bianco, A., Kostarelos, K. and Prato, M. (2005) Applications of carbon nanotubes in drug delivery. *Current Opinion in Chemical Biology*, **9**, 674–9.

15 Bianco, A., Kostarelos, K., Partidos, C.D. and Prato, M. (2005) Biomedical applications of functionalised carbon nanotubes. *Chemical Communications (Cambridge, England)*, 571–7.

16 Klumpp, C., Kostarelos, K., Prato, M. and Bianco, A. (2006) Functionalized carbon nanotubes as emerging nanovectors for the delivery of therapeutics. *Biochimica et Biophysica Acta*, **1758**, 404–12.

17 Prato, M., Kostarelos, K. and Bianco, A. (2008) Functionalized carbon nanotubes in drug design and discovery. *Accounts of Chemical Research*, **41**, 60–8.

18 Liu, Z., Tabakman, S., Welsher, K. and Dai, H. (2009) Carbon nanotubes in biology and medicine: *in vitro* and *in vivo* detection, imaging and drug delivery. *Nano Research*, **2**, 85–120.

19 Foldvari, M. and Bagonluri, M. (2008) Carbon nanotubes as functional excipients for nanomedicines: II. Drug delivery and biocompatibility issues. *Nanomedicine*, **2008** (4), 183–200.

20 Chattopadhyay, J., de Jesus Cortez, F., Chakraborty, S., Slater, N.K.H. and Billups, W.E. (2006) Synthesis of water-soluble PEGylated single-walled carbon nanotubes. *Chemistry of Materials*, **18**, 5864–8.

21 Haggenmueller, R., Rahatekar, S.S., Fagan, J.A., Chun, J., Becker, M.L., Naik, R.R., Krauss, T., Carlson, L., Kadla, J.F., Trulove, P.C., Fox, D.F., Delong, H.C., Fang, Z., Kelley, S.O. and Gilman, J.W. (2008) Comparison of the quality of aqueous dispersions of single wall carbon nanotubes using surfactants and biomolecules. *Langmuir*, **24**, 5070–8.

22 Karajanagi, S.S., Yang, H., Asuri, P., Sellitto, E., Dordick, J.S. and Kane, R.S. (2006) Protein-assisted solubilization of single-walled carbon nanotubes. *Langmuir*, **22**, 1392–5.

23 Fu, K. and Sun, Y.P. (2003) Dispersion and solubilization of carbon nanotubes. *Journal of Nanoscience and Nanotechnology*, **3**, 351–64.

24 Georgakilas, V., Tagmatarchis, N., Pantarotto, D., Bianco, A., Briand, J.P. and Prato, M. (2002) Amino acid functionalisation of water soluble carbon nanotubes. *Chemical Communications (Cambridge, England)*, 3050–1.

25 Pantarotto, D., Singh, R., McCarthy, D., Erhardt, M., Briand, J.P., Prato, M., Kostarelos, K. and Bianco, A. (2004) Functionalized carbon nanotubes for plasmid DNA gene delivery. *Angewandte Chemie International Edition*, **43**, 5242–6.

26 Cai, D., Mataraza, J.M., Qin, Z.H., Huang, Z., Huang, J., Chiles, T.C., Carnahan, D., Kempa, K. and Ren, Z. (2005) Highly efficient molecular delivery into mammalian cells using carbon nanotube spearing. *Nature Methods*, **2**, 449–54.

27 Liu, Z., Winters, M., Holodniy, M. and Dai, H. (2007) siRNA delivery into human T cells and primary cells with carbon-nanotube transporters. *Angewandte Chemie International Edition*, **46**, 2023–7.

28 Kam, N.W., Liu, Z. and Dai, H. (2005) Functionalization of carbon nanotubes via cleavable disulfide bonds for efficient intracellular delivery of siRNA and potent gene silencing. *Journal of the American Chemical Society*, **127**, 12492–3.

29 Herrero, M.A., Toma, F.M., Al-Jamal, K.T., Kostarelos, K., Bianco, A., Da Ros, T., Bano, F., Casalis, L., Scoles, G. and Prato, M. (2009) Synthesis and characterization of a carbon nanotube-dendron series for efficient siRNA delivery. *Journal of the American Chemical Society*, **131**, 9843–8.

30 Podesta, J.E., Al-Jamal, K.T., Herrero, M.A., Tian, B., Ali-Boucetta, H., Hegde, V., Bianco, A., Prato, M. and Kostarelos, K. (2009) Antitumor activity and prolonged survival by carbon-nanotube-mediated therapeutic siRNA silencing in a human lung xenograft model. *Small*, **5**, 1176–85.

31 Krajcik, R., Jung, A., Hirsch, A., Neuhuber, W. and Zolk, O. (2008) Functionalization of carbon nanotubes enables non-covalent binding and intracellular delivery of small interfering RNA for efficient knock-down of genes. *Biochemical and Biophysical Research Communications*, **369**, 595–602.

32 Wang, X., Ren, J. and Qu, X. (2008) Targeted RNA interference of cyclin A2 mediated by functionalized single-walled carbon nanotubes induces proliferation arrest and apoptosis in chronic myelogenous leukemia K562 cells. *ChemMedChem*, **3**, 940–5.

33 Kateb, B., Van Handel, M., Zhang, L., Bronikowski, M.J., Manohara, H. and Badie, B. (2007) Internalization of MWCNTs by microglia: possible application in immunotherapy of brain tumors. *Neuroimage*, **37** (Suppl. 1), S9–17.

34 Lacerda, L., Bianco, A., Prato, M. and Kostarelos, K. (2008) Carbon nanotube cell translocation and delivery of nucleic acids *in vitro* and *in vivo*. *Journal of Materials Chemistry*, **18**, 17–22.

35 Ahmed, M., Jiang, X., Deng, Z. and Narain, R. (2009) Cationic glyco-functionalized single-walled carbon nanotubes as efficient gene delivery vehicles. *Bioconjugate Chemistry*, **20**, 2017–22.

36 Pan, B., Cui, D., Xu, P., Ozkan, C., Feng, G., Ozkan, M., Huang, T., Chu, B., Li, Q., He, R. and Hu, G. (2009) Synthesis and characterization of polyamidoamine dendrimer-coated multi-walled carbon nanotubes and their application in gene delivery systems. *Nanotechnology*, **20**, 125101.

37 Delogu, L.G., Magrini, A., Bergamaschi, A., Rosato, N., Dawson, M.I., Bottini, N. and Bottini, M. (2009) Conjugation of antisense oligonucleotides to PEGylated carbon nanotubes enables efficient knockdown of PTPN22 in T lymphocytes. *Bioconjugate Chemistry*, **20**, 427–31.

38 Wu, Y., Phillips, J.A., Liu, H., Yang, R. and Tan, W. (2008) Carbon nanotubes protect DNA strands during cellular delivery. *ACS Nano*, **2**, 2023–8.

39 Jia, N., Lian, Q., Shen, H., Wang, C., Li, X. and Yang, Z. (2007) Intracellular delivery of quantum dots tagged antisense oligodeoxynucleotides by

functionalized multiwalled carbon nanotubes. *Nano Letters*, **7**, 2976–80.

40 Cui, D., Tian, F., Coyer, S.R., Wang, J., Pan, B., Gao, F., He, R. and Zhang, Y. (2007) Effects of antisense-myc-conjugated single-walled carbon nanotubes on HL-60 cells. *Journal of Nanoscience and Nanotechnology*, **7**, 1639–46.

41 Rege, K., Viswanathan, G., Zhu, G., Vijayaraghavan, A., Ajayan, P.M. and Dordick, J.S. (2006) *In vitro* transcription and protein translation from carbon nanotube-DNA assemblies. *Small*, **2**, 718–22.

42 Lanner, J.T., Bruton, J.D., Assefaw-Redda, Y., Andronache, Z., Zhang, S.J., Severa, D., Zhang, Z.B., Melzer, W., Zhang, S.L., Katz, A. and Westerblad, H. (2009) Knockdown of TRPC3 with siRNA coupled to carbon nanotubes results in decreased insulin-mediated glucose uptake in adult skeletal muscle cells. *FASEB Journal*, **23**, 1728–38.

43 Zhang, Z., Yang, X., Zhang, Y., Zeng, B., Wang, S., Zhu, T., Roden, R.B., Chen, Y. and Yang, R. (2006) Delivery of telomerase reverse transcriptase small interfering RNA in complex with positively charged single-walled carbon nanotubes suppresses tumor growth. *Clinical Cancer Research*, **12**, 4933–9.

44 Wang, X., Song, Y., Ren, J. and Qu, X. (2009) Knocking-down cyclin A(2) by siRNA suppresses apoptosis and switches differentiation pathways in K562 cells upon administration with doxorubicin. *PLoS ONE*, **4**, e6665.

45 Kam, N.W., Liu, Z. and Dai, H. (2006) Carbon nanotubes as intracellular transporters for proteins and DNA: an investigation of the uptake mechanism and pathway. *Angewandte Chemie International Edition*, **45**, 577–81.

46 Pantarotto, D., Briand, J.P., Prato, M. and Bianco, A. (2004) Translocation of bioactive peptides across cell membranes by carbon nanotubes. *Chemical Communications (Cambridge, England)*, 16–17.

47 Pantarotto, D., Partidos, C.D., Graff, R., Hoebeke, J., Briand, J.P., Prato, M. and Bianco, A. (2003) Synthesis, structural characterization, and immunological properties of carbon nanotubes functionalized with peptides. *Journal of the American Chemical Society*, **125**, 6160–4.

48 Pantarotto, D., Partidos, C.D., Hoebeke, J., Brown, F., Kramer, E., Briand, J.P., Muller, S., Prato, M. and Bianco, A. (2003) Immunization with peptide-functionalized carbon nanotubes enhances virus-specific neutralizing antibody responses. *Chemistry and Biology*, **10**, 961–6.

49 Wu, W., Wieckowski, S., Pastorin, G., Benincasa, M., Klumpp, C., Briand, J.P., Gennaro, R., Prato, M. and Bianco, A. (2005) Targeted delivery of amphotericin B to cells by using functionalized carbon nanotubes. *Angewandte Chemie International Edition*, **44**, 6358–62.

50 Feazell, R.P., Nakayama-Ratchford, N., Dai, H. and Lippard, S.J. (2007) Soluble single-walled carbon nanotubes as longboat delivery systems for platinum(IV) anticancer drug design. *Journal of the American Chemical Society*, **129**, 8438–9.

51 Ali-Boucetta, H., Al-Jamal, K.T., McCarthy, D., Prato, M., Bianco, A. and Kostarelos, K. (2008) Multiwalled carbon nanotube-doxorubicin supramolecular complexes for cancer therapeutics. *Chemical Communications (Cambridge, England)*, 459–61.

52 Chen, J., Chen, S., Zhao, X., Kuznetsova, L.V., Wong, S.S. and Ojima, I. (2008) Functionalized single-walled carbon nanotubes as rationally designed vehicles for tumor-targeted drug delivery. *Journal of the American Chemical Society*, **130**, 16778–85.

53 Yinghuai, Z., Peng, A.T., Carpenter, K., Maguire, J.A., Hosmane, N.S. and Takagaki, M. (2005) Substituted carborane-appended water-soluble single-wall carbon nanotubes: new approach to boron neutron capture therapy drug delivery. *Journal of the American Chemical Society*, **127**, 9875–80.

54 Liu, Z., Sun, X., Nakayama-Ratchford, N. and Dai, H. (2007) Supramolecular chemistry on water-soluble carbon nanotubes for drug loading and delivery. *ACS Nano*, **1**, 50–6.

55 Kostarelos, K., Lacerda, L., Pastorin, G., Wu, W., Wieckowski, S., Luangsivilay, J., Godefroy, S., Pantarotto, D., Briand, J.P., Muller, S., Prato, M. and Bianco, A. (2007) Cellular uptake of functionalized carbon nanotubes is independent of functional group and cell type. *Nature Nanotechnology*, **2**, 108–13.

56 Shi Kam, N.W., Jessop, T.C., Wender, P.A. and Dai, H. (2004) Nanotube molecular transporters: internalization of carbon nanotube-protein conjugates into Mammalian cells. *Journal of the American Chemical Society*, **126**, 6850–1.

57 Lopez, C.F., Nielsen, S.O., Moore, P.B. and Klein, M.L. (2004) Understanding nature's design for a nanosyringe. *Proceedings of the National Academy of Sciences of the United States of America*, **101**, 4431–4.

58 Jin, H., Heller, D.A. and Strano, M.S. (2008) Single-particle tracking of endocytosis and exocytosis of single-walled carbon nanotubes in NIH-3T3 cells. *Nano Letters*, **8**, 1577–85.

59 Zeineldin, R., Al-Haik, M. and Hudson, L.G. (2009) Role of polyethylene glycol integrity in specific receptor targeting of carbon nanotubes to cancer cells. *Nano Letters*, **9**, 751–7.

60 Liu, Z., Cai, W., He, L., Nakayama, N., Chen, K., Sun, X., Chen, X. and Dai, H. (2007) *In vivo* biodistribution and highly efficient tumour targeting of carbon nanotubes in mice. *Nature Nanotechnology*, **2**, 47–52.

61 Chen, R.J., Bangsaruntip, S., Drouvalakis, K.A., Kam, N.W., Shim, M., Li, Y., Kim, W., Utz, P.J. and Dai, H. (2003) Noncovalent functionalization of carbon nanotubes for highly specific electronic biosensors. *Proceedings of the National Academy of Sciences of the United States of America*, **100**, 4984–9.

62 Ryan, S.M., Mantovani, G., Wang, X., Haddleton, D.M. and Brayden, D.J. (2008) Advances in PEGylation of important biotech molecules: delivery aspects. *Expert Opinion on Drug Delivery*, **5**, 371–83.

63 van Vlerken, L.E., Vyas, T.K. and Amiji, M.M. (2007) Poly(ethylene glycol)-modified nanocarriers for tumor-targeted and intracellular delivery. *Pharmaceutical Research*, **24**, 1405–14.

64 Yang, S.T., Fernando, K.A., Liu, J.H., Wang, J., Sun, H.F., Liu, Y., Chen, M., Huang, Y., Wang, X., Wang, H. and Sun, Y.P. (2008) Covalently PEGylated carbon nanotubes with stealth character in vivo. *Small*, **4**, 940–4.

65 Liu, Z., Davis, C., Cai, W., He, L., Chen, X. and Dai, H. (2008) Circulation and long-term fate of functionalized, biocompatible single-walled carbon nanotubes in mice probed by Raman spectroscopy. *Proceedings of the National Academy of Sciences of the United States of America*, **105**, 1410–15.

66 Prencipe, G., Tabakman, S.M., Welsher, K., Liu, Z., Goodwin, A.P., Zhang, L., Henry, J. and Dai, H. (2009) PEG branched polymer for functionalization of nanomaterials with ultralong blood circulation. *Journal of the American Chemical Society*, **131**, 4783–7.

67 Shim, M., Kam, N.W.S., Chen, R.J., Li, Y.M. and Dai, H.J. (2002) Functionalization of carbon nanotubes for biocompatibility and biomolecular recognition. *Nano Letters*, **2**, 285–8.

68 Chen, R.J., Zhang, Y., Wang, D. and Dai, H. (2001) Noncovalent sidewall functionalization of single-walled carbon nanotubes for protein immobilization. *Journal of the American Chemical Society*, **123**, 3838–9.

69 Welsher, K., Liu, Z., Daranciang, D. and Dai, H. (2008) Selective probing and imaging of cells with single walled carbon nanotubes as near-infrared fluorescent molecules. *Nano Letters*, **8**, 586–90.

70 Nakayama-Ratchford, N., Bangsaruntip, S., Sun, X., Welsher, K. and Dai, H. (2007) Noncovalent functionalization of carbon nanotubes by fluorescein-polyethylene glycol: supramolecular conjugates with pH-dependent absorbance and fluorescence. *Journal of the American Chemical Society*, **129**, 2448–9.

71 Kawasaki, H., Takeda, Y. and Arakawa, R. (2007) Mass spectrometric analysis for high molecular weight synthetic polymers using ultrasonic degradation

and the mechanism of degradation. *Analytical Chemistry*, **79**, 4182–7.
72 Maeda, H. (2001) The enhanced permeability and retention (EPR) effect in tumor vasculature: the key role of tumor-selective macromolecular drug targeting. *Advances in Enzyme Regulation*, **41**, 189–207.
73 Carmeliet, P. and Jain, R.K. (2000) Angiogenesis in cancer and other diseases. *Nature*, **407**, 249–57.
74 Alessi, P., Ebbinghaus, C. and Neri, D. (2004) Molecular targeting of angiogenesis. *Biochimica et Biophysica Acta*, **1654**, 39–49.
75 Sibilia, M., Kroismayr, R., Lichtenberger, B.M., Natarajan, A., Hecking, M. and Holcmann, M. (2007) The epidermal growth factor receptor: from development to tumorigenesis. *Differentiation*, **75**, 770–87.
76 Citri, A. and Yarden, Y. (2006) EGF-ERBB signalling: towards the systems level. *Nature Reviews Molecular Cell Biology*, **7**, 505–16.
77 Linggi, B. and Carpenter, G. (2006) ErbB receptors: new insights on mechanisms and biology. *Trends in Cell Biology*, **16**, 649–56.
78 Zhang, H., Berezov, A., Wang, Q., Zhang, G., Drebin, J., Murali, R. and Greene, M.I. (2007) ErbB receptors: from oncogenes to targeted cancer therapies. *Journal of Clinical Investigation*, **117**, 2051–8.
79 Wieduwilt, M.J. and Moasser, M.M. (2008) The epidermal growth factor receptor family: biology driving targeted therapeutics. *Cellular and Molecular Life Sciences*, **65**, 1566–84.
80 Weng, X., Wang, M., Ge, J., Yu, S., Liu, B., Zhong, J. and Kong, J. (2009) Carbon nanotubes as a protein toxin transporter for selective HER2-positive breast cancer cell destruction. *Molecular Biosystems*, **5**, 1224–31.
81 Xiao, Y., Gao, X., Taratula, O., Treado, S., Urbas, A., Holbrook, R.D., Cavicchi, R.E., Avedisian, C.T., Mitra, S., Savla, R., Wagner, P.D., Srivastava, S. and He, H. (2009) Anti-HER2 IgY antibody-functionalized single-walled carbon nanotubes for detection and selective destruction of breast cancer cells. *BMC Cancer*, **9**, 351.
82 Sorkin, A. and Goh, L.K. (2008) Endocytosis and intracellular trafficking of ErbBs. *Experimental Cell Research*, **315**, 683–96.
83 Bhirde, A.A., Patel, V., Gavard, J., Zhang, G., Sousa, A.A., Masedunskas, A., Leapman, R.D., Weigert, R., Gutkind, J.S. and Rusling, J.F. (2009) Targeted killing of cancer cells *in vivo* and *in vitro* with EGF-directed carbon nanotube-based drug delivery. *ACS Nano*, **3**, 307–16.
84 Shao, N., Lu, S., Wickstrom, E. and Panchapakesan, B. (2007) Integrated molecular targeting of IGF1R and HER2 surface receptors and destruction of breast cancer cells using single wall carbon nanotubes. *Nanotechnology*, **18**, 315101.
85 Liu, J.X., Zhou, W.J., Gong, J.L., Tang, L., Zhang, Y., Yu, H.Y., Wang, B., Xu, X.M. and Zeng, G.M. (2008) An electrochemical sensor for detection of laccase activities from *Penicillium simplicissimum* in compost based on carbon nanotubes modified glassy carbon electrode. *Bioresource Technology*, **99**, 8748–51.
86 Kelemen, L.E. (2006) The role of folate receptor alpha in cancer development, progression and treatment: cause, consequence or innocent bystander? *International Journal of Cancer*, **119**, 243–50.
87 Zhao, X., Li, H. and Lee, R.J. (2008) Targeted drug delivery via folate receptors. *Expert Opinion on Drug Delivery*, **5**, 309–19.
88 Russell-Jones, G., McTavish, K., McEwan, J., Rice, J. and Nowotnik, D. (2004) Vitamin-mediated targeting as a potential mechanism to increase drug uptake by tumours. *Journal of Inorganic Biochemistry*, **98**, 1625–33.
89 Turk, M.J., Waters, D.J. and Low, P.S. (2004) Folate-conjugated liposomes preferentially target macrophages associated with ovarian carcinoma. *Cancer Letters*, **213**, 165–72.
90 Xia, W., Hilgenbrink, A.R., Matteson, E.L., Lockwood, M.B., Cheng, J.X. and Low, P.S. (2009) A functional folate

receptor is induced during macrophage activation and can be used to target drugs to activated macrophages. *Blood*, **113**, 438–46.

91 Puig-Kroger, A., Sierra-Filardi, E., Dominguez-Soto, A., Samaniego, R., Corcuera, M.T., Gomez-Aguado, F., Ratnam, M., Sanchez-Mateos, P. and Corbi, A.L. (2009) Folate receptor beta is expressed by tumor-associated macrophages and constitutes a marker for M2 anti-inflammatory/regulatory macrophages. *Cancer Research*, **69**, 9395–403.

92 Zhang, L., Hou, S., Mao, S., Wei, D., Song, X. and Lu, Y. (2004) Uptake of folate-conjugated albumin nanoparticles to the SKOV3 cells. *International Journal of Pharmaceutics*, **287**, 155–62.

93 Hong, S., Leroueil, P.R., Majoros, I.J., Orr, B.G., Baker, J.R., Jr and Banaszak Holl, M.M. (2007) The binding avidity of a nanoparticle-based multivalent targeted drug delivery platform. *Chemistry and Biology*, **14**, 107–15.

94 Lee, R.J. and Low, P.S. (1994) Delivery of liposomes into cultured KB cells via folate receptor-mediated endocytosis. *Journal of Biological Chemistry*, **269**, 3198–204.

95 Dhar, S., Liu, Z., Thomale, J., Dai, H. and Lippard, S.J. (2008) Targeted single-wall carbon nanotube-mediated Pt(IV) prodrug delivery using folate as a homing device. *Journal of the American Chemical Society*, **130**, 11467–76.

96 Zhang, X., Meng, L., Lu, Q., Fei, Z. and Dyson, P.J. (2009) Targeted delivery and controlled release of doxorubicin to cancer cells using modified single wall carbon nanotubes. *Biomaterials*, **30**, 6041–7.

97 Shi, X., Wang, S.H., Shen, M., Antwerp, M.E., Chen, X., Li, C., Petersen, E.J., Huang, Q., Weber, W.J. and Baker, J.R. (2009) Multifunctional dendrimer-modified multiwalled carbon nanotubes: synthesis, characterization, and *in vitro* cancer cell targeting and imaging. *Biomacromolecules*, **10**, 1744–50.

98 Galanzha, E.I., Shashkov, E.V., Kelly, T., Kim, J.W., Yang, L. and Zharov, V.P. (2009) *In vivo* magnetic enrichment and multiplex photoacoustic detection of circulating tumour cells. *Nature Nanotechnology*, **4**, 855–60.

99 Sabharanjak, S. and Mayor, S. (2004) Folate receptor endocytosis and trafficking. *Advanced Drug Delivery Reviews*, **56**, 1099–109.

100 Mayor, S., Sabharanjak, S. and Maxfield, F.R. (1998) Cholesterol-dependent retention of GPI-anchored proteins in endosomes. *EMBO Journal*, **17**, 4626–38.

101 Chatterjee, S., Smith, E.R., Hanada, K., Stevens, V.L. and Mayor, S. (2001) GPI anchoring leads to sphingolipid-dependent retention of endocytosed proteins in the recycling endosomal compartment. *EMBO Journal*, **20**, 1583–92.

102 Yang, J., Chen, H., Vlahov, I.R., Cheng, J.X. and Low, P.S. (2006) Evaluation of disulfide reduction during receptor-mediated endocytosis by using FRET imaging. *Proceedings of the National Academy of Sciences of the United States of America*, **103**, 13872–7.

103 Yang, J., Chen, H., Vlahov, I.R., Cheng, J.X. and Low, P.S. (2007) Characterization of the pH of folate receptor-containing endosomes and the rate of hydrolysis of internalized acid-labile folate-drug conjugates. *Journal of Pharmacology and Experimental Therapeutics*, **321**, 462–8.

104 Bae, Y., Jang, W.D., Nishiyama, N., Fukushima, S. and Kataoka, K. (2005) Multifunctional polymeric micelles with folate-mediated cancer cell targeting and pH-triggered drug releasing properties for active intracellular drug delivery. *Molecular Biosystems*, **1**, 242–50.

105 Bae, Y., Nishiyama, N. and Kataoka, K. (2007) *In vivo* antitumor activity of the folate-conjugated pH-sensitive polymeric micelle selectively releasing adriamycin in the intracellular acidic compartments. *Bioconjugate Chemistry*, **18**, 1131–9.

106 Desgrosellier, J.S. and Cheresh, D.A. (2010) Integrins in cancer: biological implications and therapeutic opportunities. *Nature Reviews Cancer*, **10**, 9–22.

107 Meyer, A., Auernheimer, J., Modlinger, A. and Kessler, H. (2006) Targeting RGD recognizing integrins: drug development, biomaterial research,

tumor imaging and targeting. *Current Pharmaceutical Design*, **12**, 2723–47.
108 Villa, C.H., McDevitt, M.R., Escorcia, F.E., Rey, D.A., Bergkvist, M., Batt, C.A. and Scheinberg, D.A. (2008) Synthesis and biodistribution of oligonucleotide-functionalized, tumor-targetable carbon nanotubes. *Nano Letters*, **8**, 4221–8.
109 Ou, Z., Wu, B., Xing, D., Zhou, F., Wang, H. and Tang, Y. (2009) Functional single-walled carbon nanotubes based on an integrin alpha v beta 3 monoclonal antibody for highly efficient cancer cell targeting. *Nanotechnology*, **20**, 105102.
110 Zavaleta, C., de la Zerda, A., Liu, Z., Keren, S., Cheng, Z., Schipper, M., Chen, X., Dai, H. and Gambhir, S. (2008) S. Noninvasive Raman spectroscopy in living mice for evaluation of tumor targeting with carbon nanotubes. *Nano Letters*, **8**, 2800–5.
111 Xiang, L., Yuan, Y., Xing, D., Ou, Z., Yang, S. and Zhou, F. (2009) Photoacoustic molecular imaging with antibody-functionalized single-walled carbon nanotubes for early diagnosis of tumor. *Journal of Biomedical Optics*, **14**, 021008.
112 Sancey, L., Garanger, E., Foillard, S., Schoehn, G., Hurbin, A., Albiges-Rizo, C., Boturyn, D., Souchier, C., Grichine, A., Dumy, P. and Coll, J.L. (2009) Clustering and internalization of integrin alphavbeta3 with a tetrameric RGD-synthetic peptide. *Molecular Therapy*, **17**, 837–43.
113 Chakravarty, P., Marches, R., Zimmerman, N.S., Swafford, A.D., Bajaj, P., Musselman, I.H., Pantano, P., Draper, R.K. and Vitetta, E.S. (2008) Thermal ablation of tumor cells with antibody-functionalized single-walled carbon nanotubes. *Proceedings of the National Academy of Sciences of the United States of America*, **105**, 8697–702.
114 Marches, R., Chakravarty, P., Musselman, I.H., Bajaj, P., Azad, R.N., Pantano, P., Draper, R.K. and Vitetta, E.S. (2009) Specific thermal ablation of tumor cells using single-walled carbon nanotubes targeted by covalently-coupled monoclonal antibodies. *International Journal of Cancer*, **125**, 2970–7.

115 McDevitt, M.R., Chattopadhyay, D., Kappel, B.J., Jaggi, J.S., Schiffman, S.R., Antczak, C., Njardarson, J.T., Brentjens, R. and Scheinberg, D.A. (2007) Tumor targeting with antibody-functionalized, radiolabeled carbon nanotubes. *Journal of Nuclear Medicine*, **48**, 1180–9.
116 Cato, M.H., D'Annibale, F., Mills, D.M., Cerignoli, F., Dawson, M.I., Bergamaschi, E., Bottini, N., Magrini, A., Bergamaschi, A., Rosato, N., Rickert, R.C., Mustelin, T. and Bottini, M. (2008) Cell-type specific and cytoplasmic targeting of PEGylated carbon nanotube-based nanoassemblies. *Journal of Nanoscience and Nanotechnology*, **8**, 2259–69.
117 Modak, S. and Cheung, N.K. (2007) Disialoganglioside directed immunotherapy of neuroblastoma. *Cancer Investigation*, **25**, 67–77.
118 Wang, C.H., Huang, Y.J., Chang, C.W., Hsu, W.M. and Peng, C.A. (2009) *In vitro* photothermal destruction of neuroblastoma cells using carbon nanotubes conjugated with GD2 monoclonal antibody. *Nanotechnology*, **20**, 315101.
119 Cerchia, L., Giangrande, P.H., McNamara, J.O. and de Franciscis, V. (2009) Cell-specific aptamers for targeted therapies. *Methods in Molecular Biology*, **535**, 59–78.
120 Chaudhuri, P., Harfouche, R., Soni, S., Hentschel, D.M. and Sengupta, S. (2010) Shape effect of carbon nanovectors on angiogenesis. *ACS Nano*, **4**, 574–82.
121 Yang, F., Hu, J., Yang, D., Long, J., Luo, G., Jin, C., Yu, X., Xu, J., Wang, C., Ni, Q. and Fu, D. (2009) Pilot study of targeting magnetic carbon nanotubes to lymph nodes. *Nanomedicine*, **4**, 317–30.
122 Freitas, R.A. Jr (2003) Particle clearance from the lymphatics, in *Nanomedicine, Volume IIA: Biocompatibility*, Landes Bioscience, Austin, TX, pp. 191–4.
123 West, K.R. and Otto, S. (2005) Reversible covalent chemistry in drug delivery. *Current Drug Discovery Technologies*, **2**, 123–60.
124 Brooks, H., Lebleu, B. and Vives, E. (2005) Tat peptide-mediated cellular delivery: back to basics. *Advanced Drug Delivery Reviews*, **57**, 559–77.

125 Torchilin, V.P. (2008) Cell penetrating peptide-modified pharmaceutical nanocarriers for intracellular drug and gene delivery. *Biopolymers*, **90**, 604–10.

126 Yang, S. and May, S. (2008) Release of cationic polymer-DNA complexes from the endosome: A theoretical investigation of the proton sponge hypothesis. *Journal of Chemical Physics*, **129**, 185105.

127 Fogueri, L.R. and Singh, S. (2009) Smart polymers for controlled delivery of proteins and peptides: a review of patents. *Recent Patents on Drug Delivery Formulation*, **3**, 40–8.

128 Pouton, C.W., Wagstaff, K.M., Roth, D.M., Moseley, G.W. and Jans, D.A. (2007) Targeted delivery to the nucleus. *Advanced Drug Delivery Reviews*, **59**, 698–717.

129 Allen, B.L., Kichambare, P.D., Gou, P., Vlasova, I.I., Kapralov, A.A., Konduru, N., Kagan, V.E. and Star, A. (2008) Biodegradation of single-walled carbon nanotubes through enzymatic catalysis. *Nano Letters*, **8**, 3899–903.

130 Allen, B.L., Kotchey, G.P., Chen, Y., Yanamala, N.V., Klein-Seetharaman, J., Kagan, V.E. and Star, A. (2009) mechanistic investigations of horseradish peroxidase-catalyzed degradation of single-walled carbon nanotubes. *Journal of the American Chemical Society*, **131**, 17194–205.

15
Application of Carbon Nanotubes to Brain Tumor Therapy

Dongchang Zhao and Behnam Badie

15.1
Introduction

The treatment of malignant brain tumors remains difficult despite recent advances in surgery, radiotherapy, and chemotherapy. Currently, no optimal treatment option is available for glioblastoma multiforme, and the patients typically survive for less than two years [1–3]. The poor prognosis is partially because of the inability to deliver chemotherapeutic agents across the blood–brain barrier (BBB) and the low tumor response to radiation. Therefore, novel and more effective strategies are warranted, and true advances may emerge from the increasing understanding in immunology, molecular biology, and nanotechnology [4–9].

One of the key advantages of carbon nanotubes (CNTs) in biomedical applications is that they can be easily internalized by cells [7–14]. The possibility of incorporating CNT-based nanomaterials into living systems has opened the way for the investigation of their potential applications in the emerging field of nanomedicine [15]. Moreover, their unique electrical, thermal and spectroscopic properties in a biological context offer further advances in the detection, monitoring and therapy of diseases, and therefore CNTs can act as delivery vehicles for a variety of molecules relevant to therapy in the clinic [16–18].

In an attempt to overcome this formidable neoplasm, CNTs targeting tumor strategies have recently undergone extensive investigation. Functionalized CNTs have been synthesized and their characteristics explored by many academic and industrial laboratories worldwide [8, 15, 19]. The aim of this chapter is to review the progress of application of CNTs to brain tumor therapy. Initially, the current status and challenge of brain tumor therapy will be summarized, after which the characteristics of CNTs for biological application will be reviewed. The opportunities and strategies for using CNTs to target tumor macrophages for brain tumor therapy will then be discussed, as will be the toxicity issues associated with the biological applications of CNTs.

15.2
The Current Challenge of Brain Tumor Therapy

15.2.1
Current Status of Clinical Practice in Brain Tumor Therapy

Primary or metastatic brain tumors affect nearly 200 000 patients in the United States annually [20]. Furthermore, in recent years, the incidence of central nervous system (CNS) metastasis has increased due to more effective modern systemic therapies, an increased availability of advanced imaging detection techniques, and vigilant surveillance protocols. For malignant gliomas there have been no underlying causes identified so far. Based on the World Health Organization (WHO) classification [21], the four main types of glioma are astrocytomas, oligodendrogliomas, ependymomas, and mixed gliomas (usually oligo-astrocytomas). Moreover, four prognostic grades have been identified, including grade I (pilocyticastrocytoma), grade II (diffuse astrocytoma), grade III (anaplastic astrocytoma), and grade IV (glioblastoma). Grade III and IV tumors are considered malignant gliomas.

The current treatment of malignant brain tumors includes maximal surgical resection, plus radiotherapy and chemotherapy [2, 3]. Surgical debulking reduces the symptoms from mass effect, and provides tissue samples for both histological diagnosis and molecular studies. Advances such as magnetic resonance imaging (MRI)-guided neuronavigation, intraoperative MRI, functional MRI, intraoperative mapping [1, 3], and fluorescence-guided surgery [3, 22] have each led to improvements in the safety of surgery, and also have increased the extent of tumor resection. Unfortunately, malignant gliomas cannot be completely eliminated surgically because of their infiltrative nature, and patients should undergo maximal surgical resection whenever possible. Moreover, the addition of radiotherapy to surgery simply increases survival among patients with glioblastomas, from 3–4 months to 7–12 months [23].

15.2.2
The Progress of Investigational Therapies for Brain Tumors

15.2.2.1 Targeted Molecular Therapy
Most malignant gliomas have the co-activation of multiple tyrosine kinases [3, 24, 25], as well as redundant signaling pathways. These aberrantly activated signaling pathways promote tumor proliferation and invasion. Recent experimental strategies have focused on inhibitors that target receptor tyrosine kinases, such as epidermal growth factor receptor (EGFR) [26], platelet-derived growth factor receptor (PDGFR) [27], and vascular endothelial growth factor receptor (VEGFR) [28], as

well as on the signal-transduction inhibitors mammalian target of rapamycin (mTOR) [29, 30], farnesyltransferase [31], phosphatidylinositol 3-kinase (PI3K), and Akt, Met, Raf, Src, and transforming growth factor-beta (TGF-β). The efficacy of these agents has been modest, however, with response rates of 0% to 15%, and no prolongation of six-month progression-free survival [3, 25, 32]. These disappointing results are due to several factors: (i) a single agent could not inhibit the multiple aberrant signal pathways; (ii) many of these agents have poor penetration across the BBB; and (iii) the tumor microenvironment influences the role of the agents.

15.2.2.2 Anti-Angiogenic Therapy

Vascular proliferation, or *neoangiogenesis*, is a distinct histopathological characteristic of malignant brain tumors [33]. Vascular endothelial growth factor(VEGF) is a key factor involved in the angiogenetic process that can elicit several responses, such as endothelial cell proliferation, extracellular matrix (ECM) degradation, cell migration, and the expression of other proangiogenic factors (e.g., matrix metalloproteinase-1, urokinase-type plasminogen activator and its receptor, plasminogen activator inhibitor-1). In preliminary clinical studies, treatment with the combination of bevacizumab (a humanized anti-VEGF monoclonal antibody) and irinotecan (a toposiomerase 1 inhibitor) was associated with a low incidence of hemorrhage and response rates of 57–63% among patients with malignant gliomas [34, 35]. However, the efficacy of these agents in prolonging patient survival has been limited.

15.2.2.3 Immunotherapy

Malignant brain tumors display an ability to evade and suppress the immune system [4, 6, 36–38]. Strategies in glioma immunotherapy broadly include cytokine therapy, passive immunotherapy, and active immunotherapy. A number of factors have been proposed to influence the role of immunotherapy:

- The passage of therapeutic agents from the circulation through the BBB favors small, uncharged, lipid-soluble molecules. The efficient use of these agents including cytokines or (MAbs) against brain tumors presents unique challenges [4].

- The binding site barrier is another challenge to antibody penetration to gliomas. To diffuse into the tumor core, the agents would have to flow against a significant concentration gradient, leaving the core of a solid tumor ineffectively treated [4].

- In particular, the microenvironment of malignant glioma poses a significant problem for the immunotherapy, such as the production of a variety of immunosuppressive substances, such as TGF-β1, prostaglandin E2 (PGE2), interleukin (IL)-10, gangliosides, STAT3, and macrophage chemoattractive protein (MCP)-1 [37, 39–42]. Moreover, the glioma-associated immunosuppressive leukocytes including microglia/macrophage cells and regulatory T cells comprise a significant component of the tumor microenvironment, and that they actively participate in angiogenesis, invasion and metastasis [43–47].

Recently, convection-enhanced delivery (CED) has emerged as an attractive drug-delivery method that may be instrumental to the development of novel therapeutic modalities [4, 48]. CED is suitable for enhancing the delivery of both large and small molecules by physical-pressure processes. In CED, the therapeutic agent is infused at high pressure through an intracranial catheter and, as opposed to diffusion, CED-mediated drug delivery relies on bulk flow. Although a number of passive immunotherapy clinical trials have utilized CED as the method of drug delivery, a major challenge with CED remains that the bulk flow produced may be slow and insufficient to deliver immunoglobulin G (IgG)-based molecules throughout a typical treatment volume. Other delivery techniques, including intraventricular infusion, intratumoral and intracavitary injections, and biodegradable polymers [e.g., biodegradable 1,3-bis(2-chloroethyl)-1-nitrosourea (BCNU) Gliadel wafers], rely on diffusion. Thus, current challenges for immunotherapy are still seeking a targeting delivery system to reduce systemic toxicity and to increase intratumoral cytokine concentrations.

15.2.2.4 Gene Therapy

Gene therapy (also referred to as "gene delivery") consists of the insertion or modification of genes into an individual's cell, in order to treat a disease. Gene delivery can be accomplished using different vectors, including viral vectors, cell-based transfer, and synthetic vectors.

Recently, retrovirus and adenovirus have been the most studied vectors for brain tumors [49–51], but the benefits to survival have been limited. The poor clinical response may be due to a poor retrovirus transfection efficiency, and to the inability of the adenovirus to penetrate as well as spread within the tumor. In addition, viral gene therapies have certain limitations, such as insertional mutagenesis, high immunogenicity, viral vector toxicity, and ethical concerns.

With the discovery of stem cells and the molecular mechanisms that mediate their tumor-tropism, cell-based gene therapies have attracted increasing attention. To date, neural stem cells (NSC) [52], neural progenitor cells (NPC) [53], embryonic stem cells (ESC)-derived astrocytes [54], bone marrow-derived stem cells [55], mesenchymal stem cells [56], endothelial progenitor cells [57], and fibroblasts [58] have each been used for brain tumor gene therapy in animal models. Nonetheless, this promising therapy requires much further investigation to evaluate its clinical efficacy.

As the field of nanotechnology continues to advance, synthetic vectors are becoming attractive delivery vehicles. These nanoparticles, which are sized between 1 and 100 nm, are of major scientific interest as they represent effectively a bridge between bulk materials and atomic or molecular structures. In general, the current synthetic vectors include liposomes [59–62], dendrimers [63, 64] and polymers [65], and nanotubes [7, 9, 12–14, 16, 19], all of which have shown promising results in animal models. In this chapter, attention will be focus on the application of CNTs.

15.3
The Characteristics of CNTs for Biological Applications

15.3.1
Single-Walled and Multi-Walled Carbon Nanotubes

Carbon nanotubes, which were discovered during the 1960s but described only in 1991 by Iijima, are comprised of graphene sheets rolled up to form a cylinder that is capped at its extremities by a hemi-fullerene [13, 14, 16]. CNTs may be composed either of a single plane of graphene (single-walled carbon nanotubes; SWNTs) or by multiple concentric layers (multi-walled carbon nanotubes; MWNTs). Most commonly, SWNTs have diameters ranging from 0.4 to 3.0 nm, and lengths that span from a few nanometers to a few microns. In contrast, MWNTs are larger, with diameters reaching 100 nm and a lengths ranging from 1 μm to several microns, or even longer (i.e., several millimeters) [13, 14, 16]. In terms of the differences between SWNTs and MWNTs – and especially in the field of biomedical applications – it is not still evident whether one system is more advantageous than the other [16]. Several methodologies for the production of both types of nanotube have been reported; these include arc discharge, laser ablation, chemical vapor deposition (CVD), and the gas-phase catalytic process (HiPCO).

15.3.2
Functionalization of CNTs

Pristine CNTs have highly hydrophobic surfaces, and are not soluble in aqueous solutions. Hence, for biomedical applications, both pristine SWNTs and MWNTs must be *functionalized* to afford water solubility and biocompatibility. The development of a proper surface functionalization on CNTs is therefore the most critical step. At present, CNTs may be functionalized in two ways [13, 14, 19]. The first approach is based on a covalent functionalization of the nanotube surface by grafting various chemical groups directly onto the backbone; the second approach involves the noncovalent coating of nanotubes with amphiphilic molecules (e.g., lipids and polymers) [8, 14, 17, 18].

15.3.2.1 Covalent Surface Modification
Currently, the covalent functionalization of CNTs employs either cycloaddition reactions to attach ammonium groups, or strong acid treatment to generate carboxylic acid groups by covalent surface modification. Use of the versatile 1,3-dipolar cycloaddition mechanism is the most widely used approach to attach molecular moieties covalently to CNTs for the delivery of therapeutic molecules [13, 14, 17, 19]. The cycloaddition reaction occurs on the aromatic sidewalls, instead of the nanotube ends and defects, as in the oxidation case. The resultant amine-terminated reactive CNT intermediates serve as the reactive centers for the subsequent attachment of functional moieties. In recent years, this has become

the general synthetic strategy, allowing for the easy preparation of a common intermediate with a reactive group, for subsequent functionalization to the desired compound.

The direct treatment of pristine CNTs with oxidizing agents represents another widely used type of covalent surface modification [66, 67]. In this process, carboxyl groups are formed at the ends of tubes, as well as at any defects on the sidewalls. Use of these carboxylic acid groups further functionalizes the nanotubes via amidation, esterification, or through the zwitterionic $COO-NH^{3+}$ formation. Subsequently, various lipophilic and hydrophilic dendrons can be attached to the CNTs via amide or ester linkages, and these offer the advantage of improving the solubility of CNTs in either organic or aqueous solvents [13, 66, 67].

15.3.2.2 Noncovalent Surface Modification

In contrast to covalent surface modification, the noncovalent functionalization of CNTs can be carried out with amphiphilic surfactant molecules or polymers, by noncovalent surface modification [13, 14, 19]. Currently, several methods are available to generate noncovalent functionalized CNTs by noncovalent surface modification. First, by taking the advantage of the $\pi-\pi$ interaction between pyrene and the nanotube surface, some proteins can be immobilized onto SWNTs functionalized by an amine-reactive pyrene derivative. In addition, single-stranded DNA (ssDNA) molecules have been widely used to solubilize SWNTs, by virtue of the $\pi-\pi$ stacking between aromatic DNA base units and the nanotube surface. Moreover, fluorescein isothiocyanate (FITC)-terminated poly(ethylene glycol) (PEG) chains are able to solubilize SWNTs, with the aromatic FITC domain $\pi-\pi$ stacked on the nanotube surface, thus yielding SWNTs having a visible fluorescence that is useful for both biological detection and imaging [68]. Second, various amphiphiles have been used to suspend CNTs in aqueous solutions, with hydrophobic domains attached to the nanotube's surface via van der Waals forces and hydrophobic effects, and polar heads to provide water-solubility [8, 10, 12]. One popular approach for generating the noncovalent functionalization of SWNTs is by using polyethylene glycol (PEG-ylated) phospholipids (PL-PEG). In this process, the two hydrocarbon chains of the lipid are strongly anchored onto the nanotube surface, with the hydrophilic PEG chain extending into the aqueous phase, thus imparting water-solubility and biocompatibility. Unlike nanotubes suspended by typical surfactants, PEGylated SWNTs prepared by this method are highly stable in various biological solutions, including serum.

15.3.3
CNT Delivery System

The unique physical and chemical features of CNTs—such as the ultra-high surface area ($1300 \, m^2 g^{-1}$), ultra-light weight, and tremendous strength—qualify them as potential delivery vehicles into biological systems. After appropriate surface modification by either covalent or noncovalent methods, functional CNTs can be used to deliver many therapeutic molecules, including antibodies, peptides, small interfering RNA (siRNA), DNA, CpG-oligonucleotides,[1] vaccines, and chemical drugs.

15.3.3.1 Delivery of Antibodies and Peptides

The use of antibodies and peptides is a common strategy for tumor targeting, and the same approach can be used to improve the selectivity of CNTs. For example, Liu and colleagues thiolated anti-Herceptin or a peptide RGD with Traut's reagent to generate active thiol groups in these targeting proteins [8, 14, 69]. Subsequently, maleimide groups were introduced onto the SWNTs by reacting PL-PEG-amine-functionalized SWNTs with a sulfosuccinimidyl 4-N-maleimidomethyl cyclohexane-1-carboxylate (Sulfo-SMCC) bifunctional linker. Finally, the activated CNTs were reacted with thiolated antibodies or peptides, obtaining the targeted CNTs bioconjugates.

15.3.3.2 Delivery of siRNA

It has been shown that siRNA is able to silence specific gene expression via RNA interference (RNAi), and this has led to a great deal of interest in both basic and applied biology. Although, viral-based siRNA delivery methods have shown promise in animal models and also in clinical trials, the safety concern of viral vectors is significant; it is important, therefore, to develop nonviral vectors for siRNA delivery. Recently, Liu and colleagues have synthesized a "smart" siRNA CNT delivery system to target CXCR4 [8, 12, 14], a chemokine receptor that has an important role in the entry of the human immunodeficiency virus (HIV) into human T cells. For this, the CXCR4 siRNAs were first thiolated to produce thiol groups, after which pyridyl disulfide groups were introduced onto the SWNTs by reacting PL-PEG-amine-functionalized SWNTs with a sulfosuccinimidyl 6-(3′-[2-pyridyldithio]-propionamido) hexanoate (Sulfo-LC-SPDP) bi-functional linker. Finally, the pyridyl disulfide group could be coupled to the thiolated siRNA to create a disulfide linkage, through a thiol exchange reaction. Once transported into the cells through endocytosis, the siRNA would be is released from the SWNTs by sulfide cleavage, and then bind to the CXCR4 mRNA so as to inhibit its translation.

15.3.3.3 Delivery of DNA Molecules

CNTs can be modified with positive charges to bind DNA plasmids for gene transfection. For example, Pantarotto *et al.* and Singh *et al.* [70–72] used amine-terminated SWNTs and MWNTs functionalized by 1,3-dipolar cycloaddition to bind DNA plasmids, and have subsequently achieved a reasonable transfection efficiency. In the study conducted by Gao *et al.* [73], amine groups were

---

1) CpG-oligodeoxynucleotides (CpG-ODN) are short, single-stranded synthetic DNA molecules that contain a cytosine "C" followed by a guanine "G." CpG-ODN can be taken up into cells by endocytosis, to activate a Toll-like receptor 9 on B lymphocytes, dendritic and natural killer (NK) cells; this results in an increased innate immunity and antibody-dependent cell cytotoxicity.

introduced to oxidized MWNTs for DNA binding and transfection, thereby successfully expressing green fluorescent protein (GFP) in mammalian cells.

15.3.3.4 Delivery of CpG

Bianco et al. [74] demonstrated that the presence of cationic functional CNTs in the delivery of synthetic oligodeoxynucleotides containing CpG motifs (CpG-ODN) could improve the immunostimulatory properties of ODN-CpGs in vitro. Synthetic oligonucleotides containing CpGs are reported to confer nonspecific protection against cellular pathogens, and to enhance antigen-specific immune responses. In order to evaluate the immunostimulatory properties of CNTs, various ratios of CNT–CpG-ODN complexes were incubated with splenocytes. The efficiency of the process was monitored by the expression of IL-6 and other proinflammatory cytokines, the production of which is stimulated by ODN-CpG. The study results showed that an increase in the immunostimulatory capacity of CpG, when complexed with CNT, was not accompanied by any enhanced secretion of IL-6, a proinflammatory cytokine that may cause harm to the host. Interestingly, IL-6 secretion was decreased in cultures stimulated with CNT–CpG-ODN complexes.

15.3.3.5 Delivery of Vaccines

Although numerous preclinical studies in mouse models have demonstrated the efficacy of peripheral vaccination against intracranial gliomas [75], therapeutic vaccines face a substantial challenge in glioma patients because they must overcome a variety of immunoregulatory mechanisms that have already established the immune escape of tumors. The basic idea of using CNTs in vaccine delivery entails linking an antigene to CNTs, without losing its conformation, thereby inducing an antibody response with the correct specificity. However, it is equally important that the incorporated CNTs do not possess any intrinsic immunogenicity and, hence, trigger an immune response. CNTs therefore act as templates, upon which chiral molecules are attached, which in turn act as centers for molecular recognition. Pantarotto et al. [76] reported the use of CNTs in eliciting an improved immune response. Peptides derived from the VP1 protein of the foot-and-mouth-disease virus (FMDV) were first coupled to SWNTs. Serum samples from inoculated (by injection with CNT conjugates) Balb/c mice were collected and analyzed (using an enzyme-linked immunosorbent assay) for the presence of antipeptide antibodies. Peptide–CNT complexes were shown to elicit a greater immune response against the peptides, with no detectable crossreactivity to the CNTs, thus confirming the nonimmunogenicity of the carrier. The CNT–protein complex was also observed to have enhanced an immune response when attached to an antigen, thus strengthening the possibility of incorporating CNTs in vaccines.

15.3.3.6 Delivery of Chemical Drugs

Fluorescent dyes and small drug molecules cargos can be simultaneously linked to 1,3-dipolar cycloaddition CNTs via amide bonds [77]. Besides covalent conjugation, a novel noncovalent supramolecular chemistry has been uncovered, for loading aromatic drug molecules to functionalized SWNTs by π–π stacking. For example, doxorubicin–a commonly used cancer chemotherapy drug–can be loaded onto the surface of PEGylated SWNTs with a remarkably high loading (up to 4 g per gram SWNT), owing to the ultrahigh surface area of the SWNTs. The supramolecular approach of drug loading on CNTs opens new opportunities for drug delivery [11]. Besides drug conjugation and loading outside nanotubes, the hollow structure of the CNTs may also allow the encapsulation of drug molecules inside nanotubes for drug delivery. Examples of materials loaded inside CNTs include fullerene balls, metal ions, and small compounds such as metallocenes.

15.4
Strategies of Application of CNTs to Brain Tumor Therapy

15.4.1
CNTs Targeting Brain Tumor-Macrophages

The results of recent studies have suggested that most tumors have devised mechanisms to escape the host immune system, which not only leads to the immune cells being ineffective in generating an antitumor response, but also exploits them to promote tumor growth. Macrophages have been proposed to play a role in tumor tissue homeostasis [78]; indeed, the infiltration of both microglia and macrophages has been well documented in malignant gliomas, the most common and malignant primary brain tumor [79–82]. Despite a large influx of immune cells, the ability of gliomas to escape the host immune system is associated with a poor prognosis and a lack of response to conventional therapy [83]. Glioma immune evasion is thought to be mediated through the secretion of immunosuppressive factors such as TGF-β, IL-10 and PGE2 into the tumor microenvironment [84, 85]. Studies conducted by the present authors and others have demonstrated that, in addition to tumor cells, tumor-infiltrating microglia/macrophages are also responsible for the production of these immunosuppressive factors [36, 86–88]. The immunosuppressed status of the microglia/macrophages suggests that targeted therapy to modulate microglia/macrophage function may represent an effective immunotherapeutic approach for gliomas.

15.4.1.1 Internalization of CNTs by BV2 Microglia Cells *in vitro*
In order to evaluate CNTs for immune modulation in brain tumors, CNT uptake by microglia and macrophages was recently evaluated [9]. For this, pristine CNTs were initially dispersed into pluronic F108 (a relatively nontoxic nonionic surfactant with primary hydroxyl groups). Functionalized CNTs were first labeled with PKH26, a nontoxic and hydrophobic red fluorescent dye, and then incubated with BV2 MG cells. Subsequently, single CNTs were found to have penetrated the cell surface within 2 h of their addition, while by 6 and 24 h all of the microglia

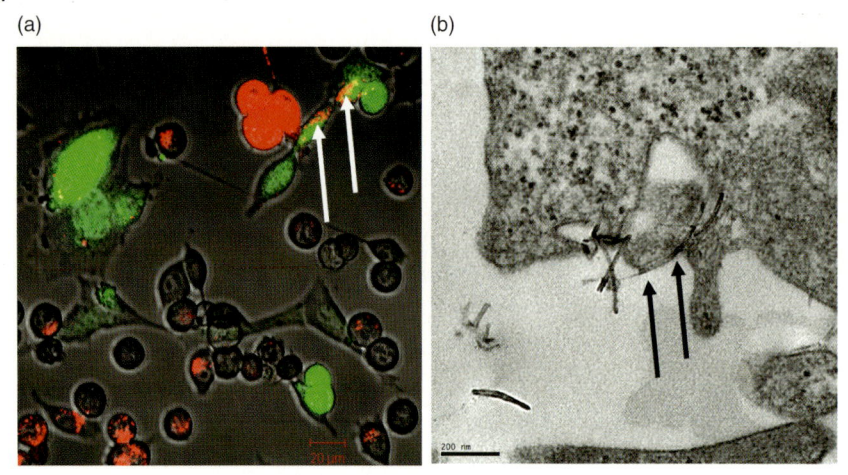

Figure 15.1 *In vitro* CNT internalization by microglia and glioma cells. (a) MWNTs-PKH (2.3 μg) were incubated (24 h) with a mixed culture of GL261-eGFP murine glioma (green cells) and BV-2 microglia cell lines. The CNTs were more efficiently taken up by microglia (red particles) than glioma cells (arrows); (b) Transmission electron microscopy demonstrating CNT uptake (arrows) by a microglia.

had internalized CNTs. These results suggested a promising application for CNTs, in the transportation of cargo across the microglia/macrophage cell membranes in the CNS (Figure 15.1).

15.4.1.2 Preferential Uptake of CNTs by Macrophages in a Glioma Model

As discussed above, the tumor microenvironment – including microglia/macrophages and regulatory T lymphocytes – participate actively in angiogenesis, invasion and metastasis, and represent (at least potentially) a target for brain tumor therapy. Thus, the initial step when considering CNTs as a delivery system would be first, to test their uptake in brain tumor models. The present authors' group recently characterized CNT uptake in an intracranial glioma model [7], whereby CNTs labeled with PKH26 were injected into either intracranial gliomas or normal control mice. All of the mice were found to tolerate the injections well, and showed no signs of toxicity, but a transient increase in microglia/macrophages infiltration, both in the glioma model and in normal control mice, was observed within 24 h of CNT injection, but returned to baseline by 72 h. It was also found that, following injection, the CNTs were colocalized to the microglia/macrophages throughout the tumor, and at the tumor edge (Figure 15.2). Furthermore, a flow cytometry intracellular analysis showed that the internalization of the CNTs varied by cell type in the glioma model. Among the brain tumor tissues, the tumor macrophages were seen to phagocytose the CNTs significantly more efficiently than did the other cell types (i.e., microglia, T cells or

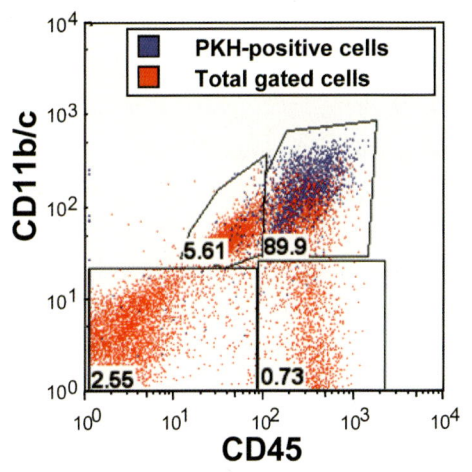

Figure 15.2 *In vivo* uptake of CNTs by macrophages in a glioma model. Intracranial GL261 mouse gliomas were injected once with CNT labeled with PKH. After 24 h, CNT internalization was assessed using flow cytometry. Most of the CNTs (blue events) were detected in tumor macrophages (CD45 high, CD 11b/c high).

tumor cells). At 24 h after the CNT injection, 70–90% of the tumor macrophages became CNT-positive. Likewise, over time the proportion of macrophages decreased to almost 50% over seven days. Overall, these findings highlighted the potential application of CNTs as a selective delivery vehicle into the tumor macrophages in gliomas.

15.4.1.3 Phosphatidylserine-Coated CNTs Targeting Microglia/Macrophages

Macrophage recognition and the uptake of apoptotic cells (also termed "efferocytosis") represents an important type of cell–cell interaction that regulates inflammation. The recognition of apoptotic cells by macrophages is largely dependent on the surface appearance of phosphatidylserine (PS), an anionic phospholipid that is normally confined to the cytosolic leaflet of the plasma membrane. According to this principle, Konduru and colleagues recently generated a PS-coated CNT delivery system [89] in which the PS-coated CNTs were efficiently taken up by a number of phagocytic cells, including murine RAW264.7 macrophages, primary monocyte-derived human macrophages, dendritic cells, and rat brain microglia *in vitro*. *In vivo* studies also showed that the aspiration of PS-coated CNTs by mice stimulated their uptake by lung alveolar macrophages. Moreover, the loading of PS-coated CNTs with pro-apoptotic cargo (cytochrome c) allowed for a targeted killing of RAW264.7 macrophages. In particular, the macrophage uptake of PS-coated CNTs altered the pattern of pro- and anti-inflammatory cytokine secretion, for instance, inhibiting tumor necrosis factor-α (TNF-α) and stimulating TGF-β and IL-10 production. Thus, PS-coating can be utilized for the targeted delivery of

CNTs with specified cargoes into professional phagocytes, and hence for the therapeutic regulation of specific populations of immune-competent cells.

15.4.2
CNTs Targeting Tumor Cells and Preliminary Efforts Towards *In Vivo* Cancer Therapy

In order to use CNTs for potential cancer treatment, an ability for the CNTs to target tumor cells is highly desirable. At present, the three approaches to tumor therapy include: (i) active targeting guided by tumor-targeting ligands; (ii) passive targeting, which relies on the enhanced permeability and retention (EPR) effect of cancerous tumors; and (iii) the thermal effect of CNTs.

15.4.2.1 CNTs Actively Targeting Tumor Cells

In order to improve the efficacy of a therapeutic modality, it is first necessary specifically to direct anti-tumor agents towards the tumors. It is well known that an integrin, $\alpha_v\beta_3$, is upregulated on various solid tumors (including those of the brain). It has been widely suggested that the $\alpha_v\beta_3$ integrin is correlated to tumor angiogenesis and metastasis, a proposal that Liu and colleagues followed up to generate an efficient CNT that was conjugated with an integrin-binding peptide (RGD) that recognized the $\alpha_v\beta_3$ integrin [69]. Liu *et al.* showed the RGD-conjugated SWNTs to exhibit a high tumor uptake of 13% injected dose per gram tissue (ID g^{-1}), compared to 4–5% ID g^{-1} in plain SWNTs without RGD in glioblastoma U87MG tumor-bearing mice. It was also found that an efficient tumor targeting could be realized only when the SWNTs were coated with long-chain PEG (SWNT PEG5400 RGD), but not with short-chain PEG. Because the SWNT PEG2000 RGD was found to have a shorter blood circulation time than SWNT PEG54000 RGD, it suggested that the short-chain PEG would have a lesser probability of being trapped in tumors, or of binding to the tumor receptors. Subsequently, other groups have improved the tumor CNT uptake by conjugating functional CNTs with anti-CD20 antibody [90], with tumor-lysate proteins [91], and with the anti-cancer agent cisplatin combined with epidermal growth factor (EGF) [92].

15.4.2.2 CNTs Passively Targeting Tumor Cells

In contrast to actively targeting the tumor cells, the passive targeting of tumor cells by CNTs can be carried out using small-molecule drugs or therapeutic siRNA. The first *in vivo* delivery of therapeutic siRNA into cancer through CNTs was achieved by Zhang *et al.* [93], who found that functional SWNTs could deliver the telomerase reverse transcriptase (TERT) siRNA very effectively, and also inhibit the expression of TERT in a mouse tumor model. Subsequently, Liu and colleagues used paclitaxel (PTX), a common chemotherapeutic agent, to conjugate to branched PEG-functionalized SWNTs via a cleavable ester bond [10]. Intravenous injection of the SWNT–PTX complex into a 4T1 murine breast cancer model led to an improved treatment efficacy compared to the clinical Cremophor-based PTX formulation, Taxol®. Moreover, both pharmacokinetics and biodistribution studies revealed a longer blood circulation half-life and a higher tumor uptake of SWNT–PTX than for simple PEGylated PTX and Taxol®. Taken together, these results suggested that the high biological effect of SWNT–PTX was most likely due to the EPR effect of cancerous tumors. Recently, Podesta and colleagues compared the efficacy of *in vivo* siRNA delivery between CNTs and liposomes [94] by constructing a proprietary toxic siRNA sequence (siTOX) that was conjugated with the functionalized MWNTs, or with a cationic liposome delivery system (DOTAP:cholesterol). The results showed that only the MWNT–siRNA complex, when administered intratumorally, abrogated the tumor growth and increased the survival of xenograft-bearing animals. The results of these studies further supported the therapeutic capacity of functionalized CNTs to deliver siRNA directly into cancerous target cells.

15.4.2.3 CNTs Thermal Effects on Tumor Cells

Thermal ablation is achieved when cells are heated above a temperature threshold, typically 55 °C. This treatment induces coagulative necrosis, a form of cell death that involves protein denaturation and membrane lysis. CNTs have the intrinsic characteristic of adsorbing energy in a near-infrared (NIR) light and in a radiofrequency field. The absorption of light induces a local increase in temperature, with deleterious consequences for the malignant cells, tissues and organs which have incorporated the CNTs. Thus, another strategy for using CNTs to treat cancer has been based on the capacity of nanotubes to convert electromagnetic radiation into heat.

Recently, several groups have demonstrated the use of CNTs conjugated with tumor-specific marker to selectively induce thermal damage to tumor cells, without harming the adjacent normal cells.

The group of Dai group [95] constructed a functional SWNT conjugated with folic acid (FA), which had a high affinity with the folate receptors (FR)s expressed in most tumor cells. In this case, FR-positive HeLa cells or normal cells without abundant FRs were incubated in a medium containing the functional SWNT–FA complex. As a consequence, extensive cell death was noted in the FR+ cells during 12–18 h after cell culture, whereas the normal cells remained intact and exhibited a normal proliferation behavior over two weeks.

Chakravarty and colleagues [96] conjugated anti-human CD22 or CD25 antibody to functional CNTs, and subsequently demonstrated the specific binding of antibody-coupled CNTs to tumor cells *in vitro*, followed by highly specific ablation with NIR light. The CD22 + CD25-Daudi cells (a human Burkitt's lymphoma cell line) were shown to bind selectively to those CNTs coupled to the anti-CD22 mAb, whereas the CD22-CD25+-activated peripheral blood mononuclear cells bound only to those CNTs coupled to the anti-CD25 mAb. Most importantly, only the specifically targeted cells were killed after exposure to NIR light. Burke and colleagues recently set up a therapeutic system using MWNTs stimulated by low-power NIR, that resulted in the eradication of mouse renal cancer [97]. In this case, tumor regression was seen to depend on the dose of MWNTs delivered to the tumor, with complete tumor regression – without recurrence for three months – being observed in 80% of the mice at a dose of 100 µg MWNTs. Overall, these reports suggested that a combination of MWNTs and NIR photothermal treatment would represent a viable approach for cancer therapy.

15.5
Toxicity Issues of CNTs in Brain Tumor Therapy

Toxicity studies of CNTs, which have been conducted largely in mice, have yielded several observations ranging from no induction of abnormalities or measurable inflammation [98], to granuloma formation in lungs [99], induced inflammation including substantial lung neutrophil influx [100], and mortality at high doses [101] following intratracheal administration. Similar discrepancies concerning cytotoxicity have emerged when CNTs were administered both intravenously and intraperitoneally. Recently, the present authors assessed the effect of MWNTs on the tumor milieu, by performing fluorescence-activated cell sorting (FACS) and quantitative reverse-transcription-polymerase chain reaction (RT-PCR) to analyze the inflammatory response and cytokine profile in a murine glioma model following a single MWNT injection [7, 9]. A transient influx of macrophages was seen in both normal and tumor-bearing brains, in response to MWNT treatment. Whereas, no significant changes in cytokine expression were noted in the normal brain, in tumor-bearing mice both IL-6 and IL-10 mRNA levels rose transiently within 48 h, while IL-12 expression was briefly suppressed. The observed increase in both IL-10 and IL-6 expression in tumors corresponded to the timing of macrophage influx into the tumors. This was in contrast to other reports, which showed elevations in pro-inflammatory cytokine expression in alveolar macrophages following CNT inhalational exposure [99]. The variations in these findings indicate that MWNT can affect the cell types in different ways, and highlights the importance of the careful selection of size, dispersion method, mode of administration, and the need for a standardized system when analyzing cytotoxicity when developing CNTs for biological applications.

Vittorio et al. [102] investigated the effects of various physico-chemical features of MWNTs on toxicity and biocompatibility in cultured human neuroblastoma cells by using MTT [3-(4,5-dimethylthiazole-2-yl)-2,5-diphenyl tetrazolium bromide], WST-1 [2-(4-odophenyl)-3-(4-nitrophenyl)-5-(2, 4-disulfophenyl)-2H tetrazolium salt], and Hoechst (in which a dye is used to visualize the nuclei and mitochondria) methods. In vitro experiments confirmed that, after three days of incubation with three different types of CNT dispersed in Pluronic F127 solution, cell viability was not affected and apoptosis and the production of reactive oxygen species (ROS) were not induced in the SH-SY5Y cells. With prolonged cultures, the cell loss proved to be minimal for preparations of 99% purity, whereas significant adverse effects were detected with preparations of 97% purity or which were acid-treated. Clearly, the CNT purity can have a direct influence on cell toxicity.

Recently, Bardi et al. [103] reported that Pluronic-coated CNTs did not induce the degeneration of cortical neurons either in vivo or in vitro. Although low concentrations of Pluronic F127 (i.e., 0.01%) induced apoptosis in mouse primary cortical neurons *in vitro* within 24 h, the presence of MWNTs prevented PF127-induced apoptosis. Furthermore, intracerebral injection of MWNTs coated with PF127 did not result in any neuronal degeneration. Taken together, these results suggested that MWNTs could reduce the toxicity associated with PF127 in cortical neurons.

In summary, although adverse reactions following the *in vivo* administration of CNTs have not yet been reported, long-term animal safety studies are still required to monitor this situation. Clearly, before any clinical applications are developed for CNTs, it is essential that their environmental and biological safety and toxicity is carefully assessed.

15.6
Conclusions and Future Directions

In this chapter, the current research strategies in the application of CNTs for brain tumor therapy have been reviewed. Surface functionalization chemistry, including various covalent and noncovalent modifications, is essential for biological application, while further modifications will be required to optimize the surface chemistry of CNTs, so as to further enhance their biocompatibility. CNTs targeting both macrophages/microglia and tumor cells represent a promising approach in brain tumor therapy. In this case, the conjugation of targeting ligands on appropriately coated CNTs may help to enhance cellular uptake via receptor-mediated endocytosis. The further development of suitable bioconjugation chemistry on nanotubes may also lead to the creation of versatile CNT-based bioconjugates for active targeting *in vivo* drug, and for gene delivery. Yet, issues of toxicity must not be ignored. While CNT anti-tumor technologies continue to generate much excitement, they require further refinement before being translated to the clinical situation.

Acknowledgments

These studies were supported by a NIH grant (R21CA131765-01A2), the American Cancer Society Research Scholar Grant (RSG-03-142-01-CNE), and a Think Cure Inc. grant.

References

1 Asthagiri, A.R., Pouratian, N., Sherman, J., Ahmed, G. and Shaffrey, M.E. (2007) Advances in brain tumor surgery. *Neurologic Clinics*, **25**, 975–1003, viii–ix.

2 Wen, P.Y. and Kesari, S. (2008) Malignant gliomas in adults. *New England Journal of Medicine*, **359**, 492–507.

3 Sathornsumetee, S., Rich, J.N. and Reardon, D.A. (2007) Diagnosis and treatment of high-grade astrocytoma. *Neurologic Clinics*, **25**, 1111–39.

4 Okada, H., Kohanbash, G., Zhu, X., Kastenhuber, E.R., Hoji, A., Ueda, R. and Fujita, M. (2009)

Immunotherapeutic approaches for glioma. *Critical Reviews in Immunology*, **29**, 1–42.

5 Ueda, R., Fujita, M., Zhu, X., Sasaki, K., Kastenhuber, E.R., Kohanbash, G., McDonald, H.A., Harper, J., Lonning, S. and Okada, H. (2009) Systemic inhibition of transforming growth factor-beta in glioma-bearing mice improves the therapeutic efficacy of glioma-associated antigen peptide vaccines. *Clinical Cancer Research*, **15**, 6551–9.

6 Fujita, M., Zhu, X., Ueda, R., Sasaki, K., Kohanbash, G., Kastenhuber, E.R., McDonald, H.A., Gibson, G.A., Watkins, S.C., Muthuswamy, R., Kalinski, P. and Okada, H. (2009) Effective immunotherapy against murine gliomas using type 1 polarizing dendritic cells – significant roles of CXCL10. *Cancer Research*, **69**, 1587–95.

7 VanHandel, M., Alizadeh, D., Zhang, L., Kateb, B., Bronikowski, M., Manohara, H. and Badie, B. (2009) Selective uptake of multi-walled carbon nanotubes by tumor macrophages in a murine glioma model. *Journal of Neuroimmunology*, **208**, 3–9.

8 Liu, Z., Tabakman, S.M., Chen, Z. and Dai, H. (2009) Preparation of carbon nanotube bioconjugates for biomedical applications. *Nature Protocols*, **4**, 1372–82.

9 Kateb, B., Van Handel, M., Zhang, L., Bronikowski, M.J., Manohara, H. and Badie, B. (2007) Internalization of MWCNTs by microglia: possible application in immunotherapy of brain tumors. *Neuroimage*, **37** (Suppl. 1), S9–17.

10 Liu, Z., Chen, K., Davis, C., Sherlock, S., Cao, Q., Chen, X. and Dai, H. (2008) Drug delivery with carbon nanotubes for in vivo cancer treatment. *Cancer Research*, **68**, 6652–60.

11 Liu, Z., Sun, X., Nakayama-Ratchford, N. and Dai, H. (2007) Supramolecular chemistry on water-soluble carbon nanotubes for drug loading and delivery. *ACS Nano*, **1**, 50–6.

12 Liu, Z., Winters, M., Holodniy, M. and Dai, H. (2007) siRNA delivery into human T cells and primary cells with carbon-nanotube transporters.

Angewandte Chemie International Edition, **46**, 2023–7.

13 Pastorin, G. (2009) Crucial functionalizations of carbon nanotubes for improved drug delivery: a valuable option? *Pharmaceutical Research*, **26**, 746–69.

14 Liu, Z., Tabakman, S., Welsher, K. and Dai, H. (2009) Carbon nanotubes in biology and medicine: *in vitro* and *in vivo* detection, imaging and drug delivery. *Nano Research*, **2**, 85–120.

15 Prato, M., Kostarelos, K. and Bianco, A. (2008) Functionalized carbon nanotubes in drug design and discovery. *Accounts of Chemical Research*, **41**, 60–8.

16 Bianco, A., Kostarelos, K. and Prato, M. (2008) Opportunities and challenges of carbon-based nanomaterials for cancer therapy. *Expert Opinion on Drug Delivery*, **5**, 331–42.

17 Foldvari, M. and Bagonluri, M. (2008) Carbon nanotubes as functional excipients for nanomedicines: II. Drug delivery and biocompatibility issues. *Nanomedicine*, **4**, 183–200.

18 Foldvari, M. and Bagonluri, M. (2008) Carbon nanotubes as functional excipients for nanomedicines: I. Pharmaceutical properties. *Nanomedicine*, **4**, 173–82.

19 Kostarelos, K., Bianco, A. and Prato, M. (2009) Promises, facts and challenges for carbon nanotubes in imaging and therapeutics. *Nature Nanotechnology*, **4**, 627–33.

20 Fisher, J.L., Schwartzbaum, J.A., Wrensch, M. and Wiemels, J.L. (2007) Epidemiology of brain tumors. *Neurologic Clinics*, **25**, 867–90.

21 Louis, D.N., Ohgaki, H., Wiestler, O.D., Cavenee, W.K., Burger, P.C., Jouvet, A., Scheithauer, B.W. and Kleihues, P. (2007) The 2007 WHO classification of tumours of the central nervous system. *Acta Neuropathologica*, **114**, 97–109.

22 Stummer, W., Pichlmeier, U., Meinel, T., Wiestler, O.D., Zanella, F. and Reulen, H.J. (2006) Fluorescence-guided surgery with 5-aminolevulinic acid for resection of malignant glioma: a randomised controlled multicentre phase III trial. *The Lancet Oncology*, **7**, 392–401.

23 Stupp, R., Mason, W.P., van den Bent, M.J., Weller, M., Fisher, B., Taphoorn, M.J., Belanger, K., Brandes, A.A., Marosi, C., Bogdahn, U., Curschmann, J., Janzer, R.C., Ludwin, S.K., Gorlia, T., Allgeier, A., Lacombe, D., Cairncross, J.G., Eisenhauer, E. and Mirimanoff, R.O. (2005) Radiotherapy plus concomitant and adjuvant temozolomide for glioblastoma. *New England Journal of Medicine*, 352, 987–96.

24 Stommel, J.M., Kimmelman, A.C., Ying, H., Nabioullin, R., Ponugoti, A.H., Wiedemeyer, R., Stegh, A.H., Bradner, J.E., Ligon, K.L., Brennan, C., Chin, L. and DePinho, R.A. (2007) Coactivation of receptor tyrosine kinases affects the response of tumor cells to targeted therapies. *Science*, 318, 287–90.

25 Sathornsumetee, S., Reardon, D.A., Desjardins, A., Quinn, J.A., Vredenburgh, J.J. and Rich, J.N. (2007) Molecularly targeted therapy for malignant glioma. *Cancer*, 110, 13–24.

26 Rich, J.N., Reardon, D.A., Peery, T., Dowell, J.M., Quinn, J.A., Penne, K.L., Wikstrand, C.J., Van Duyn, L.B., Dancey, J.E., McLendon, R.E., Kao, J.C., Stenzel, T.T., Ahmed Rasheed, B.K., Tourt-Uhlig, S.E., Herndon, J.E., 2nd, Vredenburgh, J.J., Sampson, J.H., Friedman, A.H., Bigner, D.D. and Friedman, H.S. (2004) Phase II trial of gefitinib in recurrent glioblastoma. *Journal of Clinical Oncology*, 22, 133–42.

27 Wen, P.Y., Yung, W.K., Lamborn, K.R., Dahia, P.L., Wang, Y., Peng, B., Abrey, L.E., Raizer, J., Cloughesy, T.F., Fink, K., Gilbert, M., Chang, S., Junck, L., Schiff, D., Lieberman, F., Fine, H.A., Mehta, M., Robins, H.I., DeAngelis, L.M., Groves, M.D., Puduvalli, V.K., Levin, V., Conrad, C., Maher, E.A., Aldape, K., Hayes, M., Letvak, L., Egorin, M.J., Capdeville, R., Kaplan, R., Murgo, A.J., Stiles, C. and Prados, M.D. (2006) Phase I/II study of imatinib mesylate for recurrent malignant gliomas: North American Brain Tumor Consortium Study 99-08. *Clinical Cancer Research*, 12, 4899–907.

28 Batchelor, T.T., Sorensen, A.G., di Tomaso, E., Zhang, W.T., Duda, D.G., Cohen, K.S., Kozak, K.R., Cahill, D.P., Chen, P.J., Zhu, M., Ancukiewicz, M., Mrugala, M.M., Plotkin, S., Drappatz, J., Louis, D.N., Ivy, P., Scadden, D.T., Benner, T., Loeffler, J.S., Wen, P.Y. and Jain, R.K. (2007) AZD2171, a pan-VEGF receptor tyrosine kinase inhibitor, normalizes tumor vasculature and alleviates edema in glioblastoma patients. *Cancer Cell*, 11, 83–95.

29 Galanis, E., Buckner, J.C., Maurer, M.J., Kreisberg, J.I., Ballman, K., Boni, J., Peralba, J.M., Jenkins, R.B., Dakhil, S.R., Morton, R.F., Jaeckle, K.A., Scheithauer, B.W., Dancey, J., Hidalgo, M. and Walsh, D.J. (2005) Phase II trial of temsirolimus (CCI-779) in recurrent glioblastoma multiforme: a North Central Cancer Treatment Group Study. *Journal of Clinical Oncology*, 23, 5294–304.

30 Chang, S.M., Wen, P., Cloughesy, T., Greenberg, H., Schiff, D., Conrad, C., Fink, K., Robins, H.I., De Angelis, L., Raizer, J., Hess, K., Aldape, K., Lamborn, K.R., Kuhn, J., Dancey, J. and Prados, M.D. (2005) Phase II study of CCI-779 in patients with recurrent glioblastoma multiforme. *Investigational New Drugs*, 23, 357–61.

31 Cloughesy, T.F., Wen, P.Y., Robins, H.I., Chang, S.M., Groves, M.D., Fink, K.L., Junck, L., Schiff, D., Abrey, L., Gilbert, M.R., Lieberman, F., Kuhn, J., DeAngelis, L.M., Mehta, M., Raizer, J.J., Yung, W.K., Aldape, K., Wright, J., Lamborn, K.R. and Prados, M.D. (2006) Phase II trial of tipifarnib in patients with recurrent malignant glioma either receiving or not receiving enzyme-inducing antiepileptic drugs: a North American Brain Tumor Consortium Study. *Journal of Clinical Oncology*, 24, 3651–6.

32 Chi, A.S. and Wen, P.Y. (2007) Inhibiting kinases in malignant gliomas. *Expert Opinion on Therapeutic Targets*, 11, 473–96.

33 Furnari, F.B., Fenton, T., Bachoo, R.M., Mukasa, A., Stommel, J.M., Stegh, A., Hahn, W.C., Ligon, K.L., Louis, D.N., Brennan, C., Chin, L., DePinho, R.A. and Cavenee, W.K. (2007) Malignant astrocytic glioma: genetics, biology, and

paths to treatment. *Genes and Development*, **21**, 2683–710.

34 Vredenburgh, J.J., Desjardins, A., Herndon, J.E., 2nd, Dowell, J.M., Reardon, D.A., Quinn, J.A., Rich, J.N., Sathornsumetee, S., Gururangan, S., Wagner, M., Bigner, D.D., Friedman, A.H. and Friedman, H.S. (2007) Phase II trial of bevacizumab and irinotecan in recurrent malignant glioma. *Clinical Cancer Research*, **13**, 1253–9.

35 Vredenburgh, J.J., Desjardins, A., Herndon, J.E., 2nd, Marcello, J., Reardon, D.A., Quinn, J.A., Rich, J.N., Sathornsumetee, S., Gururangan, S., Sampson, J., Wagner, M., Bailey, L., Bigner, D.D., Friedman, A.H. and Friedman, H.S. (2007) Bevacizumab plus irinotecan in recurrent glioblastoma multiforme. *Journal of Clinical Oncology*, **25**, 4722–9.

36 Zhang, L., Alizadeh, D., Van Handel, M., Kortylewski, M., Yu, H. and Badie, B. (2009) Stat3 inhibition activates tumor macrophages and abrogates glioma growth in mice. *Glia*, **57**, 1458–67.

37 Badie, B. and Schartner, J. (2001) Role of microglia in glioma biology. *Microscopy Research and Technique*, **54**, 106–13.

38 Han, S.J., Kaur, G., Yang, I. and Lim, M. (2010) Biologic principles of immunotherapy for malignant gliomas. *Neurosurgery Clinics of North America*, **21**, 1–16.

39 Badie, B., Schartner, J., Prabakaran, S., Paul, J. and Vorpahl, J. (2001) Expression of Fas ligand by microglia: possible role in glioma immune evasion. *Journal of Neuroimmunology*, **120**, 19–24.

40 Hussain, S.F., Yang, D., Suki, D., Aldape, K., Grimm, E. and Heimberger, A.B. (2006) The role of human glioma-infiltrating microglia/ macrophages in mediating antitumor immune responses. *Neuro-oncology*, **8**, 261–79.

41 Schartner, J.M., Hagar, A.R., Van Handel, M., Zhang, L., Nadkarni, N. and Badie, B. (2005) Impaired capacity for upregulation of MHC class II in tumor-associated microglia. *Glia*, **51**, 279–85.

42 Watters, J.J., Schartner, J.M. and Badie, B. (2005) Microglia function in brain tumors. *Journal of Neuroscience Research*, **81**, 447–55.

43 Fecci, P.E., Mitchell, D.A., Whitesides, J.F., Xie, W., Friedman, A.H., Archer, G.E., Herndon, J.E., 2nd, Bigner, D.D., Dranoff, G. and Sampson, J.H. (2006) Increased regulatory T-cell fraction amidst a diminished CD4 compartment explains cellular immune defects in patients with malignant glioma. *Cancer Research*, **66**, 3294–302.

44 Andaloussi, A.E., Han, Y. and Lesniak, M.S. (2008) Progression of intracranial glioma disrupts thymic homeostasis and induces T-cell apoptosis *in vivo*. *Cancer Immunology, Immunotherapy*, **57**, 1807–16.

45 Heimberger, A.B., Abou-Ghazal, M., Reina-Ortiz, C., Yang, D.S., Sun, W., Qiao, W., Hiraoka, N. and Fuller, G.N. (2008) Incidence and prognostic impact of FoxP3+ regulatory T cells in human gliomas. *Clinical Cancer Research*, **14**, 5166–72.

46 El Andaloussi, A., Han, Y. and Lesniak, M.S. (2006) Prolongation of survival following depletion of CD4+CD25+ regulatory T cells in mice with experimental brain tumors. *Journal of Neurosurgery*, **105**, 430–7.

47 Grauer, O.M., Nierkens, S., Bennink, E., Toonen, L.W., Boon, L., Wesseling, P., Sutmuller, R.P. and Adema, G.J. (2007) CD4+FoxP3+ regulatory T cells gradually accumulate in gliomas during tumor growth and efficiently suppress antiglioma immune responses *in vivo*. *International Journal of Cancer*, **121**, 95–105.

48 Ferguson, S. and Lesniak, M.S. (2007) Convection enhanced drug delivery of novel therapeutic agents to malignant brain tumors. *Current Drug Delivery*, **4**, 169–80.

49 Immonen, A., Vapalahti, M., Tyynela, K., Hurskainen, H., Sandmair, A., Vanninen, R., Langford, G., Murray, N. and Yla-Herttuala, S. (2004) AdvHSV-tk gene therapy with intravenous ganciclovir improves survival in human malignant glioma: a randomised, controlled study. *Molecular Therapy*, **10**, 967–72.

50 Ram, Z., Culver, K.W., Oshiro, E.M., Viola, J.J., DeVroom, H.L., Otto, E.,

Long, Z., Chiang, Y., McGarrity, G.J., Muul, L.M., Katz, D., Blaese, R.M. and Oldfield, E.H. (1997) Therapy of malignant brain tumors by intratumoral implantation of retroviral vector-producing cells. *Nature Medicine*, **3**, 1354–61.

51 Smitt, P.S., Driesse, M., Wolbers, J., Kros, M. and Avezaat, C. (2003) Treatment of relapsed malignant glioma with an adenoviral vector containing the herpes simplex thymidine kinase gene followed by ganciclovir. *Molecular Therapy*, **7**, 851–8.

52 Ehtesham, M., Kabos, P., Gutierrez, M.A., Chung, N.H., Griffith, T.S., Black, K.L. and Yu, J.S. (2002) Induction of glioblastoma apoptosis using neural stem cell-mediated delivery of tumor necrosis factor-related apoptosis-inducing ligand. *Cancer Research*, **62**, 7170–4.

53 Arnhold, S., Hilgers, M., Lenartz, D., Semkova, I., Kochanek, S., Voges, J., Andressen, C. and Addicks, K. (2003) Neural precursor cells as carriers for a gene therapeutical approach in tumor therapy. *Cell Transplantation*, **12**, 827–37.

54 Benveniste, R.J., Keller, G. and Germano, I. (2005) Embryonic stem cell-derived astrocytes expressing drug-inducible transgenes: differentiation and transplantation into the mouse brain. *Journal of Neurosurgery*, **103**, 115–23.

55 Lee, J., Elkahloun, A.G., Messina, S.A., Ferrari, N., Xi, D., Smith, C.L., Cooper, R., Jr, Albert, P.S. and Fine, H.A. (2003) Cellular and genetic characterization of human adult bone marrow-derived neural stem-like cells: a potential antiglioma cellular vector. *Cancer Research*, **63**, 8877–89.

56 Nakamura, K., Ito, Y., Kawano, Y., Kurozumi, K., Kobune, M., Tsuda, H., Bizen, A., Honmou, O., Niitsu, Y. and Hamada, H. (2004) Antitumor effect of genetically engineered mesenchymal stem cells in a rat glioma model. *Gene Therapy*, **11**, 1155–64.

57 Moore, X.L., Lu, J., Sun, L., Zhu, C.J., Tan, P. and Wong, M.C. (2004) Endothelial progenitor cells' "homing" specificity to brain tumors. *Gene Therapy*, **11**, 811–18.

58 Okada, H., Lieberman, F.S., Walter, K.A., Lunsford, L.D., Kondziolka, D.S., Bejjani, G.K., Hamilton, R.L., Torres-Trejo, A., Kalinski, P., Cai, Q., Mabold, J.L., Edington, H.D., Butterfield, L.H., Whiteside, T.L., Potter, D.M., Schold, S.C., Jr and Pollack, I.F. (2007) Autologous glioma cell vaccine admixed with interleukin-4 gene transfected fibroblasts in the treatment of patients with malignant gliomas. *Journal of Translational Medicine*, **5**, 67.

59 Yoshida, J., Mizuno, M., Fujii, M., Kajita, Y., Nakahara, N., Hatano, M., Saito, R., Nobayashi, M. and Wakabayashi, T. (2004) Human gene therapy for malignant gliomas (glioblastoma multiforme and anaplastic astrocytoma) by *in vivo* transduction with human interferon beta gene using cationic liposomes. *Human Gene Therapy*, **15**, 77–86.

60 Mabuchi, E., Shimizu, K., Miyao, Y., Kaneda, Y., Kishima, H., Tamura, M., Ikenaka, K. and Hayakawa, T. (1997) Gene delivery by HVJ-liposome in the experimental gene therapy of murine glioma. *Gene Therapy*, **4**, 768–72.

61 Voges, J., Reszka, R., Gossmann, A., Dittmar, C., Richter, R., Garlip, G., Kracht, L., Coenen, H.H., Sturm, V., Wienhard, K., Heiss, W.D. and Jacobs, A.H. (2003) Imaging-guided convection-enhanced delivery and gene therapy of glioblastoma. *Annals of Neurology*, **54**, 479–87.

62 de Fougerolles, A.R. (2008) Delivery vehicles for small interfering RNA *in vivo*. *Human Gene Therapy*, **19**, 125–32.

63 Agrawal, A., Min, D.H., Singh, N., Zhu, H., Birjiniuk, A., von Maltzahn, G., Harris, T.J., Xing, D., Woolfenden, S.D., Sharp, P.A., Charest, A. and Bhatia, S. (2009) Functional delivery of siRNA in mice using dendriworms. *ACS Nano*, **3**, 2495–504.

64 Waite, C.L. and Roth, C.M. (2009) PAMAM-RGD conjugates enhance siRNA delivery through a multicellular spheroid model of malignant glioma. *Bioconjugate Chemistry*, **20**, 1908–16.

65 Sarin, H. (2009) Recent progress towards development of effective systemic chemotherapy for the treatment of malignant brain tumors. *Journal of Translational Medicine*, **7**, 77.

66 Tasis, D., Tagmatarchis, N., Bianco, A. and Prato, M. (2006) Chemistry of carbon nanotubes. *Chemical Reviews*, **106**, 1105–36.

67 Schipper, M.L., Nakayama-Ratchford, N., Davis, C.R., Kam, N.W., Chu, P., Liu, Z., Sun, X., Dai, H. and Gambhir, S.S. (2008) A pilot toxicology study of single-walled carbon nanotubes in a small sample of mice. *Nature Nanotechnology*, **3**, 216–21.

68 Nakayama-Ratchford, N., Bangsaruntip, S., Sun, X., Welsher, K. and Dai, H. (2007) Noncovalent functionalization of carbon nanotubes by fluorescein-polyethylene glycol: supramolecular conjugates with pH-dependent absorbance and fluorescence. *Journal of the American Chemical Society*, **129**, 2448–9.

69 Liu, Z., Cai, W., He, L., Nakayama, N., Chen, K., Sun, X., Chen, X. and Dai, H. (2007) In vivo biodistribution and highly efficient tumour targeting of carbon nanotubes in mice. *Nature Nanotechnology*, **2**, 47–52.

70 Pantarotto, D., Singh, R., McCarthy, D., Erhardt, M., Briand, J.P., Prato, M., Kostarelos, K. and Bianco, A. (2004) Functionalized carbon nanotubes for plasmid DNA gene delivery. *Angewandte Chemie International Edition*, **43**, 5242–6.

71 Singh, R., Pantarotto, D., Lacerda, L., Pastorin, G., Klumpp, C., Prato, M., Bianco, A. and Kostarelos, K. (2006) Tissue biodistribution and blood clearance rates of intravenously administered carbon nanotube radiotracers. *Proceedings of the National Academy of Sciences of the United States of America*, **103**, 3357–62.

72 Singh, R., Pantarotto, D., McCarthy, D., Chaloin, O., Hoebeke, J., Partidos, C.D., Briand, J.P., Prato, M., Bianco, A. and Kostarelos, K. (2005) Binding and condensation of plasmid DNA onto functionalized carbon nanotubes: toward the construction of nanotube-based gene delivery vectors. *Journal of the American Chemical Society*, **127**, 4388–96.

73 Gao, L., Nie, L., Wang, T., Qin, Y., Guo, Z., Yang, D. and Yan, X. (2006) Carbon nanotube delivery of the GFP gene into mammalian cells. *Chembiochem*, **7**, 239–42.

74 Bianco, A., Hoebeke, J., Godefroy, S., Chaloin, O., Pantarotto, D., Briand, J.P., Muller, S., Prato, M. and Partidos, C.D. (2005) Cationic carbon nanotubes bind to CpG oligodeoxynucleotides and enhance their immunostimulatory properties. *Journal of the American Chemical Society*, **127**, 58–9.

75 Parney, I.F., Farr-Jones, M.A., Chang, L.J. and Petruk, K.C. (2000) Human glioma immunobiology *in vitro*: implications for immunogene therapy. *Neurosurgery*, **46**, 1169–78.

76 Pantarotto, D., Partidos, C.D., Graff, R., Hoebeke, J., Briand, J.P., Prato, M. and Bianco, A. (2003) Synthesis, structural characterization, and immunological properties of carbon nanotubes functionalized with peptides. *Journal of the American Chemical Society*, **125**, 6160–4.

77 Pastorin, G., Wu, W., Wieckowski, S., Briand, J.P., Kostarelos, K., Prato, M. and Bianco, A. (2006) Double functionalization of carbon nanotubes for multimodal drug delivery. *Chemical Communications (Cambridge)*, 1182–4.

78 de Visser, K.E., Eichten, A. and Coussens, L.M. (2006) Paradoxical roles of the immune system during cancer development. *Nature Reviews Cancer*, **6**, 24–37.

79 Badie, B. and Schartner, J.M. (2000) Flow cytometric characterization of tumor-associated macrophages in experimental gliomas. *Neurosurgery*, **46**, 957–62.

80 Roggendorf, W., Strupp, S. and Paulus, W. (1996) Distribution and characterization of microglia/macrophages in human brain tumors. *Acta Neuropathologica*, **92**, 288–93.

81 Shinonaga, M., Chang, C.C., Suzuki, N., Sato, M. and Kuwabara, T. (1988) Immunohistological evaluation of macrophage infiltrates in brain tumors.

Correlation with peritumoral edema. *Journal of Neurosurgery*, **68**, 259–65.

82 Streit, W.J. (1994) Cellular immune response in brain tumors. *Neuropathology and Applied Neurobiology*, **20**, 205–6.

83 Prados, M.D. and Levin, V. (2000) Biology and treatment of malignant glioma. *Seminars in Oncology*, **27**, 1–10.

84 Parney, I.F., Hao, C. and Petruk, K.C. (2000) Glioma immunology and immunotherapy. *Neurosurgery*, **46**, 778–92.

85 Yang, L., Ng, K.Y. and Lillehei, K.O. (2003) Cell-mediated immunotherapy: a new approach to the treatment of malignant glioma. *Cancer Control*, **10**, 138–47.

86 Badie, B., Schartner, J.M., Hagar, A.R., Prabakaran, S., Peebles, T.R., Bartley, B., Lapsiwala, S., Resnick, D.K. and Vorpahl, J. (2003) Microglia cyclooxygenase-2 activity in experimental gliomas: possible role in cerebral edema formation. *Clinical Cancer Research*, **9**, 872–7.

87 Lewis, C.E. and Pollard, J.W. (2006) Distinct role of macrophages in different tumor microenvironments. *Cancer Research*, **66**, 605–12.

88 Pollard, J.W. (2004) Tumour-educated macrophages promote tumour progression and metastasis. *Nature Reviews Cancer*, **4**, 71–8.

89 Konduru, N.V., Tyurina, Y.Y., Feng, W., Basova, L.V., Belikova, N.A., Bayir, H., Clark, K., Rubin, M., Stolz, D., Vallhov, H., Scheynius, A., Witasp, E., Fadeel, B., Kichambare, P.D., Star, A., Kisin, E.R., Murray, A.R., Shvedova, A.A. and Kagan, V.E. (2009) Phosphatidylserine targets single-walled carbon nanotubes to professional phagocytes *in vitro* and *in vivo*. *PLoS ONE*, **4**, e4398.

90 McDevitt, M.R., Chattopadhyay, D., Kappel, B.J., Jaggi, J.S., Schiffman, S.R., Antczak, C., Njardarson, J.T., Brentjens, R. and Scheinberg, D.A. (2007) Tumor targeting with antibody-functionalized, radiolabeled carbon nanotubes. *Journal of Nuclear Medicine*, **48**, 1180–9.

91 Meng, J., Meng, J., Duan, J., Kong, H., Li, L., Wang, C., Xie, S., Chen, S., Gu, N., Xu, H. and Yang, X.D. (2008) Carbon nanotubes conjugated to tumor lysate protein enhance the efficacy of an antitumor immunotherapy. *Small*, **4**, 1364–70.

92 Bhirde, A.A., Patel, V., Gavard, J., Zhang, G., Sousa, A.A., Masedunskas, A., Leapman, R.D., Weigert, R., Gutkind, J.S. and Rusling, J.F. (2009) Targeted killing of cancer cells *in vivo* and *in vitro* with EGF-directed carbon nanotube-based drug delivery. *ACS Nano*, **3**, 307–16.

93 Zhang, Z., Yang, X., Zhang, Y., Zeng, B., Wang, S., Zhu, T., Roden, R.B., Chen, Y. and Yang, R. (2006) Delivery of telomerase reverse transcriptase small interfering RNA in complex with positively charged single-walled carbon nanotubes suppresses tumor growth. *Clinical Cancer Research*, **12**, 4933–9.

94 Podesta, J.E., Al-Jamal, K.T., Herrero, M.A., Tian, B., Ali-Boucetta, H., Hegde, V., Bianco, A., Prato, M. and Kostarelos, K. (2009) Antitumor activity and prolonged survival by carbon-nanotube-mediated therapeutic siRNA silencing in a human lung xenograft model. *Small*, **5**, 1176–85.

95 Kam, N.W., O'Connell, M., Wisdom, J.A. and Dai, H. (2005) Carbon nanotubes as multifunctional biological transporters and near-infrared agents for selective cancer cell destruction. *Proceedings of the National Academy of Sciences of the United States of America*, **102**, 11600–5.

96 Chakravarty, P., Marches, R., Zimmerman, N.S., Swafford, A.D., Bajaj, P., Musselman, I.H., Pantano, P., Draper, R.K. and Vitetta, E.S. (2008) Thermal ablation of tumor cells with antibody-functionalized single-walled carbon nanotubes. *Proceedings of the National Academy of Sciences of the United States of America*, **105**, 8697–702.

97 Burke, A., Ding, X., Singh, R., Kraft, R.A., Levi-Polyachenko, N., Rylander, M.N., Szot, C., Buchanan, C., Whitney, J., Fisher, J., Hatcher, H.C., D'Agostino, R., Jr, Kock, N.D., Ajayan, P.M., Carroll, D.L., Akman, S., Torti, F.M. and Torti, S.V. (2009) Long-term survival following a single treatment of kidney tumors with multiwalled carbon nanotubes and near-infrared radiation. *Proceedings of the*

National Academy of Sciences of the United States of America, **106**, 12897–902.

98 Lacerda, L., Bianco, A., Prato, M. and Kostarelos, K. (2006) Carbon nanotubes as nanomedicines: from toxicology to pharmacology. *Advanced Drug Delivery Reviews*, **58**, 1460–70.

99 Shvedova, A.A., Kisin, E.R., Mercer, R., Murray, A.R., Johnson, V.J., Potapovich, A.I., Tyurina, Y.Y., Gorelik, O., Arepalli, S., Schwegler-Berry, D., Hubbs, A.F., Antonini, J., Evans, D.E., Ku, B.K., Ramsey, D., Maynard, A., Kagan, V.E., Castranova, V. and Baron, P. (2005) Unusual inflammatory and fibrogenic pulmonary responses to single-walled carbon nanotubes in mice. *American Journal of Physiology. Lung Cellular and Molecular Physiology*, **289**, L698–708.

100 Nemmar, A., Hoet, P.H., Vandervoort, P., Dinsdale, D., Nemery, B. and Hoylaerts, M.F. (2007) Enhanced peripheral thrombogenicity after lung inflammation is mediated by platelet-leukocyte activation: role of P-selectin. *Journal of Thrombosis and Haemostasis*, **5**, 1217–26.

101 Lam, C.W., James, J.T., McCluskey, R. and Hunter, R.L. (2004) Pulmonary toxicity of single-wall carbon nanotubes in mice 7 and 90 days after intratracheal instillation. *Toxicological Sciences*, **77**, 126–34.

102 Vittorio, O., Raffa, V. and Cuschieri, A. (2009) Influence of purity and surface oxidation on cytotoxicity of multiwalled carbon nanotubes with human neuroblastoma cells. *Nanomedicine*, **5**, 424–31.

103 Bardi, G., Tognini, P., Ciofani, G., Raffa, V., Costa, M. and Pizzorusso, T. (2009) Pluronic-coated carbon nanotubes do not induce degeneration of cortical neurons *in vivo* and *in vitro*. *Nanomedicine*, **5**, 96–104.

16
Carbon Nanotubes in Cancer Therapy, including Boron Neutron Capture Therapy (BNCT)

Amartya Chakrabarti, Hiren Patel, John Price, John A. Maguire and Narayan S. Hosmane

16.1
Introduction

Since their discovery in 1991 [1], carbon nanotubes (CNTs) have found many applications, including nanoelectronic and field emission devices, energy storage systems, and composites [2]. More recently, however, surface functionalization that has imparted water solubility to CNTs has opened the possibilities of their medicinal activity, especially in the area of cancer therapy [3]. In this chapter, the role of CNTs in several forms of cancer therapy, including boron neutron capture therapy (BNCT), will be discussed.

Initially, the recent advancements in the use of CNTs in the treatment of cancer, excluding BNCT, will be discussed, and this will be followed by a general discussion of BNCT and its development through the progression of nanotechnology. The role of CNTs in BNCT will be highlighted, and some comments made regarding the future outlook of this procedure.

16.2
Carbon Nanotubes in the Treatment of Cancer

Currently, cancer is one of the most devastating and widespread diseases, with an estimated 12 million new cases having been diagnosed worldwide in the year 2008 [4]. At present, the curative measures for killing cancer cells are still in their developmental phase, and a perfect cure has yet to be achieved. The most common treatments currently available include surgery, chemotherapy and/or radiotherapy although, unfortunately, both chemotherapy and radiotherapy will have adverse effects on healthy cells as well as cancerous cells. Moreover, in many cases surgery is not feasible. Today, a great deal of the research conducted relates specifically to the targeting of cancer cells without posing any harm to their healthy counterparts. Indeed, CNTs may play an important role in identifying such direct-path therapeutic vectors.

Nanomaterials for the Life Sciences Vol.9: Carbon Nanomaterials. Edited by Challa S. S. R. Kumar
Copyright © 2011 WILEY-VCH Verlag GmbH & Co. KGaA, Weinheim
ISBN: 978-3-527-32169-8

Carbon nanotubes found their way into cancer therapy by virtue of their nanodimensionality and chemical inertness, which allows them to penetrate and accumulate in neoplastic cells through an enhanced penetration and retention (EPR) effect. In addition, the quasi one-dimensional CNTs exhibit photoluminescence in the near-infrared (NIR) region [5], thus expanding the possibility of their application in cell imaging and probing. Further, the photothermal properties of CNTs have prompted their use in cancer cell destruction [6]. The use of CNTs in areas of drug delivery, imaging and probing, and in photoacoustical and photothermal therapy, will be discussed in the following subsections.

16.2.1
Drug Delivery

Advancements in cancer therapy depend primarily on the development of new drugs and drug-delivery systems [7, 8]. A major challenge in drug delivery is to increase the efficiency by which an agent can deliver maximum amounts of therapeutic agents to tumor cells, while keeping any adverse effects to normal cells at a minimum [9]. Fortunately, CNTs have demonstrated great promise in striking a balance between these two factors [10].

The targeted delivery of the antibacterial drug, amphotericin B (AmB), via the multiple functionalization of CNTs, was reported in 2005 by Wu and coworkers [11]. The group showed that the drug molecule could be covalently linked to both single-walled carbon nanotubes (SWNTs) or multi-walled carbon nanotubes (MWNTs), and taken up by mammalian cells (e.g., Jurkat lymphoma T cells) without any pronounced toxic effects. The composites were shown to enter the cells by a mechanism that was not mediated by endocytosis; rather, they seemed to behave like nanoneedles that penetrated the cell membrane, without causing cell death. Consequently, when it was also shown that the bound AmB was more effective than the free drug against the fungi, *Candida parapsilosos*, *Cryptococcus neoformans*, and *Candida albicans*, this provided a strong indication that the CNTs could be used as effective drug carriers.

When the SWNTs were conjugated to the drug molecule, paclitaxel (PTX) (Figure 16.1a) via phospholipid-branched poly(ethylene glycol) (PEG) side-wall attached units [12], the SWNT–PTX conjugates were found to be more effective against tumor growth in a murine 4TI breast cancer mice model than was the drug Taxol.

In a more recent report, the same group studied the *in vivo* cancer therapeutic efficiency of the drug doxorubicin (DOX), which was loaded onto PEG-functionalized SWNT via supramolecular π–π stacking [13]. A schematic depiction of the drug complex, along with some structural characterization data are presented in Figure 16.2. While DOX is highly toxic, the SWNT–DOX complex was shown to have not only a better therapeutic efficiency but also a reduced toxicity when compared to either free DOX or a DOX-loaded liposome (DOXIL). One advantage of this method was that a much greater drug loading was possible

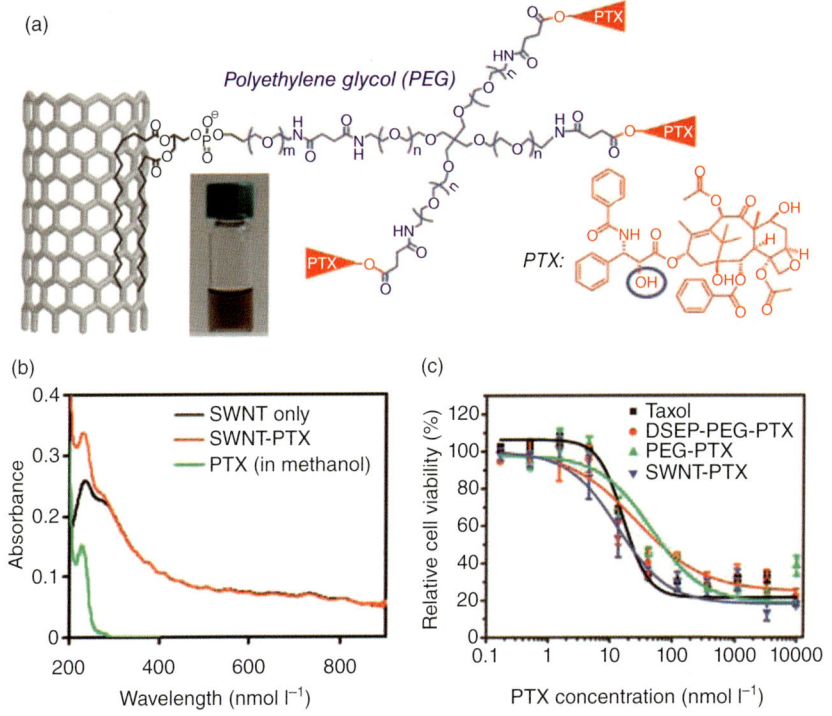

Figure 16.1 Carbon nanotube for paclitaxel (PTX) delivery. (a) Schematic illustration of PTX conjugation to SWNT functionalized by phospholipids with branched PEG chains; (b) UV-visible-NIR spectra of SWNT before (black curve) and after PTX conjugation (red curve). The absorbance peak of PTX at 230 nm (green curve) was used to measure the PTX loading on nanotubes, and the result was confirmed by radiolabel-based assay; (c) Cell survival versus concentration of PTX for 4T1 cells treated with Taxol, PEG–PTX, DSEP–PEG–PTX, or SWNT–PTX for 3 days. The PTX concentrations to cause 50% cell viability inhibition (IC_{50} values) were determined by sigmoidal fitting to be 16.4 ± 1.7 nmol l^{-1} for Taxol, 23.5 ± 1.1 nmol l^{-1} for DSPE–PEG–PTX, 28.4 ± 3.4 nmol l^{-1} for PEG–PTX, and 13.4 ± 1.8 nmol l^{-1} for SWNT–PTX [12].

compared to the specific functional group attachment of DOX; typically, up to 4 g of DOX per gram of SWNT could be achieved. Moreover, this same delivery method could be applied equally to other lipophilic aromatic drugs.

It had been shown that the anticancer drug cisplatin could be conjugated to SWNTs and used in the targeted *in vivo* killing of cancer cells. Bhirde *et al.* recently reported a successful targeting of squamous cancer cells with a bioconjugate consisting of cisplatin and the specific receptor ligand, epidermal growth factor (EGF), conjugated with oxidized SWNTs [14]. Both, *in vitro* and *in vivo* studies confirmed that the EGF-targeted bioconjugate was much more effective than were untargeted controls containing the same drug.

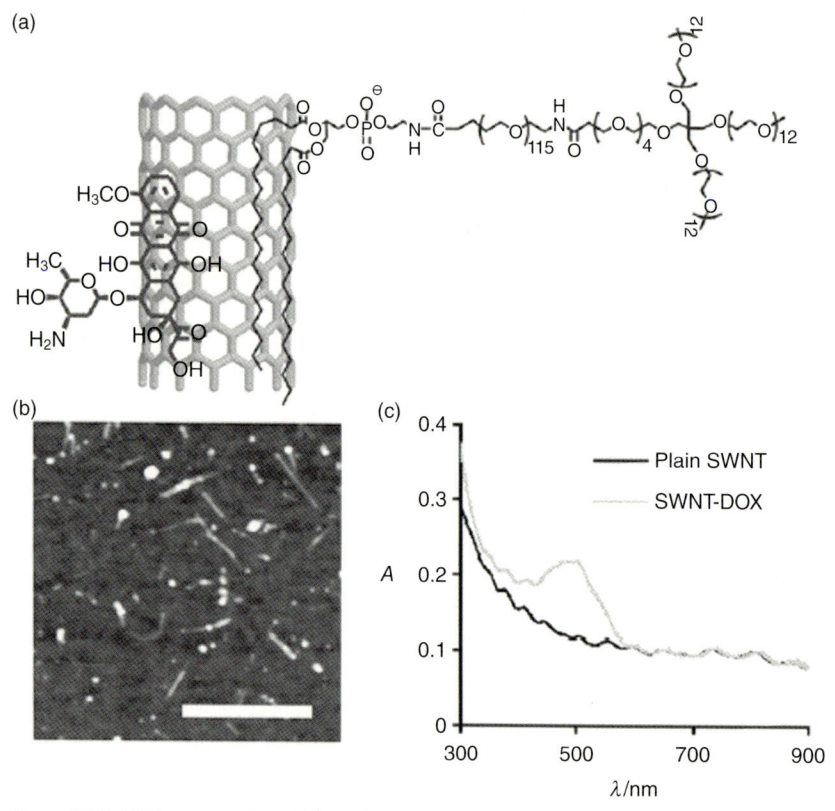

Figure 16.2 (a) Representation of the SWNT–DOX complex; (b) Atomic force microscopy image of SWNT–DOX complexes. The SWNT–DOX conjugates have an average length of 100 nm and a diameter of 2–3 nm. Scale bar: 250 nm; (c) UV/visible/NIR spectra of plain SWNTs and SWNT–DOX [13].

16.2.2
Imaging and Probing

The SWNTs are semiconducting in nature, have small band gaps on the order of about 1 eV, and they exhibit photoluminescence in the NIR region. This latter property of SWNTs has prompted their use in selective cell imaging and the probing of cell-surface receptors. For example, Cherukuri *et al.* have demonstrated the detection of SWNTs incorporated into peritoneal macrophage-like cells, at very low concentrations, with the help of a spectrofluorometer and a fluorescence microscope, modified for NIR imaging [15]. Notably, the ingested nanotubes did not exhibit any toxic effect towards the cells, and also retained their fluorescence (Figure 16.3).

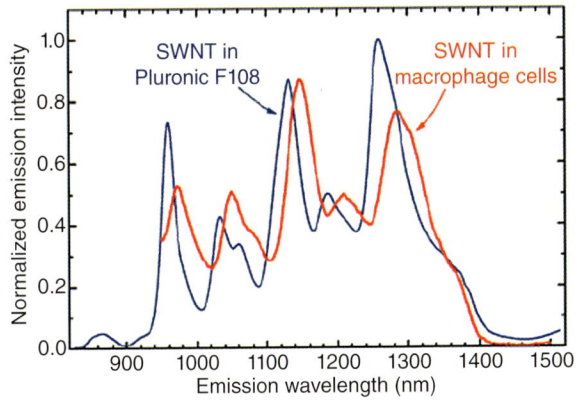

Figure 16.3 SWNT emission spectra in an aqueous Pluronic F108 suspension (blue trace) and in macrophage cells incubated in SWNT suspension and then washed (red trace). Samples were excited at 660 nm. Intensities have been scaled to aid comparison. Emission beyond 1350 nm is strongly attenuated by H_2O absorption [15].

Welsher and coworkers investigated the use of SWNTs as NIR fluorescent tags for the probing of cell-surface receptors [16]. These NIR fluorescent molecules functioned with very high specificity, as well as sensitivity, which allowed the detection of their selective binding to cell-surface receptors. The surface modification of SWNTs with phospholipid-PEG-amine functionalizations was used to conjugate two different antibodies: (i) Rituxan, which recognizes the CD20 cell-surface receptor; and (ii) Herceptin, which identifies the HER2/neu receptor on certain breast cancer cells. These procedures led to the establishment of SWNTs as NIR fluorophores for the sensitive detection of binding to cell receptors.

16.2.3
Photothermal and Photoacoustic Therapy

The strong absorbance of CNTs in the NIR region (700–1100 nm) has provided a potential application in cancer therapy. It was suggested by Kam *et al.* that the nanotubes could be optically simulated inside the living cells, and that a continuous irradiation with NIR could cause cell death due to a localized heating of the CNTs [6]. Subsequently, Huang *et al.* provided strong evidence for such a photothermal therapy using gold nanorods [17]. Notably, the functionalized CNTs, when irradiated with a noninvasive radiofrequency field, exhibited the thermal destruction of cancer cells [18].

Moon and coworkers recently reported their results on the photothermal therapy of CNTs for the *in vivo* destruction of tumor cells [19]. In this case, nude mice bearing human epidermoid mouth carcinoma KB tumor cells on their backs were treated, separately, with PEG-functionalized SWNTs and a phosphate- buffered

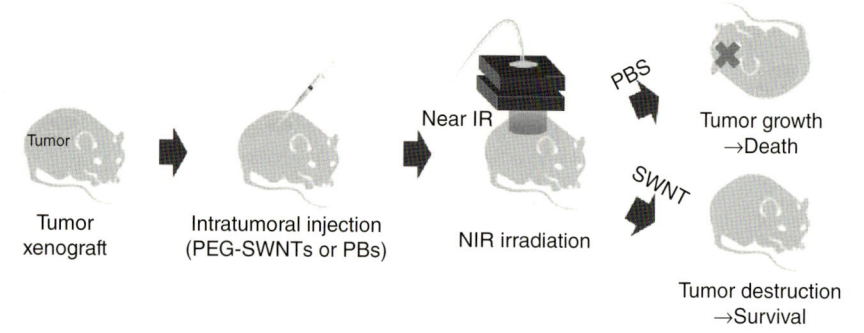

Figure 16.4 Photothermal treatments for *in vivo* tumor ablation using PEG–SWNTs. Schematic view of the procedure and results of PEG–SWNT-mediated photothermal treatment of tumors in mice [19].

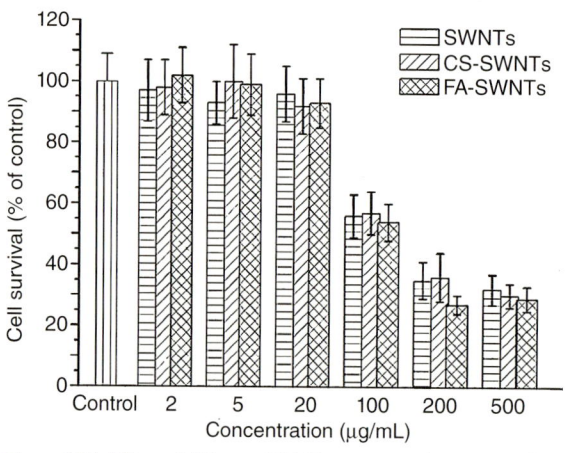

Figure 16.5 Effect of different SWNTs at various levels on cell proliferation [20].

saline (PBS) solution. The procedure and outcome are depicted schematically in Figure 16.4. The SWNT-treated mice not only survived, but also remained healthy, refraining from any toxicity or tumor recurrence over the next six months, whereas those in the PBS control group experienced cancer growth and, ultimately, death.

Photoacoustic therapy, using CNTs, has been reported by Kang *et al.* [20]. The SWNTs are known to exhibit a large photoacoustic effect upon irradiation with a 1064 nm Q-switched millisecond pulsed laser, the effect being sufficiently large to trigger an explosion of the SWNT. As a consequence, CNTs functionalized with chitosan (CS–SWNT) and conjugated with folic acid (FA) were used to create a bomb-like agent that could penetrate preferentially into the cancer cells. The lack of any negative (killing) effect of the CNTs below a concentration of 50 μg ml^{-1} is shown in Figure 16.5. The FA–SWNTs were found to be absorbed preferentially by cancer cells such that, on irradiation with a 1064 nm millisecond pulsed laser, 85% of the cancer cells were killed within 20 s while 90% of the normal cells (which did not contain the SWNTs) remained alive. Moreover, as the temperature changes during the irradiation period did not exceed ±3 °C, the cell death could not be attributed to any thermal effects. The potential of this therapeutic approach remains to be determined, however.

16.3
BNCT and Its Development through Nanotechnology

Currently, BNCT represents one of the most promising methods being investigated for cancer treatment. This binary therapy involves the selective introduction of ^{10}B-containing compounds into cancer cells, followed by irradiation with low-energy (thermal) neutrons [21]. BNCT is focused mainly on the treatment of malignant brain tumors, such as glioblastoma multiforme (GBM), which is virtually untreatable by other means. A brief overview of BNCT, and its progress as driven by nanotechnology, is provided in the following subsections.

16.3.1
BNCT: A Brief Overview

In BNCT, compounds containing ^{10}B nuclei are delivered preferentially to the tumor cells, followed by irradiation with thermal neutrons. Upon exposure to the radiation, the ^{10}B nucleus absorbs a neutron to form an excited ^{11}B state which rapidly undergoes a fission reaction, producing a high-energy α-particle, a recoil ^{7}Li ion, and a low-energy γ ray (Figure 16.6). The linear energy transfer (LET) of these charged particles has a trajectory of about 10 μm, which is essentially equivalent to a cell diameter. This assures that radiation damage is confined to the cell that originally contained the ^{10}B nucleus. Thus, if ^{10}B-containing compounds are preferentially absorbed into the cancer cells, the neoplastic cells can easily be killed, without affecting the surrounding cells. Boron-10 is attractive in that its nuclear capture cross-section is three orders of magnitude larger than the nuclei of elements such as C, O, and N, which constitute the major part of biological tissues. Boron-10 is also nonradioactive, and can be incorporated into hydrolytically stable molecules.

One of the major issues in BNCT is the synthesis of boron-containing compounds that can be delivered preferentially to cancer cells in sufficient concentrations to make the procedure viable [22]. It has been estimated that the required boron concentration for effective BNCT is approximately 10^9 atoms of ^{10}B per tumor cell, which translates into approximately 35 μg of ^{10}B per gram of tissue [23]. Despite intensive investigations having been conducted into possible boron drug-delivery agents, including carbohydrates [24], polyamines [25], porphyrins [26], and amino acids [27], only two low-molecular-weight compounds, namely sodium borocaptate (BSH) and boronophenylalanine (BPA) (Figure 16.7) are currently

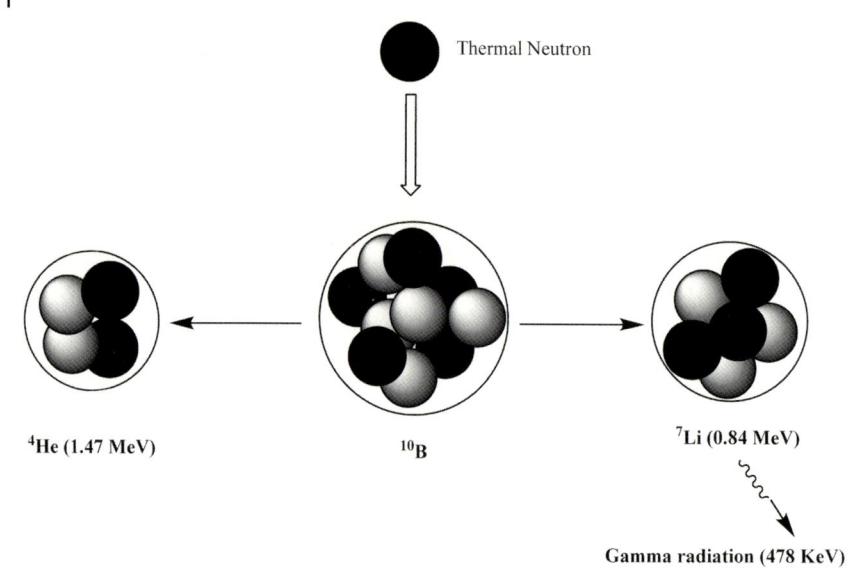

Figure 16.6 Schematic representation of boron neutron capture therapy (BNCT).

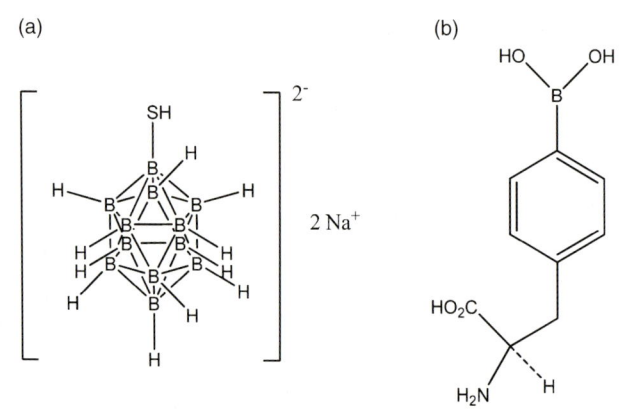

^{4}He (1.47 MeV) ^{10}B ^{7}Li (0.84 MeV)

Thermal Neutron

Gamma radiation (478 KeV)

Figure 16.7 Structures of (a) sodium borocaptate (BSH) and (b) 4-borono-L-phenyl alanine (L-BPA).

undergoing clinical trials [28]. However, to date these compounds, though nontoxic, have failed to provide any promising results due to their very low tumor:blood and tumor:brain ratios [29].

Progress in BNCT goes hand in hand with the advancement of nanotechnology [30]. Nanostructured materials could, in theory, lead to the incorporation of much greater numbers of boron atoms, providing for lower amounts of the required drug to be delivered to the cancer cells. A wide variety of nanomaterials, including liposomes, magnetic nanoparticles and nanotubes, along with dendritic macromolecules, have been tested as potential drug-delivery systems in BNCT, and their use is described in the following subsections.

16.3.2
Liposomes

Liposomes are naturally occurring nanomaterials composed of a hydrophilic head and a hydrophobic tail. They are natural components of cell walls, and as such are nontoxic, biocompatible, and can easily be tailored in accordance with the required chemical and physical characteristics. The diverse and unique properties of liposomes have led to them being an excellent choice as potential drug carriers to tumor cells. The main benefit of liposomes is that their surfaces can be modified so as to contain molecules that take advantage of receptor-mediated endocytosis. Indeed, it has been shown that the uptake of drugs encapsulated in liposomes with such modified surfaces is much greater by tumor cells than by the surrounding normal cells, in which the receptors are not overexpressed [31].

High boron-containing liposomes were prepared by Nakamura *et al.*, using double-chain nido-carboranes [32], after which the stability of liposomes in fetal bovine serum (FBS) was examined; a high incorporation of the nido-carboranes into distearoylphosphatidylcholine (DSPC) liposomes has also been reported. Similar studies were conducted with unilamellar liposomes formed through dual-chain nido-carboranes (Figure 16.8) [33], whereby the new lipid component showed a very high bilayer incorporation efficiency (98%) during liposome formation.

Another interesting approach was in the preparation of cationic liposomes, which have a strong ability to target the cell nuclei [34]. The cationic liposomes, when loaded with sugar-derived carboranes, can be considered as potential drug carriers for BNCT. More recently, ^{10}B-PEG-binding liposomes were found to act as growth suppressants for AsPC-1 tumors in nude mice implanted with human pancreatic carcinoma xenografts [35]. Moreover, the conjugate was shown capable of prolonging its circulatory time in the body by avoiding phagocytosis via the reticuloendothelial system (RES).

16.3.3
Dendritic Macromolecules

Dendrimers are spherical nanostructures with repeated branches, and are susceptible to surface functionalization [36]. Dendrimers can also be engineered to

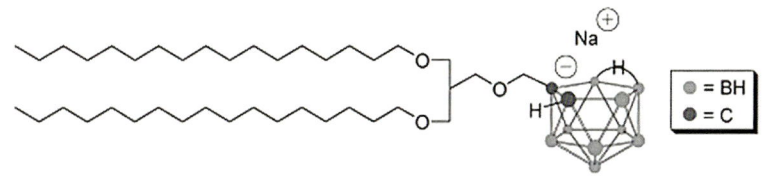

Figure 16.8 Structure of a dual-chain nido-carborane [33].

deliver either hydrophobic or hydrophilic molecules, thus opening up a wide range of applicability in the area of therapeutic drug delivery [37]. One of the most commonly used dendrimers is polyamidoamine (PAMAM), a water-soluble material that carries amine functional groups on its periphery. Positive – and also promising – results have been achieved for PAMAM loaded with anticancer drugs [38], and the possibility of PAMAM being used as a drug-delivery agent has also been reviewed [39].

As the drug molecules can be attached covalently to the dendrimer periphery, with an ease of loading and releasing, several dendrimeric systems have been developed to date exhibiting positive results [37]. Although a major problem arose when the drug molecules needed to be encapsulated inside the dendritic core, Gilles et al.. were able to combat such difficulties by developing a new drug-release methodology that involved polyethylene oxide (PEO) hybrids and dendrimers tagged with pH-sensitive hydrophobic acetal groups in the periphery [37]. In this case, when hydrolyzed under slightly acidic conditions, the hydrophobic acetal groups would vanish, ensuring a proper release of the encapsulated drug molecules.

Another major issue with dendrimers has been their toxicity, due to the presence of amine groups on their periphery. In order to reduce such cytotoxicity, a biocompatible aliphatic dendrimer has been synthesized with hydroxyl groups on the periphery [40]. At present, a compound with 16 carborane cages is undergoing investigation for its plausible application as a BNCT agent (Figure 16.9).

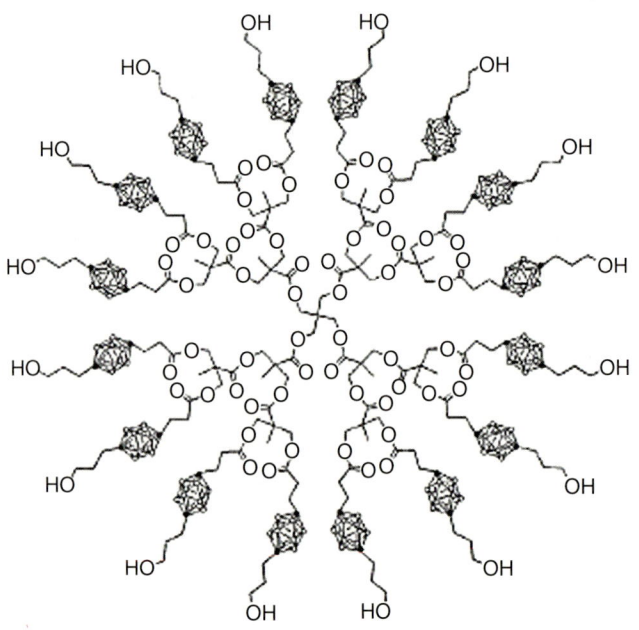

Figure 16.9 Aliphatic polyester dendrimer with 16 carborane cages [40].

16.3.4
Magnetic Nanoparticles

Most of the drugs currently used in chemotherapy are nonspecific, and attack both cancerous and healthy cells alike, which can lead to deleterious side effects. In magnetically targeted therapy, a biocompatible magnetic nanoparticle carrier is attached to the particular drug molecules being injected into a patient. When the particles enter the bloodstream, an external high-gradient magnetic field is used to concentrate the drug complexes at a specific target site. Then, when the desired concentration of complex has been achieved the drug can be released, either via enzymatic cleavage or by altering the physiological conditions [41]. This process is especially advantageous in that it leads to a reduction in the amount of cytotoxic drugs that needs to be delivered and, in turn, will (eventually) reduce the side effects encountered.

Previously, it has been shown that particles as large as 1–2 µm in dimension could be effectively concentrated in this way in intracerebral rat glima-2 (RG-2) tumors [42], and a later report indeed assessed the efficacy of 20 nm magnetic particles in targeting these tumors in rats [43]. Currently, the use of magnetic drug-delivery methods is becoming of increasing importance in medicine.

16.4
The Role of Carbon Nanotubes in BNCT

The chemical modification of CNTs through sidewall functionalization has opened new approaches to the possible application of CNTs. Most importantly, it has led to the solubilization of CNTs in aqueous media, and allowed their use in medical applications. The details of several successful fuctionalizations have been reported for both SWNTs and MWNTs, and these are very well discussed by Sun *et al.* and Bahr *et al.*, in their respective reviews [44]. Pantarotto and coworkers showed that CNTs could be used as delivery agent to the cells, based on their ability to cross the cell membrane and accumulate in the cytoplasm of 3T6, 3T3 fibroblasts and phagocytic cells, without exhibiting any toxic effects [45]. Moreover, the CNTs are able to penetrate cell walls in the same way as nanoneedles and to accumulate in the cytoplasm, but they appear unable to penetrate the nuclei of the cells. In another report, the carrier properties of SWNTs towards human promyelocytic leukemia (HL60) cells and human T cells (Jurkat) were described [46]. Together, these discoveries have prompted the idea of using CNTs as boron drug carriers for BNCT.

High-boron content, water-soluble SWNTs have been recently synthesized, and their biodistributions measured [47]. The side-wall functionalization of SWNTs was achieved via a nitrene cycloaddition, as shown in Scheme 16.1. The loading of the carborane cage per gram of nanotube was found to be 0.73 and 0.81 mmol for compounds IVa and IVb; the attached carboranes could then be decapitated to produce the corresponding nido-cages, the charges of which imparted water-solubility to the compounds.

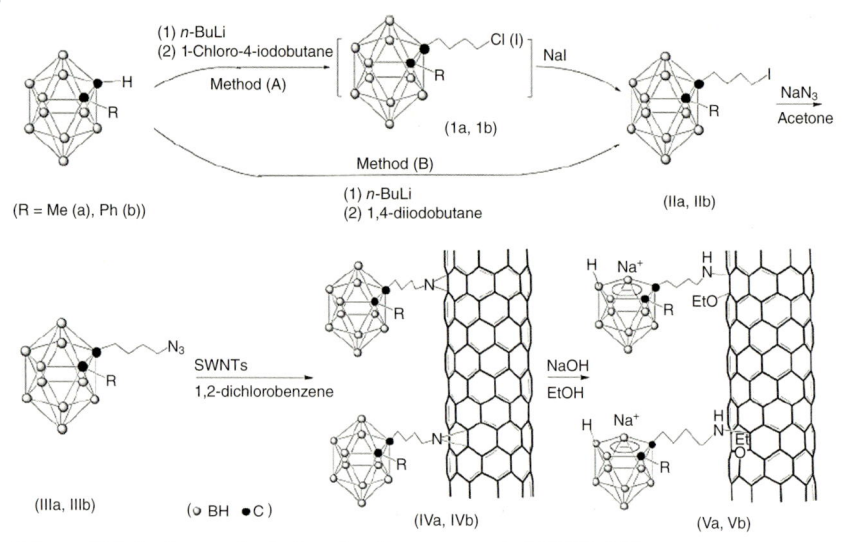

Scheme 16.1 Syntheses of substituted carborane-appended SWNTs [47].

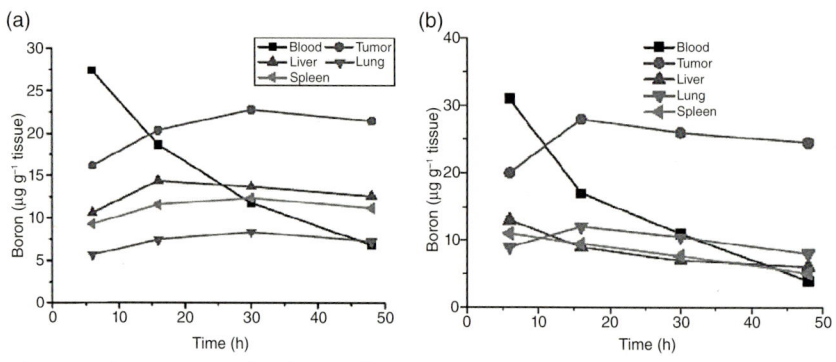

Figure 16.10 Boron tissue distribution of Va in (a) saline and (b) dimethylsulfoxide [47].

The applicability of these functionalized SWNTs in BNCT was examined through a biodistribution study on different tissues. When the conjugate Va (see Scheme 16.1) was injected into mice bearing EMT6 tumor cells, a highly favorable tumor:blood ratio of 3.12 and a boron concentration of 21.5 μg per gram of tumor cells were achieved after about 30 h (Figure 16.10). The SWNTs were also shown to be concentrated more in the tumor tissue than in the liver, lung, or spleen (Figure 16.10).

Besides sidewall attachment, it is also possible to encapsulate carboranes into the hollow spaces of CNTs [48]. For example, Morgan et al. reported the insertion of *ortho*-carborane molecules inside the nanotubes (Figure 16.11) [49]. Although the estimated filling yield was very low, this may be developed into a new methodology for the preparation of boron carriers.

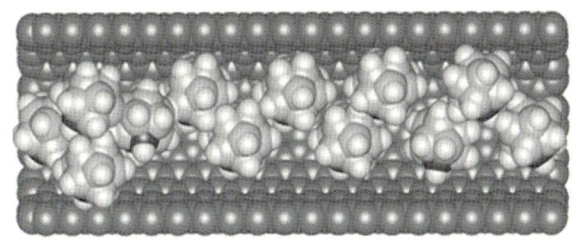

Figure 16.11 Schematic representation of *ortho*-carboranes inside SWNTs [49].

16.5
Summary and Future Outlook

In recent years, BNCT has emerged as a promising technique for combating cancer, which is recognized as one of the most deadly diseases worldwide. Yet, although several promising results have been achieved, a perfect remedy has not been identified, and new drug-delivery systems are required that involve novel therapeutic agents. Whilst nanocarriers such as liposomes, dendrimers and nanotubes have been used to raise the boron content in cancer cells, they are unable to provide the selectively delivered, high boron concentrations that are required. Consequently other nuclei, such as gadolinium, must be explored more fully as possible neutron-capture agents. Unfortunately, at present the only suitable neutron sources are nuclear reactors, and if neutron-capture therapy is to become a standard technique then more efficient, less hazardous neutron sources that can be sited within the hospital environment must be developed. The immediate future of BNCT, therefore, lies in the chemistry laboratory, and unless newer, more effective therapeutic molecules can be created it will remain a promising treatment modality that is waiting to come to fruition.

References

1 Iijima, S. (1991) Helical microtubules of graphitic carbon. *Nature*, **354**, 56–8.

2 (a) Ajayan, P.M. and Zhou, O.Z. (2001) Applications of carbon nanotubes, in *Carbon Nanotubes, Topics Appl. Phys.*, vol. 80 (eds M.S. Dresselhaus, G. Dresselhaus and P. Avouris), Springer-Verlag, Berlin Heidelberg, pp. 391–425; (b) Baughman, R.H., Zakhidov, A.A. and de Heer, W.A. (2002) Carbon nanotubes – the route toward applications. *Science*, **297**, 787–92.

3 (a) Martin, C.R. and Kohli, P. (2003) The emerging field of nanotube biotechnology. *Nature Reviews Drug Discovery*, **2**, 29–37; (b) Bianco, A. and Prato, M. (2003) Can carbon nanotubes be considered useful tools for biological applications? *Advanced Materials*, **15**, 1765–8; (c) Bianco, A., Kostarelos, K. and Prato, M. (2005) Applications of carbon nanotubes in drug delivery. *Current Opinion in Chemical Biology*, **9**, 674–9.

4 Boyle, P. and Levin, B. (2008) *World Cancer Report 2008*, World Health Organization Press.

5 O'Connell, M.J., Bachilo, S.M., Huffman, C.B., Moore, V.C., Strano, M.S., Haroz, E.H., Rialon, K.L., Boul, P.J., Noon, W.H.,

Kittrell, C., Ma, J.P., Hauge, R.H., Weisman, R.B. and Smalley, R.E. (2002) Band gap fluorescence from individual single-walled carbon nanotubes. *Science*, **297**, 593–6.

6 Kam, N.W.S., O'Connell, M., Wisdom, J.A. and Dai, H. (2005) Carbon nanotubes as multifunctional biological transporter and near-infrared agents for selective cancer cell destruction. *Proceedings of the National Academy of Sciences of the United States of America*, **102**, 11600–5.

7 Ferrari, M. (2005) Cancer nanotechnology: opportunities and challenges. *Nature Reviews Cancer*, **5**, 161–71.

8 Yih, T.C. and Al-Fandi, M. (2006) Engineered nanoparticles as precise drug delivery systems. *Journal of Cellular Biochemistry*, **97**, 1184–90.

9 Langer, R. (1998) Drug delivery and targeting. *Nature*, **392** (Suppl.), 5–10.

10 Pastorin, G. (2009) Crucial functionalizations of carbon nanotubes for improved drug delivery: a valuable option? *Pharmaceutical Research*, **26**, 746–69.

11 Wu, W., Wieckowski, S., Pastorin, G., Benincasa, M., Klumpp, C., Briand, J.-P., Gennaro, R., Prato, M. and Bianco, A. (2005) Targeted delivery of amphotericin b to cells by using functionalized carbon nanotubes. *Angewandte Chemie International Edition*, **44**, 6358–62.

12 Liu, Z., Chen, K., Davis, C., Sherlock, S., Cao, Q., Chen, X. and Dai, H. (2008) Drug delivery with carbon nanotubes for in vivo cancer treatment. *Cancer Research*, **68**, 6652–60.

13 (a) Liu, Z., Fan, A.C., Rakhra, K., Sherlock, S., Goodwin, A., Chen, X., Yang, Q., Felsher, D.W. and Dai, H. (2009) Supramolecular stacking of doxorubicin on carbon nanotubes for in vivo cancer therapy. *Angewandte Chemie International Edition*, **48**, 7668–72; (b) Liu, Z., Sun, X.M., Nakayama-Ratchford, N. and Dai, H. (2007) Supramolecular chemistry on water-soluble carbon nanotubes for drug loading and delivery. *ACS Nano*, **1**, 50–6.

14 Bhirde, A.A., Patel, V., Gavard, J., Zhang, G., Sousa, A.A., Masedunskas, A., Leapman, R.D., Weigert, R., Gutkind, J.S. and Rusling, J.F. (2009) Targeted killing of cancer cells in vivo and in vitro with EGF-directed carbon nanotube-based drug delivery. *ACS Nano*, **3**, 307–16.

15 Cherukuri, P., Bachilo, S.M., Litovsky, S.H. and Weisman, R.B. (2004) Near-infrared fluorescence microscopy of single-walled carbon nanotubes in phagocytic cells. *Journal of the American Chemical Society*, **126**, 15638–9.

16 Welsher, K., Liu, Z., Daranciang, D. and Dai, H. (2008) Selective probing and imaging of cells with single walled carbon nanotubes as near-infrared fluorescent molecules. *Nano Letters*, **8**, 586–90.

17 Huang, X., El-Sayed, I.H., Qian, W. and El-Sayed, M.A. (2006) Cancer cell imaging and photothermal therapy in the near-infrared region by using gold nanorods. *Journal of the American Chemical Society*, **128**, 2115–20.

18 Gannon, C.J., Cherukuri, P., Yakobson, B.I., Cognet, L., Kanzius, J.S., Carter, K.R., Weisman, B., Pasquali, M., Schmidt, H.K., Smalley, R.E. and Curley, S.A. (2007) Carbon nanotube-enhanced thermal destruction of cancer cells in a noninvasive radiofrequency field. *Cancer*, **110**, 2654–65.

19 Moon, H.K., Lee, S.H. and Choi, H.C. (2009) In vivo near-infrared mediated tumor destruction by photothermal effect of carbon nanotubes. *ACS Nano*, **3**, 3707–13.

20 Kang, B., Yu, D., Dai, Y., Chang, S., Chen, D. and Ding, Y. (2009) Cancer-cell targeting and photoacoustic therapy using carbon nanotubes as "bomb" agents. *Small*, **5**, 1292–301.

21 Locher, G.L. (1936) Biological effects and therapeutic possibilities of neutrons. *American Journal of Roentgenology and Radium Therapy*, **36**, 1–13.

22 Hawthorne, M.F. and Lee, M.W. (2003) A critical assessment of boron target compounds for boron neutron capture therapy. *Journal of Neuro-Oncology*, **62**, 33–45.

23 Barth, R.F., Soloway, A.H. and Brugger, R.M. (1996) Boron neutron capture therapy of brain tumors: past history, current status, and future potential. *Cancer Investigation*, **14**, 534–50.

24 (a) Tjarks, W., Anisuzzaman, A.K.M., Liu, L., Soloway, A.H., Barth, R.F., Perkins,

D.J. and Adams, D.M. (1992) Synthesis and *in vitro* evaluation of boronated uridine and glucose derivatives for boron neutron capture therapy. *Journal of Medicinal Chemistry*, **35**, 1628–33; (b) Maurer, J.L., Serino, A.J. and Hawthorne, M.F. (1988) Hydrophilically augmented glycosyl carborane derivatives for incorporation in antibody conjugation reagents. *Organometallics*, **7**, 2519–24.

25 Hariharan, J.R., Wyzlic, I.M. and Soloway, A.H. (1995) Synthesis of novel boron-containing polyamines – agents for DNA targeting in neutron capture therapy. *Polyhedron*, **14**, 823–9.

26 Kattan, G.F.-E., Lesnikowski, Z.J., Yao, S., Tanious, F., Wilson, W.D. and Schinazi, R.F. (1994) Carboranyl oligonucleotides. 2. Synthesis and physicochemical properties of dodecathymidylate containing 5-(O-carboran-1-yl)-2′-deoxyuridine. *Journal of the American Chemical Society*, **116**, 7494–501.

27 Malmquist, J. and Sjöberg, S. (1996) Asymmetric synthesis of *p*-carboranylalanine (*p*-Car) and 2-methyl-*o*-carboranylalanine (Me-*o*-Car). *Tetrahedron*, **52**, 9207–18.

28 Soloway, A.H., Tjarks, W., Barnum, B.A., Rong, F.-G., Barth, R.F., Codogni, I.M. and Wilson, J.G. (1998) The chemistry of neutron capture therapy. *Chemical Reviews*, **98**, 1515–62.

29 (a) Elowitz, E.H., Bergland, R.M., Coderre, J.A., Joel, D.D., Chadha, M. and Chanana, A.D. (1998) Biodistribution of p-boronophenylalanine in patients with glioblastoma multiforme for use in boron neutron capture therapy. *Neurosurgery*, **42**, 463–9; (b) Coderre, J.A., Elowitz, E.H., Chadha, M., Bergland, R., Capala, J., Joel, D.D., Liu, H.B., Slatkin, D.N. and Chanana, A.D. (1997) Boron neutron capture therapy for glioblastoma multiforme using p-boronophenylalanine and epithermal neutrons: trial design and early clinical results. *Journal of Neuro-Oncology*, **33**, 141–52; (c) Soloway, A.H., Hatanaka, H. and Davis, M.A. (1967) Penetration of brain and brain tumor. vii. Tumor-binding sulfhydryl boron compounds. *Journal of Medicinal Chemistry*, **10**, 714–17.

30 Yinghuai, Z., Yan, K.C., Maguire, J.A. and Hosmane, N.S. (2007) Recent developments in boron neutron capture therapy (BNCT) driven by nanotechnology. *Current Chemical Biology*, **1**, 141–9.

31 Gabizon, A., Price, D.C., Huberty, J., Bresalier, R.S. and Papahadjopoulos, D. (1990) Effect of liposome composition and other factors on the targeting of liposomes to experimental tumors: biodistribution and imaging studies. *Cancer Research*, **50**, 6371–8.

32 Nakamura, H., Miyajima, Y., Takei, T., Kasaoka, S. and Maruyama, K. (2004) Synthesis and vesicle formation of a nido-carborane cluster lipid for boron neutron capture therapy. *Chemical Communications*, 1910–11.

33 Li, T., Hamdi, J. and Hawthorne, M.F. (2006) Unilamellar liposomes with enhanced boron content. *Bioconjugate Chemistry*, **17**, 15–20.

34 Ristori, S., Oberdisse, J., Grillo, I., Donati, A. and Spalla, O. (2005) Structural characterization of cationic liposomes loaded with sugar-based carboranes. *Biophysical Journal*, **88**, 535–47.

35 Yanagie, H., Maruyama, K., Takizawa, T., Ishida, O., Ogura, K., Matsumoto, T., Sakurai, Y., Kobayashi, T., Shinohara, A., Rant, J., Skvarc, J., Ilic, R., Kuhne, G., Chiba, M., Furuya, Y., Sugiyama, H., Hisa, T., Ono, K., Kobayashi, H. and Eriguchi, M. (2006) Application of boron-entrapped stealth liposomes to inhibition of growth of tumour cells in the *in vivo* boron neutron-capture therapy model. *Biomedicine and Pharmacotherapy*, **60**, 43–50.

36 Newkome, G.R., Moorefield, C.N. and Vogtle, F. (1996) *Dendritic Macromolecules: Concepts, Synthesis, Perspectives*, Wiley-VCH Verlag GmbH, Weinheim, Germany.

37 Gillies, E.R. and Frechet, J. (2005) Dendrimers and dendritic polymers in drug delivery. *Drug Discovery Today*, **10**, 35–43.

38 Esfand, R. and Tomalia, D.A. (2001) Polyamidoamine (PAMAM) dendrimers: from biomimicry to drug delivery and biomedical applications. *Drug Discovery Today*, **6**, 427–36.

39 Kukowska-Latallo, J.F., Candido, K.A., Cao, Z., Nigavekar, S.S., Majoros, I.J., Thomas, T.P., Balogh, L.P., Khan, M.K. and Baker, J.R., Jr (2005) Nanoparticle targeting of anticancer drug improves therapeutic response in animal model of human epithelial cancer. *Cancer Research*, **65**, 5317–24.

40 Parrott, M.C., Marchington, E.B., Valliant, J.F. and Adronov, A. (2005) Synthesis and properties of carborane-functionalized aliphatic polyester Dendrimers. *Journal of the American Chemical Society*, **127**, 12081–9.

41 Alexiou, C., Arnold, W., Klein, R.J., Parak, F.G., Hulin, P., Bergemann, C., Erhardt, W., Wagenpfeil, S. and Lübbe, A.S. (2000) Locoregional cancer treatment with magnetic drug targeting. *Cancer Research*, **60**, 6641–8.

42 Pulfer, S.K., Ciccotto, S.L. and Gallo, J.M. (1999) Distribution of small magnetic particles in brain tumor-bearing rats. *Journal of Neuro-Oncology*, **41**, 99–105.

43 Sincai, M., Ganga, D., Ganga, M., Argherie, D. and Bica, D. (2005) Antitumor effect of magnetite nanoparticles in cat mammary adenocarcinoma. *Journal of Magnetism and Magnetic Materials*, **293**, 438–41.

44 (a) Sun, Y.-P., Fu, K., Lin, Y. and Huang, W. (2002) Functionalized carbon nanotubes: properties and applications. *Accounts of Chemical Research*, **35**, 1096–104.; (b) Bahr, J.L. and Tour, J.M. (2002) Covalent chemistry of single-wall carbon nanotubes. *Journal of Materials Chemistry*, **12**, 1952–8.

45 Pantarotto, D., Briand, J.-P., Prato, M. and Bianco, A. (2004) Translocation of bioactive peptides across cell membranes by carbon nanotubes. *Chemical Communications*, 16–17.

46 Kam, N.W.S., Jessop, T.C., Wender, P.A. and Dai, H. (2004) Nanotube molecular transporters: internalization of carbon nanotube-protein conjugates into mammalian cells. *Journal of the American Chemical Society*, **126**, 6850–1.

47 Yinghuai, Z., Peng, A.T., Carpenter, K., Maguire, J.A., Hosmane, N.S. and Takagaki, M. (2005) Substituted carborane-appended water-soluble single-wall carbon nanotubes: new approach to boron neutron capture therapy drug delivery. *Journal of the American Chemical Society*, **127**, 9875–80.

48 Dai, H. (2002) Carbon nanotubes: synthesis, integration and properties. *Accounts of Chemical Research*, **35**, 1035–44.

49 Morgan, D.A., Sloan, J. and Green, M.L.H. (2002) Direct imaging of o-carborane molecules within single walled carbon nanotubes. *Chemical Communications*, 2442–3.

17
Fullerenes in Photodynamic Therapy

Sulbha K. Sharma, Ying-Ying Huang, Pawel Mroz, Tim Wharton, Long Y. Chiang and Michael R. Hamblin

17.1
Introduction

The fullerene molecule, the third natural allotropic variation of carbon, was discovered in 1985 by Robert Curl, Harold Kroto, and Richard Smalley [1], who named C_{60} as Buckminsterfullerene, based on its structural similarity to geodesic structures widely credited to the architect R. Buckminster Fuller. As a result of these studies, Kroto, Smalley, and Curl were awarded the Nobel Prize in Chemistry in 1996. Unlike graphite or diamond, fullerenes are closed-cage carbon molecules, consisting of a number of five-membered rings combined with six-membered rings. The stable form of this molecule is composed of 60 carbon atoms, arranged in a soccer ball-shaped structure. Because of their unique structures and properties, fullerenes have attracted much attention from physicists, chemists and engineers, who still continue in their attempts to identify potential applications for these new carbon structures. Recently, these nanostructures have also been studied for their biological activities, with a view towards using them for biomedical applications. This rapidly growing interest in the medical application of fullerenes has drawn attention to their possible use as mediators of photodynamic therapy (PDT), on the basis of several particular properties. Some favorable PDT characteristics that fullerenes demonstrate include: (i) the presence of extended π-conjugation, giving a significant absorption of visible light; (ii) a high triplet yield and a long triplet lifetime; and (iii) the ability to generate reactive oxygen species (ROS) upon illumination. Depending on the functional groups introduced into the molecule, fullerenes are known to effectively photoinactivate pathogenic microbial cells, viruses and malignant cancer cells [2–4]. A schematic outline of the PDT applications that have been reported for fullerenes, either pristine or functionalized with various solubilizing groups, is shown in Figure 17.1. In this chapter, some basics of PDT will first be provided, followed by a brief description of the use of fullerenes as photosensitizers. The photochemical properties of fullerenes, fullerene biocompatibility, strategies to overcome any unfavorable PDT characteristics of fullerene,

Nanomaterials for the Life Sciences Vol.9: Carbon Nanomaterials. Edited by Challa S. S. R. Kumar
Copyright © 2011 WILEY-VCH Verlag GmbH & Co. KGaA, Weinheim
ISBN: 978-3-527-32169-8

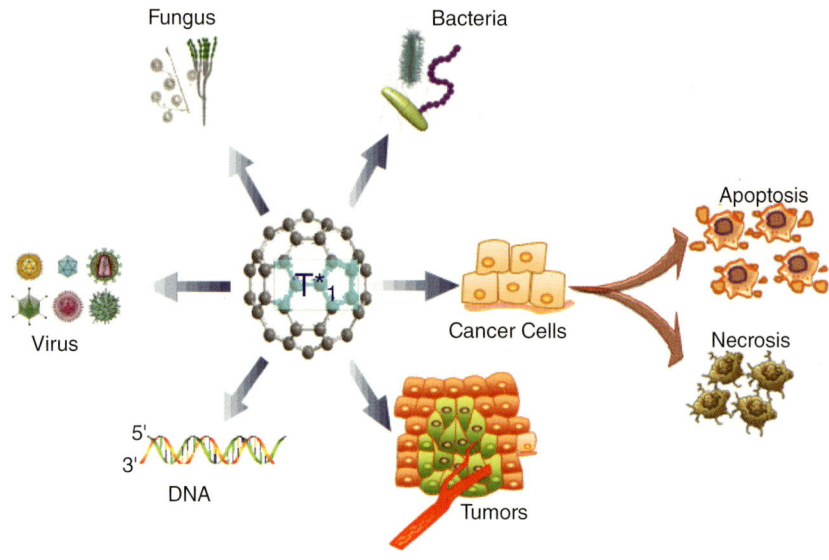

Figure 17.1 Schematic outline of the possible applications of fullerenes as photodynamic therapy sensitizers.

and existing data on the application of fullerenes in PDT, will then be discussed in greater detail.

17.2
Photodynamic Therapy

Photodynamic therapy is a nonsurgical, minimally invasive approach, which has been used for the treatment of solid tumors and many nonmalignant diseases. Previously, PDT was approved by the United States Food and Drug Administration (FDA) for use in endobronchial and endo-esophageal treatments [5], and also as a treatment for premalignant and early malignant diseases of the skin (actinic keratoses), bladder, breast, stomach, and oral cavity [6]. PDT is based on a nonthermal photochemical reaction, which requires the simultaneous presence of a photosensitizing drug (the photosensitizer, PS), oxygen, and visible light (Figure 17.2). It is principally a two-step procedure that involves the administration of a PS, followed by activation of the drug with the appropriate wavelength of light [6–9]. The selective anticancer action of PDT is the result of a low-to-moderate selective degree of PS uptake by proliferative malignant cells and the spatial confinement of light to the desired area. The photoactivation of the drug generates singlet oxygen and other ROS; this causes oxidative stress and membrane damage in the treated cells, and ultimately leads to cell death by direct cytotoxicity in addition to a dramatic anti-vascular action that impairs the blood supply to the area exposed to light [10].

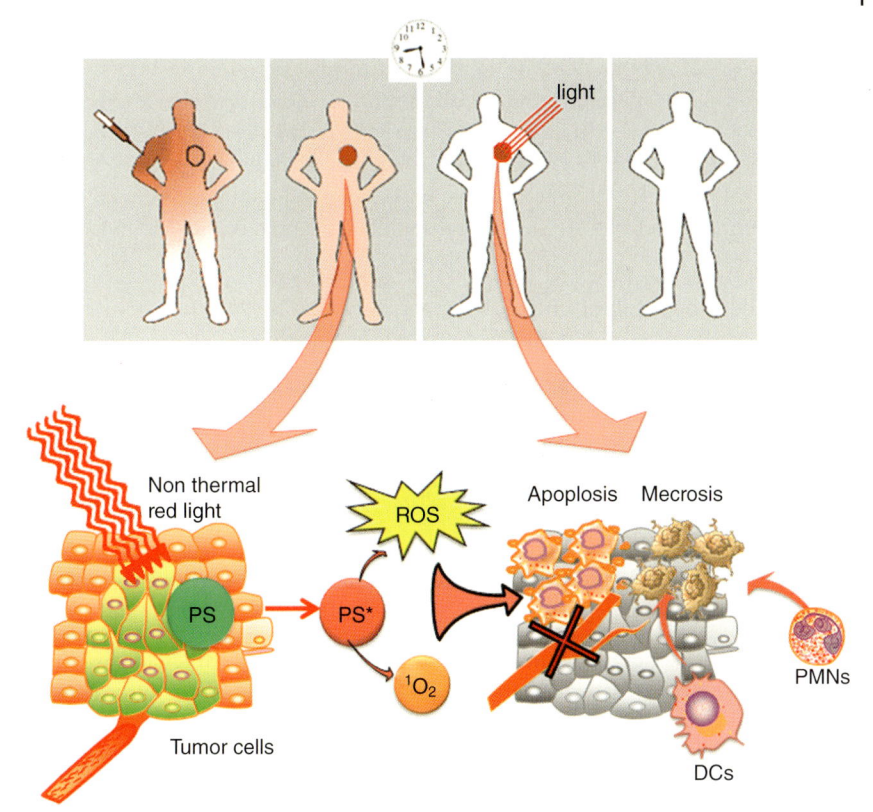

Figure 17.2 Photodynamic therapy for cancer. Typically the photosensitizer (PS) is injected intravenously and, after waiting a certain time for the PS to localize in the tumor, red light is delivered to the tumor area. The PS localized in the tumor interacts with light to produce cytotoxic reactive oxygen species (ROS), which causes a tumoricidal effect. This effect is due to a combination of direct cell killing by a mixture of apoptosis and necrosis, a rapid destruction of the tumor's blood supply, and an influx of host immune cells.

17.2.1
Traditional Photosensitizers

The structure of the PS is considered to be a critical element for the PDT effect. To date, many chemical structures have been employed as the PS for PDT purposes, and these will be briefly described at this point. The PS is characterized by an ability to absorb light of a specific wavelength and, in turn, to transfer the absorbed energy to the oxygen. The characteristics of the ideal PSs have been discussed elsewhere [11, 12], and include:

- It should produce low levels of dark toxicity to both humans and experimental animals.
- It should show a low incidence of side effects, such as hypotension (reduced blood pressure) or allergic reaction(s).

- For sufficient penetration into the tissue, the absorption of light should be in the red or far-red wavelengths. Absorption bands should be in the so called "optical window" (600–900 nm), as absorption at shorter wavelengths has less tissue penetration and is more likely to lead to skin photosensitivity (the power in sunlight drops off at $\lambda > 600$ nm). Absorption bands at high wavelengths (>800 nm) mean that the photons will not be energetic enough for the PS triplet state to transfer energy to the ground-state oxygen molecule to excite it to the singlet state.

- It should have relatively high absorption bands (>20 000–30 000 $M^{-1} cm^{-1}$) to minimize the dose required to achieve the desired effect.

- Synthesis of the PS should be relatively easy, and the starting materials readily available to make large-scale production feasible.

- It should be a pure compound with a constant composition and a stable shelf-life; ideally, it should also be water-soluble or soluble in a harmless aqueous solvent mixture.

- It should not aggregate unduly in biological environments, as this will reduce its photochemical efficiency.

- The pharmacokinetic elimination of the PS from the patient should be rapid (i.e., less than one day) in order to avoid the need for post-treatment protection from light exposure and prolonged skin photosensitivity.

- A short interval between the PS injection and illumination is desirable to facilitate outpatient treatment that is both patient-friendly and cost-effective.

- Pain on treatment is undesirable, as PDT does not usually require anesthesia or heavy sedation.

The majority of PSs, whether used clinically or experimentally, are derived from the tetrapyrrole aromatic nucleus found in many naturally occurring pigments such as heme, chlorophyll, and bacteriochlorophyll. The phthalocyanines (PC) represent a second widely studied structural group of PSs and, to a lesser extent, their related cousins, the naphthalocyanines. Another broad class of potential PSs includes completely synthetic, non-naturally occurring, conjugated pyrrolic ring systems. These comprise such structures as texaphyrins [12], porphycenes [13], and sapphyrins [14]. A final class of compounds that have been studied as PSs are nontetrapyrrole-derived naturally occurring or synthetic dyes. An example of the first group is hypericin [15], while the second group includes toluidine blue [16] and Rose Bengal [17].

17.2.2
Photophysics and Photochemistry in PDT

Light absorption and energy transfer are at the heart of PDT, the process of which is shown schematically in Figure 17.3. The ground-state PS (the "singlet state")

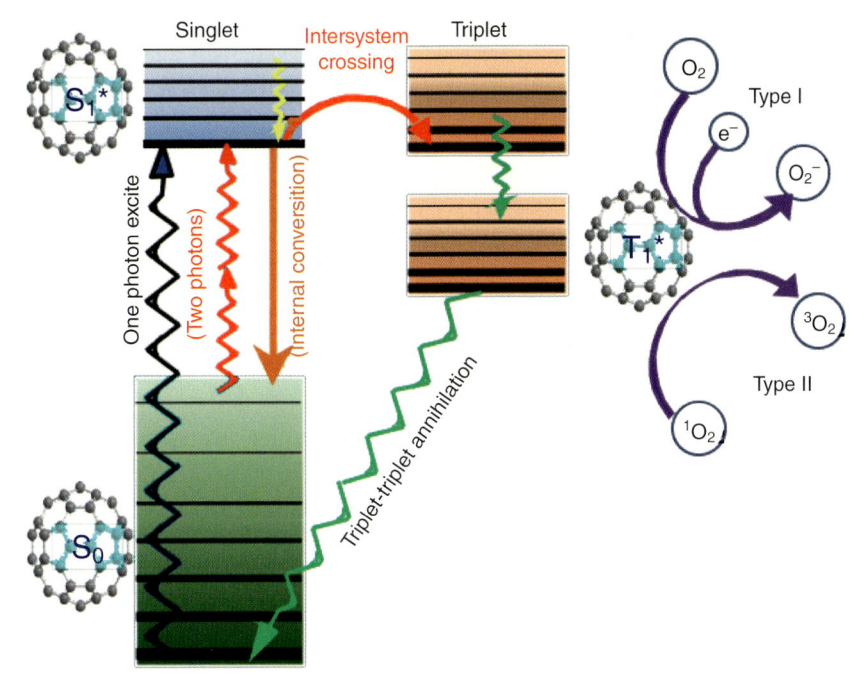

Figure 17.3 Jablonski diagram illustrating the absorption of a photon by the ground state of the singlet photosensitizer, that gives rise to the short-lived excited singlet state. This can lose energy by fluorescence (this is negligible in the case of fullerenes), by an internal conversion to heat, or by intersystem crossing to the long-lived triplet state. Fullerene triplet states are efficiently quenched by molecular oxygen (a triplet state) to give Type 2 (singlet oxygen) and Type I (hydroxyl radical) reactive oxygen species (ROS). In the absence of oxygen, the fullerene triplet states lose energy by triplet–triplet annihilation.

has two electrons with opposite spins in the low-energy molecular orbital. Following the absorption of light (photons), one of these electrons is boosted into a high-energy orbital but retains its spin (the first excited singlet state). This short-lived (nanoseconds) species loses its energy either by emitting light (fluorescence) or by an internal conversion into heat. The excited singlet-state PS may also undergo a process known as "intersystem crossing," whereby the spin of the excited electron inverts to form the relatively long-lived (microseconds to milliseconds) excited triplet state that has electron spins parallel. The long lifetime of the PS triplet state is explained by the fact that the loss of energy by emission of light (phosphorescence) is a "spin-forbidden" process, as the PS would move directly from a triplet to a singlet state. The PS excited triplet can undergo three broad types of reaction that are usually referred to as Type I, Type II, and Type III.

In a Type I reaction, the triplet PS can gain an electron from a neighboring reducing agent, such that the PS is now a radical anion bearing an additional unpaired electron. Alternatively, two triplet PS molecules can react together, involving an electron transfer to produce a pair that consists of a radical cation and a radical anion. The radical anions may further react with oxygen (with electron transfer) to produce ROS in a particular superoxide anion. A third possibility is that the PS triplet may directly transfer an electron to the ground-state oxygen, thus forming the PS radical cation and a superoxide. The radical cation may then accept an electron from a reducing agent, regenerating the PS for further reactions.

In a Type II reaction, the triplet PS can transfer its energy directly to molecular oxygen (itself a triplet in the ground state), to form excited-state singlet oxygen. Both, Type I and Type II reactions can occur simultaneously, and the ratio between these processes depends on the type of PS used, and the concentrations of substrate and oxygen.

In the less-common Type III reaction, the triplet state PS reacts directly with a biomolecule, destroying the PS and damaging the biomolecules. Type II processes are thought to best conserve the PS molecular structure in a photoactive state, and in some circumstances a single PS molecule can generate 10 000 molecules of singlet oxygen. Singlet oxygen, which is generated in Type II photochemical reactions, is generally considered as the most important mediator of PDT-induced damage, although $O_2^{-\bullet}$, H_2O_2, $OH^{\bullet}$, $NO^{\bullet}$ and other ROS are also detected in cells and tissues exposed to light and PSs [18]. Several studies have provided evidence for the active role of these metabolites in cytotoxic effects induced by PDT.

The ROS, together with singlet oxygen produced via the Type II pathway, serve as oxidizing agents that can react directly with many biological molecules. Amino acid residues in proteins are important targets that include cysteine, methionine, tyrosine, histidine, and tryptophan [19, 20]. Unsaturated lipids typically undergo ene-type reactions to produce lipid hydroperoxides (LOOHs), derived from phospholipids and cholesterol [21–24].

Because of the high reactivity and short half-life of singlet oxygen and the hydroxyl radicals, only those molecules and structures that are proximal to the area of its production (areas of PS localization) will be directly affected by PDT. The half-life of singlet oxygen in a biological system is typically $<1\,\mu s$, and therefore the radius of the action of singlet oxygen is of the order of $1\,\mu m$ [25].

17.2.3
Anticancer Mechanism of PDT

17.2.3.1 Cellular Effects
It is known that, depending on the parameters involved, when used *in vitro* PDT can kill cancer cells via apoptosis, necrosis, or autophagy:

- *Apoptosis* is a strictly controlled, energy-consuming process of suicidal cell death. It involves the activation of hydrolytic enzymes such as proteases and nucleases, leading to DNA fragmentation and the degradation of intracellular structures [26].

- *Necrosis* is a violent and quick form of degeneration that affects extensive cell populations. It is characterized by cytoplasmic swelling, the destruction of organelles and disruption of the plasma membrane, leading to a release of the intracellular contents and inflammation.

- *Autophagy* is a process used by cells to recycle their intracellular components during times of starvation. It has also been described as a process that helps cells to dispose of the damaged organelles and to replace them with new ones in order to survive [27].

17.2.3.2 *In Vivo* Effects

The direct killing effect of PDT on malignant cancer cells that has been studied in detail *in vitro* (as described above) also clearly applies *in vivo*, although in addition two separate *in vivo* mechanisms leading to PDT-mediated tumor destruction have been described. The first of these mechanisms, which has been shown to operate during the PDT of tumors, is the *vascular effect*, in which the vascular damage that occurs after completion of the PDT treatment contributes to long-term tumor control [28]. The second mechanism, which applies when tumors *in vivo* are treated with PDT, is an activation of the host immune system [29].

17.2.4
Antimicrobial Mechanism of PDT

Since the introduction of PDT during the early part of the last century, it has been known that certain microorganisms are killed by a combination of dyes and light *in vitro* [30]. Subsequently, many reports have described bacteria that have been killed or inactivated by using various combinations of PS and light [31]. During the 1990s, a fundamental difference was observed in the susceptibility to PDT of Gram-positive and Gram-negative bacteria. It was found that, in general, neutral or anionic PS molecules were bound efficiently to, and would photodynamically inactivate, Gram-positive bacteria. However, in the case of Gram-negative bacteria, the PS molecules would be bound (to a greater or lesser extent) only to the outer membrane of the bacterial cells, and the bacteria would not be inactivated following illumination [32]. The high susceptibility of the Gram-positive species to PDT was explained on a physiological basis, namely that their cytoplasmic membrane is surrounded by a relatively porous layer of peptidoglycan and lipoteichoic acid through which the PS can cross.

17.3
Fullerenes as Photosensitizers

During recent years, fullerenes have attracted considerable attention as possible PDT mediators, due largely to the rapidly growing interest in medical applications of nanotechnology. Fullerenes are considered to be potential PSs on the basis of certain favorable PDT characteristics that they possess. Although pristine C_{60} is highly insoluble in water and biological media, it is not photoactive [33] due to the formation of nanoaggregates in aqueous solvents [34]. In a 1991 editorial (in *Science*), Culotta and Koshland described C_{60} (the molecule of the year) as "Buckyballs: a wide open playing field for chemists" [35]. Since then, new PSs based on structures known as "functionalized fullerenes" that have hydrophilic or amphiphilic side chains or fused-ring structures attached to the spherical C_{60} core, have been shown to exhibit high efficiency in the production of singlet oxygen, hydroxyl radicals and superoxide anion, and have been proposed as effective PDT mediators in several applications. When compared to other, more traditional molecular PSs, fullerenes have numerous advantages and disadvantages as PSs for PDT [36]:

- Fullerenes are comparatively photostable, and manifest less photobleaching compared to tetrapyrroles. Due to the chemical structure of porphyrin-type PSs they are often good reactive substrates for the ROS they produce on illumination (particularly singlet oxygen); moreover, the photoproducts formed on oxidation of the PSs have the chromophore structure disrupted and can no longer act as PSs. By contrast, the carbon cage of the fullerene backbone is less reactive to singlet oxygen in a short timescale; consequently, these PSs can carry out many more catalytic cycles, thus dramatically extending the duration of time for which they can respond to illumination.

- Fullerenes show both types of photochemistry, comprising Types I and II. They are particularly effective in the formation of superoxide and oxygen radicals (Type I photochemistry; see below), and also in the formation of singlet oxygen (Type II photochemistry) compared to tetrapyrroles [37].

- Fullerenes can be chemically modified very easily, for tuning the drug's partition coefficient (log P for [drug in *n*-octanol]/[drug in un-ionized H_2O]) and pK_a values for the variation of *in vivo* lipophilicity, and the prediction of their distribution in a biological system.

- To enhance the overall quantum yield and the ROS production yield, light-harvesting antenna can be chemically attached onto C_{60}.

- The molecular self-assembly of fullerene cages into vesicles allows multivalent drug delivery; this allows the production of self-assembled nanoparticles that may have different tissue-targeting properties.

17.3.1
Photophysics of Fullerenes

The absorption spectra of a typical set of mono-substituted, bis-substituted and tris-substituted fullerenes show an almost monotonic decay between 300 and 700 nm; also notable is the difference that each succeeding substitution into the fullerene core makes in reducing the visible absorption, due to the successive removal of each double bond. When C_{60} is irradiated with visible light, it is excited from the S_0 ground state to a short-lived (~1.3 ns) S_1 excited state (E_S 46.1 kcal mol^{-1}).

The S_1 state rapidly decays at a rate of $5.0 \times 10^8\,s^{-1}$ and a *triplet* quantum yield (φ_T) of 1.0 to a lower-lying triplet state T_1 (E_T 37.5 kcal mol^{-1}) with a long lifetime of 50–100 µs (17.1). The $S_1 \rightarrow T_1$ decay is formally a spin-forbidden intersystem crossing (ISC), but is driven by an efficient spin–orbit coupling. In the presence of dissolved molecular oxygen ($^3O_2$), which exists as a triplet in its ground state, the fullerene T_1 state is quenched (as a consequence of the quenching, its lifetime is reduced to ~330 ns) to generate singlet oxygen ($^1O_2^*$) by energy transfer at a rate of $2 \times 10^9\,M^{-1}\,s^{-1}$ (17.2). The singlet oxygen quantum yield, φ_Δ for this process (at 532 nm excitation) has been reported to be near theoretical maximum, 1.0 [38].

17.3.2
Photochemistry of Fullerenes

It is being increasingly realized that, as compared to the standard Type 2 ROS (singlet oxygen), the ROS produced during PDT with fullerenes are biased towards Type 1 photochemical products (superoxide, hydroxyl radical, lipid hydroperoxides, hydrogen peroxide). It is known that both pristine and functionalized C_{60} fullerenes are able to catalyze the formation of ROS after illumination [39]. However, fullerenes dissolved in organic solvents in the presence of oxygen seem to preferentially produce reactive singlet oxygen (17.1) [38]. By contrast, in polar solvents – and especially those containing reducing agents (such as NADH) at concentrations found in cells (17.2) – the illumination of various fullerenes will generate different Type 1 ROS, such as superoxide anion and hydroxyl radical [40]:

$$^1C^{60} + h\nu \rightarrow {}^1C_{60}^* \rightarrow {}^3C_{60}^* + {}^3O_2 \rightarrow {}^1C_{60} + {}^1O_2^*. \tag{17.1}$$

$$2C_{60}^* + NADH \rightarrow 2C_{60}^{-\bullet} + NAD^+ + H^+ \tag{17.2}$$

$$C_{60}^{-\bullet} + O_2 \rightarrow C_{60} + O_2^{-\bullet} \tag{17.3}$$

$$2O_2^{-\bullet} + 2H^+ \rightarrow O_2 + H_2O_2 \tag{17.4}$$

$$H_2O_2 + Fe^{2+} \rightarrow OH^\bullet + OH^- + Fe^{3+} \tag{17.5}$$

$$Fe^{3+} + O_2^{-\bullet} \rightarrow Fe^{2+} + O_2 \tag{17.6}$$

Fullerenes are known to be excellent electron acceptors due their triply degenerate, low-lying lowest unoccupied molecular orbital (LUMO); typically, they are capable of accepting as many as six electrons [41]. There is, however, some evidence that fullerene excited states (in particular the triplet) are even better electron acceptors than the ground state [42, 43]. It is thought that the reduced fullerene triplet or radical anion can transfer an electron to molecular oxygen, forming the superoxide anion radical $O_2^{-\bullet}$. (17.3). By using various scavengers of ROS, physico-chemical methods [e.g., electron paramagnetic resonance (EPR), radical trapping and near-infrared spectrometry] and chemical methods (nitro blue tetrazolium reaction with superoxide), it has been shown that, whereas $^1O_2$ was generated effectively by photoexcited C_{60} in nonpolar solvents such as benzene and benzonitrile, $O_2^{-\bullet}$ and

OH$^\bullet$ were produced only in polar solvents such as water. Hydrogen peroxide is formed by the dismutation of superoxide anion (17.4). Fenton chemistry (using small amounts of ferrous iron found in cells) is able to produce hydroxyl radicals from hydrogen peroxide (17.5), while the Haber–Weiss reaction is able to reduce ferric iron back to ferrous iron in order to continue the cycle (17.6).

Miyata *et al.* observed the formation of $O_2^{-\bullet}$, with visible-light irradiation on polyvinylpyrrolidine (PVP)-solubilized C_{60} in water in the presence of NADH as a reductant and molecular oxygen [44]. The formation of $O_2^{-\bullet}$ was also evidenced by the direct observation of a characteristic signal of $O_2^{-\bullet}$ by the use of a low-temperature EPR technique at 77 K. On the other hand, no formation of $^1O_2$ was observed by the use of 2,2,6,6-tetramethyl-4-piperidone (TEMP) as a $^1O_2$ trapping agent. No near-infrared (NIR) luminescence of $^1O_2$ was observed in the aqueous C_{60}/PVP/O_2 system. These results suggest that the photoinduced bioactivities of the PVP-solubilized fullerene are caused not by $^1O_2$, but by reduced oxygen species such as $O_2^{-\bullet}$, which are generated by the electron-transfer reaction of O_{60}^{-} with molecular oxygen.

Recently, it has been has been shown in some *in vitro* studies that the phototoxicity produced by C_{60} solubilized in gamma-cyclodextrin (γ-CyD-C_{60}) was mediated by singlet oxygen (Type II), with a minor contribution of free radicals (Type I) in HaCaT keratinocytes, while $C_{60}(OH)_{24}$ water-soluble fullerene phototoxicity was mainly due to the superoxide [45]. The role of singlet oxygen in monomeric γ-CyD-C_{60} fullerene-mediated phototoxicity has been also shown in human lens epithelial cells, where the phototoxicity was manifested as apoptosis [46].

It has also been shown [47] that, under biological conditions where mild reducing agents are ubiquitous (and the environment is polar and aqueous), illuminated fullerenes produce superoxide (and possible hydroxyl radical), having switched from a Type II to a Type I mechanism. In this case, electron spin resonance (ESR) spin traps were used for superoxide, and 1270 nm luminescence measurements for singlet oxygen.

One seeming contradiction that arises in this area needs to be addressed. It is well known that fullerenes can act as antioxidants, and that C_{60} and derivatives can act as scavengers of ROS. These antioxidant effects of C_{60} have been studied in the absence of light. Both, C_{60} and polyhydroxylated fullerene derivatives have been recognized an efficient free-radical scavengers, in human neuroblastoma cells. The polyhydroxylated fullerenes have also been shown to act as mitochondrial protective antioxidants, with direct radical-scavenging and indirect antioxidant-inducing activities. C_{60} has been observed to protect the rodent liver against free-radical damage, while fullerenes have also been shown to reduce the excitotoxic and apoptotic death of cultured cortical neurons through a free radical-scavenging mechanism [48–52]. The question then, is how can the demonstrated ability of fullerenes to scavenge ROS and act as antioxidants be reconciled with the demonstrated ability of fullerenes to act as efficient producers of ROS under illumination with the correct light parameters? It was assumed that the double bonds of the fullerene cage reacted with ROS, forming covalent bonds and thereby blocking the ability of the ROS to react with sensitive biomolecules, similar to those damaged during PDT. If this assumption were true, it would be difficult to explain how fullerenes could act as efficient generators of ROS during PDT; likewise, it would be difficult to explain why fullerenes could be considered as particularly photostable photosensitizers compared to most other photosensitizers, based on the tetrapyrrole skeleton, such as porphyrins, chlorins, and phthalocyanines. One clue that might explain this paradox was proposed in 2009, by Andrievsky *et al.* [53], who showed that the main mechanism by which hydrated C_{60} can inactivate the highly reactive ROS, hydroxyl radical, not by covalently scavenging the radicals but rather by action of the coat of "ordered water" that was associated with the fullerene nanoparticle [54]. Andrievsky *et al.* claimed that the ordered water coat could slow down or trap the hydroxyl radicals for a sufficient time for two of the radicals to react with each other, thus producing the less-reactive ROS, hydrogen peroxide.

17.3.3
Interactions of Fullerenes with DNA

One of the first biological applications of photoactivated fullerenes was to effect the cleavage of DNA strands after illumination. The cleavage of supercoiled pBR322 DNA was observed after incubation with a fullerene carboxylic acid under visible light irradiation, but not in the dark [4]. Both, nicked circular and linear duplex forms of DNA were observed, and there was considerable selectivity for cleavage at guanine bases. The photoinduced action was more pronounced in D_2O, in which singlet oxygen has a longer lifetime.

An and coworkers [55] prepared a covalent conjugate between an oligodeoxynucleotide and either a fullerene or eosin (a traditional PS). Cleavage of the target complementary 285-base single-stranded DNA (ssDNA) was observed at guanosine residues in both cases upon illumination, although the fullerene conjugate was more efficient in cleavage than the eosin conjugate. Moreover, cleavage was not quenched by azide or increased by deuterium oxide, as was found for the eosin conjugate, which suggested that the mechanism followed a Type I pathway.

Boutorine *et al.* [56] described a fullerene–oligonucleotide that could bind either ssDNA or double-stranded DNA (dsDNA), and which also cleaved the strand(s) proximal to the fullerene moiety upon exposure to light. Nakanishi *et al.* [57] also observed DNA cleavage by functionalized C_{60}, while Yamakoshi *et al.* investigated the biological activities of fullerenes under illumination, including DNA cleavage, hemolysis, mutagenicity, and cell toxicity [58]. For this, the authors prepared a conjugate between a fullerene and an acridine molecule as a DNA-intercalating agent, and compared its DNA-photocleavage capacity on pBR322 supercoiled plasmid with pristine fullerene, solubilized in PVP. Subsequently, this compound showed a much more effective DNA-cleaving activity in the presence of NADH than did pure C_{60} [59]. Liu and coworkers [60] employed a water-soluble conjugate between anthryl-cyclodextrin and C_{60} to carry out the photocleavage of pGEX5X2 DNA. Ikeda *et al.* [61] used functionalized liposomes incorporating both C_{60} and C_{70} fullerenes into the lipid bilayer to carry out the photocleavage of *ColE1* supercoiled plasmid DNA using $\lambda > 350\,\text{nm}$ light. Interestingly, C_{70} was significantly better (3.5-fold) than C_{60} in photocleaving DNA.

It has been also shown that fullerene-mediated PDT may lead to mutagenic effects. For example, PVP-solubilized fullerene was found to be mutagenic in *Salmonella* strains TA102, TA104 and YG3003 in the presence of rat liver microsomes, when irradiated with visible light [62]. Notably, the mutagenicity was elevated in strain YG3003, a repair enzyme-deficient mutant of TA102, but was reduced in the presence of beta-carotene and *para*-bromophenacyl bromide, a scavenger of singlet oxygen and inhibitor, respectively, of phospholipase. These results suggested that singlet oxygen was generated by irradiating the C_{60} with visible light, and that the mutagenicity was due to the presence of oxidized phospholipids in the rat liver microsomes. The linoleate fraction isolated using high-performance liquid chromatography (HPLC) proved to be a major component and to play an important role in the mutagenic effect. The results of an ESR spectrum analysis suggested the generation of radicals at the guanine base, but not at the thymine, cytosine and adenine bases, and also the formation of 8-hydroxydeoxyguanosine (8-OH-dG). The mechanism was proposed to involve an indirect action of singlet oxygen due to the lipid peroxidation of linoleate that caused oxidative DNA damage.

17.3.4
Drug-Delivery Strategies for Fullerenes

The extreme hydrophobicity of fullerenes, and their inherent tendency to aggregate, show little promise for the application of these materials as biomedical drug-type compounds. Nevertheless, many investigations have been conducted into drug-delivery strategies for other hydrophobic drugs such as anticancer compounds (e.g., taxol), antimicrobial compounds (e.g., amphotericin B), and steroids (e.g., estrogen). Some of these strategies used to solubilize fullerenes are shown schematically in Figure 17.4.

Liposomes are vesicles composed of phospholipid bilayers that contain an aqueous interior. Hydrophobic drugs can be localized in the hydrophobic region within the lipid bilayers of liposomes or, alternatively, hydrophilic drugs can be encapsulated within the aqueous interior. A group from Japan has reported the encapsulation of C_{60} into liposomes consisting of hydrogenated lecithin and glycine soja sterol [63] (Figure 17.4a), and their use as a topical application to prevent ultraviolet (UV)-induced skin damage [64]. When Doi *et al.* [65] prepared water-soluble fullerenes, by employing lipid-membrane-incorporated fullerenes C_{60} and C_{70}, and used a fullerene exchange method from a γ-CyD cavity to vesicles, both fullerenes were shown to demonstrate a PDT-mediated killing of HeLa cells after illumination with $350\,\text{nm}$ light. Later, the same group [66] used crosslinked liposomes, termed "cerasomes," to encapsulate C_{70}, and demonstrated a similar degree of PDT-mediated killing of HeLa cells as found with the lipid membrane-incorporated C_{70} liposomes [67].

Micelles are molecular clusters formed from detergent molecules, with the hydrophobic head group at the center and the hydrophilic tail group at the outside.

(a) liposome
(b) micelle
(c) pegylated fullerene
(d) dendrimer conjugate
(e) cyclodextrin complex
(f) nanoemulsion

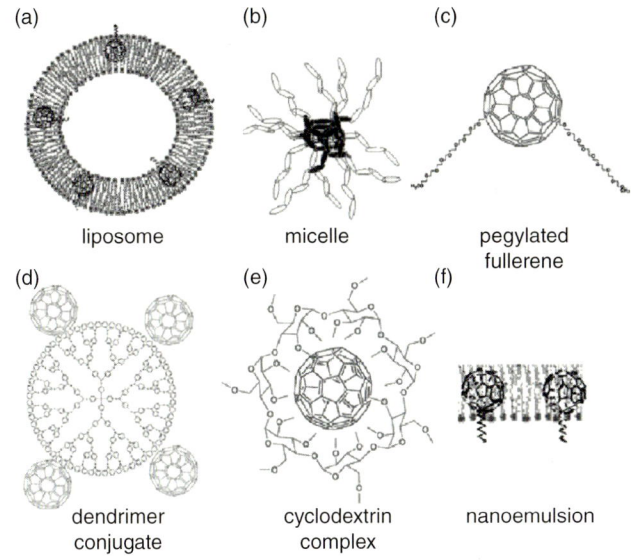

Figure 17.4 Drug-delivery strategies used to solubilize fullerenes. (a) Liposomes; (b) Micelle; (c) Dendrimer conjugate; (d) PEGylated fullerene; (e) Cyclodextrin complex; (f) Nanoemulsion.

Micelles are able to solubilize fullerenes in the hydrophobic central region of the cluster (Figure 17.4b). Yan [68] and coworkers used micelles formed from tocopheryl poly(ethylene glycol) (PEG) succinate to solubilize C_{60}, while Akiyama et al. [69] used micelles composed of a range of PEGylated block copolymers to deliver C_{60} to HeLa cells and carry out PDT.

Dendrimers (sometimes known as "starburst dendrimers") are branched molecules, the size of which depends on the number of "generation" steps used in their preparation. Dendrimers can be used to solubilize fullerenes as both noncovalent complexes and covalent conjugates (Figure 17.4c). Kojima et al. [70] synthesized polyamidoamine dendrimers having both β-CD and PEG to solubilize C_{60} as a noncovalent complex. To date, covalent polyamidoamine dendrimer conjugates have been used only to solubilize multi-walled carbon nanotubes (MWNTs) [71].

PEGylation involves the covalent attachment of PEG polymer chains to another molecule, normally a drug or therapeutic protein. Such a procedure may provide water-solubility; alternatively, it can "mask" the agent from the host's immune system, and/or also increase its hydrodynamic size (in solution), prolonging the agent's circulatory time by reducing its renal clearance. The hydrophilic nature of the PEG chains has led to their use in solubilizing hydrophobic fullerenes, by covalent attachment [72] (Figure 17.4d). With regards to a therapeutic role, PEGylated fullerenes have been used *in vivo* to mitigate arthritis via the downregulation of chondrocyte catabolic activity and the inhibition of cartilage degeneration

[73]. Similarly, studies of *in vivo* PDT using fullerenes have employed PEGylated structures as the PS [74–76].

Cyclodextrins (CDs) (Figure 17.4e) are ring structures formed from oligosaccharides, and are composed of five or more alpha-D-glucopyranoside units linked 1 → 4 (as in amylase, a fragment of starch). Rings with six glucose units are termed α-CDs, seven-membered rings as β-CDs, and eight-membered rings γ-CDs. In the past, CDs have been used to solubilize hydrophobic drugs and, indeed, they can be used to solubilize fullerenes either as noncovalent complexes or as covalent conjugates. Previously, Filippone *et al.* [77] had synthesized a novel 2 : 1 (permethylated-β-CD)–fullerene conjugate with a log *P* value of 1.58, which was in the suitable range for biological studies [78]. One of the group used C_{60} and $C_{60}(OH)_{24}$ fullerenes bi-capped with γ-cyclodextrin to carry out the *in vitro* PDT of HaCaT cells, using illumination with a band-pass filter centered at 366 nm.

Self-nanoemulsifying systems (SNES) are essentially an isotropic blend of oils, surfactants, and/or cosolvents that emulsify spontaneously to produce an oil-in-water nanoemulsion when introduced into an aqueous phase under gentle agitation (Figure 17.4f) [79]. However, whilst nanoemulsions have been shown to be superior drug-delivery vehicles for a number of hydrophobic pharmaceuticals [80–82], they have not yet been tested with fullerenes.

17.3.5
Strategies to Overcome the Unfavorable Spectral Absorption of Fullerenes

Whilst the main absorption of fullerenes occurs in the blue and green regions of the spectrum, the absorption of tetrapyrrole PSs (other than porphyrins, such as chlorins, bacteriochlorins, and phthalocyanines) has been designed to demonstrate substantial peaks in the red or far-red regions. This major disadvantage of fullerenes as PSs can, however, be overcome in several ways:

- The chemical attachment of one or two red-wavelength absorptive antenna on a C_{60} can cause ultrafast intramolecular photoinduced energy-transfer processes. In this case, the antenna absorbs the red light photons at high efficiency, and then passes the absorbed photon energy to the C_{60} cage within 250 ps, thus simulating the photoexcitation of fullerenes by red light.

- The absorption of many light photons simultaneously to supply fullerenes with a sufficiently high accumulative photo energy can be achieved by the attachment of multiple light-harvesting antennae on one C_{60} cage.

- The main application of these PSs may be confined to situations where the target is relatively superficial in nature; examples include bladder or skin cancers.

- Two-photon excitation may be used. Femtosecond-pulsed lasers at twice the wavelength of the photons to be absorbed (e.g., 800 nm) will have a dramatically increased tissue penetration, but will have a small spot size that must be scanned or rastered over the surface of the tissue to be treated.

17.3.5.1 Covalent Attachment of Light-Harvesting Antennae to Fullerenes

As porphyrins and fullerenes are attracted spontaneously to each other, this new supramolecular recognition element can be used to construct discrete host–guest complexes, as well as ordered arrays of interleaved porphyrins and fullerenes. The fullerene–porphyrin interaction underlies the successful chromatographic separation of fullerenes, and promising applications have been identified in the areas of porous framework solids and photovoltaic devices [83]. The structure of a typical fullerene–porphyrin dyad is shown in Figure 17.5a. Several reports have been made on the synthesis and characterization of fullerene–porphyrin dyads for use as artificial photosynthesis mimics [84–87].

Rancan *et al.* [88] used this approach to overcome the need to use UV or short-wavelength visible light to photoactivate fullerenes. In this case, two new fullerene-bis-pyropheophorbide a derivatives were synthesized: a mono-(FP1) and a hexa-adduct (FHP1). Whilst the C_{60}-hexa-adduct FHP1 had a significant phototoxic activity (58% cell death, after a dose of $400\,mJ\,cm^{-2}$ of 688 nm light), the

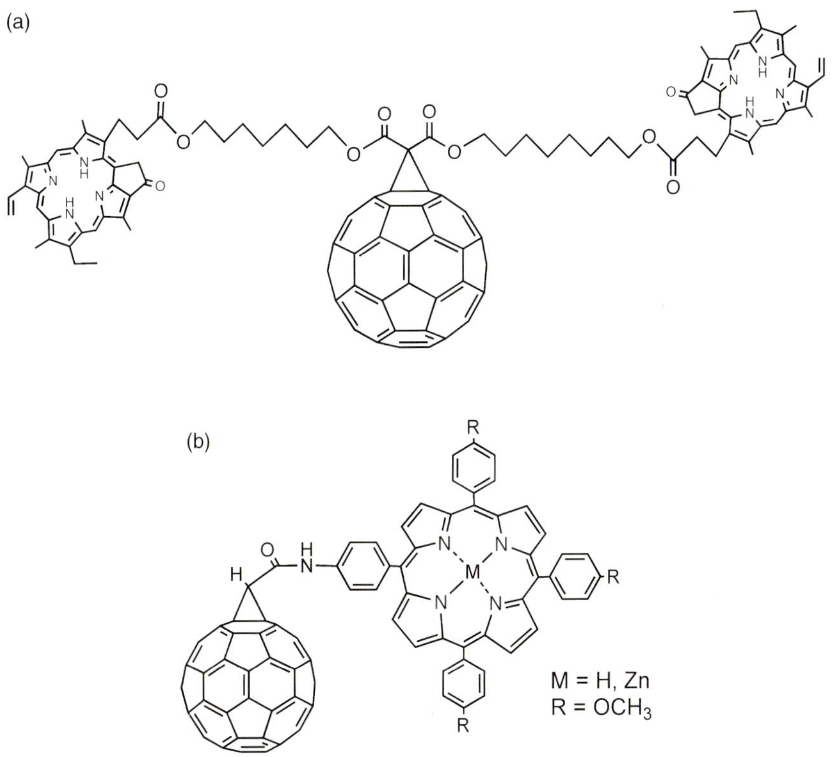

Figure 17.5 (a) Chemical structure of a dyad composed of C60 fullerene attached by linkers to two pheophorbide molecules; (b) Chemical structure of a dyad between C60 fullerene and meso-tetraphenyl porphyrin as a free base or a zinc complex.

M = H, Zn
R = OCH$_3$

mono-adduct FP1 had a very low phototoxicity, and only at higher light doses. Nevertheless, the activity of both adducts was less than that of pure pyropheophorbide a, most likely due to a lower cellular uptake of the adducts.

A group from Argentina has also studied the phototoxicity produced by tetrapyrrole–fullerene conjugates. Milanesio et al. [89] compared PDT with a porphyrin-C_{60} dyad (P-C_{60}) and its metal complex with Zn(II) (ZnP-C_{60}) (Figure 17.5b), and with 5-(4-acetamidophenyl)-10,15,20-tris(4-methoxyphenyl) porphyrin (P), both in an homogeneous medium containing photooxidizable substrates and in vitro on the Hep-2 human larynx carcinoma cell line. The $^1O_2$ yields (Φ_Δ) were determined using 9,10-dimethylanthracene (DMA). The values of Φ_Δ were heavily dependent on the solvent polarity, while cell survival after irradiation with visible light was dependent on the level of light exposure. A greater phototoxic effect was observed for P-C_{60}, which inactivated 80% of the cells after a 15-min period of irradiation. Moreover, both dyads retained a high photoactivity, even under an argon atmosphere.

Zhang et al. prepared multifunctionalized carbon nanotubules for both PDT and a photohyperthermia (PHT) approach to cancer therapy that used a single laser. In this case, zinc phthalocyanine (ZnPc) was loaded onto single-wall carbon nanohorns with holes opened (SWNHox), and a protein (bovine serum albumin; BSA) was attached to the carboxyl groups of SWNHox. Here, the ZnPc acted as the PDT agent, the SWNHox as the PHT agent, and the BSA enhanced biocompatibility. A double phototherapy effect was confirmed both in vitro and in vivo, with the tumors almost disappearing on 670 nm laser irradiation following the injection of ZnPc–SWNHox–BSA into tumors transplanted subcutaneously into mice. By contrast, the tumors continued to grow when ZnPc or SWNHox–BSA were injected alone [90].

In another study, a novel PDT agent was prepared combining a three-part system that incorporated a human α-thrombin-recognizing aptamer, a PS (chlorin e6; Ce6), and a single-walled carbon nanotube (SWNT). The PDT was based on first covalently linking a PS with an aptamer, with the product then being wrapped onto the surface of the SWNTs such that the PS would be activated by light only upon target binding. The SWNTs were found to be efficient quenchers of singlet oxygen generation (SOG). In the presence of its target, thrombin, the aptamer structure was changed dramatically, and this disturbed the DNA interaction with the SWNTs. This, in turn, caused the DNA aptamer to fall from the SWNT surface, removing the PDT quenching and restoring the SOG. Taken together, these results led to a validation of the potential of the design as a novel PDT agent, with regulation by target molecules, and an enhanced specificity and efficacy of therapeutic function, that led to the development of PDT being safer and more selective [91].

17.3.5.2 Two-Photon PDT

Although two-photon absorption (2PA) was originally proposed by Maria Göppert-Mayer in 1931 [92], the first experimental verification was provided by Werner Kaiser in 1961. This latter advance was facilitated by the recent development of the laser, which was required for excitation because of the intrinsically low intensity of 2PA. As 2PA is a third-order nonlinear optical process, it is most efficient at very high intensities. In a centrosymmetric molecule, one-photon- and two-photon-allowed transitions are mutually exclusive. In quantum mechanical terms, this difference results from the need to conserve spin. As photons have a spin of ±1, one-photon absorption requires excitation to involve an electron changing its molecular orbital to one with a spin which is different by ±1. 2PA requires a change of +2, 0, or −2. In nonresonant 2PA, two photons combine to bridge an energy gap which is larger than the energies of each photon individually. If there were an intermediate state in the gap, this could occur via two separate one-photon transitions in a process described as "resonant 2PA," "sequential 2PA," or "1 + 1 absorption." In nonresonant 2PA, the transition occurs in the absence of an intermediate state; this can be viewed as being due to a "virtual" state created by the interaction of the photons with the molecule.

Long-wavelength NIR light ($\lambda > 780$ nm) has relatively low photon energy (>1.5 eV), and is generally too low to activate most PSs by one-photon excitation. Therefore, one-photon absorption (1PA) will fail in the phototherapeutic window 780–950 nm, where tissues have a maximum transparency to light. Sensitization by simultaneous 2PA [93] combines the energy of two identical photons arriving simultaneously at the PS, and can provide the energy of a single photon of half the wavelength, which is sufficient to excite the PS to the first excited singlet state. It was not until the 1990s that rational design principles for the construction of two-photon-absorbing molecules began to be developed, in response to a need from imaging and data storage technologies, and aided by the rapid increases in computer power that allowed quantum calculations to be carried out. The accurate quantum mechanical analysis of two-photon absorbance is orders of magnitude more computationally intensive than that of one-photon absorbance.

The most important features of molecules with a strong 2PA were found to be a long conjugation system (analogous to a large antenna), and substitution by strong donor and acceptor groups (this can be thought of as inducing nonlinearity in the system and increasing the potential for charge-transfer). Therefore, many push–pull olefins exhibit high 2PA transitions, up to several thousand GM (1 GM = 10^{-50} cm^4 s photon^{-1} molecule^{-1}). *In vitro* studies with the 2PA of Photofrin® demonstrated the killing of vascular endothelial cells, but this still required a high pulse energy and long illumination times [94]. It has also been reported that two-photon PDT can be used to selectively close small blood vessels in the chicken chorioallantoic membrane [95, 96]; this is accepted as a model of choroidal neovascularization, as seen in age-related macular degeneration patients. The results of a recent study [97] showed that two-photon PDT, using a specially designed porphyrin construct with a high two-photon cross-section and a femtosecond laser, could destroy a tumor even when the laser beam had passed through the entire body thickness of the mouse to reach the lesion.

Some preliminary studies have also been conducted [98, 99] on the design and synthesis of fullerene derivatives, using an appropriate 2PA that would allow their use in two-photon PDT with NIR light in the 700–800 nm region.

17.4
Anticancer Effects of Fullerenes

17.4.1
In Vitro PDT with Fullerenes

Fullerene-mediated phototoxicity in cancer cells was first demonstrated in 1993, when Tokuyama *et al.* [4] used carboxylic acid-functionalized fullerenes at 6 μM and white light to produce growth inhibition in human HeLa cancer cells. However, the same group later reported that other carboxylic acid derivatives of C_{60} and C_{70} were completely without any photoactivity as PDT agents at 50 μM [100]. Burlaka *et al.* [101] used pristine C_{60} at 10 μM with visible light from a mercury lamp to produce some phototoxicity in Ehrlich carcinoma cells or rat thymocytes, and used EPR spin-trapping techniques to demonstrate the formation of ROS.

The cytotoxic and photocytotoxic effects of two water-soluble fullerene derivatives – a dendritic C_{60} mono-adduct and the malonic acid C_{60} tris-adduct – were tested on Jurkat cells when irradiated with UVA or UVB light [102]. The cell death was mainly caused by membrane damage, and shown to be UV dose-dependent. The tris-malonic acid fullerene was shown to be more phototoxic than the dendritic derivative, a result which contrasted with the singlet oxygen quantum yields determined for the two compounds.

Three C_{60} derivatives with two to four malonic acid groups (DMA C_{60}, TMA C_{60}, and QMA C_{60}) were each prepared and their phototoxicity against HeLa cells was determined using the MTT assay [(3-(4,5-dimethylthiazol-2-yl)-2,5-diphenyltetrazolium bromide-based colorimetric assay, in which the tetrazole is reduced by mitochondria of living cells to a purple formazan] and cell cycle analysis [103]. The relative phototoxicity of these compounds was DMA C_{60} > TMA C_{60} > QMA C_{60}. The hydroxyl radical quencher mannitol (10 mM) was unable to protect the cells against damage induced by the irradiated DMA C_{60}. The latter compound, together with irradiation, was shown to reduce the number of cells in the G_1 phase of the cell cycle, from 63% to 42%, and to increase the total number of cells in the G_2 and mitotic phases of the cell cycle, from 6% to 26%.

Ikeda and coworkers [67] employed a series of liposomal preparations of C_{60} containing cationic or anionic lipids together. Illumination with 136 J cm^{-2}, 350–500 nm produced 85% cell killing in the case of cationic liposomes, and apoptosis was also demonstrated.

Akiyama *et al.* solubilized an unmodified C_{60} of high stability using various typed of PEG-based block copolymer micelles. As a result, the cationic block copolymer micelles delivered C_{60} into human cervical cancer HeLa cells, depending on their surface densities, and also showed cytotoxic effects under photoirradiation [69].

In another study, the direct and short-term uptake (within 10 min) of fullerene into the cell membrane, using an exchange reaction from a fullerene–γ-CD complex, was monitored, and the resultant photodynamic activity for cancer cells demonstrated [104]. Doi *et al.* showed the PDT activity of cerasome-encapsulated C_{70} in HeLa cells to be similar to that for a surface-cross-linked liposome C_{70}; this

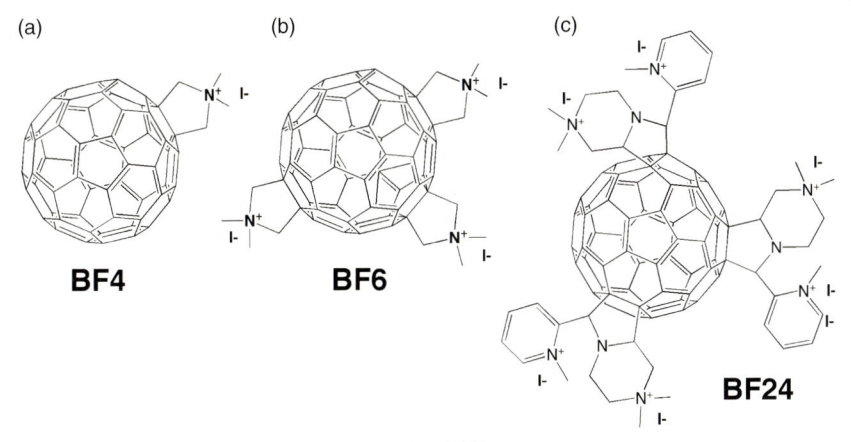

Figure 17.6 Chemical structures of functionalized fullerenes with constitutive cationic charges provided by quaternary ammonium groups. (a) BF4; (b) BF6; (c) BF24.

indicated that C_{70} could serve as a PS, without its release from the cerasome membranes [66].

More recently, the present authors have tested the hypothesis that fullerenes would be capable of killing cancer cells by PDT *in vitro*. In this case, a panel of six functionalized fullerenes was prepared in two groups, each of three compounds. The first series was constructed with one, two or three polar di-serinol groups, while the second series had one, two, or three quaternary dimethylpyrrolidinium groups. Subsequently, the C_{60} molecule mono-substituted with a single pyrrolidinium group (BF4) (Figure 17.6a) was shown to be a remarkably efficient PS that could mediate the killing of a panel of mouse cancer cells at low concentration (2 µM), and with only modest exposure to white light. Moreover, by measuring the increase in fluorescence with an intracellular probe capable of monitoring the formation of ROS, it was shown for the first time that photoactive fullerenes were indeed taken up into the cells.

The induction of apoptosis by fullerene–PDT in cancer cells at 4–6 h after illumination was also demonstrated. This was not a surprising finding, as many reports have been made of apoptosis occurring after *in vitro* PDT with conventional PSs, such as Photofrin [105], benzoporphyrin derivatives [106], and the phthalocyanine, Pc4 [107]. The relatively rapid induction of apoptosis after illumination might suggest that the fullerenes are localized in subcellular organelles such as mitochondria, as shown previously for the afore-mentioned PS. In fact, PSs that localize in lysosomes tend to cause apoptosis more slowly than do mitochondrial PS, due mainly to the release of lysosomal enzymes that subsequently activate the cytoplasmic caspases [108].

Among the six candidates, the mono-pyrrolidinium-substituted fullerene (BF4) was the most effective PS by a considerable margin. The reason for this was most likely linked to its relative hydrophobicity, as shown by a log P value >2. It has

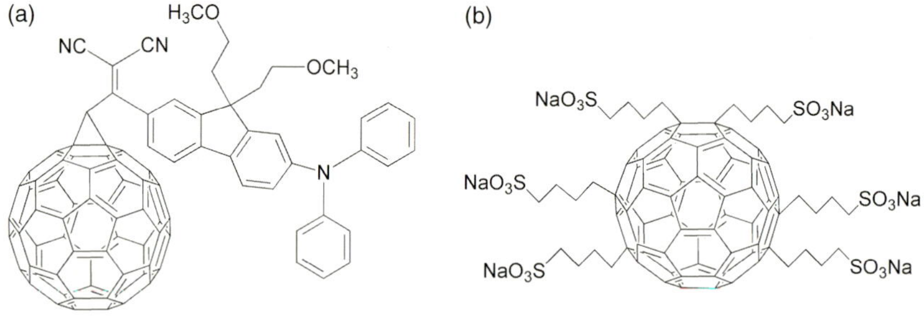

C60(>CPAF-OMe)

FC$_4$S

Figure 17.7 Chemical structures of functionalized fullerenes. (a) C60 (>CPAF-OMe); (b) FC$_4$S.

been established, in structure–function relationship studies, that the more hydrophobic a PS is (albeit up to a certain limit, when insolubility and aggregation become problematic), the more effective it is for cell killing [109–111]. The reason behind this relationship is thought to be a combination of a higher cellular uptake and a greater localization in the intracellular membrane organelles, such as mitochondria. The single cationic charge possessed by this compound is also likely to play an important role in determining its relative phototoxicity. Many lipophilic monocations have been shown to localize fairly specifically in mitochondria [112, 113] and, indeed, this property has been proposed as a strategy for targeting drugs to the mitochondria [114].

Chiang et al. [115] recently reported the synthesis of two new photoresponsive diphenylaminofluorene nanostructures, and investigated their intramolecular photoinduced energy and electron-transfer phenomena. The structural modification was made by a chemical conversion of the keto group in C_{60}(>DPAF–OMe) to a stronger electron-withdrawing 1,1-dicyanoethylenyl unit, leading to C_{60}(>CPAF–OMe) (Figure 17.7a); this resulted in an increased electronic polarization of the molecule. The modification also led to a large red shift of the major absorption band in the visible spectrum up to 600 nm. Remarkably, the $^1O_2$ quantum yield of C_{60}(>CPAF–OMe) was found to be almost sixfold higher than that of C_{60}(>DPAF–OMe), showing that the large light-harvesting enhancement of the CPAF–OMe moiety would lead to a more efficient triplet state generation of the C_{60}> cage moiety. Notably, C_{60}(>CPAF–OMe) was significantly better than C_{60}(>DPAF–OMe) for the light-mediated killing of human cancer cells.

17.4.2
In Vivo PDT with Fullerenes

The first report of using fullerenes for the PDT of actual tumors was made by Tabata, in 1997 [76]. In this case, the water-insoluble C_{60} was chemically modified with PEG, not only to render it water-soluble but also to increase its molecular size. When injected intravenously into mice carrying a subcutaneous tumor on the back, the C_{60}–PEG conjugate exhibited a greater accumulation and a more prolonged retention in tumor tissues than in normal tissues, and was excreted without showing accumulation in any specific organ. Following intravenous injection of the C_{60}–PEG conjugate or Photofrin (a recognized PS) to tumor-bearing mice, coupled with exposure of the tumor site to visible light, the volume increase of the tumor mass was suppressed, with the C_{60} conjugate exhibiting a stronger suppressive effect than Photofrin. A subsequent histological examination revealed that the conjugate injection plus light irradiation had strongly induced tumor necrosis, but had caused no damage to the overlying normal skin. Moreover, the conjugate's antitumor effect was improved in line with the increasing light fluence delivered and the C_{60} dose, with cure achieved at a conjugate dose level of 424 µg kg^{-1} and a fluence of 107 J cm^{-2}.

Liu and others [75] conjugated PEG to C_{60} (C_{60}–PEG), after which diethylenetriaminepenta-acetic acid (D2PA) was introduced to the terminal group of PEG. The C_{60}–PEG–D2PA prepared was subsequently mixed with a gadolinium acetate solution to obtain the Gd^{3+}-chelated compound, C_{60}–PEG–D2PA–Gd. Following the intravenous injection of C_{60}–PEG–D2PA–Gd into tumor-bearing mice, the antitumor effect of the PDT was evaluated, and magnetic resonance imaging (MRI) of the tumor conducted. A similar generation of superoxide upon illumination was observed with or without Gd^{3+} chelation. Intravenous injection of C_{60}–PEG–D2PA–Gd into tumor-bearing mice, plus exposure to light (400–500 nm, 53.5 J cm^{-2}), resulted in a significant anti-tumor PDT effect which was seen to depend on the timing of the light irradiation, and in turn correlated with an accumulation of the compound in the tumor, as indicated by an enhanced intensity of the MRI signal.

Chiang and coworkers reported [116] a preliminary *in vivo* study PDT using hydrophilic nanospheres formed from hexa(sulfo-*n*-butyl)-C_{60} (FC$_4$S) (Figure 17.7b) against mouse tumors. In this case, ICR mice bearing sarcoma 180 subcutaneous tumors received either an intraperitoneal or intravenous injection (5 mg kg^{-1} body weight) of water-soluble FC$_4$S dissolved in phosphate-buffered saline (PBS). The tumor site was subsequently irradiated with an argon ion laser beam at a wavelength of 515 nm, or an argon-pumped dye-laser at 633 nm with the beam focused to a diameter of 7–8 mm; the total light dose was adjusted to 100 J cm^{-2} in each experiment. Consistently, the inhibition of tumor growth was more effective using the low-wavelength, better-absorbed 515 nm laser than the 633 nm laser. Notably, the intraperitoneal administration of FC$_4$S to mice led to a slightly better inhibition of tumor growth than did the intravenous injection.

The first fullerene-based clinical treatment of a human patient with rectal adenocarcinoma was attempted by Andrievsky *et al.* in 2000 [117].

17.5
Fullerenes for Antimicrobial Photoinactivation

17.5.1
Photoinactivation of Viruses

Photodynamic reactions induced by photoactivated fullerenes have been shown to inactivate enveloped viruses [3]. Buffered solutions containing pristine C_{60} and either Semliki Forest virus (Togaviridae) or vesicular stomatitis virus (VSV) (Rhabdoviridae), when illuminated with visible light for up to 5 h, resulted in an up to seven-log loss of infectivity. The viral inactivation was oxygen-dependent and equally efficient in solutions containing protein [3]. Hirayama et al. [118] used a methoxy-PEG-conjugated fullerene at 400 µM in combination with 120 J cm^{-2} white light to destroy more than five logs of plaque-forming units of VSV. The inactivation of VSV was inhibited by oxygen removal, or by the addition of sodium azide (a known singlet oxygen quencher). The substitution of H_2O by D_2O, which is known to prolong the lifetime of singlet oxygen, promoted the virucidal activity. These results indicated that singlet oxygen may play a major role in VSV photoinactivation by the water-soluble fullerene derivative. Notably, the concentration required for virus inactivation was greater than that of other sensitizers, such as methylene blue.

Lin and coworkers [119] compared the light-dependent and light-independent inactivation of dengue-2 and other enveloped viruses by the two regioisomers of carboxyfullerene, and found the asymmetric isomer to have a greater dark activity (albeit at much higher concentrations than were needed for its PDT effect); this effect was considered due to the interaction of carboxyfullerene with the lipid envelope of the virus.

Lee et al. synthesized four novel hexakis C_{60} derivatives with varying functionalities, and evaluated (quantitatively) their photochemical properties and photodynamic disinfection efficiencies. The cationic aminofullerene hexakis, which most likely exerted an electrostatic attraction, exhibited an exceptionally rapid virus inactivation, even when compared to a commercial nano-TiO_2 photo catalyst [120].

17.5.2
Photoinactivation of Bacteria and Other Pathogens

The effectiveness of the various PSs proposed for antimicrobial PDT can be judged on several criteria. Typically, these PSs should be able to kill multiple classes of microbes at relatively low concentrations, and at low fluences of light. The PS should also be reasonably nontoxic in the dark, and should demonstrate selectivity for microbial cells over mammalian cells. The PS should, ideally, have a large extinction coefficient in the red part of the spectrum, and demonstrate high triplet and singlet oxygen quantum yields.

It has been shown, in a series of reported experiments, that cationic fullerenes fulfill many – but not all – of the aforementioned criteria. The present authors' group was the first to show that the soluble functionalized fullerenes described above – especially the cationic compound BF6 (see Figure 17.6b) – were efficient antimicrobial PSs and were capable of mediating the photodynamic inactivation (PDI) of various classes of microbial cells [2]. In this case, a broadband light source equipped with a band-pass filter was used; this provided an output of the entire visible spectrum (400–700 nm) to excite the fullerenes that maximized the absorption.

In a recent study by Spesia et al. [121], a novel N,N-dimethyl-2-(4'-N,N,N-trimethylaminophenyl)fulleropyrrolidinium iodide (DT-C_{60} (2+)) was synthesized by 1,3-dipolar cycloaddition, using 4-(N,N-dimethylamino) benzaldehyde, N-methylglycine and fullerene C_{60} and quaternization with methyl iodide. The PDI produced by these fullerene derivatives was investigated *in vitro* in *Escherichia coli*. The photosensitized inactivation of *E. coli* cellular suspensions by DT-C_{60} (2+) exhibited a approximately 3.5 log decrease in cell survival when the cultures were treated with 1 µM of sensitizer and irradiated for 30 min. This photosensitized inactivation remained high, even after one washing step. The photodynamic activity was also confirmed by a growth delay of the *E. coli* cultures. Growth was arrested when *E. coli* was exposed to 2 µM of cationic fullerene and irradiated, whereas a negligible effect was found for the non-charged MA-C_{60}.

Lee et al. synthesized four novel hexakis C_{60} derivatives with varying functionalities, and their photochemical properties and photodynamic disinfection efficiencies were quantitatively evaluated. The C_{60} derivatives were found to efficiently inactivate both *E. coli* and MS-2 bacteriophage [120].

Recently, the use has been demonstrated of innovative cationic fullerenes as broad-spectrum, light-activated antimicrobials. Following the synthesis and characterization of several new cationic fullerenes, their relative efficacies as broad-spectrum antimicrobial PSs against Gram-positive and Gram-negative bacteria, and also a fungal yeast, was determined by determining the quantitative structure–function relationships [122]. The most effective compound overall against the various classes of microbial cells had the hexacationic structure shown as BF24 in Figure 17.6c.

17.6
Summary and Future Perspectives

The combination of fullerene molecules illuminated with harmless visible light that can be absorbed leads to ROS generation that can be used to efficiently inactivate viruses and kill both cancer cells and pathogenic microbial cells. Today, the rapid growth of research efforts in nanotechnology, and more particularly in the biomedical applications of nanostructures, shows no sign of coming to a halt. Moreover, as the biomedical applications of PDT mediated by traditional PS also continue to grow, it seems certain that an increasing number of studies describing PDT mediated by various formulations of pristine or derivatized fullerenes will be conducted. Although fullerenes have certain disadvantages in terms of their optical absorption spectrum, rational strategies to overcome these deficiencies are currently in existence and will surely be further developed in the future. The preponderance of Type 1 photochemical mechanisms reported for fullerene PDT distinguishes this approach from PDT mediated by more traditional PSs, which generate singlet oxygen via a Type 2 mechanism.

Acknowledgments

The present authors' studies were supported by US NIH grants R01CA/AI838801 and R01AI050875 to M.R.H.; by grants R44AI068400 and R44CA103177 to T.W.; and by grants R01CA137108 to L.Y.C.

References

1 Kroto, H.W. (1985) C60: buckminsterfullerene. *Nature*, **318**, 162–3.
2 Tegos, G.P. *et al.* (2005) Cationic fullerenes are effective and selective antimicrobial photosensitizers. *Chemistry and Biology*, **12** (10), 1127–35.
3 Kasermann, F. and Kempf, C. (1997) Photodynamic inactivation of enveloped viruses by buckminsterfullerene. *Antiviral Research*, **34** (1), 65–70.
4 Tokuyama, H., Yamago, S. and Nakamura, E. (1993) Photoinduced biochemical activity of fullerene carboxylic acid. *Journal of the American Chemical Society*, **115**, 7918–19.
5 Dougherty, T.J. (2002) An update on photodynamic therapy applications. *Journal of Clinical Laser Medicine and Surgery*, **20** (1), 3–7.
6 Dolmans, D.E., Fukumura, D. and Jain, R.K. (2003) Photodynamic therapy for cancer. *Nature Reviews, Cancer*, **3** (5), 380–7.
7 Castano, A.P., Demidova, T.N. and Hamblin, M.R. (2004) Mechanisms in photodynamic therapy: part one–photosensitizers, photochemistry and cellular localization. *Photodiagnosis and Photodynamic Therapy*, **1** (4), 279–93.
8 Castano, A.P., Demidova, T.N. and Hamblin, M.R. (2005) Mechanisms in photodynamic therapy: part two -cellular signalling, cell metabolism and modes of cell death. *Photodiagnosis and Photodynamic Therapy*, **2** (1), 1–23.
9 Castano, A.P., Demidova, T.N. and Hamblin, M.R. (2005) Mechanisms in photodynamic therapy: part three -photosensitizer pharmacokinetics, biodistribution, tumor localization and modes of tumor destruction. *Photodiagnosis and Photodynamic Therapy*, **2** (1), 91–106.
10 Henderson, B.W. and Dougherty, T.J. (1992) How does photodynamic therapy work? *Photochemistry and Photobiology*, **55** (1), 145–57.
11 Allison, R.R. *et al.* (2004) Photosensitizers in clinical PDT. *Photodiagnosis and Photodynamic Therapy*, **1**, 27–42.
12 Detty, M.R., Gibson, S.L. and Wagner, S.J. (2004) Current clinical and preclinical photosensitizers for use in photodynamic therapy. *Journal of Medicinal Chemistry*, **47** (16), 3897–915.
13 Szeimies, R.M. *et al.* (1996) 9-Acetoxy-2,7,12,17-tetrakis-(beta-methoxyethyl)-porphycene (ATMPn), a novel photosensitizer for photodynamic therapy: uptake kinetics and intracellular localization. *Journal of Photochemistry and Photobiology B*, **34** (1), 67–72.
14 Kral, V. *et al.* (2002) Synthesis and biolocalization of water-soluble sapphyrins. *Journal of Medicinal Chemistry*, **45** (5), 1073–8.
15 Agostinis, P. *et al.* (2002) Hypericin in cancer treatment: more light on the way. *International Journal of Biochemistry and Cell Biology*, **34** (3), 221–41.
16 Stockert, J.C. *et al.* (1996) Photodynamic damage to HeLa cell microtubules induced by thiazine dyes. *Cancer Chemotherapy and Pharmacology*, **39** (1–2), 167–9.
17 Bottiroli, G. *et al.* (1997) Enzyme-assisted cell photosensitization: a proposal for an efficient approach to tumor therapy and diagnosis. The rose bengal fluorogenic substrate. *Photochemistry and Photobiology*, **66** (3), 374–83.
18 Hariharan, P.V., Courtney, J. and Eleczko, S. (1980) Production of hydroxyl radicals in cell systems exposed to haematoporphyrin and red light.

International Journal of Radiation Biology and Related Studies in Physics, Chemistry, and Medicine, **37** (6), 691–4.

19 Grune, T. *et al.* (2001) Protein oxidation and proteolysis by the nonradical oxidants singlet oxygen or peroxynitrite. *Journal of Free Radicals in Biology and Medicine*, **30** (11), 1243–53.

20 Midden, W.R. and Dahl, T.A. (1992) Biological inactivation by singlet oxygen: distinguishing $O_2(1$ delta g) and $O_2(1$ sigma g+). *Biochimica et Biophysica Acta*, **1117** (2), 216–22.

21 Bachowski, G.J., Pintar, T.J. and Girotti, A.W. (1991) Photosensitized lipid peroxidation and enzyme inactivation by membrane-bound merocyanine 540: reaction mechanisms in the absence and presence of ascorbate. *Photochemistry and Photobiology*, **53** (4), 481–91.

22 Bachowski, G.J., Korytowski, W. and Girotti, A.W. (1994) Characterization of lipid hydroperoxides generated by photodynamic treatment of leukemia cells. *Lipids*, **29** (7), 449–59.

23 Girotti, A.W. (1985) Mechanisms of lipid peroxidation. *Journal of Free Radicals in Biology and Medicine*, **1** (2), 87–95.

24 Girotti, A.W. (1983) Mechanisms of photosensitization. *Photochemistry and Photobiology*, **38** (6), 745–51.

25 Moan, J. and Berg, K. (1991) The photodegradation of porphyrins in cells can be used to estimate the lifetime of singlet oxygen. *Photochemistry and Photobiology*, **53** (4), 549–53.

26 Reed, M.W. *et al.* (1997) Synthesis of 125I-labeled oligonucleotides from tributylstannylbenzamide conjugates. *Bioconjugate Chemistry*, **8** (2), 238–43.

27 Buytaert, E., Dewaele, M. and Agostinis, P. (2007) Molecular effectors of multiple cell death pathways initiated by photodynamic therapy. *Biochimica et Biophysica Acta*, **1776** (1), 86–107.

28 Abels, C. (2004) Targeting of the vascular system of solid tumours by photodynamic therapy (PDT). *Photochemistry and Photobiological Sciences*, **3** (8), 765–71.

29 Castano, A.P., Mroz, P. and Hamblin, M.R. (2006) Photodynamic therapy and anti-tumour immunity. *Nature Reviews, Cancer*, **6** (7), 535–45.

30 Hamblin, M.R. and Hasan, T. (2004) Photodynamic therapy: a new antimicrobial approach to infectious disease? *Photochemistry and Photobiological Sciences*, **3** (5), 436–50.

31 Jori, G. *et al.* (2006) Photodynamic therapy in the treatment of microbial infections: Basic principles and perspective applications. *Lasers in Surgery and Medicine*, **38** (5), 468–81.

32 Nitzan, Y. *et al.* (1992) Inactivation of gram-negative bacteria by photosensitized porphyrins. *Photochemistry and Photobiology*, **55** (1), 89–96.

33 Hotze, E.M. *et al.* (2008) Mechanisms of photochemistry and reactive oxygen production by fullerene suspensions in water. *Environmental Science and Technology*, **42** (11), 4175–80.

34 Duncan, L.K., Jinschek, J.R. and Vikesland, P.J. (2008) C60 colloid formation in aqueous systems: effects of preparation method on size, structure, and surface charge. *Environmental Science and Technology*, **42** (1), 173–8.

35 Culotta, L. and Koshland, D.E., Jr (1991) Buckyballs: wide open playing field for chemists. *Science*, **254** (5039), 1706–9.

36 Mroz, P. *et al.* (2007) Photodynamic therapy with fullerenes. *Photochemistry and Photobiological Sciences*, **6** (11), 1139–49.

37 Martin, J.P. and Logsdon, N. (1987) Oxygen radicals are generated by dye-mediated intracellular photooxidations: a role for superoxide in photodynamic effects. *Archives of Biochemistry and Biophysics*, **256** (1), 39–49.

38 Arbogast, J.W. *et al.* (1991) Photophysical properties of C60. *The Journal of Physical Chemistry. A: Molecules, Spectroscopy, Kinetics, Environment and General Theory*, **95** (1), 11–12.

39 Foote, C.S. (1994) Photophysical and photochemical properties of fullerenes. *Topics in Current Chemistry*, **169**, 347–63.

40 Yamakoshi, Y. *et al.* (2003) Active oxygen species generated from photoexcited fullerene (C60) as potential medicines: $O_2^{-\ast}$ versus $^1O_2$. *Journal of the American Chemical Society*, **125** (42), 12803–9.

41 Koeppe, R. and Sariciftci, N.S. (2006) Photoinduced charge and energy transfer involving fullerene derivatives. *Photochemical & Photobiological Sciences*, **5** (12), 1122–31.

42 Guldi, D.M. and Prato, M. (2000) Excited-state properties of C(60) fullerene derivatives. *Accounts of Chemical Research*, **33** (10), 695–703.

43 Arbogast, J.W., Foote, C.S. and Kao, M. (1992) Electron-transfer to triplet C-60. *Journal of the American Chemical Society*, **114** (6), 2277–9.

44 Miyata, N., Yamakoshi, Y. and Nakanishi, I. (2000) Reactive species responsible for biological actions of photoexcited fullerenes. *Journal of the Pharmaceutical Society of Japan*, **120** (10), 1007–16.

45 Zhao, B. et al. (2008) Photo-induced reactive oxygen species generation by different water-soluble fullerenes (C) and their cytotoxicity in human keratinocytes. *Photochemistry and Photobiology*, **84** (5), 1215–23.

46 Zhao, B. et al. (2009) Difference in phototoxicity of cyclodextrin complexed fullerene [(gamma-CyD)2/C60] and its aggregated derivatives toward human lens epithelial cells. *Chemical Research in Toxicology*, **22** (4), 660–7.

47 Mroz, P. et al. (2007) Functionalized fullerenes mediate photodynamic killing of cancer cells: Type I versus Type II photochemical mechanism. *Journal of Free Radicals in Biology and Medicine*, **43** (5), 711–19.

48 Lens, M., Medenica, L. and Citernesi, U. (2008) Antioxidative capacity of C(60) (buckminsterfullerene) and newly synthesized fulleropyrrolidine derivatives encapsulated in liposomes. *Biotechnology and Applied Biochemistry*, **51** (Pt 3), 135–40.

49 Spohn, P. et al. (2009) C60 fullerene: a powerful antioxidant or a damaging agent? The importance of an in-depth material characterization prior to toxicity assays. *Environmental Pollution*, **157** (4), 1134–9.

50 Cai, X. et al. (2008) Polyhydroxylated fullerene derivative C(60)(OH)(24) prevents mitochondrial dysfunction and oxidative damage in an MPP(+)-induced cellular model of Parkinson's disease. *Journal of Neuroscience Research*, **86** (16), 3622–34.

51 Gharbi, N. et al. (2005) [60]fullerene is a powerful antioxidant in vivo with no acute or subacute toxicity. *Nano Letters*, **5** (12), 2578–85.

52 Dugan, L.L. et al. (1996) Buckminsterfullerenol free radical scavengers reduce excitotoxic and apoptotic death of cultured cortical neurons. *Neurobiology of Disease*, **3** (2), 129–35.

53 Andrievsky, G.V. et al. (2009) Peculiarities of the antioxidant and radioprotective effects of hydrated C60 fullerene nanostuctures *in vitro* and *in vivo*. *Journal of Free Radicals in Biology and Medicine*, **47** (6), 786–93.

54 Weiss, D.R., Raschke, T.M. and Levitt, M. (2008) How hydrophobic buckminsterfullerene affects surrounding water structure. *Journal of Physical Chemistry B*, **112** (10), 2981–90.

55 An, Y.-Z. et al. (1996) Sequence-specific modification of guanosine in DNA by a C60-linked deoxyoligonucleotide: evidence for a non-singlet oxygen mechanism. *Tetrahedron*, **52**, 5179–89.

56 Boutorine, A.S.T., Takasugi, H., Isobe, M., Nakamura, H., Helene, E. and Angew, C. (1994) Fullerene-oligonucleotide conjugates: photo-induced sequence-specific DNA cleavage. *Angewandte Chemie International Edition in English*, **33** (23/24), 2462–5.

57 Nakanishi, I. et al. (2001) DNA cleavage via electron transfer from NADH to molecular oxygen photosensitized by g-cyclodextrin-bicapped C60, in *Fullerenes for the New Millennium* (ed. P.V. Kamat, D.M. Guldi and D.M. Kadish), The Electrochemical Society, pp. 138–51.

58 Yamakoshi, Y., Sueyoshi, S. and Miyata, N. (1999) [Biological activity of photoexcited fullerene]. *Kokuritsu Iyakuhin Shokuhin Eisei Kenkyusho Hokoku* (117), 50–60.

59 Yamakoshi, Y.N. et al. (1996) Acridine adduct of [60]fullerene with enhanced DNA-cleaving activity. *Journal of Organic Chemistry*, **61** (21), 7236–7.

60 Liu, Y. et al. (2005) Water-soluble β-cyclodextrin derivative possessing a

fullerene tether as an efficient photodriven DNA-cleavage reagent. *Tetrahedron Letters*, **46** (14), 2507–11.

61 Ikeda, A. *et al.* (2007) An extremely effective DNA photocleavage utilizing functionalized liposomes with a fullerene-enriched lipid bilayer. *Journal of the American Chemical Society*, **129** (14), 4140–1.

62 Sera, N., Tokiwa, H. and Miyata, N. (1996) Mutagenicity of the fullerene C60-generated singlet oxygen dependent formation of lipid peroxides. *Carcinogenesis*, **17** (10), 2163–9.

63 Kato, S. *et al.* (2009) Biological safety of liposome-fullerene consisting of hydrogenated lecithin, glycine soja sterols, and fullerene-C60 upon photocytotoxicity and bacterial reverse mutagenicity. *Toxicology and Industrial Health*, **25** (3), 197–203.

64 Kato, S. *et al.* (2009) Defensive effects of fullerene-C60/liposome complex against UVA-induced intracellular reactive oxygen species generation and cell death in human skin keratinocytes HaCaT, associated with intracellular uptake and extracellular excretion of fullerene-C60. *Journal of Photochemistry and Photobiology B*, **98** (2), 144–51.

65 Doi, Y. *et al.* (2008) Intracellular uptake and photodynamic activity of water-soluble [60]- and [70]fullerenes incorporated in liposomes. *Chemistry*, **14** (29), 8892–7.

66 Ikeda, A. *et al.* (2009) Photodynamic activity of C70 caged within surface-cross-linked liposomes. *Chemistry–An Asian Journal*, **4** (1), 199–205.

67 Ikeda, A. *et al.* (2007) Induction of cell death by photodynamic therapy with water-soluble lipid-membrane-incorporated [60]fullerene. *Organic and Biomolecular Chemistry*, **5** (8), 1158–60.

68 Yan, A. *et al.* (2007) Tocopheryl polyethylene glycol succinate as a safe, antioxidant surfactant for processing carbon nanotubes and fullerenes. *Carbon N Y*, **45** (13), 2463–70.

69 Akiyama, M. *et al.* (2008) Solubilisation of [60]fullerenes using block copolymers and evaluation of their photodynamic activities. *Organic and Biomolecular Chemistry*, **6** (6), 1015–19.

70 Kojima, C. *et al.* (2008) Aqueous solubilization of fullerenes using poly(amidoamine) dendrimers bearing cyclodextrin and poly(ethylene glycol). *Bioconjugate Chemistry*, **19** (11), 2280–4.

71 Pan, B. *et al.* (2009) Synthesis and characterization of polyamidoamine dendrimer-coated multi-walled carbon nanotubes and their application in gene delivery systems. *Nanotechnology*, **20** (12), 125101.

72 Hooper, J.B., Bedrov, D. and Smith, G.D. (2008) Supramolecular self-organization in PEO-modified C60 fullerene/water solutions: influence of polymer molecular weight and nanoparticle concentration. *Langmuir*, **24** (9), 4550–7.

73 Yudoh, K. *et al.* (2007) Water-soluble C60 fullerene prevents degeneration of articular cartilage in osteoarthritis via down-regulation of chondrocyte catabolic activity and inhibition of cartilage degeneration during disease development. *Arthritis and Rheumatism*, **56** (10), 3307–18.

74 Nitta, N. *et al.* (2008) Is the use of fullerene in photodynamic therapy effective for atherosclerosis? *Cardiovascular and Interventional Radiology*, **31** (2), 359–66.

75 Liu, J. *et al.* (2007) Preparation of PEG-conjugated fullerene containing Gd3+ ions for photodynamic therapy. *Journal of Controlled Release*, **117** (1), 104–10.

76 Tabata, Y., Murakami, Y. and Ikada, Y. (1997) Photodynamic effect of polyethylene glycol-modified fullerene on tumor. *Japanese Journal of Cancer Research*, **88** (11), 1108–16.

77 Filippone, S., Heimann, F. and Rassat, A. (2002) A highly water-soluble 2:1 beta-cyclodextrin-fullerene conjugate. *Chemical Communications (Cambridge, England)*, (14), 1508–9.

78 Zhao, B. *et al.* (2008) Pristine (C60) and hydroxylated [C60(OH)24] fullerene phototoxicity towards HaCaT keratinocytes: type I vs type II mechanisms. *Chemical Research in Toxicology*, **21** (5), 1056–63.

79 Bansal, T. *et al.* (2008) Solid self-nanoemulsifying delivery systems as a

80 Bali, V., Ali, M. and Ali, J. (2010) Novel nanoemulsion for minimizing variations in bioavailability of ezetimibe. *Journal of Drug Targeting*, **18** (7), 506–19.

81 Amani, A. *et al.* (2009) Factors affecting the stability of nanoemulsions – use of artificial neural networks. *Pharmaceutical Research*, **27** (1), 37–45.

82 Shakeel, F. and Faisal, M.S. (2010) Nanoemulsion: a promising tool for solubility and dissolution enhancement of celecoxib. *Pharmaceutical Development and Technology*, **15** (1), 53–6.

83 Boyd, P.D. and Reed, C.A. (2005) Fullerene-porphyrin constructs. *Accounts of Chemical Research*, **38** (4), 235–42.

84 El-Khouly, M.E. *et al.* (2005) Spectral, electrochemical, and photophysical studies of a magnesium porphyrin-fullerene dyad. *Physical Chemistry Chemical Physics*, **7** (17), 3163–71.

85 Imahori, H. (2004) Porphyrin-fullerene linked systems as artificial photosynthetic mimics. *Organic and Biomolecular Chemistry*, **2** (10), 1425–33.

86 Schuster, D.I. *et al.* (2004) Design, synthesis, and photophysical studies of a porphyrin-fullerene dyad with parachute topology; charge recombination in the marcus inverted region. *Journal of the American Chemical Society*, **126** (23), 7257–70.

87 Vail, S.A. *et al.* (2006) Energy and electron transfer in beta-alkynyl-linked porphyrin-[60]fullerene dyads. *Journal of Physical Chemistry B*, **110** (29), 14155–66.

88 Rancan, F. *et al.* (2005) Fullerene-pyropheophorbide a complexes as sensitizer for photodynamic therapy: uptake and photo-induced cytotoxicity on Jurkat cells. *Journal of Photochemistry and Photobiology B*, **80** (1), 1–7.

89 Milanesio, M.E. *et al.* (2005) Porphyrin-fullerene C60 dyads with high ability to form photoinduced charge-separated state as novel sensitizers for photodynamic therapy. *Photochemistry and Photobiology*, **81** (4), 891–7.

90 Zhang, M. *et al.* (2008) Fabrication of ZnPc/protein nanohorns for double photodynamic and hyperthermic cancer phototherapy. *Proceedings of the National Academy of Sciences of the United States of America*, **105** (39), 14773–8.

91 Zhu, Z. *et al.* (2008) Regulation of singlet oxygen generation using single-walled carbon nanotubes. *Journal of the American Chemical Society*, **130** (33), 10856–7.

92 Goeppert-Mayer, M. (1931) Über Elementarakte mit zwei Quantensprüngen. *Annals of Physics*, **9**, 273–95.

93 Bhawalkar, J.D. *et al.* (1997) Two-photon photodynamic therapy. *Journal of Clinical Laser Medicine and Surgery*, **15** (5), 201–4.

94 Karotki, A. *et al.* (2006) Simultaneous two-photon excitation of photofrin in relation to photodynamic therapy. *Photochemistry and Photobiology*, **82** (2), 443–52.

95 Samkoe, K.S. *et al.* (2007) Complete blood vessel occlusion in the chick chorioallantoic membrane using two-photon excitation photodynamic therapy: implications for treatment of wet age-related macular degeneration. *Journal of Biomedical Optics*, **12** (3), 034025.

96 Samkoe, K.S. and Cramb, D.T. (2003) Application of an ex ovo chicken chorioallantoic membrane model for two-photon excitation photodynamic therapy of age-related macular degeneration. *Journal of Biomedical Optics*, **8** (3), 410–17.

97 Starkey, J.R. *et al.* (2008) New two-photon activated photodynamic therapy sensitizers induce xenograft tumor regressions after near-IR laser treatment through the body of the host mouse. *Clinical Cancer Research*, **14** (20), 6564–73.

98 Verma, S. *et al.* (2005) Self-assembled photoresponsive amphiphilic diphenylaminofluorene-C60 conjugate vesicles in aqueous solution. *Langmuir*, **21** (8), 3267–72.

99 Chiang, L.Y. *et al.* (2002) Synthesis of C60-diphenylaminofluorene dyad with large 2PA cross-sections and efficient intramolecular two-photon energy transfer. *Chemical Communications (Cambridge, England)*, (17), 1854–5.

100 Irie, K. et al. (1996) Photocytotoxicity of water-soluble fullerene derivatives. *Bioscience, Biotechnology, and Biochemistry*, **60** (8), 1359–61.

101 Burlaka, A.P. et al. (2004) Catalytic system of the reactive oxygen species on the C60 fullerene basis. *Experimental Oncology*, **26** (4), 326–7.

102 Rancan, F. et al. (2002) Cytotoxicity and photocytotoxicity of a dendritic C(60) mono-adduct and a malonic acid C(60) tris-adduct on Jurkat cells. *Journal of Photochemistry and Photobiology B*, **67** (3), 157–62.

103 Yang, X.L., Fan, C.H. and Zhu, H.S. (2002) Photo-induced cytotoxicity of malonic acid [C(60)]fullerene derivatives and its mechanism. *Toxicology in Vitro*, **16** (1), 41–6.

104 Ikeda, A. et al. (2009) Direct and short-time uptake of [70]fullerene into the cell membrane using an exchange reaction from a [70]fullerene-gamma-cyclodextrin complex and the resulting photodynamic activity. *Chemical Communications (Cambridge, England)*, (12), 1547–9.

105 He, X.Y. et al. (1994) Photodynamic therapy with photofrin II induces programmed cell death in carcinoma cell lines. *Photochemistry and Photobiology*, **59** (4), 468–73.

106 Granville, D.J. et al. (1998) Rapid cytochrome c release, activation of caspases 3, 6, 7 and 8 followed by Bap31 cleavage in HeLa cells treated with photodynamic therapy. *FEBS Letters*, **437** (1–2), 5–10.

107 Gupta, S., Ahmad, N. and Mukhtar, H. (1998) Involvement of nitric oxide during phthalocyanine (Pc4) photodynamic therapy-mediated apoptosis. *Cancer Research*, **58** (9), 1785–8.

108 Kessel, D. et al. (2000) Determinants of the apoptotic response to lysosomal photodamage. *Photochemistry and Photobiology*, **71** (2), 196–200.

109 Ben-Dror, S. et al. (2006) On the correlation between hydrophobicity, liposome binding and cellular uptake of porphyrin sensitizers. *Photochemistry and Photobiology*, **82** (3), 695–701.

110 Cauchon, N. et al. (2005) Structure-photodynamic activity relationships of substituted zinc trisulfophthalocyanines. *Bioconjugate Chemistry*, **16** (1), 80–9.

111 Potter, W.R. et al. (1999) Parabolic quantitative structure-activity relationships and photodynamic therapy: application of a three-compartment model with clearance to the in vivo quantitative structure-activity relationships of a congeneric series of pyropheophorbide derivatives used as photosensitizers for photodynamic therapy. *Photochemistry and Photobiology*, **70** (5), 781–8.

112 Ross, M.F. et al. (2006) Accumulation of lipophilic dications by mitochondria and cells. *Biochemical Journal*, **400** (1), 199–208.

113 Rottenberg, H. (1984) Membrane potential and surface potential in mitochondria: uptake and binding of lipophilic cations. *Journal of Membrane Biology*, **81** (2), 127–38.

114 Murphy, M.P. and Smith, R.A. (2007) Targeting antioxidants to mitochondria by conjugation to lipophilic cations. *Annual Review of Pharmacology and Toxicology*, **47**, 629–56.

115 Chiang, L.Y. et al. (2010) Synthesis and characterization of highly fullerenyl dyads with a close chromophore antenna-C60 cage contact and effective photodynamic. *Journal of Materials Chemistry*, **20**, 5280–93.

116 Chi, Y. et al. (1999) Hexa(sulfobutyl) fullerene-induced photodynamic effect on tumors in vivo and toxicity study in rats. *Proceedings of the Electrochemical Society*, **99**, 234–49.

117 Andrievsky, G. et al. (2000) First clinical case of treatment of patient (volunteer) with rectal adenocarcinoma by hydrated fullerenes. Natural course of the disease or non-specific anticancer activity? Poster presentation No. 0377, Session N9. Proceedings, Biochemical and Pharmaceutical Aspects of Fullerene Materials, 197th Meeting of The Electrochemical Society, Toronto, Canada, 14-19 May 2000.

118 Hirayama, J. et al. (1999) Photoinactivation of vesicular stomatitis

virus with fullerene conjugated with methoxy polyethylene glycol amine. *Biological and Pharmaceutical Bulletin*, **22** (10), 1106–9.

119 Lin, C.P., Lynch, M.C. and Kochevar, I.E. (2000) Reactive oxidizing species produced near the plasma membrane induce apoptosis in bovine aorta endothelial cells. *Experimental Cell Research*, **259** (2), 351–9.

120 Lee, I. *et al*. (2009) Photochemical and antimicrobial properties of novel C60 derivatives in aqueous systems. *Environmental Science and Technology*, **43** (17), 6604–10.

121 Spesia, M.B., Milanesio, M.E. and Durantini, E.N. (2008) Synthesis, properties and photodynamic inactivation of *Escherichia coli* by novel cationic fullerene C60 derivatives. *European Journal of Medicinal Chemistry*, **43** (4), 853–61.

122 Huang, L. *et al*. (2010) Innovative cationic fullerenes as broad-spectrum light-activated antimicrobials. *Nanomedicine*, **6** (3), 442–52.

Index

a

activation energy
– bulk diffusion   14
– carbon atoms   13
– surface diffusion   14
aggregation
– carbon nanohorns (CNHs)   88f., 152
– carbon nanotubes (CNTs)   311f., 337
– fullerenes (C_{60})   262, 430
– graphene   73
– metal atoms   16
– nanodiamond (ND)   114
antibacterial therapy   140ff.
anticancer agent   333ff.
– associated with CNTs   338ff.
– cisplatin   97f., 153, 341f.
– monoclonal antibodies   334, 337, 349
aspect ratio   133f., 145, 163, 311, 349, 353
atomic force microscopy (AFM)   52f.
– CNTs   140, 188, 232
– graphene   73
– polystyrene mask   175
– SWNT-FET   232f.
Auger electron microscope   39

b

basic structural units (BSUs)   4
biocompatibility
– CNT-based substrates   173
– CNT–bioactive calcium phosphate nanoparticles (CP-NP)   180
– CNTs   79, 177
– graphene-based materials   69, 79
– nanodiamonds   6
– nanoparticles   275
– NGO–PEG sheets   80
– poly(ethylene glycol) (PEG)   80
biomedical applications of carbon nanohorns   87ff.
– CNH–H_2P nanohybrid   92
– DDS, see drug delivery systems
– SWCNHs   93f.
biomedical applications of carbon nanotubes   161ff.
biomedical applications of graphene   78ff.
– biocompatible paper   79
– drug dilvery   79ff.
– graphene–DNA hybrids   81f.
biomedical applications of magnetic carbon nanotubes/fullerenes   120ff.
biomedical applications of magnetic graphene   125
biomedical applications of magnetic graphite   116ff.
biomedical applications of nanodiamond   113ff.
biosensor
– amperometric   190f., 206ff.
– CAR-bound SWNT-FET   232
– chemiresistive   226
– CNT-based   187ff.
– CNT/metal nanoparticle hybrid glucose   197f., 205
– CNT/sol–gel   194f.
– detection limit   205, 207
– DNA–graphene   125
– dopamine   202
– durability   188, 208
– electrochemical   190
– enzymatic   203ff.
– ethanol   207f.
– field-effect (FE)   189, 217ff.
– glutamate   205ff.
– homocysteine   201
– indole acetic acid (IAA)   202
– lifetime   188, 208
– mediated-glucose   102, 193, 199, 203ff.
– mediator-free   206

- nicotinamide adenine dinucleotide (NAD)   199, 201
- nonenzymatic   199ff.
- receptors   224ff.
- response time   198
- sensitivity   198
- SWCNHs   101ff.
- SWNT-based   217ff.
- SWNT-FET   217ff.
- SWNT-FET–DNA   231
- voltammetric   191, 194, 198, 202, 208

biosensing mechanism
- adsorption   219
- electrostatic gating effect   218f.
- Schottky barrier modulation   218ff.

blood brain barrier (BBB)   381
boron neutron capture therapy (BNCT)   403ff.
boron nitride (BN) nanotube   118f.
brain tumor therapy   381f., 389ff.

c

cancer therapy
- boron neutron capture therapy, see BNCT
- brain tumor-macrophages   389ff.
- chemotherapy   145ff.
- CNTs   309ff.
- cryoablation   316
- heated particles   334f.
- high-intensity focused ultrasound   316
- hyperthermia   312f., 393
- in vitro   318, 405
- in vivo   319, 392f., 405, 408
- microwave ablation   315
- RFA, see radiofrequency ablation
- targeted   349ff.

carbon
- activated   6
- allotropes   6, 62, 87, 111ff.
- amorphous   18, 77, 89
- black   6, 11

carbon nanofibers
- fishbone   17
- platelet   17
- vapor-grown   5, 11ff.

carbon nanohorns (CNHs)   3, 87ff.
- bud-like   87
- dahlia-like   87f.
- double-wall (DWCNHs)   88

carbon nanohorns (CNHs) functionalization
- chemical surface modification   90
- covalent   91f.
- noncovalent   92f., 103

carbon nanohorns (CNHs)
- heat-treated SWCNHs (hSWCNHs)   90
- multi-walled (MWCNHs)   88
- pristine   89f.
- single-wall (SWCNHs)   87f., 152
- structure   88

carbon nanorods (CNRs)
- AuNRs   274

carbon nanotubes
- aligned   17f., 167f., 195f.
- armchair   14f.
- bioactive calcium phosphate nanoparticles (CP-NP)   178ff.
- bundle   64
- cancer therapy   309ff.
- chirality   14f., 21, 163, 313
- curvature   22, 24
- diameter   14
- DNA-wrapped   138
- double-walled (DWNTs)   311
- energy-minimized structure   282ff.
- ferromagnetic filled-   122f.
- free-standing matrix   175ff.
- fishbone   17
- formation on nickel   13

carbon nanotube functionalization
- bioactive molecules   23f.
- br-PEG–DPCC functionalized Irinotecan   292ff.
- chemical   163
- covalent   163, 189, 221f., 272, 274f., 385f.
- defect-mediated   22, 163
- endohedral   163
- enzyme   193ff.
- external   133
- fluorescein   355
- hybrid method   189
- nitrogen   60f., 134
- noncovalent   20, 163, 189, 222f., 274, 277, 322, 334f., 385f.
- oxygen   24, 59f., 221
- selective   24

carbon nanotubes growth   163
- rate   14, 18
- root-growth   16f.
- super-growth technique   18
- tip-growth   16f.

carbon nanotubes
- half-life   354, 360, 393
- hexagonally aligned   168
- hydrophobic   222, 322
- imaging   323f.
- interconnected   170
- length   355

– localization to malignant tissues 321f.
– matrix enhancement 164
– metal-capped 196
– metallic 16, 313
– multi-walled (MWNTs) 5, 9f., 13f., 133ff.
– nickel-coated MWNTs 120
– nucleation 18
– periodically aligned 168, 170
carbon nanotubes properties
– chemical 189, 311
– electrical 188f., 311
– electrochemical 189ff.
– intrinsic 324
– mechanical 3, 164f., 188, 311
carbon nanotubes
– pyramid-like 171f.
– semiconducting 14, 16, 21, 313
– single-wall (SWNTs) 9f., 13ff.
– three-dimensional cavity network 170ff.
– three-dimensional matrices 161, 166ff.
– tissue engineering 161ff.
– uptake of CNTs, see cellular uptake of CNTs
– vertical aligned (VACNTs) 167f., 197
– zigzag 14f., 118
carbon
– soots 9, 11, 19
– vapor-grown fibers (VGCFs) 11
catalyst
– biocompatible 18
– deactivation 5, 12
– density 17
– heterogeneous 16
– life time 18
– macrophages 390f.
– metal particle 14f.
– shape 6
– solvent effect 7
– steam-reforming 13
CDDP (cisplatin) 96ff.
cellular uptake of CNTs 354ff.
– nonspecific 356, 360, 369
– protein binding 356
– receptor-mediated 369
central nervous system (CNS) 382
charge-coupled device (CCD) 41f., 62
charge transport 189f.
chemotherapy
– CNTs 309, 312, 320
– hypothermic drug delivery 146ff.
– MWNTs 145ff.
cisplatin, see anticancer agent
cluster 6, 9
CNTs, see carbon nanotubes

coating
– carbon 11, 116f., 120
– CNTs 121, 356, 391f.
– phosphatidylserine- 391f.
complex
– CNT–cisplatin-EGF 340ff.
– CNT–DOX 323, 404, 406
– CNT/polymer 123, 165f., 340, 392
– combinatorial 285
– graphene–surfactant 73
– MWNT–GnRH 152, 337
– MWNT–hydroxycamptothecin 340f.
– SWNT–doxorubicin
– SWNT–folic acid (FA) 322, 334, 336f., 393, 408
– SWNT–Pt(IV) 322
– SWNT–PTX 277, 280f., 392f., 404f.
composite
– CNH/nanoparticle binary nanomaterial 93
– CNT/sol–gel electrode 194f.
– CNT/Nafion 193
– collagen–CNT hybrids 174
– MWNT–NF–PMG film 208
– MWNT–PE layer-by-layer 176
– PPy/CNT/GOx 193ff.
– SWCNH-based binary nanomaterial 93
– UNCD-based 9
confocal microscopy 324f., 338
core–shell structures 116f.
coulomb
– blockade 119ff.
– staircase 119ff.
current–voltage characteristics 82
cytotoxicity
– catalyst particles 19
– CNT–DOX complexes 323
– drugs 357
– PEG–DXR–oxSWCNHs 97f.
– radiofrequency 338
– SWNT–folic acid 322
– SWNT–Pt(IV) 322
– surfaces 79

d

Debye screening length 225, 228
defect
– adatom–vacancy pair 119
– carbon nanohorns (CNHs) 88
– carbon nanotubes (CNTs) 21f., 118
– C–C bond change 118
– free graphene 69
– free graphite 59
– free MWNT 51

– pyramidalization angle 22f.
– Stone–Wales (SW) 22, 118
– structural 24
– topological 22f., 119
– vacancy 23, 118f.
– wall 311
density functional theory (DFT) 92, 116, 124
density of states (DOS) 55f., 124
diamonds
– gem quality 113
– nanodiamond (ND) 6f., 22, 113ff.
– soot 8
– ultra-dispersed (UDD) 8
– ultra-nanocrystalline (UNCD) 6, 8f.
diamondoids 6
differential scanning calorimetry (DSC) 314
diffusion
– bulk 14
– carbon atoms 13f.
– Fe surface 14
– Franz diffusion-type cell 145
DNA
– based molecular diagnostics 231
– encased MWNTs 317ff.
– functionalized SWNTs 324
– graphene hybrids 81f.
– hybridization 221
– lable-free 81
– single-stranded DNA (ssDNA) 81, 296ff.
drug delivery systems (DDS) in boron neutron capture therapy (BNCT)
– dendritic macromolecules 411f.
– liposomes 411
– magnetic nanoparticles 413
drug delivery systems (DDS) of carbon nanohorns (CNHs)
– CDDP@hSWCNH 96
– CDDP@oxSWCNH 96f.
– LAOx–SWCNHs–BSA complex 98
– MAGoxSWCNHs 101f.
– MWNTs 133ff.
– oxSWCNHs 95f.
– PEG–DXR–oxSWCNHs 97, 99
– SWCNHs 95ff.
– VCM–oxSWCNHs–PEG 99
– ZnPc–oxSWCNHs—BSA 99
drug delivery systems (DDS) of carbon nanotubes (CNTs)
– antibodies 387
– chemical drugs 388f.
– controlled release 301, 370
– CpG 388

– DNA molecules 387f.
– ferromagnetic nanomaterial-coated CNTs 121
– hypothermic 146ff.
– peptides 387
– polymeric 320
– si-RNA 387
– size-limited drug efflux pump 353
– supramolecular approach 389
– vaccines 388
drug delivery systems (DDS) of fullerenes 430ff.
– cyclodextrins (CDs) 432
– dendrimers 431
– liposomes 430
– micelles 430f.
– PEGylation 431f.
– self-nanoemulsifying systems (SNES) 432
drug delivery systems (DDS) of nanodiamond 114

e
electrical double-layer 220, 225f.
electrode
– aligned-CNT 195f.
– basal plane pyrolytic graphite (BPPG) 191
– CNT/metal nanoparticle 197ff.
– CNT-modified needle 191, 195, 206
electrode fabrication
– adsorption technique 190f.
– covalent bonding 192
– high-temperature pyrolysis technique 195f.
– hybrid 197ff.
– in-situ 195
– polymer entrapment 192f.
electrode
– GCE, *see* glassy carbon electrode
– MWNT/gold 192
– surfaces 192
electron diffraction 36f.
electron energy-loss spectroscopy (EELS) 39, 41, 55f., 58
– electron energy loss near-edge structure (ELNES) 56f.
– extended energy loss fine structure (EXELFS) 56
– graphene 76ff.
– scanning transmission electron microscopy (TEM) 49
electropermeabilization 151
electrostatic gating effect 218

encapsulating
- CNTs   353
- liposome   277
- nanocapsules   272, 274
- transition metal particles   12, 16, 116
energy-dispersive X-ray spectroscopy (EDX)   39, 53f.
- CNT–bioactive calcium phosphate nanoparticles (CP-NP)   179f.
- scanning transmission electron microscopy (TEM)   49
enhanced permeability and retention (EPR) effect   154, 311, 313, 320f., 357, 392
epitaxial growth, see graphene
exfoliation, see synthesis
extracellular matrix (ECM)   161f.

f

fabrication
- bottom-up   188
- CNT-based electrode, see electrode
- fullerene   261f., 264
- top-down   188
filling
- capillary   136
- intertube   133ff.
- quantum dots (QDs)   136f.
fluorescence microscopy
- SWNT   316
- graphene paper   79
Food and Drug Administration (FDA)   276, 420
full width at half-maximum (FWHM)   56, 59
fullerene   3, 6, 9f., 239ff.
- antimicrobial photoinactivation   439ff.
- covalent attachment of light-harvesting antennae   433f.
- emission sources   254, 259f., 263
- environmental impact   239f., 262ff.
- fabrication   261f., 264
- photochemistry of   427ff.
- photodynamic therapy (PDT)   419ff.
- photophysics of   426f.
- spectral absorption   432ff.
- suspensions   239ff.
- usages   260
functionalization of CNTs, see CNTs
functionalization of graphene, see graphene

g

gene therapy   138f.
glassy carbon electrode (GCE)
- CNTs   191, 204, 208, 218

- SWCNH-modified   103
graphene
- amine (GA) sheets   81
- bi-layer   74, 76
- defect-free   69
- epitaxial growth   71, 125
- few-layer   74
- flakes   71, 73
- free-standing   70
graphene functionalization
- NGO sheets   80f.
- NGO–PEG sheets   80f.
- NGO–PEG–SN-38 sheets   81
graphene
- nanoscale GO (NGO)   80f.
- oxide (GO)   71, 80f.
- oxygen-free   77
- sheets   4f., 14, 71, 80f., 163
- single-layer   71, 74, 76
- synthesis   70
- thermal conductivity   69
graphite
- exfoliation   70f., 73
- foamy-like   116
- highly oriented pyrolytic (HOPG)   59f., 70, 73, 78, 115f., 190
- onion-like   8f.
- oxidation–exfoliation reduction   71
- oxide   71
- sonication   71ff.
graphitization   4, 63

h

heat-shock proteins (HSPs)   315
heteroatoms   4, 22
hole
- opening method   96
- thermal closing   90
Hückel method   124
hydrocarbons
- decomposition   5, 10f.
- dissociation   14
hyperthermia, see cancer therapy
hypothermic chemoperfusion (HC)   148
- intraperitoneal (IPHC)   148
- nanotube-induced   149
- radiofrequency   149, 151

i

immobilization
- abrasive   190f.
- biomolecules   221ff.
- glucose oxidase (GOx)   220
- metal particles   223f.

- soybean peroxidase (SBP) 102
- ssDNA 231
impurities
- carbon 15, 17, 19
- graphite 8
- inorganic 19f.
- metal oxide 8
infrared spectroscopy (IR) 64
- absorption 146, 149
- fourier transform (FT-IR) 65, 78
- graphene 77f.
- laser stimulation 148, 150
- near-infrared (NIR) 100, 146, 275, 316f., 406f.
- SWCNHs 100
interaction
- arene–CNT 20
- biomolecular 220
- CNT–CNT 64
- CNT–ssDNA 296ff.
- electron–electron exchange 124
- electron–phonon 73
- electron–specimen 38f.
- extracellular matrix (ECM) 161
- fullerenes–DNA 429f.
- low-interaction supports 16
- high-interaction counterparts 16
- hydrophobic 222
- metal–substrate 16f.
- MWNTs–cell membranes 133
- nanoparticle–drug 275
- π–π 222
- van der Waals 20, 26, 64, 191, 277, 284
interface
- CNT/analyte 189
- electrode/solution 190, 207
- metal–support 13
internalization
- cellular 311, 322, 358
- CNTs 359, 389f.
- DNA–SWNTs 354f.
ion implantation 112

k

kinetics
- pharmaco- 352
- pyrolysis 10f.
- reaction 14
- solid carbon formation 6

l

layer-by-layer (LbL)-assembled 165, 175f., 278
Lennard–Jones potential 64

m

magic bullet 349
magnetic carbon materials
- bio-inspired 111ff.
- carbon nanotubes/fullerenes 117ff.
- graphene 124ff.
- graphite 115ff.
- nanodiamond 113ff.
magnetic resonance imaging (MRI) 100f., 111, 116f.
- SWCNHs 100f.
- thermometry 149
- transition metals (TM) 116f.
magnetic tunnel junctions (MTJs) 120f.
magnetoresistance (MR) sensors
- carbon nanotubes/fullerenes 119
- high-temperature-operating Hall device 114
- nanodiamond 114f.
- tunnel (TMR) 120f.
macrophage receptor with collagenous structure (MARCO) 134
mean square displacement, see simulation techniques
micellization process 275
minimum inhibitory concentration (MIC) 142
molecular dynamics (MDs), see simulation techniques
molecule-by-molecule-assembled 277
Monte Carlo (MC), see simulation techniques
most probable number (MPN) 233f.

n

nanocapsules, see encapsulating
nanodiamonds (NDs), see diamonds
nanoelectrode ensemble (NEE) 194f., 197
nanofilaments, see carbon nanotubes
nanohorns, see carbon nanohorns
nanoparticles (NP)
- accumulation 321
- bioactive calcium phosphate (CP-NP) 178ff.
nanorods (NR), see carbon nanorods
nanosphere lithography (NSL), see synthesis
natural organic matter (NOM) 262
neutron diffraction (ND) 36f.

o

optical microscopy
- SWCNHs 102
- SWNTs 324, 337
own N-layered integrated molecular orbital and molecular mechanics (ONIOM) 92

p

π–π stacking   26, 90, 93, 222f., 323, 340f.
– supramolecular   404
PEG   80f., 97f., 276f., 280ff.
– nontoxic   276
– PEGylation   276f., 280ff.
pharmacokinetics   352
photoacoustic effect   349, 407ff.
photodynamic therapy (PDT)   419ff.
– antimicrobial mechanism   425
– cellular effects   424f.
– Fablonski diagram   423
– in vivo   425, 438f.
– in vitro   435ff.
– two-photon   434f.
photoinactivation
– bacteria   440f.
– dynamic (PDI)   440f.
– viruses   439f.
photosensitizer (PS)
– C_{70} 437
– excited triplets   423f.
– fullerenes   425f.
– intersystem crossing   423
– traditional   421f.
photothermal effect   311, 316f., 407ff.
phototoxicity   438
porosity
– carbon nanohorns (CNHs)   89f.
– interstitial   89f.
powder diffraction file (PDF)   36
precipitation, see synthesis
precursor
– hydrocarbon   4f., 10ff.
– polyaromatic   10
proton relaxation rate   148
paclitaxel (PTX)
– functionalized-branched lipophilic species   280ff.
– loaded conjugate   280f., 340f.
– hydrophobicity   277
purification
– antibody   228
– CNTs   19f., 196f.
– fullerenes   8f.
– liquid chromatography   9
– protein   228
pyrolysis   4f., 10ff.

q

quantum dots (QDs)   135ff.
quenched   7

r

radial distribution function, see simulation techniques
radiation therapy   309, 312
radiofrequency ablation (RFA)   314f., 317, 320, 333
– noninvasive field   338, 407
Raman spectroscopy   62ff.
– carbon nanohorns (CNHs)   89
– fingerprint spectra   73
– graphene   73f.
– graphite   74, 77f.
– infrared scattering   275
– multi-color   363
– MWNTs   24, 63
– radial breathing mode (RBM)   63f.
– SWNTs   62f.
reaction
– Boudouard   11f.
– carbothermal   18
– coupling   25f.
– cycloaddition   22, 25, 91f., 323, 388
– kinetics   14
– oxidation   24
– polymerase chain (PCR)   231
– rate   11, 15
– reverse CVD   15
– reverse-transcription polymerase chain (RT-PCR)   231
– steam-reforming   12f.
receptors
– aptamers   224ff.
– enzymes   229f.
– fragment antibodies   227ff.
– immunoglobins   227
– proteins   229f.
rehybridization   163
reticuloendothelial system (RES)   276, 320f.
retransformation   7
reactive oxygen species (ROS)   421, 424, 427, 429, 437

s

scaffolds   161f.
scanning electron microscopy (SEM)   39ff.
– backscattered electrons (BSEs)   39ff.
– CNT network   173
– core–shell structures   117
– environmental (ESEM)   40
– field emission (FESEM)   200
– MWNTs   19, 21, 163
– secondary electrons   38f.
– SWNTs   19, 21
scanning probe microscopy   49

scanning transmission electron microscopy (STEM)   49
– high-angle annular dark field (HAADF)   49
scanning tunneling microscopy (STM)   49ff.
– energy band gaps   118
self-assembled monolayer (SAM), *see* synthesis
silicon wafers   17f.
simulation techniques
– amorphous cell (AC)   279
– atomistic modelling   278
– condensed-phase   279
– Discover technique   279
– dissipative particles dynamics (DPD)   274, 279
– Forcite analysis   281, 293
– mean square displacement (MSD)   288, 290
– MesoDyn   279
– mesoscale methods   279, 300
– molecular dynamics (MDs)   14, 276, 280ff.
– Monte Carlo (MC)   278
– radial distribution function (RDF)   284, 288ff.
– torsion   282, 294f.
solubility
– CNT   163, 312
– fullerene   239ff.
– nanodiamond   114
superconducting quantum interference device (SQUID) analysis
– nickel-filled CNTs   124
superstructure   88
surface area
– biosensor   189, 199
– carbon nanohorns (CNHs)   90
– carbon nanotubes (CNTs)   22, 386
– graphene   69
– high-   16
– nanodiamond   115
– oxides   16
surface-to-volume ratio   217
synthesis
– arc-discharge   9f., 19, 72, 163, 311, 385
– arc-in-liquid   88
– catalytic chemical vapor deposition (CCVD)   10ff.
– catalyzed   4f., 10, 72
– chemical vapor deposition (CVD)   10f., 72, 163, 196, 311, 385
– double plasma-enhanced hot filament chemical vapor deposition   121

– electrochemical decoration technique   228
– electron beam evaporation   167
– epitaxial growth   71, 125
– exfoliation   70f.
– explosive detonation   7f.
– floating catalyst CCVD   15ff.
– gas-phase catalytic process (HiPCO)   15, 385
– graphene   69ff.
– high-pressure high-temperature (HPHT)   7f.
– hot filament chemical vapor deposition (HFCVD)   114
– immobilized catalyst CCVD   16f.
– in-situ growth   122
– laser ablation   9, 163, 385
– laser chemical vapor deposition (LCVD)   121
– microwave plasma-enhanced chemical vapor deposition (MPCVD)   198ff.
– nanodiamond (ND)   7
– nanosphere lithography (NSL)   167, 175
– pattern growth technique   228
– plasma-enhanced CVD (PECVD)   18, 166, 168, 311
– platinum-catalyzed cyclodehydrogenation   10
– precipitation   13, 16, 154
– reactive-ion etching   176
– self-assembly method   227, 229
– shock-wave   7f.
– Soxhlet extraction   9
– substrate-free gas-phase   72
– three-step CCVD mechanism   14
– uncatalyzed   4f.

t

targeting of cancer
– active   358, 370, 392
– biotin receptor   365f.
– brain tumor therapy   382ff.
– disialoganglioside (GD2)   368
– erbB familiy members   360ff.
– folate receptor α   363
– integrins   366f.
– markers for lymphomas or leukemias   367f.
– molecular therapy   383f.
– passive   357f., 370, 392
– trafficking of targeted drug-delivery vehicles   358f., 361f.
temperature-driven rearrangement   4
temperature-programmed desorption (TPD)   25

thermal
- decomposition, see pyrolysis
- lethality   315
- transition   314
thermogravimetry-differential thermal analysis (TD-DTA)   89
time-of-flight mass spectroscopy (TOF-MS)
- fullerenes   262
tissue engineering   161ff.
toxicity of CNHs   94f.
toxicity of CNT
- aggregates   312
- brain tumor therapy   394f.
- drug delivery systems   320f.
- metal-induced   19f., 26
- MWNT   174
- PAH-induced   19f.
- reduction   142
- SWNTs   140
toxicity of fullerenes
- aqueous suspensions   246
- C_{60} on algae   251, 255
- C_{60} on bacteria   251, 256ff.
- C_{60} on fish   247ff.
- C_{60} on invertebrates   250, 252f.
- C_{60} on soil microbes   251, 256ff.
- dark   421
transepithelial electrical resistance (TEER)   145
transformation
- UNCDs–graphite   8
- thermal   314
transition metal (TM)   5, 16, 116f., 122ff.
- encapsulation   116f., 122ff.
- filled CNTs   116f., 122f.
transmission electron microscopy (TEM)   41ff.
- aberration correction   45f.
- analytical   42
- atomic-resolution direct TEM image   82f.
- carbon nanofiber (CNF)   12
- carbon nanohorns (CNHs)   88
- diffraction contrast   43
- graphene   70, 75ff.
- high resolution (HR-TEM)   5, 12, 19f., 43ff.
- imaging   42f., 82f.
- in-situ   14
- inverse contrast   153

- iron carbide (Fe_3C)   13
- mass–thickness contrast   43
- MWNTs   5, 12, 14, 19f., 133, 141
- phase contrast   43f.
- phase-contrast transfer function (CTF)   44f.
- selected area electron diffraction (SAED)   41
- SWNTs   19
- tomography   46ff.
- vapor-grown fibers (VGCF)   12

u

UDD, see diamond
ultracentrifugation   21f., 73
UNCD, see diamond
UV
- induced skin damage   430
- visible–NIR spectra   405f.

v

vapor–liquid–solid (VLS) theory   13
voltammetry
- cyclic (CV)   191, 194, 198, 202
- differential pulse   208

w

wound healing
- hydrogels   143
- MWNTs 142ff.
- SWNT–chitosan film   144, 199
- SWNT–HA film   144
- therapies   143

x

X-ray absorption spectroscopy (XAS)   57f.
- near-edge structure (XANES)   58
- near-edge X-ray absorption fine structure (NEXAFS)   58
X-ray diffraction (XRD)   36f.
- carbon nanohorns (CNHs)   89
- iron carbide (Fe_3C)   13
X-ray photoelectron spectroscopy (XPS)   58ff.
- electron spectroscopy for chemical analysis (ESCA)   58
- graphene   78
- in situ   14
- iron particles   14
- nickel particles   14